W9-BPN-858

✳

Race, Class, and Gender

An Anthology

Race, Class, and Gender

An Anthology

EIGHTH EDITION

MARGARET L. ANDERSEN
University of Delaware

PATRICIA HILL COLLINS
University of Maryland

WADSWORTH
CENGAGE Learning·

Australia • Brazil • Japan • Korea • Mexico • Singapore • Spain • United Kingdom • United States

WADSWORTH
CENGAGE Learning

Race, Class, and Gender: An Anthology, **Eighth Edition**
Margaret L. Andersen and
Patricia Hill Collins

Publisher: Linda Ganster

Acquisitions Editor: Erin Mitchell

Assistant Editor: Mallory Ortberg

Media Editor: John Chell

Marketing Program Manager:
Janay Pryor

Marketing Communications
Manager: Tami Strang

Art and Cover Direction,
Production Management, and
Composition: PreMediaGlobal

Manufacturing Planner: Judy
Inouye

Rights Acquisitions Specialist:
Roberta Broyer

Text Researcher: Pablo D'Stair

Cover Designer: Norman Baugher

Cover Image: © Fototaras/
Shutterstock

© 2013, 2010 Wadsworth, Cengage Learning

ALL RIGHTS RESERVED. No part of this work covered by the copyright herein may be reproduced, transmitted, stored, or used in any form or by any means graphic, electronic, or mechanical, including but not limited to photocopying, recording, scanning, digitizing, taping, Web distribution, information networks, or information storage and retrieval systems, except as permitted under Section 107 or 108 of the 1976 United States Copyright Act, without the prior written permission of the publisher.

For product information and
technology assistance, contact us at **Cengage Learning
Customer & Sales Support, 1-800-354-9706**.

For permission to use material from this text or product,
submit all requests online at **www.cengage.com/permissions**.
Further permissions questions can be e-mailed to
permissionrequest@cengage.com.

Library of Congress Control Number: 2011945950

ISBN-13: 978-1-111-83094-6

ISBN-10: 1-111-83094-0

Wadsworth
20 Davis Drive
Belmont, CA 94002-3098
USA

Cengage Learning is a leading provider of customized learning solutions with office locations around the globe, including Singapore, the United Kingdom, Australia, Mexico, Brazil, and Japan. Locate your local office at **www.cengage.com/global**.

Cengage Learning products are represented in Canada by Nelson Education, Ltd.

To learn more about Wadsworth, visit
www.cengage.com/wadsworth.

Purchase any of our products at your local college store or at our preferred online store **www.cengagebrain.com**.

Printed in the United States of America
2 3 4 5 6 7 16 15 14 13

Contents

✳

Preface

We write this preface at a time when the social dynamics of class—and its relationship to gender and race is—changing. In 2009, a global economic crisis battered the U.S. housing and financial markets and left many people without jobs. People who had worked their entire lives saw their life's savings disappear as their houses dropped in value and their savings accounts for sending their kids to college or for their own retirement shrank. Many families faced foreclosure, unable to pay their mortgages. Some families became homeless. The global financial crisis affected nearly everyone in some way, but it didn't affect everyone in the same way. Race, class, and gender shaped people's experiences with this global crisis. People who were more vulnerable before the crisis were hit harder. Racial/ethnic families, poor White families and/or families maintained by single mothers were hit harder than others. Moreover, middle class families also found themselves with shrinking bank accounts and no jobs. This crisis shows how race, class, and gender continue to shape the American opportunity structure.

To explain this global economic crisis, you need to know how race, class, and gender—and their relationship to other social factors—are systematically interrelated.

That is the theme of this book: race, class, and gender simultaneously shape social issues and experiences in the United States. Central to the book is the idea that race, class, and gender are interconnected and that they must be understood as operating together if you want to understand the experiences of diverse groups and particular issues and events in society. We want this book to help students see how the lives of different groups develop in the context of their race, class, and gender location in society.

Since the publication of the first edition of this book, the study of race, class, and gender has become much more present in people's thinking and teaching. Yet, as our introductory essay will argue, not all of the work on race, class, and gender is centered in the intersectional framework that is central to this book. Often, people continue to treat one or the other of these social factors in isolation. Many studies of race, class, and gender also treat race, class, and gender as if they were equivalent experiences. Although we see them as interrelated—and sometimes similar in how they work—we also understand that each has its own dynamic, but a dynamic that can only be truly understood in relationship to the others. Moreover, with the growth of race, class, and gender studies, other social factors, such as sexuality, nationality, age, and disability, are also connected to the structural intersections of race, class, and gender. We hope that this book helps students understand how these structural phenomena—that is, the social forces of race, class, and gender and their connection with other social variables—are deeply embedded in the social structure of society.

We hope we achieve this by presenting an anthology that is more than a collection of readings. Our book is strongly centered in an analytical framework about the interconnections among race, class, and gender. In this edition we continue our efforts to further develop a framework of the *intersectionality* of race, class, and gender, or as Patricia Hill Collins has labeled it, the *matrix of domination*. The organization of the book features this framework. Our introductory essay distinguishes what we mean by this intersectional framework from other models of studying "difference." The four parts of the book are intended to help students see the importance of this intersectional framework, to engage critically the core concepts on which the framework is based, and to analyze different social institutions and current social issues using this framework, including being able to apply it to understanding processes of social change.

The four major parts of the book reflect these goals. We open with a section and essay entitled "Why Race, Class, and Gender Still Matter." As in past editions, we include essays in Part I that engage students in personal narratives, as a way of helping them step beyond their own social location and to see how race, class, gender, sexuality, and other social factors shape people's lives differently. We also intend for this section to show students the very different experiences that anchor the study of race, class, and gender. Being able to center oneself in the lived experience of others is, we think, important to grasping the importance of race, class, gender, and other social factors, such as sexuality, ethnicity, religion, and so forth. Even when dominant groups may claim that race, class, and gender no longer matter in society, we know this is not true. We begin our book with essays that show their continuing, even if changing, significance.

The title of Part II of this book, "Systems of Power and Inequality," reflects how we conceptualize race, class, gender, ethnicity, and sexuality as linked through systems of power and inequality. The essays in Part II link ethnicity, nationality, and sexuality to the study of race, class, and gender. Here we want students to understand that the individual identities and personal narratives of Part I, "Why Race, Class, and Gender Still Matter," are structured by intersecting systems of power. Although we caution readers against thinking that all forms of difference are the same, we know that race, class, and gender are interconnected with other systems of power and inequality.

Part III, "The Structure of Social Institutions," focuses on how race, class, and gender structure social institutions—now and historically. In this edition we have particularly strengthened the section on "Work and Economic Transformation" to reflect the growing significance of economic issues in the lives of our students. Finally, we again revised Part IV, "Pulling It All Together." Many anthologies use their final section to show how students can make a difference in society, once they understand the importance of race, class, and gender. We think this is a tall order for students who may have had only a few weeks to even begin to understand how race, class, and gender matter—and matter together. Thus, rather than using our final section to make it seem that change can be easily accomplished, we use this section to show the different ways that one can *apply* understanding the intersectionality of race, class, and gender in different social sites.

Throughout, our book is grounded in a sociological perspective, although the articles come from different perspectives, disciplines, and experiences. Several articles provide a historical foundation for understanding how race, class, and gender have emerged. We also include materials that bring a global dimension to the study of race, class, and gender—not just by looking comparatively at other cultures but also by analyzing how the process of globalization is shaping life in the United States.

As in earlier editions, we have selected articles based primarily on two criteria: (1) accessibility to undergraduate readers and the general public, not just highly trained specialists, and (2) articles that are grounded in race *and* class *and* gender— in other words, intersectionality. Not all articles accomplish this as much as we would like, but we try not to select articles that focus exclusively on one issue while ignoring the others. In this regard, our book differs significantly from other anthologies on race, class, and gender that include many articles on each issue, but do less to show how they are connected. We also distinguish our book from those that are centered in a multicultural perspective. Although multiculturalism is important, we think that race, class, and gender go beyond the appreciation of cultural differences. Rather, we see race, class, and gender as embedded in the structure of society and significantly influencing group cultures. But race, class, and gender are also structures of group opportunity, power, and privilege. We searched for articles that are conceptually and theoretically informed and at the same time accessible to undergraduate readers. Although it is important to think of race, class, and gender as analytical categories, we do not want to lose sight of how they affect human experiences and feelings; thus, we have included personal narratives that are reflective and analytical. We think that personal accounts generate empathy and also help us connect personal experiences to social structural conditions.

ORGANIZATION OF THE BOOK

As noted previously, we have organized the book to reflect how we approach the study of race, class, and gender. The first part establishes the framework within which we approach the study of race, class, and gender. We introduce each of the four parts with an essay we wrote to analyze the issues raised by the reading selections. These essays are an important part of this book because they establish the conceptual foundation that we use to think about race, class, and gender.

Part I, "Why Race, Class, and Gender Still Matter," contains several pieces that reflect on how race, class, and gender shape individual and collective experiences. These different accounts are not meant to represent everyone who has been excluded, but to ground readers in the practice of thinking about the lives of people who rarely occupy the center of dominant group thinking. When excluded groups are placed in the forefront of thought, they are typically stereotyped and distorted. When you look at society through the different perspectives of persons reflecting on their location within systems of race, class, and gender oppression, you begin to see just how much race, class, and gender really matter—and continue to matter, despite claims to the contrary. Because several of the essays in this part are grounded in personal experiences, they capture student interest, but we also use them analytically to help students think beyond any one person's life.

Part II, "Systems of Power and Inequality," provides the conceptual foundation for understanding how race, class, and gender are linked together and how they link with other systems of power and inequality, especially ethnicity and sexuality. We treat each of these separately here, not because we think they stand alone, but to show students how each operates so they can better see their interlocking nature. The introductory essay provides working definitions for these major concepts and presents some of the contemporary data through which students can see how race, class, and gender stratify contemporary society.

In Part III, "The Structure of Social Institutions," we examine how intersecting systems of race, class, and gender shape the organization of social institutions and how, as a result, these institutions affect group experience. Social scientists routinely document how Latinos, African Americans, women, workers, and other distinctive groups are affected by institutional structures. We know this is true but want to go beyond these analyses to scrutinize how institutions are constructed through race, class, and gender relations. The articles in Part III show how we should not relegate the study of racial-ethnic groups, the working class, and women only to subjects marked explicitly as race, class, or gender studies. As categories of social experience, race, class, and gender shape all social institutions and systems of meaning; thus, it is important to think about people of color, different class experiences, and women in analyzing all social institutions and belief systems.

Finally, in Part IV, "Pulling It All Together" we provide different examples of how using an intersectional perspective can help you understand the different social locations where social change and social action occur. Throughout the book, we have avoided a "social problems" approach, as if race, class, and gender are solely about victimization. People are not just victims; they are creative and visionary. The articles here show different ways that people react to the realities of race, class, and gender in their lives. We think that, together, these articles illustrate for students how to apply a race, class, and gender framework in order to understand how people negotiate race, class, and gender in their everyday lives.

We know that developing a complex understanding of the interrelationships between race, class, and gender is not easy and involves a long-term process engaging personal, intellectual, and political change. We do not claim to be models of perfection in this regard. We have been pleased by the strong response to the first seven editions of this book, and we are fascinated by how race, class, and gender studies have developed since the publication of our first edition. We know further work is needed. Our own teaching and thinking has been transformed by developing this book. We imagine many changes still to come.

CHANGES TO THE EIGHTH EDITION

We have made several changes in the seventh edition that strengthen and refresh *Race, Class, and Gender*. These include the following:

- Twenty new readings.
- A stronger focus on economic issues throughout the book, including improved readings in the section on class (in Part II), and a revised section on work and economic transformation (in Part III).
- Continued focus on the media and popular culture, with material on youth, social networking and the Internet.
- A stronger section on sexuality, including a new essay on sexuality and the media, the connections among racism, sexism, and homophobia, as well as work on young people and heterosexuality, sex trafficking, and connections among race, ethnicity, and sexuality.
- A revised final part that helps students apply an understanding of race, class, and gender in different social locations and in the process of social change.
- Four revised introductions by us; this is one of the noted strengths of our book compared to others. These introductions articulate the analytical model we use to understand the intersections among race, class, and gender.
- New material on race, class, gender and important current issues, including the sub-prime mortgage situation, sustainability, women's rights as human rights, and hate crimes.

PEDAGOGICAL FEATURES

We realize that the context in which you teach matters. If you teach in an institution where students are more likely to be working class, perhaps how the class system works will be more obvious to them than it is for students in a more privileged college environment. Many of those who use this book will be teaching in segregated environments, given the high degree of race and class—and even gender—segregation in education. Thus, how one teaches this book should reflect the different environments where faculty work. Ideally, the material in this book should be discussed in a multiracial, multicultural atmosphere, but we realize that is not always the case. We hope that the content of the book and the pedagogical features that enhance it will help bring a more inclusive analysis to educational settings than might be there to start with.

We see this book as more than just a collection of readings. The book has an analytical logic to its organization and content, and we think it can be used to format a course. Of course, some faculty will use the articles in an order different from how we present them, but we hope the four parts will help people develop the framework for their course. We also provide pedagogical tools to help people expand their teaching and learning beyond the pages of the book.

We have included features with this edition that provide faculty with additional teaching tools. They include the following:

- *Instructor's manual.* This edition includes an instructor's manual with suggestions for classroom exercises, discussion and examination questions, and course assignments.

- *Index.* The index will help students and faculty locate particular topics in the book quickly and easily.

- *Website.* Online support is available through the Wadsworth Sociology Resource Center. We encourage people to browse the *Race, Class, and Gender* page at the Wadsworth website or related resources. The address is www.cengage.com/sociology.

A NOTE ON LANGUAGE

Reconstructing existing ways of thinking to be more inclusive requires many transformations. One transformation needed involves the language we use when referring to different groups. Language reflects many assumptions about race, class, and gender; and for that reason, language changes and evolves as knowledge changes. The term *minority,* for example, marginalizes groups, making them seem somehow outside the mainstream or dominant culture. Even worse, the phrase *non-White,* routinely used by social scientists, defines groups in terms of what they are not and assumes that Whites have the universal experiences against which the experiences of all other groups are measured. We have consciously avoided using both of these terms throughout this book, although this is sometimes unavoidable.

We have capitalized Black in our writing because of the specific historical experience, varied as it is, of African Americans in the United States. We also capitalize White when referring to a particular group experience; however, we recognize that White American is no more a uniform experience than is African American. We use *Hispanic* and *Latina/o* interchangeably, though we recognize that is not how groups necessarily define themselves. When citing data from other sources (typically government documents), we use *Hispanic* because that is usually how such data are reported.

Language becomes especially problematic when we want to talk about features of experience that different groups share. Using shortcut terms like Hispanic, Latina/o, Native American, and women of color homogenizes distinct historical experiences. Even the term *White* falsely unifies experiences across such factors as ethnicity, region, class, and gender, to name a few. At times, though, we want to talk of common experiences across different groups, so we have used labels such as Latina/o, Asian American, Native American, and women of color to do so. Unfortunately, describing groups in this way reinforces basic categories of oppression. We do not know how to resolve this problem but want readers to be aware of the limitations and significance of language as they try to think more inclusively about diverse group experiences.

ACKNOWLEDGMENTS

An anthology rests on the efforts of more people than the editors alone. This book has been inspired by our work with scholars and teachers from around the country who are working to make their teaching and writing more inclusive and sensitive to the experiences of all groups. Over the years of our own collaboration, we have each been enriched by the work of those trying to make higher education a more equitable and fair institution. In that time, our work has grown from many networks that have generated new race, class, and gender scholars. These associations continue to sustain us. Many people contributed to the development of this book.

We especially thank David Strohecker at the University of Maryland, College Park for his expert research assistance; Dave was a real partner in developing this book, and we could not have completed it on time without him. We also thank Rebecca Emel at the University of Delaware for doing all the research about contributors. We appreciate the support given by our institutions, with special thanks to Provost Tom Apple at the University of Delaware and Reeve Vanneman, Chair of the Sociology Department at the University of Maryland. Special thanks as well to Dianna DiLorenzo, Sarah Hedrick, and Joan Stock at the University of Delaware for helping to keep things from unraveling! As always, Erin Mitchell has proven to be an exceptional editor. We gratefully acknowledge her support and her willingness to work with us through our busy schedules. We also thank the anonymous reviewers who provided valuable commentary on the prior edition and thus helped enormously in the development of the eighth edition.

Developing this book has been an experience based on friendship, hard work, travel, and fun. We especially thank Roger, Valerie, Lauren, and Richard for giving us the support, love, and time we needed to do this work well. This book has deepened our friendship, as we have grown more committed to transformed ways of thinking and being. We are lucky to be working on a project that continues to enrich our work and our friendship.

About the Editors

Margaret L. Andersen (B.A. Georgia State University; M.A., Ph.D., University of Massachusetts, Amherst) is the Edward F. and Elizabeth Goodman Rosenberg Professor of Sociology at the University of Delaware where she also holds joint appointments in Black American Studies and Women's Studies and serves as the Executive Director for the President's Diversity Initiative. She has received two teaching awards at the University of Delaware. She has published numerous books and articles, including *Thinking about Women: Sociological Perspectives on Sex and Gender* (9th ed., Allyn and Bacon, 2011); *Race and Ethnicity in Society: The Changing Landscape* (edited with Elizabeth Higginbotham, 3rd ed., Cengage, 2012); *On Land and On Sea: A Century of Women in the Rosenfeld Collection* (Mystic Seaport Museum, 2007); *Living Art: The Life of Paul R. Jones, African American Art Collector* (University of Delaware Press, 2009); and *Sociology: The Essentials* (co-authored with Howard F. Taylor, Cengage, 2012). She received the American Sociological Association's Jessie Bernard Award for expanding the horizons of sociology to include the study of women and the Eastern Sociological Society's Merit Award and Robin Williams Lecturer Award. She is a past vice president of the American Sociological Association and past president of the Eastern Sociological Society.

Patricia Hill Collins (B.A., Brandeis University; M.A.T., Harvard University; Ph.D., Brandeis University) is distinguished university professor of Sociology at the University of Maryland, College Park, and Charles Phelps Taft Emeritus Professor of African American Studies and Sociology at the University of Cincinnati. She is the author of numerous articles and books including *Another Kind of Public Education: Race, Schools, the Media and Democratic Possibilities* (Beacon, 2009), *From Black Power to Hip Hop: Racism, Nationalism and Feminism* (Temple University, 2006); *Black Sexual Politics: African Americans, Gender and the New Racism* (Routledge, 2004), which won the Distinguished Publication Award from the American Sociological Association; *Fighting Words* (University of Minnesota, 1998); and *Black Feminist Thought: Knowledge, Consciousness, and the Politics of Empowerment* (Routledge, 1990, 2000), which won the Jessie Bernard Award of the American Sociological Association and the C. Wright Mills Award of the Society for the Study of Social Problems. In 2008–2009, she served as the 100th president of the American Sociological Association.

About the Contributors

Joan Acker is professor emerita of Sociology at the University of Oregon. She founded and directed the University of Oregon's Center for the Study of Women in Society and is the recipient of the American Sociological Association's Career of Distinguished Scholarship Award as well as the Jessie Bernard Award for feminist scholarship. She is the author of *Class Questions, Feminist Answers* as well as many other works in the areas of gender, institutions, and class.

Michelle Alexander is associate professor of law at the Moritz College of Law at Ohio State University. She has served as director of the Racial Justice Project for the ACLU of Northern California, and has clerked for Justice Harry A. Blackmun on the United States Supreme Court.

Teresa Amott is provost and dean of faculty at Hobart and William Smith Colleges. She is the co-author of *Race, Gender, and Work: A Multicultural Economic History of Women in the United States* (with Julie Matthaei) and *Caught in Crisis: Women in the U.S. Economy Today*.

Maxine Baca Zinn is professor emerita of Sociology, Michigan State University. Widely known for her work on Latina families and family diversity, she is the co-author (with D. Stanley Eitzen) of *Diversity in Families, In Conflict and Order, Globalization: The Transformation of Social Worlds*, and *Social Problems* and the co-editor of *Gender through the Prism of Difference: A Sex and Gender Reader* (with Pierrette Hondagneu-Sotelo and Michael A. Messner).

Janani Balasubramanian is an undergraduate student at Stanford University, studying Atmosphere and Energy Engineering, and minoring in Feminist Studies. She served as an intern with the Climate Economics Group in SEI-U.S. Center.

Fabricio E. Balcazar is associate professor in the Department of Psychology at the University of Illinois at Chicago (UIC). He is also the director of the Center for Capacity Building on Minorities with Disabilities Research at UIC.

Kristen Barber is a doctoral candidate in Sociology at the University of Southern California, focusing her research on gender and work, inequality, the body, and qualitative methods. She is author and co-author of several articles, book chapters, and encyclopedia entries.

Marianne Bertrand is Chris P. Dialynas Professor of Economics and Neubauer Family Faculty Fellow at Chicago Booth University School of Business. Her work has been published in the *Quarterly Journal of Economics*, the *Journal of Political Economy*, the *American Economic Review*, and the *Journal of Finance*, among others.

Pamela Block is associate professor at the State University of New York at Stony Brook. She is also an adjunct faculty member at the Center for Alcohol and Addiction Studies at Brown University.

Danah Boyd is a senior researcher at Microsoft Research, and an adjunct associate professor at the University of New South Wales. She is currently a fellow at the Harvard University Berkman Center for Internet and Society, and her work has contributed to the book *Hanging Out, Messing Around, and Geeking Out: Kids Living and Learning with New Media* (by Mizuko Ito, Dilan Mahendran, Katynka Z. Martínez, C. J. Pascoe, et al.).

Denise Brennan is an associate professor of Anthropology in the Sociology and Anthropology Department at Georgetown University. She is the author of several books and articles about the global sex trade, human trafficking, and women's labor, primarily in Latin America and the Caribbean. She was recently awarded an American Association of University Women fellowship to conduct field research for her next book on the process of recovery and resettlement after trafficking.

Charlotte Bunch is Board of Governor's Distinguished Service Professor of Women's and Gender Studies at Rutgers University, and the founding director and senior scholar at the Center for Women's Global Leadership (CWGL). She has written several essays and edited or co-edited nine anthologies. She was inducted into the National Women's Hall of Fame in 1996.

Erica Chito Childs is associate professor of Sociology at Hunter College, City University of New York. She is a council member for the American Sociological Association section on Racial and Ethnic Minorities.

Ward Churchill (enrolled Keetoowah Band Cherokee) is a long-time native rights activist and was professor of Ethnic Studies and Coordinator of American Indian Studies at the University of Colorado until 2007. His books include *Agents of Repression, Fantasies of the Master Race, From a Native Son,* and *A Little Matter of Genocide: Holocaust and Denial in the Americas.*

Judith Ortiz Cofer is the Franklin Professor of English and Creative Writing at the University of Georgia. She is the author of numerous books of poetry and essays, and a novel, *The Line of the Sun.* She has won the PEN/Martha Albrand Special Citation in nonfiction for her book *Silent Dancing.*

Gary David Comstock is university Protestant chaplain and visiting associate professor at Wesleyan University. He has published numerous books, including *Gay Theology Without Apology; Unrepentant, Self-Affirming, Practicing: Lesbian/ Bisexual/Gay People Within Organized Religion; Que(e)rying Religion: A Critical Anthology;* and *Violence against Lesbians and Gay Men.*

Mary Yu Danico is vice chair of the Psychology and Sociology Department at California Polytechnic State University, Pomona. Her research interests include immigration and migration, Korean American diaspora, Asian diaspora, and transnational families and communities. She is the co-editor of *Transforming the Academy: People of Color, Queers, and Women in Higher Education.*

Bonnie Thornton Dill is professor of Women's Studies, director of the Research Consortium on Gender, Race, and Ethnicity, and dean of the college of arts and humanities at the University of Maryland, College Park. Her books include *Women of Color in U.S. Society,* co-edited with Maxine Baca Zinn, and *Across the Boundaries of Race and Class: Work and Family among Black Female Domestic Servants.*

Rachel E. Dubrofsky is an assistant professor in the Department of Communication at the University of South Florida. Her research projects focus on studies of the media and women in popular culture, especially the phenomenon of reality television. She is the author of *The Surveillance of Women on Reality Television* and numerous articles on feminist media studies.

Abby Ferber is professor of sociology, director of the Matrix Center, and co-director of Women's and Ethnic Studies at the University of Colorado at Colorado Springs. She is the author of *White Man Falling: Race, Gender and White Supremacy* and *Hate Crime in America: What Do We Know?* She is co-author of *Making a Difference: University Students of Color Speak Out* and co-editor of the anthology *Privilege: A Reader* with Michael Kimmel.

Charles A. Gallagher is the chair of the Department of Sociology at LaSalle University with research specialties in race and ethnic relations, urban sociology, and inequality. He has published several articles on subjects such as colorblind political narratives, racial categories within the context of interracial marriages, and perceptions of privilege based on ethnicity.

Herbert J. Gans has been a prolific and influential sociologist for more than fifty years. His published works on urban renewal and suburbanization are intertwined with his personal advocacy and participant observation, including a stint as consultant to the National Advisory Commission on Civil Disorder. He is the author of the classic *The Urban Villagers* as well as the more recent *Democracy and the News.*

Naomi Gerstel is professor of Sociology at the University of Massachusetts Amherst. She is the author of several articles, including *Giving and Taking Family Leaves: Right or Privilege,* and *Fathering, Class and Gender: A Comparison of Physicians and EMTs* (with Carla Shows).

Rosalind Gill is professor of Subjectivity and Cultural Analysis at The Open University. She is the author of *Gender and the* Media, and *New Femininities: Post-feminism, Neoliberalism, and Subjectivity* (with Christina Scharff).

Chong-suk Han teaches in the Sociology Department at Temple University and is currently a doctoral candidate in Social Welfare at the University of Washington. His research focuses on shared experiences of marginalization and the intersections of race and sexuality.

Adia M. Harvey (Wingfield) is an associate professor of Sociology at Georgia State University specializing in race, class, and gender, work and occupations, and social theory. She is the author of *Yes We Can? White Racial Framing and the 2008 Presidential Campaign* (with Joe Feagin) and *Doing Business with Beauty: Black Women, Hair Salons, and the Racial Enclave Economy*.

Pierrette Hondagneu-Sotelo is professor of Sociology, American Studies, and Ethnicity at the University of Southern California. She is author of *Gendered Transitions: Mexican Experiences of Immigration* and *Doméstica: Immigrant Workers Cleaning and Caring in the Shadow of Affluence,* which won the Society for Social Problems. C. Wright Mills Award.

Insight Center for Community Economic Development is a national research, consulting, and legal organization dedicated to building economic health and opportunities in vulnerable communities.

Lawrence R. Jacobs is the Walter F. and Joan Mondale Chair for Political Studies and director of the Center for the Study of Politics and Governance at the University of Minnesota. He co-edits the *Chicago Series in American Politics* for the University of Chicago Press (with James Druckman), and has published a number of scholarly works.

Barbara Jensen is a community and counseling psychologist in private practice in Minneapolis-St. Paul. She also teaches sociology and psychology as an adjunct at Metropolitan State University and serves on the steering committee of the Minnesota Center for Labor and Working Class Studies.

Jonathan Ned Katz was the first tenured professor of Lesbian and Gay Studies in the United States (Department of Lesbian and Gay Studies, City College of San Francisco). He is the founder of the Queer Caucus of the College Art Association. He is also the co-founder of the activist group Queer Nation.

Haunani Kay-Trask is a Hawaiian scholar and poet and has been an activist in the Native Hawaiian community for over 20 years. She is a professor and former director of the Center for Hawaiian Studies at the University of Hawaii at Manoa.

Christopher B. Keys is a department chair and professor of Community Psychology at DePaul University.

Nazli Kibria is associate professor of Sociology at Boston University. She is the author of *Becoming Asian American: Identities of Second Generation Chinese and Korean Americans* and *Family Tightrope: The Changing Lives of Vietnamese Americans*.

Chungmei Lee is a research associate at the Civil Rights Project. She has served on several projects concerning the training and professional development of teachers, including Harvard University's Program for Professional Education.

David Leonhardt writes "Economic Scene," a regular column on economics and business in *The New York Times*. He is also a staff writer for the *New York Times Magazine,* and he has contributed to such publications as *Business Week* and *The Washington Post.*

R. L'Heureux Lewis is assistant professor of Sociology and Black Studies at the City College of New York (CUNY). His commentary has been highlighted in the US World News Report, National Public Radio, the Detroit Free Press, and others.

Arturo Madrid is Murchison Distinguished Professor of the Humanities at Trinity University and the recipient of the Charles Frankel Prize in the Humanities, National Endowment for the Humanities, in 1996. From 1984 until 1993 he served as founding president of The Tomis Rivera Center, a national institute for policy studies on Latino issues.

Julianne Malveaux holds a Ph.D. in economics and is a syndicated columnist who writes regularly for *Essence* and also appears regularly on CNN, BET, Fox News, MSNBC, and other television news shows. She is the author of *Wall Street, Main Street, and the Side Street: A Mad Economist Takes a Stroll.* She is president and CEO of her own multimedia company.

Gregory Mantsios is the director of Worker Education at Queens College, the City University of New York.

Elizabeth Martinez is a Chicana scholar and activist. She is the co-founder of the Institute for MultiRacial Justice. Her latest book is *DeColores Means All of Us: Latina Views for a Multi-Colored Century.*

Julie Matthaei is a professor of Economics at Wellesley College. Her books include *An Economic History of Women in America: Women's Work, The Sexual Division of Labor, and the Development of Capitalism,* and (with Teresa Amott) *Race, Gender, and Work: A Multicultural Economic History of Women in the United States.*

Peggy McIntosh is associate director of the Wellesley College Center for Research on Women. She is the founder and co-director of the National SEED Project on Inclusive Curriculum—a project that helps teachers make school climates fair and equitable. She is the co-founder of the Rocky Mountain Women's Institute.

Michael Messner is professor of Sociology and Gender Studies at the University of Southern California. He is the author of King of the Wild Suburb: A Memoir of Fathers, Sons, and Guns; It's All For the Kids: Gender, Families, and *Youth Sports, Men's Lives* (with Michael Kimmel) and numerous other books and articles on gender, men and masculinity, and sports.

Doug Meyer is a doctoral candidate and adjunct lecturer in Sociology at the City University of New York (CUNY).

James A. Morone is professor of Political Science and Urban Studies at Brown University and president of the Politics and History Section of the American Political Science Association. He has published eight books, including a New York Times Notable Book (for *The Democratic Wish* in 1991).

Jennifer Mueller is assistant professor of Management at the Wharton School at the University of Pennsylvania. Her research focuses on interpersonal relationships in organizations, creativity and innovation, team dynamics, and emotions at work.

Sendhil Mullainathan is professor of Economics at Harvard University. He is the co-founder of The MIT Poverty Action Lab, and of ideas42, a Harvard-based think-tank that uses human psychology to understand the economic lives of people.

Carolina Bank Muñoz is associate professor of Sociology at Brooklyn College. She is the author of *Transnational Tortillas: Race, Gender and Shop Floor Politics in Mexico and the United States,* as well as many other publications.

Carla O'Connor is associate professor and Arthur F. Thurnau Professor in the School of Education at the University of Michigan. She co-edited (with Erin McNamara Horvat) the book, *Beyond Acting White: Reframing the Debate on Black Student Achievement,* and is a founding member of the NSF-sponsored Center for the Study of Black Youth in Context.

Melvin L. Oliver is the SAGE Sara Miller McCune Dean of Social Sciences and professor of Sociology at the University of California, Santa Barbara. He also serves on the boards of the Urban Institute in Washington D.C., the William T. Grant Foundation, and is an advisory board member of the National Poverty Center at the Gerald R. Ford School of Public Policy, University of Michigan, Ann Arbor.

Gary Orfield is a professor of Education and the co-director of the Civil Rights Project at the UCLA Graduate School of Education & Information Studies. His research interests include civil rights, education policy, urban policy, and minority opportunity. He is the author of several books and articles in these areas, as well as the recipient of the 2007 "Social Justice in Education" Award by the American Educational Research Association.

Gina M. Pérez is an associate professor of Comparative American Studies at Oberlin College. Her research focuses on topics such as U.S. Latinas/os, gender, political economy, migration, transnationalism, urban anthropology, and poverty. She is the author of *The Near Northwest Side Story: Migration, Displacement, and Puerto Rican Families.*

Lillian Rubin lives and works in San Francisco. She is an internationally known lecturer and writer. Some of her books include *The Man with the Beautiful Voice, Tangled Lives, the Transcendent Child,* and *Intimate Strangers.*

Almas Sayeed is the child welfare policy coordinator at the Kansas Coalition Against Sexual and Domestic Violence.

Janny Scott is a staff writer for *The New York Times.* She is the author of numerous articles dealing with the interactions of race, class, and politics.

Thomas M. Shapiro is Pokross Chair of Law and Social Policy at Brandeis University. His latest book is *The Hidden Cost of Being African American*. His earlier work with Melvin L. Oliver, *Black Wealth/White Wealth,* has won several awards including the American Sociological Association Distinguished Scholarly Award.

Jael Silliman is associate professor of Women's Studies at the University of Iowa. She is the co-editor of *Dangerous Intersections: Feminist Perspectives on Population, Environment, and Development* and the author of *Jewish Portraits, Indian Frames: Women's Narratives from a Diaspora of Hope*.

C. Matthew Snipp is professor of Sociology at Stanford University. He is the author of *American Indians: The First of This Land and Public Policy Impacts on American Indian Economic Development*. His tribal heritage is Oklahoma Cherokee and Choctaw.

Jessica Taft is assistant professor of Sociology at Davidson College. She has written several book chapters and articles and is currently the chair of the Girls' Studies Interest Group at the National Women's Studies Association.

Ronald T. Takaki is professor of Ethnic Studies at the University of California, Berkeley. He is the author of several books, including *Iron Cages: Race and Culture in 19th Century America; Strangers from a Different Shore: A History of Asian Americans;* and *A Different Mirror: A History of Multicultural America*.

Jeanne Theoharis is professor of Political Science at Brooklyn College. She has written and edited several books and articles including her latest publication, *Hidden in Plain Sight: Southern Exceptionalism and the Civil Rights Movement in the North*. She has also served as the chairwoman of the Northern Radical Liberalism Roundtable at the American Studies Association Annual Conference (2007).

Jeremiah Torres is a graduate of Stanford University, where he studied symbolic systems. His article "Label Us Angry" appeared in the book *Asian American X,* a collection of essays about the experiences of contemporary Asian Americans.

Eisa Nefertari Ulen is a teacher of English at Hunter College in New York City as well as an essayist and a journalist. She has contributed to many anthologies and other publications, including *The Washington Post, Ms.,* and *Health*. She is the recipient of a Frederick Douglass Creative Arts Center Fellowship for Young African American Fiction Writers and a Hunter College Presidential Award for Excellence for Teaching.

Linda Trinh Vô is associate professor and chair of the Department of Asian American Studies School of Humanities at the University of California, Irvine. Her research interests include Asian American studies, race and ethnic relations, immigration theory, and social stratification and inequality. She also participates in community activism as a member of such organizations as the Orange County Asian and Pacific Islander Community Alliance (OCAPICA) and Project Moti-VATe, a mentoring program for Vietnamese American teens.

Mary C. Waters is M.E. Zukerman Professor of Sociology and Harvard College Professor at Harvard University. She is the author of *Black Identities: West Indian*

Immigrant Dreams and *American Realities; Ethnic Options: Choosing Identities in America;* and numerous articles on race, ethnicity, and immigration.

Sandra Weissinger is assistant professor of Sociology at the Southern University of New Orleans. She is the author of *What Globalization Means to Class Reproduction: How Class Reproduction is Changing in the United States* (under the advisement of Zine Magubane).

Kath Weston is a sociological anthropologist who has written several books, including *Families We Choose: Lesbians, Gays, and Kinship; Render Me, Gender Me; Long Slow Burn;* and *Gender in Real Times.*

Christine L. Williams is a professor of Sociology at the University of Texas at Austin. Her research interests include gender and sexuality; work, occupations, and organizations; qualitative methodology; and sociological theory. She is the author of *Still a Man's World: Men Who Do "Women's Work"* as well as numerous articles and book chapters.

Jennifer Wriggins is associate professor of Law at the University of Southern Maine. She specializes in the practice of civil litigation.

✳

PART I

Why Race, Class, and Gender Still Matter

MARGARET L. ANDERSEN AND
PATRICIA HILL COLLINS

The United States is a nation where people are supposed to be able to rise above their origins. Those who want to succeed, it is believed, can do so through hard work and solid effort because the nation is founded on the principle of equality. Although equality has historically been denied to many, there is now a legal framework in place that guarantees protection from discrimination and equal treatment for all citizens. Historic social movements, such as the civil rights movement and the feminist movement, raised people's consciousness about the rights of minority groups and women. Moreover, these movements have generated new opportunities for multiple groups—African Americans, Latinos, White women, disabled people, lesbian, gay, bisexual, trans-gendered (LGBT) peoples, and older people, to name some of the groups that have been beneficiaries of civil rights action and legislation. And, if you ask Americans if they support non-discrimination policies, the overwhelming number will say, "Yes." Why, then, do race, class, and gender still matter?

Race, class, and gender still matter because they continue to structure society in ways that value some lives more than others. Currently, some groups have more opportunities and resources, while other groups struggle. Race, class, and gender matter because they remain the foundations for systems of power and inequality that, despite our nation's diversity, continue to be among the most significant social facts of people's lives. Thus, despite having removed the formal barriers to opportunity, the United States is still highly stratified along lines of race, class, and gender.

In this book, we ask students to think about race, class, and gender as *systems of power*. We want to encourage readers to imagine ways to transform, rather than reproduce, existing social arrangements. This starts with shifting one's thinking so that groups who are often silenced or ignored become heard. All social groups are located in a system of power relationships wherein your social location can shape what you know—and what others know about you. As a result, dominant forms of knowledge have been constructed largely from the experiences of the most powerful—that is, those who have the most access to systems of education and communication. Thus, to acquire a more inclusive view—one that pays attention to group experiences that may differ from your own—requires that you form a new frame of vision.

You can think of this as if you were taking a photograph. For years, poor people, women, and people of color—and especially poor women of color—were totally outside the frame of vision of more powerful groups or distorted by their views. If you move your angle of sight to include those who have been overlooked, however, some accepted points of view may seem less revealing or just plain wrong. Completely new subjects can also appear. This is more than a matter of sharpening one's focus, although that is required for clarity. Instead, this new angle of vision means actually seeing things differently, perhaps even changing the lens you look through—thereby removing the filters (or stereotypes and misconceptions) that you bring to what you see and think.

DEVELOPING A RACE, CLASS, AND GENDER PERSPECTIVE

In this book we ask you to think about how race, class, and gender matter in shaping everyone's lived experiences. We focus on the United States, but increasingly the inclusive vision we present here matters on a global scale as well. Thinking from a perspective that engages race, class, and gender is not just about illuminating the experiences of oppressed groups. It also changes how we understand groups who are on both sides of power and privilege. For example, the development of women's studies has changed what we know and how we think about women; at the same time, it has changed what we know and how we think about men. This does not mean that women's studies is about "male-bashing." It means taking the experiences of women and men seriously and analyzing how race, class, and gender shape the experiences of both men and women—in different, but interrelated, ways. Likewise, the study of racial and ethnic groups begins by learning the diverse histories and experiences of these groups. In doing so, we also transform our understanding of White experiences. Rethinking class means seeing the vastly different experiences of both wealthy, middle-class, working class, and poor people in the United States and learning to think differently about privilege and opportunity. The exclusionary thinking that comes from past frames of vision simply does not reveal the intricate interconnections that exist among the different groups that comprise the U.S. society.

It is important to stress that thinking about race, class, and gender is not just a matter of studying victims. Relying too heavily on the experiences of poor people, women, and people of color can erase our ability to see race, class, and gender as an integral part of everyone's experiences. We remind students that race, class, and gender have affected the experiences of all individuals and groups. As a result, we do not think we should talk only about women when talking about gender or only about poor people when talking about class. Because race, class, and gender affect the experiences of all, it is important to study Whites when analyzing race, the experiences of the affluent when analyzing class, and to study men when analyzing gender. Furthermore, we should not forget women when studying race or think only about Whites when studying gender.

So you might ask, how does reconstructing knowledge about excluded groups matter? To begin with, knowledge is not just some abstract thing—good to have, but not all that important. There are real consequences to having partial or distorted knowledge. First, knowledge is not just about content and information; it provides an orientation to the world. What you know frames how you behave and how you think about yourself and others. If what you know is wrong because it is based on exclusionary thought, you are likely to act in exclusionary ways, thereby reproducing the racism, anti-Semitism, sexism, class oppression, and homophobia of society. This may not be because you are intentionally racist, anti-Semitic, sexist, elitist, or homophobic; it may simply be because you do not know any better. Challenging oppressive race, class, and gender relations in society requires reconstructing what we know so that we have some basis from which to change these damaging and dehumanizing systems of oppression.

Second, learning about other groups helps you realize the partiality of your own perspective; furthermore, this is true for both dominant and subordinate groups. Knowing only the history of Puerto Rican women, for example, or seeing their history only in single-minded terms will not reveal the historical linkages between the oppression of Puerto Rican women and the exclusionary and exploitative treatment of African Americans, working-class Whites, Asian

American men, and similar groups. This is discussed by Ronald T. Takaki in his essay included here ("A Different Mirror") on the multicultural history of American society.

Finally, having misleading and incorrect knowledge leads to the formation of bad social policy—policy that then reproduces, rather than solves, social problems. U.S. immigration policy has often taken a one-size-fits-all approach, failing to recognize that vast differences among groups coming to the United States privilege some and disadvantage others. Taking a broader view of social issues fosters more effective social policy.

RACE, CLASS, AND GENDER AS A MATRIX OF DOMINATION

Race, class, and gender shape the experiences of all people in the United States. This fact has been widely documented in research and, to some extent, is commonly understood. Thus, for years, social scientists have studied the consequences of race, class, and gender inequality for different groups in society. The framework of race, class, and gender studies presented here, however, explores how race, class, and gender operate *together* in people's lives. Fundamentally, race, class, and gender are *intersecting* categories of experience that affect all aspects of human life; thus, they *simultaneously* structure the experiences of all people in this society. At any moment, race, class, or gender may feel more salient or meaningful in a given person's life, but they are overlapping and cumulative in their effects.

In this volume we focus on several core features of this intersectional framework for studying race, class, and gender. First, we emphasize *social structure* in our efforts to conceptualize intersections of race, class, and gender. We use the approach of a *matrix of domination* to analyze race, class, and gender. A matrix of domination sees social structure as having multiple, interlocking levels of domination that stem from the societal configuration of race, class, and gender relations. This structural pattern affects individual consciousness, group interaction, and group access to institutional power and privileges (Collins 2000). Within this structural framework, we focus less on comparing race, class, and gender as separate systems of power than on investigating the structural patterns that join them. Because of the simultaneity of race, class, and gender in people's lives, intersections of race, class, and gender can be seen in individual stories and personal experience. In fact, much exciting work on the intersections of race, class, and gender appears in autobiographies, fiction, and personal essays. We do recognize the significance of these individual narratives and include many here, but we also emphasize social structures that provide the context for individual experiences.

Second, studying interconnections among race, class, and gender within a context of social structures helps us understand how race, class, and gender are manifested differently, depending on their configuration with the others. Thus, one might say African American men are privileged *as men,* but this may not be true when their race and class are also taken into account. Otherwise, how can we possibly explain the particular disadvantages African American men experience in the criminal justice system, in education, and in the labor market? For that matter, how can we explain the experiences that Native American women undergo—disadvantaged by the unique experiences that they have based on race, class, *and* gender—none of which is isolated from the effects of the others? Studying the connections among race, class, and gender reveals that divisions by race and by class and by gender are not as clear-cut as they may seem. White women, for example, may be disadvantaged because of gender but privileged by race and perhaps (but not necessarily) by class. And increasing class differentiation within racial-ethnic groups reminds us that race is not a monolithic category, as can be seen in the fact that White poverty is increasing more than poverty among other groups, even while some Whites are the most powerful members of society.

Third, the matrix of domination approach to race, class, and gender studies is historically grounded. We have chosen to emphasize the intersections of race, class, and gender as institutional systems that have had a special impact in the United States. Yet race, class, and gender intersect with other categories of experience, such as sexuality, ethnicity, age, ability, religion, and nationality. Historically, these intersections have taken varying forms from one society to the next; within any given society, the connections among them also shift. Thus, race is not inherently more important than gender, just as sexuality is not inherently more significant than class and ethnicity.

Given the complex and changing relationships among these categories of analysis, we ground our analysis in the historical, institutional context of the United States. Doing so means that race, class, and gender emerge as fundamental categories of analysis in the U.S. setting, so significant that in many ways they influence all of the other categories. Systems of race, class, and gender have been so consistently and deeply codified in U.S. laws that they have had intergenerational effects on economic, political, and social institutions. For example, the capitalist class relations that have characterized all phases of U.S. history have routinely privileged or penalized groups organized by gender and by race. U.S.

social institutions have reproduced economic equalities for poor people, women, and people of color from one generation to the next. Thus, in the United States, race, class, and gender demonstrate visible, long-standing, material effects that in many ways foreshadow more recently visible categories of ethnicity, religion, age, ability, and/or sexuality.

DIFFERENCE, DIVERSITY, AND MULTICULTURALISM

How does the matrix of domination framework differ from other ways of conceptualizing race, class, and gender relationships? We think this can be best understood by contrasting the *matrix of domination framework* to what might be called a *difference framework* of race, class, and gender studies as well as related frameworks that emphasize diversity and multiculturalism. A difference framework, though viewing some of the common processes in race, class, and gender relations, tends to focus on unique group experiences. Books that use a framework of difference (or diversity or multiculturalism) will likely include writings by diverse groups of people, but on closer inspection, you will see that many of these writings treat race, class, and gender separately. Although we think such studies are valuable and add to the body of knowledge about race, class, and gender, we distinguish our work by looking at the *interrelationships* among race, class, and gender, not just their unique ways of being experienced.

You might think of the distinction between the two approaches as one of thinking comparatively, which is an example of one of the core features of a difference framework, versus thinking relationally, which is the hallmark of the matrix of domination approach. For example, in the difference framework individuals are encouraged to compare their experiences with those supposedly unlike them. When you think comparatively, you might look at how different groups have, for example, encountered prejudice and discrimination or you might compare laws prohibiting interracial marriage to current debates about same-sex marriage. These are important and interesting questions, but they are taken a step further when you think beyond comparison to the structural relationships between different group experiences. In contrast, when you think relationally, you see the social structures that *simultaneously* generate unique group histories and link them together in society. You then untangle the workings of social systems that shape the experiences of different people and groups, and you move beyond just comparing (for example) gender oppression with race oppression or the oppression of gays and lesbians with that of racial groups.

Recognizing how intersecting systems of power shape different groups' experiences positions you to think about changing the system, not just documenting the effects of such systems on different people.

The language of difference encourages comparative thinking. People think comparatively when they learn about experiences other than their own and begin comparing and contrasting the experiences of different groups. This is a step beyond centering one's thinking in a single group (typically one's own), but it is nonetheless limited. For example, when students encounter studies of race, class, and gender for the first time, they often ask, "How is this group's experience like or not like my own?" This is an important question and a necessary first step, but it is not enough. For one thing, it frames one's understanding of different groups only within the context of other groups' experiences; thus, it can assume an artificial norm against which different groups are judged. Furthermore, it tends to promote ranking the oppression of one group compared to another, as if the important thing were to determine who is most victimized. Thinking comparatively tends to assume that race, class, and gender constitute separate and independent components of human experience that can be compared for their similarities and differences.

We should point out that comparative thinking can foster greater understanding and tolerance, but comparative thinking alone can also leave intact the power relations that create race, class, and gender relations. Because the concept of difference contains the unspoken question "different from what?" this framework can privilege those who are deemed to be "normal" and stigmatize people who are labeled as "different." And because it is based on comparison, the very concept of difference fosters dichotomous, either/or thinking. Some approaches to difference place people in either/or categories, as if one is either Black or White, oppressed or oppressor, powerful or powerless, normal or different when few of us fit neatly into any of these restrictive categories.

Some difference frameworks try to move beyond comparing systems of race, class, and gender by thinking in terms of an *additive* approach. The additive approach is reflected in terms such as *double* and *triple jeopardy*. Within this logic, poor African American women seemingly experience the triple oppression of race, gender, and class, whereas poor Latina lesbians encounter quadruple oppression, and so on. But social inequality cannot necessarily be quantified in this fashion. Adding together "differences" (thought to lie in one's difference from the norm) produces a hierarchy of difference that ironically reinstalls those who are additively privileged at the top while relegating those who are additively oppressed to the bottom. We do not think of race and gender oppression in the simple additive terms implied by phrases such as double and triple jeopardy. The effects of race, class, and gender do "add up," both over time and in intensity of impact, but seeing race, class, and gender only in additive terms misses the social structural connections among them and the particular ways in which different configurations of race, class, and gender affect group experiences.

Within difference frameworks, this additive thinking can foster another troubling outcome. One can begin with the concepts of race, class, and gender and continue to "add on" additional types of difference. Ethnicity, sexuality, religion, age, and ability all can be added on to race, class, and gender in ways that suggest that any of these forms of difference can substitute for others. This use of difference fosters a view of oppressions as equivalent and as being the same. Recognizing that difference encompasses more than race, class, and gender is a step in the right direction. But continuing to add on many distinctive forms of difference can be a never-ending process. After all, there are as many forms of difference as there are individuals. Ironically, this form of recognizing difference can erase the workings of power just as effectively as diversity initiatives.

When it comes to conceptualizing race, class, and gender relations, the matrix of domination approach also differs from another version of the focus on difference, namely, thinking about diversity. *Diversity* has become a catchword for trying to understand the complexities of race, class, and gender in the United States. What does *diversity* mean? Because the American public has become a more heterogeneous population, *diversity* has become a buzzword—popularly used, but loosely defined. People use *diversity* to mean cultural variety, numerical representation, changing social norms, and the inequalities that characterize the status of different groups. In thinking about diversity, people have recognized that race, class, gender, age, sexual orientation, and ethnicity matter; thus, groups who have previously been invisible, including people of color, gays, lesbians, and bisexuals, older people, and immigrants, are now in some ways more visible. At the same time that diversity is more commonly recognized, however, these same groups continue to be defined as "other"; that is, they are perceived through dominant group values, treated in exclusionary ways, and subjected to social injustice and economic inequality.

The movement to "understand diversity" has made many people more sensitive and aware of the intersections of race, class, and gender. Thinking about diversity has also encouraged students and social activists to see linkages to other categories of analysis, including sexuality, age, religion, physical disability, national identity, and ethnicity. But appreciating diversity is not the only point. The very term *diversity* implies that understanding race, class, and gender is simply a matter of recognizing the plurality of views and experiences in society—as if race, class, and gender were benign categories that foster diverse experiences instead of systems of power that produce social inequalities.

Diversity initiatives hold that the diversity created by race, class, and gender differences are pleasing and important, both to individuals and to society as a whole—so important, in fact, that diversity should be celebrated. Under diversity initiatives, ethnic foods, costumes, customs, and festivals are celebrated, and students and employees receive diversity training to heighten their multicultural awareness. Diversity initiatives also advance a notion that, despite their differences, "people are really the same." Under this view, the diversity created by race, class, and gender constitutes cosmetic differences of style, not structural opportunities.

Certainly, opening our awareness of distinct group experiences is important, but some approaches to diversity can erase the very real differences in power that race, class, and gender create. For example, diversity initiatives have asked people to challenge the silence that has surrounded many group experiences. In this framework, people think about diversity as "listening to the voices" of a multitude of previously silenced groups. This is an important part of coming to understand race, class, and gender, but it is not enough. One problem is that people may begin hearing the voices as if they were disembodied from particular historical and social conditions. This perspective can make experience seem to be just a matter of competing discourses, personifying "voice" as if the voice or discourse itself constituted lived experience. Second, the "voices" approach suggests that any analysis is incomplete unless every voice is heard. In a sense, of course, this is true, because inclusion of silenced people is one of the goals of race/class/gender work. But in a situation where it is impossible to hear every voice, how does one decide which voices are more important than others? One might ask, who are the privileged listeners within these voice metaphors?

We think that the matrix of domination model is more analytical than either the difference or diversity frameworks *because of its focus on structural systems of power and inequality*. This means that race, class, and gender involve more than either comparing and adding up oppressions or privileges or appreciating cultural diversity. The matrix of domination model requires analysis and criticism of existing systems of power and privilege; otherwise, understanding diversity becomes just one more privilege for those with the greatest access to education—something that has always been a mark of the elite class. Therefore, race, class, and gender studies mean more than just knowing the cultures of an array of human groups. Instead, studying race, class, and gender means recognizing and analyzing the hierarchies and systems of domination that permeate society and limit our ability to achieve true democracy and social justice.

Finally, the matrix of domination framework challenges the idea that race, class, and gender are important only at the level of culture—an implication of the catchword *multiculturalism*. *Culture* is traditionally defined as the "total way of life" of a group of people. It encompasses both material and symbolic components and is an important dimension of understanding human life. Analysis of culture per se, however, tends to look at the group itself rather than at the broader conditions within which the group lives. Of course, as anthropologists know, a sound analysis of culture situates group experience within these social structural conditions. Nonetheless, a narrow focus on culture tends to ignore social conditions of power, privilege, and prestige. The result is that multicultural studies often seem tangled up with notions of cultural pluralism—as if knowing a culture other than one's own is the only goal of a multicultural education.

Because we approach the study of race, class, and gender with an eye toward transforming thinking, we see our work as differing somewhat from the concepts implicit in the language of difference, diversity, or multiculturalism. Although we think it is important to see the diversity and plurality of different cultural forms, in our view this perspective, taken in and of itself, misses the broader point of understanding how racism, class relations, and sexism have shaped the experience of groups. Imagine, for example, looking for the causes of poverty solely within the culture of currently poor people, as if patterns of unemployment, unexpected health care costs, rising gas prices, and home mortgage foreclosures had no effect on people's opportunities and life decisions. Or imagine trying to study the oppression of LGBT people in terms of gay culture only. Obviously, doing this turns attention to the group itself and away from the dominant society. Likewise, studying race only in terms of Latino culture or Asian American culture or African American culture, or studying gender only by looking at women's culture, encourages thinking that blames the victims for their own oppression. For all of these reasons, the focus of this book is on the institutional, or structural, bases for race, class, and gender relations.

Thus, this book is not just about comparing differences, understanding diversity, or describing multicultural societies. Instead, we attempt to develop a structural perspective on the relationships among race, class, and gender as systems of power, the hallmark of the matrix of domination framework. We recognize that reaching these goals will require rejecting the kind of exclusionary thinking that virtually erased some groups' experiences and embracing an inclusive perspective that incorporates neglected groups and themes. Inclusive perspectives begin the recognition that the United States is a multicultural and diverse society. Population data and even casual observations reveal that obvious truth, but developing an inclusive perspective requires more than recognizing the plurality of experiences in this society. Understanding race, class, and gender means coming to see the systematic exclusion and exploitation of some groups as well as the intergenerational privileges of others. This is more than just adding in different group experiences to already established frameworks of thought. It means constructing new analyses that are focused on the centrality of race, class, and gender in the experiences of us all.

DEVELOPING AN INCLUSIVE PERSPECTIVE

We want readers to understand that race, class, and gender are linked experiences, no one of which is more important than the others; the three are interrelated and together configure the structure of U.S. society. You can begin to develop a more inclusive perspective by asking: How does the world look different if we put the experiences of those who have been excluded at the center of our thinking? At first, people might be tempted to simply assert the perspective and experience of their own group. Initially, this claiming of one's experience can be valuable and empowering, but ultimately centering exclusively in one's own experiences discourages inclusionary, relational thinking.

Developing an inclusive perspective calls for more than just seeing the world through the perspective of any one group whose views have been distorted or ignored. Remember that group membership cuts across race, class, and gender categories. For example, one may be an Asian American working-class woman or a Latino middle-class man or a gay, White working-class woman. Inclusive perspectives see the interconnections between these experiences and do not reduce a given person's or group's life to a single factor. In addition, developing an inclusive perspective entails more than just summing up the experiences of individual groups, as in the additive model discussed previously. Race, class, and gender are social structural categories. This means that they are embedded in the institutional structure of society. Understanding them requires a social structural analysis—by which we mean revealing the race, class, and gender patterns and processes that form the very framework of society.

We believe that thinking about the experiences of those who have been excluded from knowledge changes how we think about society, history, and culture. No longer do different groups seem "different," "deviant," or "exotic." Rather, specific patterns of the intersections of race, class, and gender are revealed, as are the connections that exist among groups. We then learn how our different experiences are linked, both historically and currently.

Once you understand that race, class, and gender are *simultaneous* and *intersecting* systems of relationship and meaning, you also can see the distinctive ways that other categories of experience intersect in society. Age, religion, sexual orientation, nationality, physical ability, region, and ethnicity also shape systems of privilege and inequality. We have tried to integrate these different experiences throughout the book, although we could not include as much as we would have liked.

Because analysis of the historical role of diverse groups is critical to understanding who we are as a society and a culture, we open this section with Ronald T. Takaki's "A Different Mirror." Takaki makes a point of showing the common connections in the histories of African Americans, Chicanos, Irish Americans, Jews, and Native Americans. He argues that only when we understand a multidimensional history that encompasses race, class, and gender will we see ourselves in the full complexity of our humanity. Several readings in this section rely heavily on personal accounts that reflect the diverse experiences of race, class, gender, and/or sexual orientation. We intend for the personal nature of these accounts—especially those that provide personal accounts of what exclusion means and how it feels—to build empathy among groups. We think that empathy encourages an emotional stance that is critical to relational thinking and developing an inclusive perspective.

Arturo Madrid ("Missing People and Others"), for example, shows how his experience as a young Latino student was silenced throughout his educational curriculum, leaving him to feel like an "other" in a society where he seemingly had no place, no history, no culture. For him, this involves more than just acknowledging the diverse histories, cultures, and experiences of groups who have been defined as marginal in society—what we have come to think of as "valuing diversity." But there is something more important than just valuing the diverse histories and cultures of the different groups who constitute society and that is to recognize how groups whose experiences have been vital in the formation of society and culture have also been silenced in the construction of knowledge about this society. The result is that what we know—about the experiences of both these silenced groups and the dominant culture—is distorted and incomplete. Indeed, for that matter, ignoring such experiences also gives us a distorted view of how the nation itself has developed.

How much did you learn about the history of group oppression in your formal education? You probably touched briefly on topics such as the labor movement, slavery, women's suffrage, perhaps even the Holocaust, but most likely these were brief excursions from an otherwise dominant narrative that ignored working-class people, women, and people of color, along with others. For that matter, how much of what you study now is centered in the experiences of the most dominant groups in society? Think about the large number of social science studies that routinely make general conclusions about the population when they have been based on research done primarily on middle-class college students, or on men. Or, how much of the literature you read and artistic creations that you study are the work of Native Americans; Muslim Americans; new immigrant populations; Asian Americans; Latinos/as; African Americans; or gays, lesbians, or women?

By minimizing the experiences and creations of these different groups, we communicate that their work and creativity is less important and less central to the development of culture than is the history of White American men. What false or incomplete conclusions does this exclusionary thinking generate? When you learn, for example, that democracy and egalitarianism were central cultural beliefs in the early history of the United States, how do you explain the enslavement of millions of African Americans, the genocide of Native Americans, the absence of laws against child labor, the presence of laws forbidding intermarriage between Asian Americans and White Americans?

This book asks you to think more inclusively. Without doing so you are prone to understand society, your own life within it, and the experiences of others through stereotypes and the misleading information that is all around you. For example, Jeremiah Torres ("Label Us Angry") shows us how sometimes the pain of living with stereotypes is very personal and painful. Torres describes how his seemingly trouble-free childhood changed overnight when he became the victim of a hate crime in a community known for its acceptance of multiculturalism.

What new experiences, understandings, theories, histories, and analyses do these readings inspire? What does it take for a member of one group (say a Latino male) to be willing to learn from and value the experiences of another (for example, an Indian Muslim woman)? These essays show that, although we are caught in multiple systems, we can learn to see our connection to others.

This is not just an intellectual exercise. As Haunani-Kay Trask shows ("From a Native Daughter"), there can also be a gap between dominant cultural narratives and people's actual experiences. As she, a native Hawaiian tells it, the official history she learned in schools was not what she was taught in her family and community. Dominant narratives can try to justify the oppression of different groups, but the unwritten, untold, subordinated truth can be a source for knowledge in pursuit of social justice.

Engaging oneself at the personal level is critical to this process of thinking differently about race, class, and gender. Changing one's mind is not just a matter of assessing facts and data, though that is important; it also requires examining one's feelings. We incorporate personal narratives into this opening section of the book to encourage you to think about your personal story. We each have one that is shaped by race, class and gender. Almas Sayeed's narrative ("Chappals and Gym Shorts"), for example, illustrates the challenges facing a young woman who tries to explain her growing feminist perspective to her father who loves her but who also wants her to focus only on getting married. Unlike more conventional forms of sociological data (such as surveys, interviews, and even direct observations), personal accounts such as those by Sayeed, Trask, and Torres, are more likely to elicit emotional responses. Traditionally, social science has defined emotional engagement as an impediment to objectivity. Sociology, for example, has emphasized rational thought as the basis for social action and has often discouraged more personalized reflection, but the capacity to reflect on one's experience makes us distinctly human. Personal documents tap the private, reflective dimension of life, enabling us to see the inner lives of others and, in the process, revealing our own lives more completely.

The idea that objectivity is best reached only through rational thought is a specifically Western and masculine way of thinking—one that we challenge throughout this book. Including personal narratives is not meant to limit our level of understanding only to individuals. In "White Privilege: Unpacking the Invisible Knapsack," Peggy McIntosh describes how the system of racial privilege becomes invisible to those who benefit from it, even though it structures the everyday life of both White people and people of color. McIntosh's personal narrative about examining patterns of privilege in her everyday life enabled her to develop objective knowledge about domination. As sociologists, we study individuals in groups as a way of revealing the social structures shaping collective experiences. In doing so, we discover our common experiences and see the impact of the social structures of race, class, and gender on our experiences. Much is at stake in our willingness to develop an inclusive perspective. Thinking relationally enables us to see connections that were formerly invisible. Through a discussion of eugenics, the authors of "Race, Poverty, and Disability: Three Strikes and You're Out! Or Are You?" examine how common ways of thinking about disability intersect with similar thinking about race and poverty. Proponents of eugenics believed in so-called higher and lower races and supported state-sponsored programs to control the population of the lower race. The article deals with how eugenics thinking may have disappeared from official public policy, yet its spirit persists in influencing contemporary public policy toward people of color living with disabilities in poverty. This article is a good example of how the kind of relational thinking of race, class and gender provides new angles of vision on important social issues. It also points to why race, class and gender still matter.

We hope that understanding the significance of race, class, and gender as will encourage readers to put the experiences of the United States itself into a broader context. Knowing how race, class, and gender operate within U.S. national borders should help you see beyond those borders. We hope that developing an awareness of how the increasingly global basis of society influences the configuration of race, class, and gender relationships in the United States will encourage readers to cast an increasingly inclusive perspective on the world itself.

REFERENCE

Collins, Patricia Hill. 2000. *Black Feminist Thought: Knowledge, Consciousness, and Empowerment.* New York: Routledge.

1

Missing People and Others

Joining Together to Expand the Circle

ARTURO MADRID

I am a citizen of the United States, as are my parents and as were their parents, grandparents, and great-grandparents. My ancestors' presence in what is now the United States antedates Plymouth Rock, even without taking into account any American Indian heritage I might have.

I do not, however, fit those mental sets that define America and Americans. My physical appearance, my speech patterns, my name, my profession (a professor of Spanish) create a text that confuses the reader. My normal experience is to be asked, "And where are *you* from?"

My response depends on my mood. Passive-aggressive, I answer, "From here." Aggressive-passive, I ask, "Do you mean where am I originally from?" But ultimately my answer to those follow-up questions that ask about origins will be that we have always been from here.

Overcoming my resentment I will try to educate, knowing that nine times out of ten my words fall on inattentive ears. I have spent most of my adult life explaining who I am not. I am exotic, but—as Richard Rodriguez of *Hunger of Memory* fame so painfully found out—not exotic enough ... not Peruvian, or Pakistani, or Persian, or whatever.

I am, however, very clearly *the other,* if only your everyday, garden-variety, domestic *other.* *I've* always known that I was *the other,* even before I knew the vocabulary or understood the significance of being *the other.*

I grew up in an isolated and historically marginal part of the United States, a small mountain village in the state of New Mexico, the eldest child of parents native to that region and whose ancestors had always lived there. In those vast and empty spaces, people who look like me, speak as I do, and have names like mine predominate. But the *americanos* lived among us: the descendants of those nineteenth-century immigrants who dispossessed us of our lands; missionaries who came to convert us and stayed to live among us; artists who became enchanted with our land and humanscape and went native; refugees from unhealthy climes, crowded spaces, unpleasant circumstances; and, of course, the inhabitants of Los Alamos, whose socio-cultural distance from us was moreover accentuated by the fact that they occupied a space removed from and proscribed to us. More importantly, however, they—*los americanos*—were omnipresent

SOURCE: From *Change* 20 (May/June 1988): 55–59. Reprinted by permission.

(and almost exclusively so) in newspapers, newsmagazines, books, on radio, in movies and, ultimately, on television.

Despite the operating myth of the day, school did not erase my otherness. It did try to deny it, and in doing so only accentuated it. To this day, schooling is more socialization than education, but when I was in elementary school—and given where I was—socialization was everything. School was where one became an American. Because there was a pervasive and systematic denial by the society that surrounded us that we were Americans. That denial was both explicit and implicit. My earliest memory of the former was that there were two kinds of churches: theirs and ours. The more usual was the implicit denial, our absence from the larger cultural, economic, political and social spaces—the one that reminded us constantly that we were *the other*. And school was where we felt it most acutely.

Quite beyond saluting the flag and pledging allegiance to it (a very intense and meaningful action, given that the United States was involved in a war and our brothers, cousins, uncles, and fathers were on the front lines) becoming American was learning English and its corollary—not speaking Spanish. Until very recently ours was a proscribed language—either *de jure* (by rule, by policy, by law) or *de facto* (by practice, implicitly if not explicitly; through social and political and economic pressure). I do not argue that learning English was not appropriate. On the contrary. Like it or not, and we had no basis to make any judgments on that matter, we were Americans by virtue of having been born Americans, and English was the common language of Americans. And there was a myth, a pervasive myth, that said that if we only learned to speak English well—and particularly without an accent—we would be welcomed into the American fellowship.

Senator Sam Hayakawa notwithstanding, the true text was not our speech, but rather our names and our appearance, for we would always have an accent, however perfect our pronunciation, however excellent our enunciation, however divine our diction. That accent would be heard in our pigmentation, our physiognomy, our names. We were, in short, *the other*.

Being *the other* means feeling different; is awareness of being distinct; is consciousness of being dissimilar. It means being outside the game, outside the circle, outside the set. It means being on the edges, on the margins, on the periphery. Otherness means feeling excluded, closed out, precluded, even disdained and scorned. It produces a sense of isolation, of apartness, of disconnectedness, of alienation.

Being *the other* involves a contradictory phenomenon. On the one hand being *the other* frequently means being invisible. Ralph Ellison wrote eloquently about that experience in his magisterial novel *The Invisible Man*. On the other hand, being *the other* sometimes involves sticking out like a sore thumb. What is she/he doing here?

If one is *the other,* one will inevitably be perceived unidimensionaly; will be seen stereotypically; will be defined and delimited by mental sets that may not bear much relation to existing realities. There is a darker side to otherness as well. *The other* disturbs, disquiets, discomforts. It provokes distrust and suspicion. *The other* makes people feel anxious, nervous, apprehensive, even fearful. *The other* frightens, scares.

For some of us being *the other* is only annoying; for others it is debilitating; for still others it is damning. Many try to flee otherness by taking on protective colorations that provide invisibility, whether of dress or speech or manner or name. Only a fortunate few succeed. For the majority, otherness is permanently sealed by physical appearance. For the rest, otherness is betrayed by ways of being, speaking or of doing.

I spent the first half of my life downplaying the significance and conse-quences of otherness. The second half has seen me wrestling to understand its complex and deeply ingrained realities; striving to fathom why otherness denies us a voice or visibility or validity in American society and its institutions; strug-gling to make otherness familiar, reasonable, even normal to my fellow Americans.

I am also a missing person. Growing up in northern New Mexico I had only a slight sense of our being missing persons. *Hispanos,* as we called (and call) our-selves in New Mexico, were very much a part of the fabric of the society and there were Hispano professionals everywhere about me: doctors, lawyers, school teachers, and administrators. My people owned businesses, ran organizations and were both appointed and elected public officials.

To be sure, we did not own the larger businesses, nor at the time were we permitted to be part of the banking world. Other than that, however, people who looked like me, spoke like me, and had names like mine, predominated. There was, to be sure, Los Alamos, but as I have said, it was removed from our realities.

My awareness of our absence from the larger institutional life of society became sharper when I went off to college, but even then it was attenuated by the circumstances of history and geography. The demography of Albuquerque still strongly reflected its historical and cultural origins, despite the influx of Mid-westerners and Easterners. Moreover, many of my classmates at the University of New Mexico in Albuquerque were Hispanos, and even some of my professors were.

I thought that would change at UCLA, where I began graduate studies in 1960. Los Angeles already had a very large Mexican population, and that popu-lation was visible even in and around Westwood and on the campus. Many of the groundskeepers and food-service personnel at UCLA were Mexican. But Mexican-American students were few and mostly invisible, and I do not recall seeing or knowing a single Mexican-American (or, for that matter, black, Asian, or American Indian) professional on the staff or faculty of that institution during the five years I was there.

Needless to say, persons like me were not present in any capacity at Dart-mouth College—the site of my first teaching appointment—and, of course, were not even part of the institutional or individual mind-set. I knew then that we—a "we" that had come to encompass American Indians, Asian Americans, black Americans, Puerto Ricans, and women—were truly missing persons in American institutional life.

Over the past three decades, the *de jure* and *de facto* segregations that have historically characterized American institutions have been under assault. As a consequence, minorities and women have become part of American institutional life, and although there are still many areas where we are not to be found, the missing persons phenomenon is not as pervasive as it once was.

However, the presence of *the other,* particularly minorities, in institutions and in institutional life, is, as we say in Spanish, *a flor de tierra;* spare plants whose roots do not go deep, a surface phenomenon, vulnerable to inclemencies of an economic, political, or social nature.

Our entrance into and our status in institutional life is not unlike a scenario set forth by my grandmother's pastor when she informed him that she and her family were leaving their mountain village to relocate in the Rio Grande Valley. When he asked her to promise that she would remain true to the faith and continue to involve herself in the life of the church, she assured him that she would and asked him why he thought she would do otherwise.

"Doña Trinidad," he told her, "in the Valley there is no Spanish church. There is only an American church." "But," she protested, "I read and speak English and would be able to worship there." Her pastor's response was: "It is possible that they will not admit you, and even if they do, they might not accept you. And that is why I want you to promise me that you are going to go to church. Because if they don't let you in through the front door, I want you to go in through the back door. And if you can't get in through the back door, go in the side door. And if you are unable to enter through the side door I want you to go in through the window. What is important is that you enter and that you stay."

Some of us entered institutional life through the front door; others through the back door; and still others through side doors. Many, if not most of us, came in through windows and continue to come in through windows. Of those who entered through the front door, some never made it past the lobby; others were ushered into corners and niches. Those who entered through back and side doors inevitably have remained in back and side rooms. And those who entered through windows found enclosures built around them. For despite the lip service given to the goal of the integration of minorities into institutional life, what has occurred instead is ghettoization, marginalization, isolation.

Not only have the entry points been limited: in addition, the dynamics have been singularly conflictive. Gaining entry and its corollary—gaining space—have frequently come as a consequence of demands made on institutions and institutional officers. Rather than entering institutions more or less passively, minorities have, of necessity, entered them actively, even aggressively. Rather than taking, they have demanded. Institutional relations have thus been adversarial, infused with specific and generalized tensions.

The nature of the entrance and the nature of the space occupied have greatly influenced the view and attitudes of the majority population within those institutions. All of us are put into the same box; that is, no matter what the individual reality, the assessment of the individual is inevitably conditioned by a perception that is held of the class. Whatever our history, whatever our record, whatever our validations, whatever our accomplishments, by and large we are perceived unidimensionally and are dealt with accordingly.

My most recent experience in this regard is atypical only in its explicitness. A few years ago I allowed myself to be persuaded to seek the presidency of a large and prestigious state university. I was invited for an interview and presented myself before the selection committee, which included members of the board of trustees. The opening question of the brief but memorable interview was directed at me by a member of that august body. "Dr. Madrid," he asked, "why does a one-dimensional person like you think he can be the president of a multi-dimensional institution like ours?"

If, as I happen to believe, the well-being of a society is directly related to the degree and extent to which all of its citizens participate in its institutions, we have a challenge before us. One of the strengths of our society—perhaps its main strength—has been a tradition of struggle against clubbishness, exclusivity, and restriction.

Today, more than ever, given the extraordinary changes that are taking place in our society, we need to take up that struggle again—irritating, grating, troublesome, unfashionable, unpleasant as it is. As educated and educator members of this society, we have a special responsibility for leading the struggle against marginalization, exclusion, and alienation.

Let us work together to assure that all American institutions, not just its precollegiate educational and penal institutions, reflect the diversity of our society. Not to do so is to risk greater alienation on the part of a growing segment of our society. It is to risk increased social tension in an already conflictive world. And ultimately it is to risk the survival of a range of institutions that, for all their defects and deficiencies, permit us the space, the opportunity, and the freedom to improve our individual and collective lot; to guide the course of our government, and to redress whatever grievances we have. Let us join together to expand, not to close the circle.

2

Chappals and Gym Shorts
An Indian Muslim Woman in the Land of Oz

ALMAS SAYEED

It was finals week during the spring semester of my sophomore year at the University of Kansas, and I was buried under mounds of papers and exams. The stress was exacerbated by long nights, too much coffee and a chronic, building pain in my permanently splintered shins (left over from an old sports injury). Between attempting to understand the nuances of Kant's *Critique of Pure Reason* and applying the latest game-theory models to the 1979 Iranian revolution, I was regretting my decision to pursue majors in philosophy, women's studies *and* international studies.

My schedule was not exactly permitting much down time. With a full-time school schedule, a part-time job at Lawrence's domestic violence shelter and preparations to leave the country in three weeks, I was grasping to hold onto what little sanity I had left. Wasn't living in Kansas supposed to be more laid-back than this? After all, Kansas was the portal to the magical land of Oz, where wicked people melt when doused with mop water and bright red, sparkly shoes could substitute for the services of American Airlines, providing a quick getaway. Storybook tales aside, the physical reality of this period was that my deadlines were inescapable. Moreover, the most pressing of these deadlines was completely non-school related: my dad, on his way home to Wichita, was coming for a brief visit. This would be his first stay by himself, without Mom to accompany him or act as a buffer.

Dad visited me the night before my most difficult exam. Having just returned from spending time with his family—a group of people with whom he historically had an antagonistic relationship—Dad seemed particularly relaxed in his stocky six-foot-four frame. Wearing one of the more subtle of his nineteen cowboy hats, he arrived at my door hungry, greeting me in Urdu, our mother tongue, and laden with gifts from Estée Lauder for his only daughter. Never mind that I rarely wore makeup and would have preferred to see the money spent on my electric bill or a stack of feminist theory books from my favorite used bookstore. If Dad's visit was going to include a conversation about how little I use beauty products, I was not going to be particularly receptive.

SOURCE: From *Colonize This! Young Women of Color on Today's Feminism* edited by Daisy Hernandez and Bushra Rehman, pp. 203–214. Copyright © 2002 Daisy Hernandez and Bushar Rehman. Reproduced with permission of Avalon Publishing Group.

"Almas," began my father from across the dinner table, speaking in his British-Indian accent infused with his love of Midwestern colloquialisms, "You know that you won't be a spring chicken forever. While I was in Philadelphia, I realized how important it is for you to begin thinking about our culture, religion and your future marriage plans. I think it is time we began a two-year marriage plan so you can find a husband and start a family. I think twenty-two will be a good age for you. You should be married by twenty-two."

I needed to begin thinking about the "importance of tradition" and be married by twenty-two? This, from the only Indian man I knew who had Alabama's first album on vinyl and loved to spend long weekends in his rickety, old camper near Cheney Lake, bass fishing and listening to traditional Islamic Quavali music? My father, in fact, was in his youth crowned "Mr. Madras," weightlifting champion of 1965, and had left India to practice medicine and be an American cowboy in his spare time. But he wanted *me* to aspire to be a "spring chicken," maintaining some unseen hearth and home to reflect my commitment to tradition and culture.

Dad continued, "I have met a boy that I like for you very much. Masoud's son, Mahmood. He is a good Muslim boy, tells great jokes in Urdu and is a promising engineer. We should be able to arrange something. I think you will be very happy with him!" Dad concluded with a satisfied grin.

Masoud, Dad's cousin? This would make me and Mahmood distant relatives of some sort. And Dad wants to "arrange something"? I had brief visions of being paraded around a room, serving tea to strangers in a sari or a shalwar kameez (a traditional South Asian outfit for women) wearing a long braid and chappals (flat Indian slippers), while Dad boasted of my domestic capabilities to increase my attractiveness to potential suitors. I quickly flipped through my mental Rolodex of rhetorical devices, acquired during years of women's studies classes and found the card blank. No doubt, even feminist scholar Catherine MacKinnon would have been rendered speechless sitting across the table in a Chinese restaurant speaking to my overzealous father.

It is not that I hadn't already dealt with the issue. In fact, we had been here before, ever since the marriage proposals began (the first one came when I was fourteen). Of course, when they first began, it was a family joke, as my parents understood that I was to continue my education. The jokes, however, were always at my expense: "You received a proposal from a nice boy living in our mosque. He is studying medicine," my father would come and tell me with a huge, playful grin. "I told him that you weren't interested because you are too busy with school. And anyway you can't cook or clean." My father found these jokes particularly funny, given my dislike of household chores. In this way, the eventuality of figuring out how to deal with these difficult issues was postponed with humor.

Dad's marriage propositions also resembled conversations that we had already had about my relationship to Islamic practices specific to women, some negotiated in my favor and others simply shelved for the time being. Just a year ago, Dad had come to me while I was home for the winter holidays, asking me to begin wearing *hijab,* the traditional headscarf worn by Muslim women. I categorically refused, maintaining respect for those women who chose to do so. I understood that for numerous women, as well as for Dad, hijab symbolized something much more than covering a woman's body or hair; it symbolized a way to adhere to religious and cultural traditions in order to prevent complete Western immersion. But even my sympathy for this concern didn't change my feeling that hijab constructed me as a woman first and a human being second. Veiling seemed to reinforce the fact that inequality between the sexes was a natural, inexplicable phenomenon that is impossible to overcome, and that women should cover themselves, accommodating an unequal hierarchy, for the purposes of modesty and self-protection. I couldn't reconcile these issues and refused my father's request to don the veil. Although there was tension—Dad claimed I had yet to have my religious awakening—he chose to respect my decision.

Negotiating certain issues had always been part of the dynamic between my parents and me. It wasn't that I disagreed with them about everything. In fact, I had internalized much of the Islamic perspective of the female body while simultaneously admitting to its problematic nature (to this day, I would rather wear a wool sweater than a bathing suit in public, no matter how sweltering the weather). Moreover, Islam became an important part of differentiating myself from other American kids who did not have to find a balance between two opposing cultures. Perhaps Mom and Dad recognized the need to concede certain aspects of traditional Islamic norms, because for all intents and purposes, I had been raised in the breadbasket of America.

By the time I hit adolescence, I had already established myself outside of the social norm of the women in my community. I was an athletic teenager, a competitive tennis player and a budding weightlifter. After a lot of reasoning with my parents, I was permitted to wear shorts to compete in tennis tournaments, but I was not allowed to show my legs or arms (no tank tops) outside of sports. It was a big deal for my parents to have agreed to allow me to wear shorts in the first place. The small community of South Asian Muslim girls my age, growing up in Wichita, became symbols of the future of our community in the United States. Our bodies became the sites to play out cultural and religious debates. Much in the same way that Lady Liberty had come to symbolize idealized stability in the *terra patria* of America, young South Asian girls in my community were expected to embody the values of a preexisting social structure. We were scrutinized for what we said, what we wore, being seen with boys in public and for lacking grace and piety. Needless to say, because of disproportionate muscle mass, crooked teeth, huge Lucy glasses and a disposition to walk pigeon-toed, I was not among the favored.

To add insult to injury, Mom nicknamed me "Amazon Woman," lamenting the fact that she—a beautiful, petite lady—had produced such a graceless, unfeminine creature. She was horrified by how freely I got into physical fights with my younger brother and arm wrestled boys at school. She was particularly frustrated by the fact that I could not wear her beautiful Indian jewelry, especially her bangles and bracelets, because my wrists were too big. Special occasions, when I had to slather my wrists with tons of lotion in order to squeeze my hands into her tiny bangles, often bending the soft gold out of shape, caused us both infinite amounts of grief. I was the snot-nosed, younger sibling of the Bollywood (India's Hollywood) princess that my mother had in mind as a more appropriate representation of an Indian daughter. Rather, I loved sports, sports figures and books. I hated painful makeup rituals and tight jewelry.

It wasn't that I had a feminist awakening at an early age. I was just an obnoxious kid who did not understand the politics raging around my body. I did not possess the tools to analyze or understand my reaction to this process of social conditioning and normalization until many years later, well after I had left my parent's house and the Muslim community in Wichita. By positioning me as a subject of both humiliation and negotiation, Mom and Dad had inadvertently laid the foundations for me to understand and scrutinize the process of conditioning women to fulfill particular social obligations.

What was different about my dinner conversation with Dad that night was a sense of immediacy and detail. Somehow discussion about a "two-year marriage plan" seemed to encroach on my personal space much more than had previous jokes about my inability to complete my household chores or pressure to begin wearing hijab. I was meant to understand that when it came to marriage, I was up against an invisible clock (read: social norms) that would dictate how much time I had left: how much time I had left to remain desirable, attractive and marriageable. Dad was convinced that it was his duty to ensure my long-term security in a manner that reaffirmed traditional Muslim culture in the face of an often hostile foreign community. I recognized that the threat was not as extreme as being shipped off to India in order to marry someone I had never met. The challenge was far more subtle than this. I was being asked to choose my community; capitulation through arranged marriage would show my commitment to being Indian, to being a good Muslim woman and to my parents by proving that they had raised me with a sense of duty and the willingness to sacrifice for my culture, religion and family.

There was no way to tell Dad about my complicated reality. Certain characteristics of my current life already indicated failure by such standards. I was involved in a long-term relationship with a white man, whose father was a prison guard on death row, an occupation that would have mortified my upper-middle-class, status-conscious parents. I was also struggling with an insurmountable crush on an *actress* in the Theater and Film Department. I was debating my sexuality in terms of cultural compatibility as well as gender. Moreover, there was no way to tell Dad that my social circle was supportive of these non-traditional romantic explorations. My friends in college had radically altered my perceptions of marriage and family. Many of my closest friends, including my roommates, were coming to terms with their own life-choices, having recently come out of the closet but unable to tell their families about their decisions. I felt inextricably linked to this group of women, who, like me, often had to lead double lives. The immediacy of fighting for issues such as queer rights, given the strength and beauty of my friends' romantic relationships, held far more appeal for me than the topics of marriage and security that my father broached over our Chinese dinner. There was no way to explain to my loving, charismatic, steadfastly religious father, who was inclined to the occasional violent

outburst, that a traditional arranged marriage not only conflicted with the feminist ideology I had come to embrace, but it seemed almost petty in the face of
larger, more pressing issues.

Although I had no tools to answer my father that night at dinner, feminist
theory had provided me with the tools to understand *why* my father and I were
engaged in the conversation in the first place. I understood that in his mind, Dad
was fulfilling his social obligation as father and protector. He worried about my
economic stability and, in a roundabout way, my happiness. Feminism and
community activism had enabled me to understand these things as part of a proscribed role for women. At the same time, growing up in Kansas and coming to
feminism here meant that I had to reconcile a number of different issues. I am a
Muslim, first-generation Indian, feminist woman studying in a largely homogeneous white, Christian community in Midwestern America. What sacrifices are
necessary for me to retain my familial relationships as well as a sense of personal
autonomy informed by Western feminism?

The feminist agenda in my community is centered on ending violence
against women, fighting for queer rights and maintaining women's reproductive
choices. As such, the way that I initially became involved with this community
was through community projects such as "Womyn Take Back the Night,"
attending pride rallies and working at the local domestic violence shelter. I am
often the only woman of color in feminist organizations and at feminist events.
Despite having grown up in the Bible belt, it is difficult for me to relate to stories
told by my closest friends of being raised on cattle ranches and farms, growing up
Christian by default and experiencing the strict social norms of small, religious
communities in rural Kansas. Given the context of this community—a predominantly white, middle-class, college town—I have difficulty explaining that my
feminism has to address issues like, "I should be able to wear *both* hijab *and* shorts
if I choose to." The enormity of our agenda leaves little room to debate issues
equally important but applicable only to me, such as the meaning of veiling,
arranged marriages versus dating and how the north-south divide uniquely
disadvantages women in the developing world.

It isn't that the women in my community ever turned to me and said, "Hey
you, brown girl, stop diluting our priorities." To the contrary, the majority of
active feminists in my community are eager to listen and understand my sometimes divergent perspective. We have all learned to share our experiences as
women, students, mothers, partners and feminists. We easily relate to issues of
male privilege, violence against women and figuring out how to better appreciate the sacrifices made by our mothers. From these commonalities we have
learned to work together, creating informal social networks to complete
community projects.

The difficulty arises when trying to put this theory and discussion into
practice. Like last year, when our organization, the Womyn's Empowerment
Action Coalition, began plans for the Womyn Take Back the Night march and
rally, a number of organizers were eager to include the contribution of a petite,
white belly dancer in the pre-march festivities. When I voiced my concern that
historically belly dancing had been used as a way to objectify women's bodies in

the Middle East, one of my fellow organizers (and a very good friend) laughed and called me a prude: "We're in Kansas, Almas," she said. "It doesn't mean the same thing in our culture. It is different here than over *there*." I understood what she meant, but having just returned from seven months in the West Bank, Palestine two months before, for me over there *was* over here. In the end, the dance was included while I wondered about our responsibility to women outside of the United States and our obligation to address the larger social, cultural issues of the dance itself.

To reconcile the differences between my own priorities and those of the women I work with, I am learning to bridge the gap between the Western white women (with the occasional African American or Chicana) feminist canon and my own experience as a first-generation Indian Muslim woman living in the Midwest. I struggle with issues like cultural differences, colonialism, Islam and feminism and how they relate to one another. The most difficult part has been to get past my myopic vision of simply laying feminist theory written by Indian, Muslim or postcolonial theorists on top of American-Western feminism. With the help of feminist theory and other feminists, I am learning to dissect Western models of feminism, trying to figure out what aspects of these models can be applied to certain contexts. To this end, I have had the privilege of participating in projects abroad, in pursuit of understanding feminism in other contexts.

For example, while living with my extended family in India, I worked for a micro-credit affiliate that advised women on how to get loans and start their own businesses. During this time I learned about the potential of micro-enterprise as a weapon against the feminization of poverty. Last year, I spent a semester in the West Bank, Palestine, studying the link between women and economics in transitional states and beginning to understand the importance of women's efforts during revolution. These experiences have been invaluable to me as a student of feminism and women's mobilization efforts. They have also shaped my personal development, helping me understand where the theoretical falls short of solving for the practical. In Lawrence, I maintain my participation in local feminist projects. Working in three different contexts has highlighted the amazing and unique ways in which feminism develops in various cultural settings yet still maintains certain commonalities.

There are few guidebooks for women like me who are trying to negotiate the paradigm of feminism in two different worlds. There is a delicate dance here that I must master—a dance of negotiating identity within interlinking cultural spheres. When faced with the movement's expectations of my commitment to local issues, it becomes important for me to emphasize that differences in culture and religion are also "local issues." This has forced me to change my frame of reference, developing from a rebellious tomboy who resisted parental imposition to a budding social critic, learning how to be a committed feminist and still keep my cultural, religious and community ties. As for family, we still negotiate despite the fact that Dad's two-year marriage plan has yet to come to fruition in this, my twenty-second year.

3

From a Native Daughter

HAUNANI-KAY TRASK

E noi'i wale mai nō ka haole, a,
'a'ole e pau nō hana a Hawai'i 'imi loa
Let the haole freely research us in detail
But the doings of deep delving Hawai'i
will not be exhausted.

—Kepelino
19th-century Hawaiian historian

When I was young, the story of my people was told twice: once by my parents, then again by my school teachers. From my *'ohana* (family), I learned about the life of the old ones: how they fished and planted by the moon; shared all the fruits of their labors, especially their children; danced in great numbers for long hours; and honored the unity of their world in intricate genealogical chants. My mother said Hawaiians had sailed over thousands of miles to make their home in these sacred islands. And they had flourished, until the coming of the *haole* (whites).

At school, I learned that the "pagan Hawaiians" did not read or write, were lustful cannibals, traded in slaves, and could not sing. Captain Cook had "discovered" Hawai'i and the ungrateful Hawaiians had killed him. In revenge, the Christian god had cursed the Hawaiians with disease and death.

I learned the first of these stories from speaking with my mother and father. I learned the second from books. By the time I left for college, the books had won out over my parents, especially since I spent four long years in a missionary boarding school for Hawaiian children.

When I went away I understood the world as a place and a feeling divided in two: one *haole* (white), and the other *kānaka* (Native). When I returned ten years later with a Ph.D., the division was sharper, the lack of connection more painful. There was the world that we lived in—my ancestors, my family, and my people—and then there was the world historians described. This world, they had written, was the truth. A primitive group, Hawaiians had been ruled by bloodthirsty priests and despotic kings who owned all the land and kept our people in feudal subjugation. The chiefs were cruel, the people poor.

SOURCE: From *A Native Daughter* by Haunani-Kay Trask. Common Courage Press, Monroe, Maine 1993. Reprinted by permission of the author.

But this was not the story my mother told me. No one had owned the land before the *haole* came; everyone could fish and plant, except during sacred periods. And the chiefs were good and loved their people.

Was my mother confused? What did our *kūpuna* (elders) say? They replied: Did these historians (all *haole*) know the language? Did they understand the chants? How long had they lived among our people? Whose stories had they heard?

None of the historians had ever learned our mother tongue. They had all been content to read what Europeans and Americans had written. But why did scholars, presumably well-trained and thoughtful, neglect our language? Not merely a passageway to knowledge, language is a form of knowing by itself; a people's way of thinking and feeling is revealed through its music.

I sensed the answer without needing to answer. From years of living in a divided world, I knew the historian's judgment: *There is no value in things Hawai-ian; all value comes from things haole.*

Historians, I realized, were very much like missionaries. They were a part of the colonizing horde. One group colonized the spirit; the other, the mind. Frantz Fanon had been right, but not just about Africans. He had been right about the bondage of my own people: "By a kind of perverted logic, [colonial-ism] turns to the past of the oppressed people, and distorts, disfigures, and destroys it" (1963:210). The first step in the colonizing process, Fanon had writ-ten, was the deculturation of a people. What better way to take our culture than to remake our image? A rich historical past became small and ignorant in the hands of Westerners. And we suffered a damaged sense of people and culture because of this distortion.

Burdened by a linear, progressive conception of history and by an assump-tion that Euro-American culture flourishes at the upper end of that progression, Westerners have told the history of Hawai'i as an inevitable if occasionally bittersweet triumph of Western ways over "primitive" Hawaiian ways. A few authors—the most sympathetic—have recorded with deep-felt sorrow the pass-ing of our people. But in the end, we are repeatedly told, such an eclipse was for the best.

Obviously it was best for Westerners, not for our dying multitudes. This is why the historian's mission has been to justify our passing by celebrating Western dominance. Fanon would have called this missionizing, intellectual colonization. And it is clearest in the historian's insistence that *pre-haole* Hawaiian land tenure was "feudal"—a term that is now applied, without question, in every mono-graph, in every schoolbook, and in every tour guide description of my people's history.

From the earliest days of Western contact my people told their guests that *no one* owned the land. The land—like the air and the sea—was for all to use and share as their birthright. Our chiefs were *stewards* of the land; they could not own or privately possess the land any more than they could sell it.

But the *haole* insisted on characterizing our chiefs as feudal landlords and our people as serfs. Thus, a European term which described a European practice founded on the European concept of private property—feudalism—was imposed upon a people halfway around the world from Europe and vastly different from her in every conceivable way. More than betraying an ignorance of Hawaiian culture and history, however, this misrepresentation was malevolent in design.

By inventing feudalism in ancient Hawai'i, Western scholars quickly transformed a spiritually based, self-sufficient economic system of land use and occupancy into an oppressive, medieval European practice of divine right ownership, with the common people tied like serfs to the land. By claiming that a Pacific people lived under a European system—that the Hawaiians lived under feudalism—Westerners could then degrade a successful system of shared land use with a pejorative and inaccurate Western term. Land tenure changes instituted by Americans and in line with current Western notions of private property were then made to appear beneficial to the Hawaiians. But in practice, such changes benefited the *haole,* who alienated the people from the land, taking it for themselves.

The prelude to this land alienation was the great dying of the people. Barely half a century after contact with the West, our people had declined in number by eighty percent. Disease and death were rampant. The sandalwood forests had been stripped bare for international commerce between England and China. The missionaries had insinuated themselves everywhere. And a debt-ridden Hawaiian king (there had been no king before Western contact) succumbed to enormous pressure from the Americans and followed their schemes for dividing the land.

This is how private property land tenure entered Hawai'i. The common people, driven from their birthright, received less than one percent of the land. They starved while huge haole-owned sugar plantations thrived.

And what had the historians said? They had said that the Americans "liberated" the Hawaiians from an oppressive "feudal" system. By inventing a false feudal past, the historians justify—and become complicitous in—massive American theft.

Is there "evidence"—as historians call it—for traditional Hawaiian concepts of land use? The evidence is in the sayings of my people and in the words they wrote more than a century ago, much of which has been translated. However, historians have chosen to ignore any references here to shared land use. But there is incontrovertible evidence in the very structure of the Hawaiian language. If the historians had bothered to learn our language (as any American historian of France would learn French) they would have discovered that we show possession in two ways: through the use of an "a" possessive, which reveals acquired status, and through the use of an "o" possessive, which denotes inherent status. My body (*ko'u kino*) and my parents (*ko'u mākua*), for example, take the "o" form; most material objects, such as food *(ka'u mea'ai)* take the "a" form. But land, like one's body and one's parents, takes the "o" possessive (*ko'u 'āina*). Thus, in our way of speaking, land is inherent to the people; it is like our bodies and our parents. The people cannot exist without the land, and the land cannot exist without the people.

Every major historian of Hawai'i has been mistaken about Hawaiian land tenure. The chiefs did not own the land: they *could not* own the land. My mother was right and the *haole* historians were wrong. If they had studied our language they would have known that no one owned the land. But was their failing merely ignorance, or simple ethnocentric bias?

No, I did not believe them to be so benign. As I read on, a pattern emerged in their writing. Our ways were inferior to those of the West, to those of the historians' own culture. We were "less developed," or "immature," or "authoritarian." In some tellings we were much worse. Thus, Gavan Daws (1968), the most famed modern historian of Hawai'i, had continued a tradition established earlier by missionaries Hiram Bingham (1848; reprinted, 1981) and Sheldon Dibble (1909), by referring to the old ones as "thieves" and "savages" who regularly practiced infanticide and who, in contrast to "civilized" whites, preferred "lewd dancing" to work. Ralph Kuykendall (1938), long considered the most thorough if also the most boring of historians of Hawai'i, sustained another fiction—that my ancestors owned slaves, the outcast *kauwā*. This opinion, as well as the description of Hawaiian land tenure as feudal, had been supported by respected sociologist Andrew Lind.... Finally, nearly all historians had refused to accept our genealogical dating of A.D. 400 or earlier for our arrival from the South Pacific. They had, instead, claimed that our earliest appearance in Hawai'i could only be traced to A.D.1100. Thus at least seven hundred years of our history were repudiated by "superior" Western scholarship. Only recently have archaeological data confirmed what Hawaiians had said these many centuries (Tuggle 1979).[1]

Suddenly the entire sweep of our written history was clear to me. I was reading the West's view of itself through the degradation of my own past. When historians wrote that the king owned the land and the common people were bound to it, they were saying that ownership was the only way human beings in their world could relate to the land, and in that relationship, some one person had to control both the land and the interaction between humans.

And when they said that our chiefs were despotic, they were telling of their own society, where hierarchy always results in domination. Thus any authority or elder is automatically suspected of tyranny.

And when they wrote that Hawaiians were lazy, they meant that work must be continuous and ever a burden.

And when they wrote that we were promiscuous, they meant that lovemaking in the Christian West is a sin.

And when they wrote that we were racist because we preferred our own ways to theirs, they meant that their culture needed to dominate other cultures.

And when they wrote that we were superstitious, believing in the *mana* of nature and people, they meant that the West has long since lost a deep spiritual and cultural relationship to the earth.

And when they wrote that Hawaiians were "primitive" in their grief over the passing of loved ones, they meant that the West grieves for the living who do not walk among their ancestors.

For so long, more than half my life, I had misunderstood this written record, thinking it described my own people. But my history was nowhere present. For we had not written. We had chanted and sailed and fished and built and prayed. And we had told stories through the great blood lines of memory: genealogy.

To know my history, I had to put away my books and return to the land. I had to plant *taro* in the earth before I could understand the inseparable bond between people and '*āina*. I had to feel again the spirits of nature and take gifts of plants and fish to the ancient altars. I had to begin to speak my language with our elders and leave long silences for wisdom to grow. But before anything else, I needed to learn the language like a lover so that I could rock within her and lie at night in her dreaming arms.

There was nothing in my schooling that had told me of this, or hinted that somewhere there was a longer, older story of origins, of the flowing of songs out to a great but distant sea. Only my parents' voices, over and over, spoke to me of a Hawaiian world. While the books spoke from a different world, a Western world.

And yet, Hawaiians are not of the West. We are of *Hawai'i Nei,* this world where I live, this place, this culture, this *'āina.*

What can I say, then, to Western historians of my place and people? Let me answer with a story.

A while ago I was asked to appear on a panel on the American overthrow of our government in 1893. The other panelists were all *haole.* But one was a *haole* historian from the American continent who had just published a book on what he called the American anti-imperialists. He and I met briefly in preparation for the panel. I asked him if he knew the language. He said no. I asked him if he knew the record of opposition to our annexation to America. He said there was no real evidence for it, just comments here and there. I told him that he didn't understand and that at the panel I would share the evidence. When we met in public and spoke, I said this:

There is a song much loved by our people. It was written after Hawai'i had been invaded and occupied by American marines. Addressed to our dethroned Queen, it was written in 1893, and tells of Hawaiian feelings for our land and against annexation to the United States. Listen to our lament:

Kaulana nā pua a'o Hawai'i	Famous are the children of Hawai'i
Kupa'a ma hope o ka 'āina	Who cling steadfastly to the land
Hiki mai ka 'elele o ka loko 'ino	Comes the evil-hearted with
Palapala 'ānunu me ka pākaha	A document greedy for plunder
Pane mai Hawai'i moku o Keawe	Hawai'i, island of Keawe, answers
Kokua nā hono a'o Pi'ilani	The bays of Pi'ilani [of Maui, Moloka'i, and Lana'i] help
Kako'o mai Kaua'i Mano	Kaua'i of Mano assists
Pau pu me ke one o Kakuhihewa	Firmly together with the sands of Kakuhihewa
'A'ole a'e kau i ka pūlima	Do not put the signature
Maluna o ka pepa o ka 'enemi	On the paper of the enemy
Ho'ohui 'āina kū'ai hewa	Annexation is wicked sale
I ka pono sivila a'o ke kānaka	Of the civil rights of the Hawaiian people
Mahope mākou o Lili'ūlani	We support Lili'uokalani
A loa'a 'e ka pono o ka 'āina	Who has earned the right to the land
Ha'ina 'ia mai ana ka puana	The story is told
'O ka po'e i aloha i ka 'āina	Of the people who love the land

This song, I said, continues to be sung with great dignity at Hawaiian political gatherings, for our people still share the feelings of anger and protest that it conveys.

But our guest, the *haole* historian, answered that this song, although beautiful, was not evidence of either opposition or of imperialism from the Hawaiian perspective.

Many Hawaiians in the audience were shocked at his remarks, but, in hindsight, I think they were predictable. They are the standard response of the historian who does not know the language and has no respect for its memory.

Finally, I proceeded to relate a personal story, thinking that surely such a tale could not want for authenticity since I myself was relating it. My *tūtū* (grandmother) had told my mother who had told me that at the time of the overthrow a great wailing went up throughout the islands, a wailing of weeks, a wailing of impenetrable grief, a wailing of death. But he remarked again, this too is not evidence.

And so, history goes on, written in long volumes by foreign people. Whole libraries begin to form, book upon book, shelf upon shelf. At the same time, the stories go on, generation to generation, family to family.

Which history do Western historians desire to know? Is it to be a tale of writings by their own countrymen, individuals convinced of their "unique" capacity for analysis, looking at us with Western eyes, thinking about us within Western philosophical contexts, categorizing us by Western indices, judging us by Judeo-Christian morals, exhorting us to capitalist achievements, and finally, leaving us an authoritative-because-Western record of their complete misunderstanding?

All this has been done already. Not merely a few times, but many times. And still, every year, there appear new and eager faces to take up the same telling, as if the West must continue, implacably, with the din of its own disbelief. But there is, as there has been always, another possibility. If it is truly our history Western historians desire to know, they must put down their books, and take up our practices. First, of course, the language. But later, the people, the *āina*, the stories. Above all, in the end, the stories. Historians must listen, they must hear the generational connections, the reservoir of sounds and meanings.

They must come, as American Indians suggested long ago, to understand the land. Not in the Western way, but in the indigenous way, the way of living within and protecting the bond between people and '*āina*. This bond is cultural, and it can be understood only culturally. But because the West has lost any cultural understanding of the bond between people and land, it is not possible to know this connection through Western culture. This means that the history of indigenous people cannot be written from within Western culture. Such a story is merely the West's story of itself.

Our story remains unwritten. It rests within the culture, which is inseparable from the land. To know this is to know our history. To write this is to write of the land and the people who are born from her.

NOTES

1. See also Fornander (1878—85; reprinted, 1981). Lest one think these sources antiquated, it should be noted that there exist only a handful of modern scholarly works on the history of Hawai'i. The most respected are those by Kuykendall (1938) and

Daws (1968), and a social history of the 20th century by Lawrence Fuchs (1961). Of these, only Kuykendall and Daws claim any knowledge of pre-*haole* history, while concentrating on the 19th century. However, countless popular works have relied on these two studies which, in turn, are themselves based on primary sources written in English by extremely biased, anti-Hawaiian Westerners such as explorers, traders, missionaries (e.g., Bingham [1848; reprinted, 1981] and Dibble [1909]), and sugar planters. Indeed, a favorite technique of Daws's—whose *Shoal of Time* was once the most acclaimed and recent general history—is the lengthy quotation without comment of the most racist remarks by missionaries and planters. Thus, at one point, half a page is consumed with a "white man's burden" quotation from an 1886 *Planters Monthly* article ("It is better here that the white man should rule ...," etc., p. 213). Daws's only comment is, "The conclusion was inescapable." To get a sense of such characteristic contempt for Hawaiians, one has but to read the first few pages, where Daws refers several times to the Hawaiians as "savages" and "thieves" and where he approvingly has Captain Cook thinking, "It was a sensible primitive who bowed before a superior civilization" (p. 2). See also—among examples too numerous to cite—his glib description of sacred *hula* as a "frivolous diversion," which, instead of work, the Hawaiians "would practice energetically in the hot sun for days on end ... their bare brown flesh glistening with sweat" (pp. 65–66). Daws, who repeatedly displays an affection for descriptions of Hawaiian skin color, taught Hawaiian history for some years at the University of Hawai'i. He once held the Chair of Pacific History at the Australian National University's Institute of Advanced Studies.

Postscript: Since this article was written, the first scholarly history by a Native Hawaiian was published in English: *Native Land and Foreign Desires* by Lilikalà Kame'eleihiwa (Honolulu: Bishop Museum Press, 1992).

REFERENCES

Bingham, Hiram. 1981. *A Residence of Twenty-one Years in the Sandwich Islands*. Tokyo: Charles E. Tuttle.

Daws, Gavan. 1968. *Shoal of Time: A History of the Hawaiian Islands*. Honolulu: University of Hawai'i Press.

Dibble, Sheldon. 1909. *A History of the Sandwich Islands*. Honolulu: Thrum Publishing.

Fanon, Frantz. 1963. *The Wretched of the Earth*. New York: Grove Press.

Fornander, Abraham. 1981. *An Account of the Polynesian Race, Its Origins, and Migrations and the Ancient History of the Hawaiian People to the Times of Kamehameha I*. Routledge, Vermont: Charles E. Tuttle.

Fuchs, Lawrence H. 1961. *Hawaii Pono: A Social History*. New York: Harcourt Brace & World.

Kuykendall, Ralph. 1938. *The Hawaiian Kingdom, 1778–1854: Foundation and Transformation*. Honolulu: University of Hawai'i Press.

Tuggle, H. David. 1979. "Hawai'i," in *The Prehistory of Polynesia*, ed. Jessie D. Jennings. Cambridge: Harvard University Press.

4

Label Us Angry

JEREMIAH TORRES

It hurts to know that the most painful and shocking event of my life happened in part because of my race—something I can never change. On October 23, 1998, my friend and I experienced what would forever change our perceptions of our hometown and society in general.

We both attended elementary, middle, and high school in the quiet, prosperous, seemingly sophisticated college town of Palo Alto. In the third grade, we happily sang "It's a Small World," holding hands with the children of professors, graduate students, and professionals of the area, oblivious to our diversity in race, culture, or experience. Our small world grew larger as we progressed through the school system, each year learning more about what made us different from each other. But on that October evening, the world grew too large for us to handle.

Carlos and I were ready for a night out with the boys. It was his seventeenth birthday, and we were about to celebrate at the pool hall. I pulled out of the Safeway driveway as a speeding driver delivered a jolting honk. I followed him out, speeding to catch up with him, my immediate anger getting the better of me.

We lined up at the stoplight, and the passenger, a young white man dressed for the evening, rolled down his window; I followed. He looked irritated.

"He wasn't honking at you, you stupid fuck!"

His words slapped me across the face. I opened my stunned mouth, only to deliver an empty breath, so I gave him my middle finger until I could return some angry words. He grimaced and reached under his seat to pull out a bottle of mace, spraying it directly in my face, barely missing Carlos, who witnessed the bizarre scene in shock. It burned.

"Take that you fucking lowlifes! Stupid chinks!"

Carlos instinctively bolted out the door at those words. He started pounding the white guy without a second thought, with a new anger he had never known or felt before. Pssssht! The white guy hit Carlos point blank in the face with the mace. He screamed; tires squealed; "fuck you's" were exchanged.

We spent the next ten minutes half-blind, clutching our eyes in the burning pain, cursing in raging anger that made us forget for moments the intense, throbbing fire on our faces. I crawled out of my car to follow Carlos's screams and curses, opening my eyes to the still, spectating traffic surrounding us. I stumbled

SOURCE: From Har, Arar, and John Hsu, eds. 2004. *Asian American X: An Intersection of 21st Century Asian American Voices*, pp. 15–18, Ann Arbor, MI: University of Michigan Press. Reprinted by permission.

to the sidewalk, where Carlos pounded the ground and recalled the words of the white guy. We needed water.

I stumbled further to a nearby house that had lights in the living room. I doorbelled frantically, but nobody answered. I appealed to the traffic for help. They just watched, forming a new route around my car to continue about their evening. The mucous membranes in our sinuses cut loose, and we spit every few seconds to sustain our gasping breaths. After nearly five minutes of appeals, a kind woman stopped to call the cops and give us water to quench the burning.

The cops came within minutes with advice for dealing with the mace. We tried to identify the car and the white guy who had sprayed us, and they sent out the obligatory all points bulletin. They questioned us soon after, asking if we were in a gang. I returned a blank stare with a silent "no." Apparently, two Filipino teenagers finding trouble on a Friday evening raised suspicions of a new Filipino gang in Palo Alto—yeah, all five of us.

I often ask myself if it would have been different had I been driving a BMW and dressed in an ironed polo shirt and slacks, like a typical Palo Alto kid. Maybe then the white guy would not have been afraid and called us lowlifes and chinks. I don't think so. He wasn't afraid of us; he initiated the curses and maced us from a safe distance. He reached out to hurt us because he was having a bad day and we looked different.

That night was our first encounter with overt racism that stems from a hatred of difference. We hadn't seen it through the smiles and happy songs of elementary school or the isolated cliques of middle and high school, but now we knew it was there. We hadn't seen it through the clean-cut, sophisticated facade of the Palo Alto white guy, but now we knew it was there. The "low-life," "chink," and "gangster" labels made us different, marginalizing us from the town we called home.

Those labels made us angry, but we hesitated to project that anger. At first, we didn't tell anyone except our closest friends, afraid our parents would find out and react irrationally by locking us in our rooms to keep us away from trouble. But then we realized that the trouble had found us, and we decided to voice our anger.

We wrote an anonymous article in the school newspaper narrating the incident and the underlying racism that had come to surface. We noted that the incident wasn't purely racial, or a hate crime, but proof that racist tendencies still exist, even in open-minded suburban towns like Palo Alto. Parents, students, and teachers were shocked, maybe because they knew the truth in what we were saying. Many asked if it was Carlos and me who had been maced, but I responded, "Does it matter? What matters is that some people in this town still can't accept diversity. It's sad." We confronted the community with an issue previously reserved for hypothetical classroom discussions and brought it into the open. It was the least we could do to release our anger and expose its roots, hoping for a change in those who chose to label us.

After the article, Carlos and I took different routes. I continued with my studies, complying with my regimen of high school classes and activities as my anger subsided. I tried to lay the incident aside, having exposed it and promoted self-inspection and possible change in others through writing. Carlos remained angry. Why not? He got a face full of mace and racist labels for his seventeenth birthday. He alienated himself from the white majority and returned the mean gestures of the white guy to the yuppie congregation of Palo Alto. He became an outsider. Whenever someone would look at him funny, he would stare back, sometimes too harshly.

On the day after finals, he was making his way through the front parking lot of school when a parent looked at him funny. He stared back. The parent called him a punk. Carlos exploded. He cursed and gestured all he could at the father, and when he sped away in his Suburban, Carlos followed. Carlos couldn't keep up with the Suburban, so he took a quarter from his pocket and threw it at the back window, shattering it to pieces. Carlos ran away when the cops came to school.

Within two days, students had identified Carlos as the perpetrator, and he was suspended from school as the father called his lawyer, indicting Carlos of "assault with the intent to hurt." Weeks passed until a court hearing, and Carlos attended anger management counseling, but he was still angry—angry that he was being tried over throwing a quarter and that once again "the white guys were winning." His mother scraped up the little money she had to spare to afford him a lawyer for the trial, but there was no contesting the father's accusations. Carlos was sentenced to a night in juvenile hall and two hundred hours of community service over some angry words and throwing a quarter. He became a convicted felon.

He had learned once again that he couldn't win against the labels thrown at him, the labels that hurt him more than the mace or the night in juvy, and so he became more of an outsider. In both cases, the labels distanced us from the "normal" Palo Altans: white, clean-cut, wealthy. That division didn't always exist, however; it was created by the generalizations "normal" Palo Altans made through labels. To them, we looked like lowlifes, chinks, gangsters, and punks. In truth, we were two Filipino Americans headed toward Stanford and Berkeley, living in a town that swiftly disowned us with four reckless labels after raising us for ten years. Label us angry.

5

A Different Mirror

RONALD T. TAKAKI

I had flown from San Francisco to Norfolk and was riding in a taxi to my hotel to attend a conference on multiculturalism. Hundreds of educators from across the country were meeting to discuss the need for greater cultural diversity in the curriculum. My driver and I chatted about the weather and the tourists. The sky was cloudy, and Virginia Beach was twenty minutes away. The rearview mirror reflected a white man in his forties. "How long have you been in this country?" he asked. "All my life," I replied, wincing. "I was born in the United States." With a strong southern drawl, he remarked: "I was wondering because your English is excellent!" Then, as I had many times before, I explained: "My grandfather came here from Japan in the 1880s. My family has been here, in America, for over a hundred years." He glanced at me in the mirror. Somehow I did not look "American" to him; my eyes and complexion looked foreign.

Suddenly, we both became uncomfortably conscious of a racial divide separating us. An awkward silence turned my gaze from the mirror to the passing landscape, the shore where the English and the Powhatan Indians first encountered each other. Our highway was on land that Sir Walter Raleigh had renamed "Virginia" in honor of Elizabeth I, the Virgin Queen. In the English cultural appropriation of America, the indigenous peoples themselves would become outsiders in their native land. Here, at the eastern edge of the continent, I mused, was the site of the beginning of multicultural America. Jamestown, the English settlement founded in 1607, was nearby: the first twenty Africans were brought here a year before the Pilgrims arrived at Plymouth Rock. Several hundred miles offshore was Bermuda, the "Bermoothes" where William Shakespeare's Prospero had landed and met the native Caliban in *The Tempest*. Earlier, another voyager had made an Atlantic crossing and unexpectedly bumped into some islands to the south. Thinking he had reached Asia, Christopher Columbus mistakenly identified one of the islands as "Cipango" (Japan). In the wake of the admiral, many peoples would come to America from different shores, not only from Europe but also Africa and Asia. One of them would be my grandfather. My mental wandering across terrain and time ended abruptly as we arrived at my destination. I said good-bye to my driver and went into the hotel, carrying a vivid reminder of why I was attending this conference.

SOURCE: From Ronald T. Takaki, *A Different Mirror: A History of Multicultural America*, pp. 1–17. Copyright © 1993 by Ronald Takaki. Reprinted by permission of Little, Brown & Company.

Questions like the one my taxi driver asked me are always jarring, but I can understand why he could not see me as American. He had a narrow but widely shared sense of the past—a history that has viewed American as European in ancestry. "Race," Toni Morrison explained, has functioned as a "metaphor" necessary to the "construction of Americanness": in the creation of our national identity, "American" has been defined as "white."[1]

But America has been racially diverse since our very beginning on the Virginia shore, and this reality is increasingly becoming visible and ubiquitous. Currently, one-third of the American people do not trace their origins to Europe; in California, minorities are fast becoming a majority. They already predominate in major cities across the country—New York, Chicago, Atlanta, Detroit, Philadelphia, San Francisco, and Los Angeles.

This emerging demographic diversity has raised fundamental questions about America's identity and culture. In 1990, *Time* published a cover story on "America's Changing Colors." "Someday soon," the magazine announced, "white Americans will become a minority group." How soon? By 2056, most Americans will trace their descent to "Africa, Asia, the Hispanic world, the Pacific Islands, Arabia—almost anywhere but white Europe." This dramatic change in our nation's ethnic composition is altering the way we think about ourselves. "The deeper significance of America's becoming a majority nonwhite society is what it means to the national psyche, to individuals' sense of themselves and their nation—their idea of what it is to be American."...[2]

What is fueling the debate over our national identity and the content of our curriculum is America's intensifying racial crisis. The alarming signs and symptoms seem to be everywhere—the killing of Vincent Chin in Detroit, the black boycott of a Korean grocery store in Flatbush, the hysteria in Boston over the Carol Stuart murder, the battle between white sportsmen and Indians over tribal fishing rights in Wisconsin, the Jewish-black clashes in Brooklyn's Crown Heights, the black-Hispanic competition for jobs and educational resources in Dallas, which *Newsweek* described as "a conflict of the have-nots," and the Willie Horton campaign commercials, which widened the divide between the suburbs and the inner cities.[3]

This reality of racial tension rudely woke America like a fire bell in the night on April 29, 1992. Immediately after four Los Angeles police officers were found not guilty of brutality against Rodney King, rage exploded in Los Angeles. Race relations reached a new nadir. During the nightmarish rampage, scores of people were killed, over two thousand injured, twelve thousand arrested, and almost a billion dollars' worth of property destroyed. The live televised images mesmerized America. The rioting and the murderous melee on the streets resembled the fighting in Beirut and the West Bank. The thousands of fires burning out of control and the dark smoke filling the skies brought back images of the burning oil fields of Kuwait during Desert Storm. Entire sections of Los Angeles looked like a bombed city. "Is this America?" many shocked viewers asked. "Please, can we get along here," pleaded Rodney King, calling for calm. "We all can get along. I mean, we're all stuck here for a while. Let's try to work it out."[4]

But how should "we" be defined? Who are the people "stuck here" in America? One of the lessons of the Los Angeles explosion is the recognition of the fact that we are a multiracial society and that race can no longer be defined in the binary terms of white and black. "We" will have to include Hispanics and Asians. While blacks currently constitute 13 percent of the Los Angeles population, Hispanics represent 40 percent. The 1990 census revealed that South Central Los Angeles, which was predominantly black in 1965 when the Watts rebellion occurred, is now 45 percent Hispanic. A majority of the first 5,438 people arrested were Hispanic, while 37 percent were black. Of the fifty-eight people who died in the riot, more than a third were Hispanic, and about 40 percent of the businesses destroyed were Hispanic-owned. Most of the other shops and stores were Korean-owned. The dreams of many Korean immigrants went up in smoke during the riot: two thousand Korean-owned businesses were damaged or demolished, totaling about $400 million in losses. There is evidence indicating they were targeted. "After all," explained a black gang member, "we didn't burn our community, just *their* stores."[5]

"I don't feel like I'm in America anymore," said Denisse Bustamente as she watched the police protecting the firefighters. "I feel like I am far away." Indeed, Americans have been witnessing ethnic strife erupting around the world—the rise of neo-Nazism and the murder of Turks in Germany, the ugly "ethnic cleansing" in Bosnia, the terrible and bloody clashes between Muslims and Hindus in India. Is the situation here different, we have been nervously wondering, or do ethnic conflicts elsewhere represent a prologue for America? What is the nature of malevolence? Is there a deep, perhaps primordial, need for group identity rooted in hatred for the other? Is ethnic pluralism possible for America? But answers have been limited. Television reports have been little more than thirty-second sound bites. Newspaper articles have been mostly superficial descriptions of racial antagonisms and the current urban malaise. What is lacking is historical context; consequently, we are left feeling bewildered.[6]

How did we get to this point, Americans everywhere are anxiously asking. What does our diversity mean, and where is it leading us? *How* do we work it out in the post–Rodney King era?

Certainly one crucial way is for our society's various ethnic groups to develop a greater understanding of each other. For example, how can African Americans and Korean Americans work it out unless they learn about each other's cultures, histories, and also economic situations? This need to share knowledge about our ethnic diversity has acquired new importance and has given new urgency to the pursuit for a more accurate history....

While all of America's many groups cannot be covered [here], the English immigrants and their descendants require attention, for they possessed inordinate power to define American culture and make public policy. What men like John Winthrop, Thomas Jefferson, and Andrew Jackson thought as well as did mattered greatly to all of us and was consequential for everyone. A broad range of groups [is important]: African Americans, Asian Americans, Chicanos, Irish, Jews, and Indians. While together they help to explain general patterns in our society, each has contributed to the making of the United States.

African Americans have been the central minority throughout our country's history. They were initially brought here on a slave ship in 1619. Actually, these first twenty Africans might not have been slaves; rather, like most of the white laborers, they were probably indentured servants. The transformation of Africans into slaves is the story of the "hidden" origins of slavery. How and when was it decided to institute a system of bonded black labor? What happened, while freighted with racial significance, was actually conditioned by class conflicts within white society. Once established, the "peculiar institution" would have consequences for centuries to come. During the nineteenth century, the political storm over slavery almost destroyed the nation. Since the Civil War and emancipation, race has continued to be largely defined in relation to African Americans—segregation, civil rights, the underclass, and affirmative action. Constituting the largest minority group in our society, they have been at the cutting edge of the Civil Rights Movement. Indeed, their struggle has been a constant reminder of America's moral vision as a country committed to the principle of liberty. Martin Luther King clearly understood this truth when he wrote from a jail cell: "We will reach the goal of freedom in Birmingham and all over the nation, because the goal of America is freedom. Abused and scorned though we may be, our destiny is tied up with America's destiny."[7]

Asian Americans have been here for over one hundred and fifty years, before many European immigrant groups. But as "strangers" coming from a "different shore," they have been stereotyped as "heathen," exotic, and unassimilable. Seeking "Gold Mountain," the Chinese arrived first, and what happened to them influenced the reception of the Japanese, Koreans, Filipinos, and Asian Indians as well as the Southeast Asian refugees like the Vietnamese and the Hmong. The 1882 Chinese Exclusion Act was the first law that prohibited the entry of immigrants on the basis of nationality. The Chinese condemned this restriction as racist and tyrannical. "They call us 'Chink,'" complained a Chinese immigrant, cursing the "white demons." "They think we no good! America cuts us off. No more come now, too bad!" This precedent later provided a basis for the restriction of European immigrant groups such as Italians, Russians, Poles, and Greeks. The Japanese painfully discovered that their accomplishments in America did not lead to acceptance, for during World War II, unlike Italian Americans and German Americans, they were placed in internment camps. Two-thirds of them were citizens by birth. "How could I as a 6-month-old child born in this country," asked Congressman Robert Matsui years later, "be declared by my own Government to be an enemy alien?" Today, Asian Americans represent the fastest-growing ethnic group. They have also become the focus of much mass media attention as "the Model Minority" not only for blacks and Chicanos, but also for whites on welfare and even middle-class whites experiencing economic difficulties.[8]

Chicanos represent the largest group among the Hispanic population, which is projected to outnumber African Americans. They have been in the United States for a long time, initially incorporated by the war against Mexico. The treaty had moved the border between the two countries, and the people of "occupied" Mexico suddenly found themselves "foreigners" in their "native land." As historian Albert Camarillo pointed out, the Chicano past is an integral part of America's westward expansion, also known as "manifest destiny." But while the early Chicanos were a colonized people, most of them today have immigrant roots. Many began the trek to El Norte in the early twentieth century. "As I had heard a lot about the United States," Jesus Garza recalled, "it was my dream to come here." "We came to know families from Chihuahua, Sonora, Jalisco, and Durango," stated Ernesto Galarza. "Like ourselves, our Mexican neighbors had come this far moving step by step, working and waiting, as if they were feeling their way up a ladder." Nevertheless, the Chicano experience has been unique, for most of them have lived close to their homeland—a proximity that has helped reinforce their language, identity, and culture. This migration to El Norte has continued to the present. Los Angeles has more people of Mexican origin than any other city in the world, except Mexico City. A mostly mestizo people of Indian as well as African and Spanish ancestries, Chicanos currently represent the largest minority group in the Southwest, where they have been visibly transforming culture and society.[9]

The Irish came here in greater numbers than most immigrant groups. Their history has been tied to America's past from the very beginning. Ireland represented the earliest English frontier: the conquest of Ireland occurred before the colonization of America, and the Irish were the first group that the English called "savages." In this context, the Irish past foreshadowed the Indian future. During the nineteenth century, the Irish, like the Chinese, were victims of British colonialism. While the Chinese fled from the ravages of the Opium Wars, the Irish were pushed from their homeland by "English tyranny." Here they became construction workers and factory operatives as well as the "maids" of America. Representing a Catholic group seeking to settle in a fiercely Protestant society, the Irish immigrants were targets of American nativist hostility. They were also what historian Lawrence J. McCaffrey called "the pioneers of the American urban ghetto," "previewing" experiences that would later be shared by the Italians, Poles, and other groups from southern and eastern Europe. Furthermore, they offer contrast to the immigrants from Asia. The Irish came about the same time as the Chinese, but they had a distinct advantage: the Naturalization Law of 1790 had reserved citizenship for "whites" only. Their compatible complexion allowed them to assimilate by blending into American society. In making their journey successfully into the mainstream, however, these immigrants from Erin pursued an Irish "ethnic" strategy: they promoted "Irish" solidarity in order to gain political power and also to dominate the skilled blue-collar occupations, often at the expense of the Chinese and blacks.[10]

Fleeing pogroms and religious persecution in Russia, the Jews were driven from what John Cuddihy described as the "Middle Ages into the Anglo-American world of the *goyim* 'beyond the pale.'" To them, America represented the Promised Land. This vision led Jews to struggle not only for themselves but also for other oppressed groups, especially blacks. After the 1917 East St. Louis race riot, the Yiddish *Forward* of New York compared this antiblack violence to a 1903 pogrom in Russia: "Kishinev and St. Louis—the same soil, the same people." Jews cheered when Jackie Robinson broke into the Brooklyn Dodgers in 1947. "He was adopted as the surrogate hero by many of us growing up at the time," recalled Jack Greenberg of the NAACP Legal Defense Fund. "He was the way we saw ourselves triumphing against the forces of bigotry and ignorance." Jews stood shoulder to shoulder with blacks in the Civil Rights Movement: two-thirds of the white volunteers who went south during the 1964 Freedom Summer were Jewish. Today Jews are considered a highly successful "ethnic" group. How did they make such great socioeconomic strides? This question is often reframed by neoconservative intellectuals like Irving Kristol and Nathan Glazer to read: if Jewish immigrants were able to lift themselves from poverty into the mainstream through self-help and education without welfare and affirmative action, why can't blacks? But what this thinking overlooks is the unique history of Jewish immigrants, especially the initial advantages of many of them as literate and skilled. Moreover, it minimizes the virulence of racial prejudice rooted in American slavery.[11]

Indians represent a critical contrast, for theirs was not an immigrant experience. The Wampanoags were on the shore as the first English strangers arrived in what would be called "New England." The encounters between Indians and whites not only shaped the course of race relations, but also influenced the very culture and identity of the general society. The architect of Indian removal, President Andrew Jackson told Congress: "Our conduct toward these people is deeply interesting to the national character." Frederick Jackson Turner understood the meaning of this observation when he identified the frontier as our transforming crucible. At first, the European newcomers had to wear Indian moccasins and shout the war cry. "Little by little," as they subdued the wilderness, the pioneers became "a new product" that was "American." But Indians have had a different view of this entire process. "The white man," Luther Standing Bear of the Sioux explained, "does not understand the Indian for the reason that he does not understand America." Continuing to be "troubled with primitive fears," he has "in his consciousness the perils of this frontier continent.... The man from Europe is still a foreigner and an alien. And he still hates the man who questioned his path across the continent." Indians questioned what Jackson and Turner trumpeted as "progress." For them, the frontier had a different "significance": their history was how the West was lost. But their story has also been one of resistance. As Vine Deloria declared, "Custer died for your sins."[12]

By looking at these groups from a multicultural perspective, we can comparatively analyze their experiences in order to develop an understanding of their differences and similarities. Race, we will see, has been a social construction that has historically set apart racial minorities from European immigrant groups. Contrary to the notions of scholars like Nathan Glazer and Thomas Sowell, race in America has not been the same as ethnicity. A broad comparative focus also allows us to see how the varied experiences of different racial and ethnic groups occurred within shared contexts.

During the nineteenth century, for example, the Market Revolution employed Irish immigrant laborers in New England factories as it expanded cotton fields worked by enslaved blacks across Indian lands toward Mexico.

Like blacks, the Irish newcomers were stereotyped as "savages," ruled by passions rather than "civilized" virtues such as self-control and hard work. The Irish saw themselves as the "slaves" of British oppressors, and during a visit to Ireland in the 1840s, Frederick Douglass found that the "wailing notes" of the Irish ballads reminded him of the "wild notes" of slave songs. The United States annexation of California, while incorporating Mexicans, led to trade with Asia and the migration of "strangers" from Pacific shores. In 1870, Chinese immigrant laborers were transported to Massachusetts as scabs to break an Irish immigrant strike; in response, the Irish recognized the need for interethnic working-class solidarity and tried to organize a Chinese lodge of the Knights of St. Crispin. After the Civil War, Mississippi planters recruited Chinese immigrants to discipline the newly freed blacks. During the debate over an immigration exclusion bill in 1882, a senator asked: If Indians could be located on reservations, why not the Chinese?[13]

Other instances of our connectedness abound. In 1903, Mexican and Japanese farm laborers went on strike together in California: their union officers had names like Yamaguchi and Lizarras, and strike meetings were conducted in Japanese and Spanish. The Mexican strikers declared that they were standing in solidarity with their "Japanese brothers" because the two groups had toiled together in the fields and were now fighting together for a fair wage. Speaking in impassioned Yiddish during the 1909 "uprising of twenty thousand" strikers in New York, the charismatic Clara Lemlich compared the abuse of Jewish female garment workers to the experience of blacks: "[The bosses] yell at the girls and 'call them down' even worse than I imagine the Negro slaves were in the South." During the 1920s, elite universities like Harvard worried about the increasing numbers of Jewish students, and new admissions criteria were instituted to curb their enrollment. Jewish students were scorned for their studiousness and criticized for their "clannishness." Recently, Asian American students have been the targets of similar complaints: they have been called "nerds" and told there are "too many" of them on campus.[14]

Indians were already here, while blacks were forcibly transported to America, and Mexicans were initially enclosed by America's expanding border. The other groups came here as immigrants: for them, America represented liminality—a new world where they could pursue extravagant urges and do things they had thought beyond their capabilities. Like the land itself, they found themselves "betwixt and between all fixed points of classification." No longer fastened as fiercely to their old countries, they felt a stirring to become new people in a society still being defined and formed.[15]

These immigrants made bold and dangerous crossings, pushed by political events and economic hardships in their homelands and pulled by America's demand for labor as well as by their own dreams for a better life. "By all means let me go to America," a young man in Japan begged his parents. He had calculated that in one year as a laborer here he could save almost a thousand yen—an amount equal to the income of a governor in Japan. "My dear Father," wrote an immigrant Irish girl living in New York, "Any man or woman without a family are fools that would not venture and come to this plentyful Country where no man or woman ever hungered." In the shtetls of Russia, the cry "To America!" roared like "wildfire." "America was in everybody's mouth," a Jewish immigrant recalled. "Businessmen talked [about] it over their accounts; the market women made up their quarrels that they might discuss it from stall to stall; people who had relatives in the famous land went around reading their letters." Similarly, for Mexican immigrants crossing the border in the early twentieth century, El Norte became the stuff of overblown hopes. "If only you could see how nice the United States is," they said, "that is why the Mexicans are crazy about it."[16]

The signs of America's ethnic diversity can be discerned across the continent—Ellis Island, Angel Island, Chinatown, Harlem, South Boston, the Lower East Side, places with Spanish names like Los Angeles and San Antonio or Indian names like Massachusetts and Iowa. Much of what is familiar in America's cultural landscape actually has ethnic origins. The Bing cherry was developed by an early Chinese immigrant named Ah Bing. American Indians were cultivating corn, tomatoes, and tobacco long before the arrival of Columbus. The term *okay* was derived from the Choctaw word *oke,* meaning "it is so." There is evidence indicating that the name *Yankee* came from Indian terms for the English—from *eankke* in Cherokee and *Yankwis* in Delaware. Jazz and blues as well as rock and roll have African American origins. The "Forty-Niners" of the Gold Rush learned mining techniques from the Mexicans; American cowboys acquired herding skills from Mexican *vaqueros* and adopted their range terms—such as *lariat* from *la reata, lasso* from *lazo,* and *stampede* from *estampida.* Songs like "God Bless America," "Easter Parade," and "White Christmas" were written by a Russian-Jewish immigrant named Israel Baline, more popularly known as Irving Berlin.[17]

Furthermore, many diverse ethnic groups have contributed to the building of the American economy, forming what Walt Whitman saluted as "a vast, surging, hopeful army of workers." They worked in the South's cotton fields, New England's textile mills, Hawaii's cane fields, New York's garment factories, California's orchards, Washington's salmon canneries, and Arizona's copper mines. They built the railroad, the great symbol of America's industrial triumph....

Moreover, our diversity was tied to America's most serious crisis: the Civil War was fought over a racial issue—slavery....

... The people in our study have been actors in history, not merely victims of discrimination and exploitation. They are entitled to be viewed as subjects—as men and women with minds, wills, and voices.

> In the telling and retelling
>   of their stories,
> They create communities
>   of memory.

They also re-vision history. "It is very natural that the history written by the victim," said a Mexican in 1874, "does not altogether chime with the story of the victor." Sometimes they are hesitant to speak, thinking they are only "little people." "I don't know why anybody wants to hear my history," an Irish maid said apologetically in 1900. "Nothing ever happened to me worth the tellin'."[18]

But their stories are worthy. Through their stories, the people who have lived America's history can help all of us, including my taxi driver, understand that Americans originated from many shores, and that all of us are entitled to dignity. "I hope this survey do a lot of good for Chinese people," an immigrant told an interviewer from Stanford University in the 1920s. "Make American people realize that Chinese people are humans. I think very few American people really know anything about Chinese." But the remembering is also for the sake of the children. "This story is dedicated to the descendants of Lazar and Goldie Glauberman," Jewish immigrant Minnie Miller wrote in her autobiography. "My history is bound up in their history and the generations that follow should know where they came from to know better who they are." Similarly, Tomo Shoji, an elderly Nisei woman, urged Asian Americans to learn more about their roots: "We got such good, fantastic stories to tell. All our stories are different." Seeking to know how they fit into America, many young people have become listeners; they are eager to learn about the hardships and humiliations experienced by their parents and grandparents. They want to hear their stories, unwilling to remain ignorant or ashamed of their identity and past.[19]

The telling of stories liberates. By writing about the people on Mango Street, Sandra Cisneros explained, "the ghost does not ache so much." The place no longer holds her with "both arms. She sets me free." Indeed, stories may not be as innocent or simple as they seem to be. Native-American novelist Leslie Marmon Silko cautioned:

I will tell you something about stories ...
They aren't just entertainment.
*   Don't be fooled.*

Indeed, the accounts given by the people in this study vibrantly re-create moments, capturing the complexities of human emotions and thoughts. They also provide the authenticity of experience. After she escaped from slavery, Harriet Jacobs wrote in her autobiography: "[My purpose] is not to tell you what I have heard but what I have seen—and what I have suffered." In their sharing of memory, the people in this study offer us an opportunity to see ourselves reflected in a mirror called history.[20]

In his recent study of Spain and the New World, *The Buried Mirror,* Carlos Fuentes points out that mirrors have been found in the tombs of ancient Mexico, placed there to guide the dead through the underworld. He also tells us about the legend of Quetzalcoatl, the Plumed Serpent: when this god was given a mirror by the Toltec deity Tezcatlipoca, he saw a man's face in the mirror and realized his own humanity. For us, the "mirror" of history can guide the living and also help us recognize who we have been and hence are. In *A Distant Mirror,* Barbara W. Tuchman finds "phenomenal parallels" between the "calamitous fourteenth century" of European society and our own era. We can, she observes, have "greater fellow-feeling for a distraught age" as we painfully recognize the "similar disarray," "collapsing assumptions," and "unusual discomfort."[21]

But what is needed in our own perplexing times is not so much a "distant" mirror, as one that is "different." While the study of the past can provide collective self-knowledge, it often reflects the scholar's particular perspective or view of the world. What happens when historians leave out many of America's peoples? What happens, to borrow the words of Adrienne Rich, "when someone with the authority of a teacher" describes our society, and "you are not in it"? Such an experience can be disorienting—"a moment of psychic disequilibrium, as if you looked into a mirror and saw nothing."[22]

Through their narratives about their lives and circumstances, the people of America's diverse groups are able to see themselves and each other in our common past. They celebrate what Ishmael Reed has described as a society "unique" in the world because "the world is here"—a place "where the cultures of the world crisscross." Much of America's past, they point out, has been riddled with racism. At the same time, these people offer hope, affirming the struggle for equality as a central theme in our country's history. At its conception, our nation was dedicated to the proposition of equality. What has given concreteness to this powerful national principle has been our coming together in the creation of a new society. "Stuck here" together, workers of different backgrounds have attempted to get along with each other.

> *People harvesting*
> *Work together unaware*
> *Of racial problems,*

wrote a Japanese immigrant describing a lesson learned by Mexican and Asian farm laborers in California.[23]

Finally, how do we see our prospects for "working out" America's racial crisis? Do we see it as through a glass darkly? Do the televised images of racial hatred and violence that riveted us in 1992 during the days of rage in Los Angeles frame a future of divisive race relations—what Arthur Schlesinger Jr. has fearfully denounced as the "disuniting of America"? Or will Americans of diverse races and ethnicities be able to connect themselves to a larger narrative? Whatever happens, we can be certain that much of our society's future will be influenced by which "mirror" we choose to see ourselves. America does not belong to one race or one group.... Americans have been constantly redefining their national identity from the moment of first contact on the Virginia shore. By sharing their stories, they invite us to see ourselves in a different mirror.[24]

NOTES

1. Toni Morrison, *Playing in the Dark: Whiteness in the Literary Imagination* (Cambridge, Mass., 1992), p. 47.

2. William A. Henry III, "Beyond the Melting Pot," in "America's Changing Colors," *Time*, vol. 135, no. 15 (April 9, 1990), pp. 28–31.

3. "A Conflict of the Have-Nots," *Newsweek*, December 12, 1988, pp. 28–29.

4. Rodney King's statement to the press, *New York Times*, May 2, 1992, p. 6.

5. Tim Rutten, "A New Kind of Riot," *New York Times Review of Books*, June 11, 1992, pp. 52–53; Maria Newman "Riots Bring Attention to Growing Hispanic

Presence in South-Central Area," *New York Times*, May 11, 1992, p. A10; Mike Davis, "In L.A. Burning All Illusions," *The Nation*, June 1, 1992, pp. 744–745; Jack Viets and Peter Fimrite, "S.F. Mayor Visits Riot-Torn Area to Buoy Businesses," *San Francisco Chronicle*, May 6, 1992, p. A6.

6. Rick DelVecchio, Suzanne Espinosa, and Carl Nolte, "Bradley Ready to Lift Curfew," *San Francisco Chronicle*, May 4, 1992, p. A1.

7. Abraham Lincoln, "The Gettysburg Address," in *The Annals of America*, vol. 9, *1863–1865: The Crisis of the Union* (Chicago, 1968), pp. 462–463; Martin Luther King, *Why We Can't Wait* (New York, 1964), pp. 92–93.

8. Interview with old laundryman, in "Interviews with Two Chinese," circa 1924, Box 326, folder 325, Survey of Race Relations, Stanford University, Hoover Institution Archives; Congressman Robert Matsui, speech in the House of Representatives on the 442 bill for redress and reparations, September 17, 1987, *Congressional Record* (Washington, D.C., 1987), p. 7584.

9. Albert Camarillo, *Chicanos in a Changing Society: From Mexican Pueblos to American Barrios in Santa Barbara and Southern California, 1848–1930* (Cambridge, Mass., 1979), p. 2; Juan Nepornuceno Seguín, in David J. Weber (ed.), *Foreigners in Their Native Land: Historical Roots of the Mexican Americans* (Albuquerque, N. Mex., 1973), p. vi; Jesus Garza, in Manuel Garnio, *The Mexican Immigrant: His Life Story* (Chicago, 1931), p. 15; Ernesto Galarza, *Barrio Boy: The Story of a Boy's Acculturation* (Notre Dame, Ind., 1986), p. 200.

10. Lawrence J. McCaffrey, *The Irish Diaspora in America* (Washington, D.C., 1984), pp. 6, 62.

11. John Murray Cuddihy, *The Ordeal of Civility: Freud, Marx, Levi Strauss, and the Jewish Struggle with Modernity* (Boston, 1987), p. 165; Jonathan Kaufman, *Broken Alliance: The Turbulent Times between Blacks and Jews in America* (New York, 1989), pp. 28, 82, 83–84, 91, 93, 106.

12. Andrew Jackson, First Annual Message to Congress, December 8, 1829, in James D. Richardson (ed.), *A Compilation of the Messages and Papers of the Presidents, 1789–1897* (Washington, D.C., 1897), vol. 2, p. 457; Frederick Jackson Turner, "The Significance of the Frontier in American History," in *The Early Writings of Frederick Jackson Turner* (Madison, Wis., 1938), pp. 185ff.; Luther Standing Bear, "What the Indian Means to America," in Wayne Moquin (ed.), *Great Documents in American Indian History* (New York, 1973), p. 307; Vine Deloria, Jr., *Custer Died for Your Sins: An Indian Manifesto* (New York, 1969).

13. Nathan Glazer, *Affirmative Discrimination: Ethnic Inequality and Public Policy* (New York, 1978); Thomas Sowell, *Ethnic America: A History* (New York, 1981); David R. Roediger, *The Wages of Whiteness: Race and the Making of the American Working Class* (London, 1991), pp. 134–136; Dan Caldwell, "The Negroization of the Chinese Stereotype in California," *Southern California Quarterly*, vol. 33 (June 1971), pp. 123–131.

14. Thomas Almaguer, "Racial Domination and Class Conflict in Capitalist Agriculture: The Oxnard Sugar Beet Workers' Strike of 1903," *Labor History*, vol. 25, no. 3 (summer 1984), p. 347; Howard M. Sachar, *A History of the Jews in America* (New York, 1992), p. 183.

15. For the concept of liminality, see Victor Turner, *Dramas, Fields, and Metaphors: Symbolic Action in Human Society* (Ithaca, N.Y., 1974), pp. 232, 237; and Arnold Van

Gennep, *The Rites of Passage* (Chicago, 1960). What I try to do is to apply liminality to the land called America.

16. Kazuo Ito, *Issei: A History of Japanese Immigrants in North America* (Seattle, 1973), p. 33; Arnold Schrier, *Ireland and the American Emigration, 1850–1900* (New York, 1970), p. 24; Abraham Cahan, *The Rise of David Levinsky* (New York, 1960; originally published in 1917), pp. 59–61; Mary Antin, quoted in Howe, *World of Our Fathers* (New York, 1983), p. 27; Lawrence A. Cardoso, *Mexican Emigration to the United States, 1897–1931* (Tucson, Ariz., 1981), p. 80.

17. Ronald Takaki, *Strangers from a Different Shore: A History of Asian Americans* (Boston, 1989), pp. 88–89; Jack Weatherford, *Native Roots: How the Indians Enriched America* (New York, 1991), pp. 210, 212; Carey McWilliams, *North from Mexico: The Spanish-Speaking People of the United States* (New York, 1968), p. 154; Stephan Themstrom (ed.), *Harvard Encyclopedia of American Ethnic Groups* (Cambridge, Mass., 1980), p. 22; Sachar, *A History of the Jews in America*, p. 367.

18. Weber (ed.), *Foreigners in Their Native Land*, p. vi; Hamilton Holt (ed.), *The Life Stories of Undistinguished Americans as Told by Themselves* (New York, 1906), p. 143.

19. "Social Document of Pany Lowe, interviewed by C. H. Burnett, Seattle, July 5, 1924," p. 6, Survey of Race Relations, Stanford University, Hoover Institution Archives; Minnie Miller, "Autobiography," private manuscript, copy from Richard Balkin; Tomo Shoji, presentation, Obana Cultural Center, Oakland, California, March 4, 1988.

20. Sandra Cisneros, *The House on Mango Street* (New York, 1991), pp. 109–110; Leslie MarmonSilko, *Ceremony* (New York, 1978), p. 2; Harriet A. Jacobs, *Incidents in the Life of a Slave Girl, written by herself* (Cambridge, Mass., 1987; originally published in 1857), p. xiii.

21. Carlos Fuentes, *The Buried Mirror: Reflections on Spain and the New World* (Boston, 1992), pp. 10, 11, 109; Barbara W. Tuchman, *A Distant Mirror: The Calamitous 14th Century* (New York, 1978), pp. xiii, xiv.

22. Adrienne Rich, *Blood, Bread, and Poetry: Selected Prose, 1979–1985* (New York, 1986), p. 199.

23. Ishmael Reed, "America: The Multinational Society," in Rick Simonson and Scott Walker (eds.), *Multi-cultural Literacy* (St. Paul, 1988), p. 160; Ito, *Issei*, p. 497.

24. Arthur M. Schlesinger, Jr., *The Disuniting of America: Reflections on a Multicultural Society* (Knoxville, Tenn., 1991); Carlos Bulosan, *America Is in the Heart: A Personal History* (Seattle, 1981), pp. 188–189.

6

White Privilege
Unpacking the Invisible Knapsack

PEGGY MCINTOSH

Through work to bring materials from Women's Studies into the rest of the curriculum, I have often noticed men's unwillingness to grant that they are overprivileged, even though they may grant that women are disadvantaged. They may say they will work to improve women's status, in the society, the university, or the curriculum, but they can't or won't support the idea of lessening men's. Denials which amount to taboos surround the subject of advantages which men gain from women's disadvantages. These denials protect male privilege from being fully acknowledged, lessened, or ended.

Thinking through unacknowledged male privilege as a phenomenon, I realized that since hierarchies in our society are interlocking, there was most likely a phenomenon of white privilege which was similarly denied and protected. As a white person, I realized I had been taught about racism as something which puts others at a disadvantage, but had been taught not to see one of its corollary aspects, white privilege, which puts me at an advantage.

I think whites are carefully taught not to recognize white privilege, as males are taught not to recognize male privilege. So I have begun in an untutored way to ask what it is like to have white privilege. I have come to see white privilege as an invisible package of unearned assets which I can count on cashing in each day, but about which I was "meant" to remain oblivious. White privilege is like an invisible weightless knapsack of special provisions, maps, passports, codebooks, visas, clothes, tools, and blank checks.

Describing white privilege makes one newly accountable. As we in Women's Studies work to reveal male privilege and ask men to give up some of their power, so one who writes about having white privilege must ask, "Having described it, what will I do to lessen or end it?"

After I realized the extent to which men work from a base of unacknowledged privilege, I understood that much of their oppressiveness was unconscious. Then I remembered the frequent charges from women of color that white women whom they encounter are oppressive. I began to understand why we are justly

SOURCE: "White Privilege: Unpacking the Invisible Knapsack," published in *Peace and Freedom Magazine*, July/August 1989. Copyright © 1988 by Peggy McIntosh. May not be copied without permission of the author, mmcintosh@wellesley.edu. A longer analysis and list of privileges, including heterosexual privilege, is available from Peggy McIntosh, Wellesley College, Center for Research on Women, Wellesley, MA, 02481-8203. Tel. (617) 283-2520; Fax (617) 283-2504.

seen as oppressive, even when we don't see ourselves that way. I began to count the ways in which I enjoy unearned skin privilege and have been conditioned into oblivion about its existence.

My schooling gave me no training in seeing myself as an oppressor, as an unfairly advantaged person, or as a participant in a damaged culture. I was taught to see myself as an individual whose moral state depended on her individual moral will. My schooling followed the pattern my colleague Elizabeth Minnich has pointed out: whites are taught to think of their lives as morally neutral, normative, and average, and also ideal, so that when we work to benefit others, this is seen as work which will allow "them" to be more like "us."

I decided to try to work on myself at least by identifying some of the daily effects of white privilege in my life. I have chosen those conditions which I think in my case *attach somewhat more to skin-color privilege* than to class, religion, ethnic status, or geographical location, though of course all these other factors are intricately intertwined. As far as I can see, my African American co-workers, friends and acquaintances with whom I come into daily or frequent contact in this particular time, place, and line of work cannot count on most of these conditions.

1. I can if I wish arrange to be in the company of people of my race most of the time.

2. If I should need to move, I can be pretty sure of renting or purchasing housing in an area which I can afford and in which I would want to live.

3. I can be pretty sure that my neighbors in such a location will be neutral or pleasant to me.

4. I can go shopping alone most of the time, pretty well assured that I will not be followed or harassed.

5. I can turn on the television or open to the front page of the paper and see people of my race widely represented.

6. When I am told about our national heritage or about "civilization," I am shown that people of my color made it what it is.

7. I can be sure that my children will be given curricular materials that testify to the existence of their race.

8. If I want to, I can be pretty sure of finding a publisher for this piece on white privilege.

9. I can go into a music shop and count on finding the music of my race represented, into a supermarket and find the staple foods which fit with my cultural traditions, into a hairdresser's shop and find someone who can cut my hair.

10. Whether I use checks, credit cards, or cash, I can count on my skin color not to work against the appearance of financial reliability.

11. I can arrange to protect my children most of the time from people who might not like them.

12. I can swear, or dress in secondhand clothes, or not answer letters, without having people attribute these choices to the bad morals, the poverty, or the illiteracy of my race.

13. I can speak in public to a powerful male group without putting my race on trial.

14. I can do well in a challenging situation without being called a credit to my race.

15. I am never asked to speak for all the people of my racial group.

16. I can remain oblivious of the language and customs of persons of color who constitute the world's majority without feeling in my culture any penalty for such oblivion.

17. I can criticize our government and talk about how much I fear its policies and behavior without being seen as a cultural outsider.

18. I can be pretty sure that if I ask to talk to "the person in charge," I will be facing a person of my race.

19. If a traffic cop pulls me over or if the IRS audits my tax return, I can be sure I haven't been singled out because of my race.

20. I can easily buy posters, postcards, picture books, greeting cards, dolls, toys, and children's magazines featuring people of my race.

21. I can go home from most meetings of organizations I belong to feeling somewhat tied in, rather than isolated, out-of-place, outnumbered, unheard, held at a distance, or feared.

22. I can take a job with an affirmative action employer without having cow-orkers on the job suspect that I got it because of my race.

23. I can choose public accommodation without fearing that people of my race cannot get in or will be mistreated in the places I have chosen.

24. I can be sure that if I need legal or medical help, my race will not work against me.

25. If my day, week, or year is going badly, I need not ask of each negative episode or situation whether it has racial overtones.

26. I can choose blemish cover or bandages in "flesh" color and have them more or less match my skin.

I repeatedly forgot each of the realizations on this list until I wrote it down. For me white privilege has turned out to be an elusive and fugitive subject. The pressure to avoid it is great, for in facing it I must give up the myth of meritocracy. If these things are true, this is not such a free country; one's life is not what one makes it; many doors open for certain people through no virtues of their own.

In unpacking this invisible knapsack of white privilege, I have listed conditions of daily experience which I once took for granted. Nor did I think of any of these perquisites as bad for the holder. I now think that we need a more finely differentiated taxonomy of privilege, for some of these varieties are only what one would want for everyone in a just society, and others give license to be ignorant, oblivious, arrogant and destructive.

I see a pattern running through the matrix of white privilege, a pattern of assumptions which were passed on to me as a white person. There was one main piece of cultural turf; it was my own turf, and I was among those who could control the turf. *My skin color was an asset for any move I was educated to want to make.* I could think of myself as belonging in major ways, and of making social systems work for me. I could freely disparage, fear, neglect, or be oblivious to anything outside of the dominant cultural forms. Being of the main culture, I could also criticize it fairly freely.

In proportion as my racial group was being made confident, comfortable, and oblivious, other groups were likely being made inconfident, uncomfortable, and alienated. Whiteness protected me from many kinds of hostility, distress, and violence, which I was being subtly trained to visit in turn upon people of color.

For this reason, the word "privilege" now seems to me misleading. We usually think of privilege as being a favored state, whether earned or conferred by birth or luck. Yet some of the conditions I have described here work to systematically overempower certain groups. Such privilege simply *confers dominance* because of one's race or sex.

I want, then, to distinguish between earned strength and unearned power conferred systemically. Power from unearned privilege can look like strength when it is in fact permission to escape or to dominate. But not all of the privileges on my list are inevitably damaging. Some, like the expectation that neighbors will be decent to you, or that your race will not count against you in court, should be the norm in a just society. Others, like the privilege to ignore less powerful people, distort the humanity of the holders as well as the ignored groups.

We might at least start by distinguishing between positive advantages which we can work to spread, and negative types of advantages which unless rejected will always reinforce our present hierarchies. For example, the feeling that one belongs within the human circle, as Native Americans say, should not be seen as privilege for a few. Ideally it is an *unearned entitlement.* At present, since only a few have it, it is an *unearned advantage* for them. This paper results from a process of coming to see that some of the power which I originally saw as attendant on being a human being in the U.S. consisted in *unearned advantage* and *conferred dominance.*

I have met very few men who are truly distressed about systemic, unearned male advantage and conferred dominance. And so one question for me and others like me is whether we will be like them, or whether we will get truly distressed, even outraged, about unearned race advantage and conferred dominance and if so, what we will do to lessen them. In any case, we need to do more work in identifying how they actually affect our daily lives. Many, perhaps most, of our white students in the U.S. think that racism doesn't affect them because they are not people of color; they do not see "whiteness" as a racial identity. In addition, since race and sex are not the only advantaging systems at work, we need similarly to examine the daily experience of having age advantage, or ethnic advantage, or physical ability, or advantage related to nationality, religion, or sexual orientation.

Difficulties and dangers surrounding the task of finding parallels are many. Since racism, sexism, and heterosexism are not the same, the advantaging associated with them should not be seen as the same. In addition, it is hard to disentangle aspects of unearned advantage which rest more on social class, economic class, race, religion, sex and ethnic identity than on other factors. Still, all of the oppressions are interlocking, as the Combahee River Collective Statement of 1977 continues to remind us eloquently.

One factor seems clear about all of the interlocking oppressions. They take both active forms which we can see and embedded forms which as a member of the dominant group one is taught not to see. In my class and place, I did not see myself as a racist because I was taught to recognize racism only in individual acts of meanness by members of my group, never in invisible systems conferring unsought racial dominance on my group from birth.

Disapproving of the systems won't be enough to change them. I was taught to think that racism could end if white individuals changed their attitudes. [But] a "white" skin in the United States opens many doors for whites whether or not we approve of the way dominance has been conferred on us. Individual acts can palliate, but cannot end, these problems.

To redesign social systems we need first to acknowledge their colossal unseen dimensions. The silences and denials surrounding privilege are the key political tool here. They keep the thinking about equality or equity incomplete, protecting unearned advantage and conferred dominance by making these taboo subjects. Most talk by whites about equal opportunity seems to me now to be about equal opportunity to try to get into a position of dominance while denying that *systems* of dominance exist.

It seems to me that obliviousness about white advantage, like obliviousness about male advantage, is kept strongly inculturated in the United States so as to maintain the myth of meritocracy, the myth that democratic choice is equally available to all. Keeping most people unaware that freedom of confident action is there for just a small number of people props up those in power, and serves to keep power in the hands of the same groups that have most of it already.

Though systemic change takes many decades, there are pressing questions for me and I imagine for some others like me if we raise our daily consciousness on the perquisites of being light-skinned. What will we do with such knowledge? As we know from watching men, it is an open question whether we will choose to use unearned advantage to weaken hidden systems of advantage, and whether we will use any of our arbitrarily-awarded power to try to reconstruct power systems on a broader base.

7

Race, Poverty, and Disability

Three Strikes and You're Out! Or Are You?

PAMELA BLOCK, FABRICIO E. BALCAZAR, AND CHRISTOPHER B. KEYS

INTRODUCTION

Throughout the twentieth century, theories of biology and culture presented images of race, class, and disability in terms of deficiency and dependence. Biological models represented certain ethnic and racial groups as genetically inferior. Cultural models represented these groups as trapped in an inescapable cycle of poverty. Both models represented people of color with disabilities as social victims and/or socially threatening. Policies and practices based on these theories projected pejorative images on low-income ethnic and racial groups, and on people with disabilities. Individuals belonging to more than one of these categories—or all three—were especially vulnerable to social stigma.

Such images of biological and cultural pathology have been rejected by many modern theorists who have adopted a minority group model for the analysis of groups disenfranchised on the basis of race, class, gender, sexual orientation, and/or disability. In these formulations, racism, classism, ableism and other forms of prejudice create barriers that result in social and economic marginalization for members of disenfranchised groups. African Americans and Latinos with disabilities face multiple challenges of discrimination and social barriers related to race and class, as well as to disability. However, insufficient attention has been given to the interaction of these variables.

RACE, POVERTY AND DISABILITY
AS BIOLOGICAL PATHOLOGY

From the time of the early eugenics family studies until eugenics' recent revival in academic and popular discussion, biological theories have always intersected theories of race, class, and disability. In the early 1900s, eugenics was the primary

SOURCE: "Race, Poverty and Disability: Three Strikes and You're Out! Or Are You?" by Pamela Block, Fabricio E. Balcazar, & Christopher B. Keys. *Social Policy*, Fall 2002. Reprinted by permission of the authors.

ideological framework in which policies and practices were developed to manage marginalized populations. Eugenics, the science of the genetic improvement of the human race, was used to establish race and class distinctions as "natural" and incontrovertible. The dominance of the upper class was mandated by their superior genetic heritage; the poor remained in poverty because of their degenerate genes.

Many proponents of eugenics believed in a hierarchy of "lower" and "higher" races. These theorists sought a state-controlled program to improve the quality of the human gene pool by optimizing the breeding of elite white Anglo-Saxon Protestants (positive eugenics), and controlling the reproduction of "degenerates," "morons," and those of "inferior" racial or ethnic groupings (negative eugenics). Special attention was given to those of "mixed" heritage and to foreigners from places other than Western Europe. "Mongrel" families (mixed Caucasian, African-American and Native American racial heritage) were considered unstable and degenerate. The genetic inferiority of the new wave of immigrants was widely discussed. Alarmed at the supposed decrease in the birthrate of white Anglo-Saxon Protestants, eugenicists feared "race suicide." One psychologist advocated that "morons" be allowed to serve in the army in World War I, so that fewer "normal men" would be wasted.

Eugenicists believed that racial degeneration would change the national character, subsuming a primarily Nordic-European population under a "tide" of unwanted immigrants. Adherents of the biological-pathology model developed standardized tests that they believed proved the inferior intelligence of African Americans and immigrants. The biological model highlighted multiple stigma as justification for regressive social policy. Some case studies showed how racial character, poverty, and disability interacted with other factors such as illicit pregnancy, prostitution, gambling, alcoholism, drug addiction, tuberculosis, and syphilis. When several of these factors coexisted, it was considered an indication of degeneration due to pathological inheritance. Policymakers and high profile health professionals in the United States responded by establishing programs for the institutionalization and sterilization of supposedly pathological populations. Hundreds of thousands of individuals considered "feeble-minded" were institutionalized, sometimes for life, in facilities that still exist today.

The extreme example of both positive and negative eugenics was Nazi Germany. Those with Aryan characteristics were encouraged to have large families, while thousands of others were involuntarily sterilized and millions of others were annihilated under a policy of racial purification. Advances in the science of genetics and the example of Nazi Germany served to discredit the eugenics movement in the United States. However, eugenicists continued to take an active role in many professions, including population science and social service provision for marginalized groups (e.g., people on welfare, psychiatric patients, and people with cognitive disabilities).

The term "eugenics" disappeared from professional and popular discourse when it was discredited and thus no longer considered an effective means of program implementation. However, decades after eugenics disappeared from official public policy, eugenicists continued to influence the social service system by advocating institutionalization and sterilization of people with developmental disabilities and sterilization of women on welfare. The legacy of treatment models developed by eugenicists in the early 1900s (such as segregation from the opposite sex and sterilization) is still apparent in practices, such as genetic counseling, which encourage the abortion of fetuses with disabilities like Down's Syndrome. It is also visible in the physical and organizational structures of mental institutions and in the psychological scars of people who have been institutionalized for long periods. Additionally, eugenics ideologies persisted through the subtle influence of cultural beliefs concerning marginalized groups, evidenced by contradictory representations of people with disabilities in literature, film, and television as social victims or social threats. People with disabilities are often portrayed as unable to control their sexuality and capable of erupting into random and unpredictable acts of violence. Society and those with disabilities are only considered safe when such people are locked behind the walls of an institution or when deceased.

RACE, POVERTY AND DISABILITY
AS CULTURAL PATHOLOGY

In the 1930s, biological theories began to be replaced by the theories of culture promulgated by sociologists and anthropologists. Various views of poverty were presented: as cultural disintegration and deprivation; as inspiring pathological forms of cultural structure, i.e., the "culture of poverty"; or as a result of the loss or lack of culture.

Although recognizably distinct from earlier biological theories, the new cultural theories included regressive images of people living in poverty that were disturbingly close to earlier models. Whether the explanation was biological or cultural, these differences (e.g. to be poor, a racial minority, or to have a disability) were presented as a form of self-defeating and self-perpetuating pathology.

Whether defined as a pathological lack of culture, or a culture that was pathological, the people living in poverty were considered deficient by leading theorists and in public policy. For example, the concept of "culture of poverty" impacted Department of Labor and War On Poverty thinking in the 1960s.

The prevalence of the deficiency approach has also characterized societal perceptions of people with disabilities. For example, the 1994 reauthorization of the Developmental Disabilities Assistance and Bill of Rights Act (PL 103-230), defined developmental disabilities as a series of "substantial functional limitations" in "self-care," "self-direction," "capacity for independent living," and "economic self-sufficiency." These "substantial functional limitations" were also exhibited by the "white trash" in an earlier period, with biology thought to be the root cause. Like pathological genes, the culture of poverty was reproduced from generation to generation, all but impossible to escape. Although by the 1950s and 1960s eugenics had been largely discredited, its ideas unintentionally provided a model of cultural pathology to supplant eugenics notions of biological pathology. Thus the social problems exhibited by minorities, such as teen pregnancy, alcohol and drug abuse, welfare culture, serial monogamy or multiple partners, and destabilized families, could be explained through cultural, rather than biological deficiencies. Whether in biological or cultural terms, the contours of these pathology arguments were quite comparable, focusing on the identification of individual deficits and blaming the victim for having them.

FROM MINORITY GROUP MODEL TO AN EMPOWERMENT FRAMEWORK OF ANALYSIS

Immediately after their appearance, scholars and activists began challenging the concepts of the culture of poverty and cultural deprivation, protesting that the supposed traits of people living in poverty were contemptuous distortions based on middle-class biases. Urban anthropologists wrote ethnographic accounts of African-American communities highlighting and valorizing cultural differences. Community, social, and developmental psychologists suggested that so-called "deficits" in language and learning actually reflected cultural differences, not cultural deficiency. Activists protested an approach that blamed victims for their problems and ignored the strengths of their cultures.

Like the eugenics family studies, the culture of poverty thesis functioned to justify blaming the poor for their poverty and blaming social problems on innate deficiencies. Culture became an alternative to biological explanations for the inferiority, dependency and marginality of certain social groups. Individuals from marginalized groups became categorized as mentally retarded, not because of their inherent "incompetence," but due to racial and class bias within school and social service systems.

The contrary view claimed that social problems should be addressed through the elimination of unequal power relations and re-distribution of wealth and income. This social perspective on disability is based on a minority group model that has also been adopted to explain systematic exclusion on the basis of sex, race, and sexual orientation. This model functions to liberate individuals with disabilities from long-held societal prejudices and mistaken assumptions.

Identity politics grew out of collective movements for equality on the basis of gender, race, and sexual orientation. People with disabilities are now applying this principle to their own situations. However, within the minority group model, a single issue (such as race, gender, or sexual orientation) is usually the focus for identity formation. Other issues are considered secondary or are not considered at all. Perhaps this is a response to earlier biological and cultural models that tended to use a multi-issue approach to "prove" the pathological nature (and culture) of marginalized groups. In the process of identity formation, attempts to stay as far away as possible from the earlier pejorative images of race and poverty have resulted in the inadvertent exclusion of individuals facing the "triple jeopardy" of race, poverty, and disability. Perhaps to overcome the pressures of marginalization and to rally a critical mass of support, it is more effective, especially initially, to focus on one common characteristic. But inevitably to some degree this single-focus strategy excludes those facing triple jeopardy and their multiple concerns.

Disability rights activists reject earlier models of disability, adopting instead a model that emphasizes overcoming social barriers rather than focusing on individual pathology. In contrast to static biological and cultural models, the minority group model is a dynamic view of disability resulting from the interaction of the individual and the environment. Disability-rights activists argue that social and cultural values encouraging discrimination or the segregation of people with disabilities must be changed. In this model, the emphasis shifts from the individual to the society, and from victim-blaming to strategies for social change. Education, organizing, a sense of community, and pride in identity replace earlier experiences of isolation and shame. Individuals can band together with others who face similar challenges and learn how to challenge and navigate existing systems to their advantage. A good example is the national advocacy organization of people with physical disabilities known as ADAPT (www.adapt.org). This organization originally focused on efforts to make public transportation accessible in major cities in the country. Since the passage of the Americans with Disabilities Act, the group has focused more on efforts to promote state and national legislation regarding needs for personal care attendants. This group tends to approach social change one-issue-at-a-time.

Earlier biological and cultural models focused on individual pathology, thus obscuring the significant social barriers faced by people with the multiple stigmas of race, poverty, and disability. With multiple stigmas (such as gender, race, and disability), opportunities for employment become increasingly limited and people become trapped in poverty. Disability, poverty, and minority status are linked and intensify the already negative relationship between economic status and the existence of a disability. For example, the majority of Latinos with disabilities live below the poverty level, and African Americans with disabilities are particularly disadvantaged. Biological and cultural theories cannot sufficiently explain the discrepancies between the disabled and non-disabled, minority and non-minority populations. People living with multiple stigmas have limited opportunities and resources available to them and face societal barriers and oppression that result in poverty and exclusion.

An empowerment framework, one that refers to the increased degree of control people can have over relevant aspects of their environment, better accounts for systemic inequalities faced by minorities with disabilities and attempts to overcome these obstacles.

The minority group model of disability must encompass the complex and multilayered identities held by people with disabilities. For some individuals, disability is only one of many stigmas they face. A disability-rights model that does not recognize the multiple barriers resulting from the "triple jeopardy" of race, poverty, and disability may inadvertently exclude a large number of people who deserve representation, many of whom are most likely to be left behind. The one-issue-at-a-time approach is an effective but narrow strategy for social change. An empowerment framework of analysis that incorporates an understanding of multifaceted issues can serve as a catalyst for helping people to overcome multiple stigmas and support their efforts to seek positive social change.

This framework encourages people from multiple constituencies to develop a common agenda based on shared unmet needs. It also promotes self-reliance and an increased awareness and understanding of the social forces that maintain oppression and discrimination. Empowered individuals or groups are more likely to challenge the status quo and to pursue remedy to the social inequality that characterizes the existence of low-income people of color with disabilities.

CONCLUSION

Throughout the twentieth century, several models have evolved to address the problems of people from marginalized groups, such as those living in poverty or who have disabilities. In the early decades of the twentieth century, a model of biological pathology was used to explain persistent social problems, and pathological inheritance was believed to cause mental and physical deficiency. In the decades following World War II, biological explanations gave way to models of cultural pathology. Cultural deprivation was related to mental and physical deficiency, resulting in deficits in language and learning. Both biological and cultural pathology models have been criticized for over-looking the importance of social and political inequality. The minority group model was developed in the 80s and 90s to explain the persistence of the social barriers faced by people with disabilities and other marginalized groups. Currently, an empowerment framework re-conceptualizes the problem and specifies action sequences by which individuals from marginalized groups with multiple stigmas may gain the social, political, and economic support needed to overcome barriers to their full participation in society.

Early perceptions of people with disabilities as victims and threats have been increasingly replaced by images of social actors empowered to and capable of overcoming functional limitations and social barriers. Attempts continue to be made to better understand the complex interaction of multiple variables, such as disability, gender, race, and economic status. Developing theories of social change support the promotion of empowerment. Minorities with disabilities living in poverty can have the strength, understanding, and motivation to learn about their own situation and act to change it. They can develop a critical consciousness about their oppressed status, and then plan and take action to address the impact of negative societal attitudes on them and others in similar circumstances.

The barriers faced by people of color with disabilities living in poverty should not be underestimated. Racism, ableism, and poverty severely limit opportunity. Established organizations are usually unprepared to serve minority populations with multiple needs. For example, there is a growing population of inner-city minorities who have become permanently disabled as a result of violence. Service providers have difficulty meeting their needs. In the past, such situations were addressed through single-issue movements, and service provision that divided and dealt with each issue in isolation.

What is needed now is greater implementation of empowerment approaches that address the nexus of race, poverty, and disability. With critical awareness, support and social change, individuals and groups facing multiple stigmata need not be relegated to the sidelines of society. Essential to this important mission are more indigenous leaders and grassroots groups who have experience working on these different levels to meet the needs, affirm the rights and enhance the choices of people of color with disabilities living in low-income neighborhoods. Three strikes need not mean you're out, but rather playing a game with a greater degree of difficulty.

PART II

✳

Systems of Power and Inequality

MARGARET L. ANDERSEN AND PATRICIA HILL COLLINS

O ne of the most important things to learn about race, class, and gender is that they are *systemic forms of inequality*. Although most people tend to think of them as individual characteristics (or identities), they are built into the very structure of society—and it is this social fact that drives our analysis of race, class, and gender as *intersectional systems of inequality*. This does not make them irrelevant as individual or group characteristics but points you to the analysis of *social structure* to think about how race, class, and gender operate; what they mean; and how they influence people's lives.

Locating racial oppression in the structure of social institutions provides a different frame of analysis than analyzing only individuals. Using a social structural analysis of race, class, and gender turns your attention to how they work as *systems of power*—systems that differentially advantage and disadvantage groups depending on their social location. Moreover, this means that no one of these social facts singularly predetermines where you will be situated within this system of power and social relationships. Thus, not all men are equally powerful and not all women are equally oppressed. When you focus on the intermingling structural relationships of race, class, *and* gender, you see a more complex, ever-changing, and multidimensional social order.

To repeat, neither race nor class nor gender operates alone. They do so within a system of simultaneous, interrelated social relationships—what we have earlier called the *matrix of domination*. This means that they also engage other

61

social facts—ethnicity, sexuality, age, disability, even the region where you live, and so forth. In this section we examine race, class, and gender from an institutional or structural perspective. And, later in the section, we examine how race, class, and gender also intersect with ethnicity, nationality, and sexuality. You will see that each of these systems of power and inequality intertwine—and they do so in different ways at different points in time. Thus, sexuality has become more visible as a system of social subordination; beliefs about sexuality have long served to buttress the beliefs that support racial and gender subordination. Likewise, a system of racial subordination has historically been one way that class structure was created: Some people accumulated property through the appropriation of other people's labor, even while groups who provided the labor that produced property for others were denied basic rights of citizenship, such as the right to vote, the right to marry, the right to own property, and the right to be considered a U.S. citizen.

Throughout Part II, you should keep the concept of *social structure* in mind. Race, class, and gender form a structure of social relations. This structure is supported by ideological beliefs that make things appear "normal" and "acceptable," thus clouding our awareness of how the structure operates. Philosopher Marilyn Frye likens this to a birdcage. From the bird's point of view, you might see only one wire, thus wondering why you cannot just fly away. But when you see the whole structure of the cage, you realize there is a whole structure of containment enclosing the bird (Frye 1983). For example, one of the prevailing beliefs about racism is that it is largely a thing of the past, now that formal barriers to racial discrimination have been removed. But, as Charles Gallagher ("Color-Blind Privilege: The Social and Political Functions of Erasing the Color Line in Post-Race America") points out in his formulation of *color-blind racism,* racism persists, even when it takes on new forms. Perhaps one "wire" from the cage has been removed—laws endorsing racial discrimination—but a whole structure of racism, including new belief systems, is in place.

Understanding the intersections among race, class, and gender—and how they interrelate with ethnicity and sexuality—requires knowing how to conceptualize each factor. Although we would rather not treat them separately, we do so initially here to learn what each means and how each is manifested in different group experiences. At the same time, the readings in this section examine the connections among race, class, gender, and other important social categories—namely, ethnicity and sexuality. As we review each in turn, you will notice several common themes.

First, *each is a socially constructed category*. That is, their significance stems not from some "natural" state, but from what they have become as the result of social and historical processes. Second, notice how *each tends to construct groups in binary (or polar opposite) terms:* man/woman, Black/White, rich/poor, gay/straight, or citizen/alien. These binary constructions create the "otherness" that we examined in the first part of this book. Third, *each is a category of individual and group identity, but note—and this is important—they are also social structures.* That is, they are not just about identity but are also about group location in a system of power and inequality. Thus, as we move into Part II, you will see how they are part of the institutional fabric of society. More than being about individual and group identity, race, class, and gender—and the other systems that they intersect with—shape patterns in the labor market, families, state institutions (such as the government and the law), mass media, and so forth. This is a key difference, as we have seen, in a model that focuses solely on difference and one that focuses on the matrix of domination. That is, the matrix of domination forces you to look at social structures, while models of difference often dissolve into individual or group identities. Finally, *neither race, class, nor gender is a fixed category.* Because they are social constructions, their form—and their interrelationship—changes over time. This also means that social change is possible.

As you learn about race, class, and gender, you should keep the intersectional model in mind. One way to think about their interrelationship in a social system is to imagine a typical college basketball game. This will probably seem familiar: the players on the court, the cheerleaders moving about on the side, the band playing, fans cheering, boosters watching from the best seats, and—if the team is ranked—perhaps a television crew. Everybody seems to have a place in the game. Everybody seems to be following the rules. But what explains the patterns that we see and don't see?

Race clearly matters. The predominance of young African American men on many college basketball teams is noticeable. Why do so many young Black men play basketball? Some people argue that African Americans are simply better in areas requiring physical skills such as sports, but there is another reality: for many young Black men, sports may seem the only hope for a good job, so sports, like the military, can seem like an attractive mobility route. Perceived promises of high salaries, endorsements, and merchandise can make young people believe sports are a path to success, thus, many young African American boys believe they can earn a good living playing professional sports. The odds of actually doing so are extremely slim. In truth, of the forty thousand African American boys playing high school basketball, only thirty-five will make it to the NBA (National Basketball Association) and only seven of those will be starters. This makes the odds of success 0.000175 (Eitzen 1999)! In the face of other systematic disadvantages posed by race, however, some will still think this is their best chance for success.

But does a racial analysis fully explain the "rules" of college basketball? Not really. Who benefits from college basketball? Yes, players (both Black and White, and, increasingly, immigrants) get scholarships and a chance to earn college degrees. Players reap the rewards, but who really benefits? College athletics is big business, and as amateur athletes the players are forbidden to take any payment for their skills. They may have the hope of an NBA contract or a college degree if they graduate. But few actually turn pro. Indeed, many never graduate from college. Among Division I Black basketball players, 60 percent graduate from college, compared with 82 percent of White male basketball players. However, graduation rates among Black male college athletes (including all sports) are slightly higher than among nonathletes, most likely because of the scholarship support they receive. Among women basketball players, graduation rates are virtually identical for White and Black women, at 64 percent and 62 percent respectively. Student-athletes have much higher graduation rates than nonathletes as well as higher graduation rates than male student-athletes (National Collegiate Athletic Association 2010).

Winning teams also benefit educational institutions. Winning teams garner increased admissions applications, more alumni giving, higher levels of corporate support, and television revenues. Simply put, athletics—including college athletics—is big business. Corporate sponsors want their names and products identified with winning teams and athletes; advertisers want their products promoted by winning players. Even though college athletes are forbidden to promote products, corporations create and market products in conjunction with prevailing excitement about basketball, sustained by the players' achievements. Athletic shoes, workout clothing, cars, and beer all target the consumer dollars of those who enjoy watching basketball. And how many jobs are supported by the revenues generated from the enterprise of college basketball? Of course, there are the sports reporters, team physicians, trainers, and coaches—most of them defined as professionals. But there are also the service workers—preparing and selling food, cleaning the toilets and stands, and maintaining the stadium. A class system defines one's place in this system of inequality, so much so that many of the service workers are invisible to the fans. So, while institutions profit the most—whether it be schools, product manufacturers, or advertisers—there are differential benefits depending on your "rank" within the system of basketball. And the stakes get even bigger when you move beyond college into the world of professional sports.

So, do race and class fully explain the "rules" of basketball? What about gender? Only in a few schools does women's basketball draw as large an audience as men's. And certainly in the media, men's basketball is generally the public's focal point, even though women's sports are increasingly popular. In college basketball, like the pros, most of the coaches and support personnel are men, as are the camera crew and announcers. Where are the women? A few are coaches, rarely paid what the men receive—even on the most winning teams. Those closest to the action on the court may be cheerleaders—tumbling, dancing, and being thrown into the air in support of the exploits of the athletes. Others may be in the band. Some women are in the stands, cheering the team—many of them accompanied by their husbands, partners, boyfriends, parents, and children. Many work in the concession stands, fulfilling women's roles of serving others. Still others are even more invisible, left to clean the restrooms, locker rooms, and stands after the crowd goes home.

Men's behavior reveals a gendered dimension to basketball, as well. Where else are men able to put their arms around one another, slap one another's buttocks, hug one another, or cry in public without having their "masculinity" questioned? Sportscasters, too, bring gender into the play of sports, such as when they talk about men's athletic achievements as heroic but talk about women athletes' looks or their connection to children. For that matter, look at the prominence given to men's teams in sports pages of the daily newspaper compared with sports news about women, who are typically relegated to the back pages—if their athletic accomplishments are reported at all. Sometimes what we don't see can be just as revealing as what we do see. Gender, as a feature of the game on the court, is so familiar that it may go unquestioned.

This discussion of college basketball demonstrates how each factor—race, class, and gender—provides an important, yet partial, perspective on social action. Using an intersectional model, we not only see each of them in turn but also the connections among them. In fact, race, class, and gender are so inextricably intertwined that gaining a comprehensive understanding of a basketball game requires thinking about all of them and how they work together. New questions then emerge: Why are most of those serving the food in concession stands likely to be women and men of color? How are norms of masculinity played out through sport? What class and racial ideologies are promoted through assuming that sports are a mobility route for those who try hard enough? If race, class, and gender relations are embedded in something as familiar and widespread as college basketball, to what extent are other social practices, institutions, relations, and social issues similarly structured?

Race, gender, and class divisions are deeply embedded in the structure of social institutions such as work, family, education, and the state. They shape human relationships, identities, social institutions, and the social issues that emerge from within institutions. You can see the intersections of race, class, and gender in three realms of society: the representational realm, the social interaction realm, and the social structural realm. The representational realm includes the symbols, language, and images that convey racial meanings in society; the social interaction realm refers to the norms and behaviors observable in human relationships; and the social structural realm involves the institutional sites where power and resources are distributed in society (Glenn 2002: 12).

This means that race, class, and gender affect all levels of our experience—our consciousness and ideas, our interaction with others, and the social institutions we live within. And, because they are interconnected, no one can be subsumed under the other. In this section of the book, although we focus on each one to provide conceptual grounding, keep in mind that race, class, and gender are connected and overlapping—in all three realms of society: the realm of ideas, interaction, and institutions.

You might begin by considering a few facts:

- The United States is in the midst of a sizable redistribution of wealth, with a greater concentration of wealth and income in the hands of a few than at most previous periods of time. At the same time, a declining share of income is going to the middle class—a class that finds its position slipping, relative to years past (Warren 2009).

- Within class groups, racial group experiences are widely divergent. Thus, although there has been substantial growth of an African American and Latino middle class, they have a more tenuous hold on this class status than groups with more stable footing in the middle class. Furthermore, there is significant class differentiation within different racial groups (Charles 2006; Pattillo 2007; Lacy 2007).

- Women in the top 25 percent of income groups have seen the highest wage growth of any group over the last twenty years; the lowest earning groups of women, like men, have seen wages fall, while the middle has remained flat (Mishel, Bernstein, and Allegretto 2005). Class differences within gender are hidden by thinking of women as a monolithic group.

- Women of color, including Latinas, African American women, Native American women, and Asian American women, are concentrated in the bottom rungs of the labor market along with recent immigrant women (www.bls.gov).

- Poverty has been steadily increasing in the United States since 2000; it is particularly severe among women, especially among women of color and their children; in recent years, poverty has risen most among Hispanics (DeNavas-Walt et al. 2010).

- At both ends of the economic spectrum, there is a growth of gated communities: well-guarded, locked neighborhoods for the rich and prisons for the poor—particularly Latinos and African American men. At the same time, growth in the rate of imprisonment is highest among women (Glaze 2010).

None of these facts can be explained through an analysis that focuses only on race or class or gender. Clearly, race matters. Class matters. Gender matters. And they matter together. We turn now to defining some of the basic concepts involving the systems of power and inequality that we examine in this section.

RACE AND RACISM

For many years, race and racism in the United States have been thought of in terms of Black/White relations. But changes in the racial landscape of the United States make such a framework now inadequate for thinking about all group experiences, a point made by Elizabeth Martinez (in "Seeing More than Black and White"). Martinez shows that Black/White relations have defined racism in the United States for centuries but that a rapidly changing population that includes diverse Latino groups is forcing Americans to reconsider the nature of racism. Color, she argues, has been the marker of race, but she challenges the dualistic thinking that has promoted this racist thinking. Moreover, other groups—Asian Americans, Native Americans, American Muslims, and others—are demonstrating the ethnic and racial diversity in the United States. This reality has resulted in shifting understandings of race and ethnic relations. Still, it is important to locate the study of race, racism, and ethnicity in the shifting relations of power that also characterize the treatment of different groups. This makes race and racism fundamentally about the character of U.S. social institutions.

Still, most people think about race and racism in the context of prejudice. How does prejudice differ from racism? *Prejudice* is a hostile attitude toward a person who is presumed to have alleged negative characteristics associated with a group to which he or she belongs. Racism is more systematic than this and is not the same thing as prejudice. *Prejudice* refers to people's attitudes. *Racism* is a system of power and privilege; it can be manifested in people's attitudes but is rooted in society's structure. It is reflected in the different group advantages and disadvantages, based on their location in this societal system. Racism is structured into society, not just in people's minds. As such, it is built into the very fabric of dominant institutions in the United States and has been since the founding of the nation. This is referred to as *systemic racism,* meaning the "complex array of anti-Black practices, the unjustly gained political-economic power of Whites, the continuing economic and other resource inequalities along racial lines, and the White racial ideologies and attitudes created to maintain and rationalize White privilege and power" (Feagin 2000: 6).

Understanding racism as systemic also means that people may not be individually racist but can still benefit from a system that is organized to benefit some at the expense of others, as Charles Gallagher points out in his essay on color-blind racism ("Color-Blind Privilege"). *Color-blind racism* is a new form of racism in which dominant groups assume that race no longer matters—even when society is highly racially segregated and when individual and group well-being is still strongly determined by race. Many people also believe that being nonracist means being color-blind—that is, refusing to recognize or treat as significant a person's racial background and identity. But to ignore the significance of race in a society where racial groups have distinct historical and contemporary experiences is to deny the reality of their group experience. Being color-blind in a society structured on racial privilege means assuming that everybody is "White," which is why people of color might be offended by friends who say, for example, "But I never think of you as Black." This blindness to the persistent realities of race also leads to an idea that there is nothing we should be doing about it—either individually or collectively—thus, racism is perpetuated. Gallagher's essay shows that the form of racism changes over time. Racial discrimination is no longer legal, but racism continues to structure relations among groups and to differentiate the power that different groups have. Specific racial group histories differ, but different racial groups share common experiences of racial oppression. Thus, Chinese Americans were never enslaved, but they experienced forced residential segregation and economic exploitation based on their presumed racial characteristics. Mexican Americans were never placed on federal reservations, as Native Americans have been, but in some regards both groups share the experience of colonization by White settlers. Both have experienced having their lands appropriated by White people; Native Americans were removed from their lands and forced into reservations, if not killed. Chicanos originally held land in what is now the American Southwest, but it was taken following the Mexican American War; in 1848, Mexico ceded huge parts of what are now California, New Mexico, Nevada, Colorado, Arizona, and Utah to the United States for $15 million. Mexicans living there were one day Mexicans, the next living in the United States, though without all the rights of citizens.

Most people assume that race is biologically fixed, an assumption that is fueled by arguments about the presumed biological basis for different forms of inequality. The concept of race is more social than biological, however. Scientists working on the human genome project have even found that there is no "race" gene. But you should not conclude from this that race is not "real." It is just that its reality stems from its social significance. That is, the meaning and significance of *race* stems from specific social, historical, and political contexts. It is these contexts that make race meaningful, not just whatever physical differences may exist among groups.

To understand this, think about how racial categories are created, by whom, and for what purposes. Racial classification systems reflect prevailing views of race, thereby establishing groups that are presumed to be "natural." These constructed racial categories then serve as the basis for allocating resources; furthermore, once defined, the categories frame political issues and conflicts (Omi and Winant 1994). Omi and Winant define *racial formation* as "the sociohistorical process by which racial categories are created, inhabited, transformed, and destroyed" (1994: 55). In Nazi Germany, Jews were considered to be a race—a social construction that became the basis for the Holocaust. Abby L. Ferber's essay ("What White Supremacists Taught a Jewish Scholar about Identity") shows the complexities that evolve in the social construction of race. As someone who studies White supremacist groups, she sees how White racism defines her as Jewish, even while she lives in society as White. Her reflections reveal, too, the interconnections between racism and anti-Semitism (the hatred of Jewish people), reminding us of the interplay between different systems of oppression.

In understanding racial formations, we see that societies construct rules and practices that define groups in racial terms. Moreover, racial meanings constantly change as institutions evolve and as different groups contest prevailing racial definitions. Some groups are "racialized," others are not. Where, for example, did the term *Caucasian* come from? Although many take it to be "real" and don't think about its racist connotations, the term has racist origins. It was developed in the late eighteenth century by a German anthropologist, Johann Blumenbach. He developed a racial classification scheme that put people from the Russian Caucuses at the top of the racial hierarchy because he thought Caucasians were the most beautiful and sophisticated people; darker people were put on the bottom of the list: Asians, Africans, Polynesians, and Native Americans (Hannaford 1996). It is amazing when you think about it that this term remains with us, with few questioning its racist origin and connotations.

Consider also the changing definitions of *race* in the U.S. census. Given the large number of multiracial groups and the increasing diversity brought about by immigration, we can no longer think of race in mutually exclusive terms. In 1860, only three "races" were presumed to exist—Whites, Blacks, and mulattoes. By 1890, however, these original three races had been joined by five others—quadroon, octoroon, Chinese, Japanese, and Indian. Ten short years later, this list shrank to five races—White, Black, Chinese, Japanese, and Indian—a situation reflecting the growth of strict segregation in the South (Rodriquez 2000).

People of mixed racial heritage have long presented a challenge to census classifications. The U.S. government now allows people to check multiple boxes to identify themselves as more than one race. You can check "Hispanic" as a separate category. This change in the census (which began with the 2000 census) reflects the growing number of multiracial people in the United States. The census categories are not just a matter of accurate statistics; they have significant consequences for the apportionment of societal resources. Thus, while some might argue that we should not "count" race at all, doing so is important because data on racial groups are used to enforce voting rights, to regulate equal employment opportunities, and to determine various governmental supports, among other things.

Nazli Kibria's research article ("The Contested Meanings of 'Asian American': Racial Dilemmas in the Contemporary U.S.") also shows how the concept of race is constructed in particular historical moments in time. As she shows, race is fundamentally a system of power, and racial identities are shaped within that hierarchy. For Asian Americans, changes in immigration laws have shifted how race is constructed. Groups who at one time would have had national identities develop a pan-ethnic identity that stems from their experiences in the specific racial environment of the United States. Thus, racial meaning is a shifting social construction.

The overarching structure of racial power relations means that placement in this structure leads to differences in how people see racism—or don't—and what they are willing to do about it. Herbert J. Gans explores this idea in his essay, "Race as Class." Different groups are perceived within a racial hierarchy that distributes both material and symbolic resources in society. Some groups become less "racially marked" via upward class mobility, although, as Gans shows, this does not usually occur for African Americans—indicative of the strong hold that racial beliefs have on the general American public. As we will see, even with changes in attitudes and some optimism about racial progress, marked differences by race are still evident in employment, political representation, schooling, and other basic measures of group well-being.

CLASS

Class is a major force in American society. As Janny Scott and David Leonhardt show ("Shadowy Lines That Still Divide"), class shapes the life chances of Americans even though there is a strong cultural belief that class mobility is possible. Class is not just about money, however. The consumer culture of America makes it seem that everyone is middle-class because the products associated with this class status are so ubiquitous. But despite this perception, class distinctions are not fading.

Rather than thinking of social class as a rank held by an individual, think of social class as a series of relations that pervade the entire society and shape our social institutions and relationships with one another. Although class shapes identity and individual well-being, class is a system that differentially structures group access to economic, political, cultural, and social resources. Social class is not just a matter of material difference; it is a pattern of domination in which some groups have more power than others. *Power* is the ability to influence and dominate others. This refers not just to interpersonal power but also to the structural power that some groups have because of their position in a racialized and gendered class system that is fundamental to the system of capitalism. This is well argued by Joan Acker in her article, "Is Capitalism Gendered and Racialized?" Her answer is, "Yes." Under capitalism, groups with vast amounts of wealth also influence systems such as the media and the political process in ways that less powerful groups cannot. Privilege in social class thus encompasses both a position of material advantage and the ability to control and influence others.

The class system is currently undergoing profound changes, changes that are linked to patterns of economic transformation in the political economy—changes that are both global and domestic. Jobs are being exported overseas as vast multinational corporations seek to enhance their profits by promoting new markets and cutting the costs of labor. Within the United States, there is a shift from a manufacturing-based economy to a service economy, with corresponding changes in the types of jobs available and the wages attached to these jobs. Fewer skilled, decent-paying manufacturing jobs exist today than in the past. Fewer workers are covered by job benefits and unemployment insurance. This has become more evident to people, especially in light of the economic recession and downturn that began in 2009.

With unemployment high, jobs scarce, and a shrinking safety net to provide benefits to people in need, class divisions in the United States have perhaps become more obvious. The nation now experiences greater inequality—a growing gap between the haves and have-nots—than has occurred at any other time in history. Income growth has been greatest for those at the top end of the population—both the upper 20 percent and the upper 5 percent of all income groups, regardless of race. For everyone else, income has remained flat. In the nation's cities and towns, homelessness has become increasingly apparent even to casual observers. And the wealth of some contrasted with the poverty of many makes a striking contrast in the society's distribution of resources. The consequences are real and are reflected in such basic measures of well-being as health, as explained by Lawrence R. Jacobs and James A. Morone ("Health and Wealth: Our Appalling Health Inequality Reflects and Reinforces Society's Other Gaps").

An important way to understand class inequality is to understand the different meaning and significance of income and wealth. *Income* is the amount of money brought into a household in one year. Measures of income in the United States are based on annually reported census data drawn from a sample of the population. *Median income* is the income level above and below which half of the population lies. It is the best measure of group income standing. Thus, in 2009, median income for non–Hispanic White households was $54,620 (meaning half of such households earned more than this and half earned less); this is the "middle." Black households had a median income of $32,068; Hispanic households, $38,679; Asian Americans, $64,308 (DeNavas-Walt et al. 2011). As you will see in Figure 1, family structure is also a factor in predicting income.

Something to keep in mind is that because household income results from the income of individual workers, many households need more than one earner just to reach median levels of income. Some workers also have to hold multiple jobs to make ends meet.

But income only tells part of the story. Differences in patterns of wealth are even more revealing and show how inequality is perpetuated in society. *Wealth* is determined by adding up all of one's financial assets and subtracting all of one's debt—that is, one's net worth. Income and wealth are related but they are not

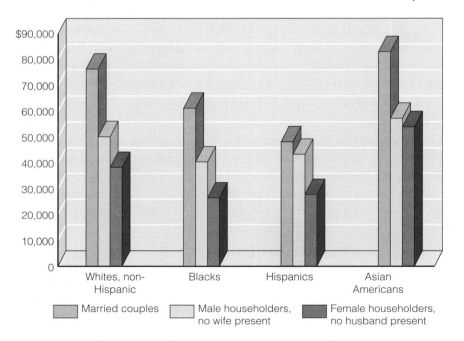

FIGURE 1   Median Annual Household Income by Race and Household Status, 2009

SOURCE: www.census.gov

the same thing. As important as income can be in determining one's class status and measuring inequality, wealth is even more significant.

Consider this: Imagine two recent college graduates. They graduate in the same year, from the same college, with the same major and the identical grade point average. Both get jobs with the same salary in the same company. One student's parents paid all college expenses and gave her a car upon graduation. The other student worked while in school and has graduated with substantial debt from student loans. This student's family has no money to help support the new worker. Who is better off? Same salary, same credentials, but one person has a clear advantage—one that will be played out many times over as the young worker buys a home, finances her own children's education, and possibly inherits additional assets. This shows you the significance of wealth—not just income—in structuring social class.

Thus, income data indicate quite dramatic differences in race, class, and gender standing. Furthermore, wealth differences are also startling. The wealthiest 1 percent of the population hold a larger share of the nation's total wealth than the bottom 90 percent combined—even while household debt has been soaring (Economic Policy Institute 2009). For most Americans, debt, not wealth, is more common: 25 percent of White households, 61 percent of Black households, and 54 percent of Hispanic households have *no financial assets at all* (Oliver and Shapiro 2006). And, there are vast differences in wealth holdings among different racial groups. The median net worth of White households is more than ten times that of African American and Latino households. This has meant that the recent economic downturn has been catastrophic for African Americans, as Melvin L. Oliver and Thomas M. Shapiro show ("Sub-Prime as a Black Catastrophe"). Whole communities have been devastated by housing foreclosures. And, adding in gender to race, the report from the Insight Center for Community Economic Development ("Lifting as We Climb: Women of Color, Wealth, and America's Future"), women and especially women of color are especially disadvantaged when it comes to wealth.

Wealth provides a *cumulative* advantage to those who have it. Wealth produces more wealth because inheritance allows people to transmit economic status from one generation to the next. Wealth helps pay for college costs for children and down payments on houses; it can cushion the impact of emergencies, such as unexpected unemployment or sudden health problems. Even small amounts of wealth can provide the cushion that averts economic disaster for families. Buying a home, investing, being free of debt, sending one's children to college, and transferring economic assets to the next generation are all instances of class advantage that add up over time and produce advantage, even beyond one's current income level. Sociologists Melvin Oliver and Thomas Shapiro (2006) have found, for example, that even Black and White Americans at the same income level, with the same educational and occupational assets, have a substantial difference in their financial assets—an average difference of $43,143 per year! This means that, even when earning the same income, the two groups are in quite different class situations—although both may be considered middle class. When you study class in relationship to race, you will see that overall there are wide differences in the class status of Whites and people of color, but you should be careful not to see all Whites and Asian Americans as well off and all African Americans, Native Americans, and Latinos as poor. White households on average possess higher accumulated wealth and have higher incomes than Black, Hispanic, and Native American households, but large numbers of White households do not. White people also account for almost half (43 percent) of the nation's poor (DeNavas-Walt et al. 2011). Class experiences even *within* racial groups can vary widely, as we have seen an expansion of the Black and Latino middle class in recent years.

These facts should caution us about conclusions based on *aggregate data* (that is, data that represent whole groups). Such data give you a broad picture of group differences, but they are not attentive to the more nuanced picture you see when taking into account race, class, and gender (along with other factors, such as age, level of education, occupation, and so forth). Aggregate data on Asian Americans, for example, show them as a group to be relatively well off. This portrayal, like the stereotypes of a model minority, however, obscures significant differences both when making comparisons between Asian Americans and other groups and between Asian American groups. So, for example, although Asian American median income is, in the aggregate, higher than for White Americans, this does not mean all Asian American families are better off than White families. Twelve percent of Asian Americans are poor, compared with 9.9 percent of White, non-Hispanic people (DeNavas-Walt et al. 2011). Recent Asian immigrants have the highest rates of poverty, including the Hmong, Laotians, and Cambodians, whose rates of poverty match that of African Americans and Native Americans—about a 25 percent poverty rate (Le 2008).

Examining poverty is a revealing way to see the intersection of race, class, and gender. Women and their children are especially hard hit by poverty. Forty percent of Black families and of Hispanic families headed by women are poor, as are 19 percent of Asian American and 20 percent of White families headed by women (see Figure 2). Poverty rates among children (those under eighteen years of age) are especially disturbing: in 2009, 20 percent of all children in the United States lived below the *poverty line* ($21,832 for a family with three children and one adult). When adding race, the figures are even more disturbing: 35 percent of African American children, 32 percent of Hispanic children, 14 percent of

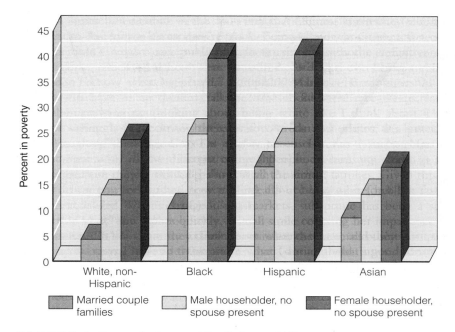

FIGURE 2 Poverty by Race and Family Status, 2009
SOURCE: www.census.gov

Asian/Pacific Islander children, and 9 percent of White (non–Hispanic) children
are poor—astonishing figures for one of the most affluent nations in the world
(DeNavas-Walt et al. 2010).

GENDER

Gender, like race, is a social construction, not a biological imperative. *Gender* is
rooted in social institutions and results in patterns within society that structure
the relationships between women and men and that give them differing positions
of advantage and disadvantage within institutions. As an identity, gender is
learned; that is, through gender socialization, people construct definitions of
themselves and others that are marked by gender. Like race, however, gender
cannot be understood at the individual level alone. Gender is structured in social
institutions, including work, families, mass media, and education.

   You can see this if you think about the concept of a gendered institution.
Gendered institution is now used to define the total patterns of gender relations
that are "present in the processes, practices, images, and ideologies, and distribution of power in the various sectors of social life" (Acker 1992: 567). This term
brings a much more structural analysis of gender to the forefront. Rather than

seeing gender only as a matter of interpersonal relationships and learned identities, this framework focuses the analysis of gender on relations of power—just as
thinking about institutional racism focuses on power relations and economic and
political subordination—not just interpersonal relations. Changing gender relations is not just a matter of changing individuals. As with race and class, change
requires transformation of institutional structures.

Gender, however, is not a monolithic category. Maxine Baca Zinn, Pierrette
Hondagneu Sotelo, and Michael Messner argue ("Gender Through the Prism of
Difference") that, although gender is grounded in specific power relations, it is
important to understand that gender is constructed differently depending on the
specific social locations of diverse groups. Thus race, class, nationality, sexual orientation, and other factors produce varying social and economic consequences that
cannot be understood by looking at gender differences alone. These authors ask us
to move beyond studying differences and instead to use multiple "prisms" to see
and comprehend the complexities of multiple systems of domination—each of
which shapes and is shaped by gender. Seen in this way, men appear as a less
monolithic and uni-dimensional group as well.

Race, class, gender, and sexuality *together* construct stereotypes. Each gains
meaning in relationship to the others (Glenn 2002). Judith Ortiz Cofer ("The
Myth of the Latin Woman: I Just Met a Girl Named Maria") shows how the
combination of her gender identity with her status as a Latina results in quite
hurtful stereotypes that she continuously encounters in everyday life. You cannot
understand her experience as a woman without also locating her in the ethnic,
racial, national, and migration experiences that are also part of her daily life.
Thus, gender oppression is maintained through multiple systems—systems that
are reflected in group stereotypes. These stereotypes also sexualize groups in different, but particular, ways.

One place where the stereotypes associated with gender are constantly reproduced is in the popular media. *Controlling images* of women (and men) in popular
culture are both racialized and sexualized, especially (but not exclusively) for
women of color. In the media, race, gender, and sexual stereotypes intertwine in
different ways. African American men are stereotyped as hypermasculine and over-
sexed, and African American women as promiscuous, bad mothers, and nurturing
"mammies" who care for everyone else, but not their own children. Latinos are
stereotyped as "macho" and, like African American men, sexually passionate, but
out of control. Latinas are stereotyped as either "hot" or virgin-like. Similarly,
White women are sexually stereotyped in dichotomous terms as "madonnas" or
"whores." Working-class women are more likely to be seen as "sluts" and
upper-class women as frigid and cold. We can see that controlling images of

sexuality are part of the architecture of race, class, and gender oppression (Collins 2000, 2007), revealing the interlocking systems of race, class, gender, and sexuality.

Gender ideologies abound in other ways, too. A common one is that women now have it made. The facts tell us otherwise. True, the gap between women's and men's income has shrunk, although most analysts agree that the narrowing of the gap reflects a drop in men's wages more than an increase in women's wages. And, women are more present in professional jobs—women who have become defined as the stereotypical "working woman." Despite this new image, though, most women remain concentrated in gender-segregated occupations with low wages, little opportunity for mobility, and stressful conditions. This is particularly true for women of color, who are more likely to be in occupations that are both race- and gender-segregated. Income and occupational data, however, do not tell the full story for women. High rates of violence against women—whether in the home, on campus, in the workplace, or on the streets—indicate the continuing devaluation of and danger for women in this society.

When people are shut out of dominant social institutions, they often form new social structures and organizations. Adia M. Harvey ("Becoming Entrepreneurs: Intersections of Race, Class, and Gender at the Black Beauty Salon") shows how African American women have created their own social spaces—spaces where they can realize their own entrepreneurial talents. Harvey's research on Black-woman-owned beauty salons shows how people actively construct their lives even in the face of various and multiple forms of oppression.

Turning the tables, Kristen Barber ("The Well-Coiffed Man" Class. Race, and Heterosexual Masculinity in the Hair Salon") studies how men construct an identity as masculine even while in women's space. Their study of metrosexual men shows how these men both transgress expected gender boundaries, while also resisting a perception of themselves as too "feminine." One point is that masculinity, like other identities we examine here, is a fluid and social construction.

The construction of femininity has long been thought to define women's gender roles. But, an inclusive perspective helps us see the different ways that women construct "womanhood" and men construct "manhood." Carla O'Connor and her co-authors argue that Black women are more likely to be socialized as independent and self-sufficient, even while still being "womanly." The life histories of the women in their study document the specific ways that gender socialization emerges among Black women, especially within the context of community expectations about what it means to be a woman.

Altogether, the articles in this section show the social basis of gender. Although identities develop in the context of individual lives, they are shaped by the particular configurations of gender that emerge in a society also structured by race and class.

ETHNICITY AND MIGRATION

Sociologists traditionally definee *ethnicity to refer to* groups who share a common culture; however, like race, ethnicity develops within the context of systems of power. Thus, the meaning and significance of ethnicity can shift over time and in different social and political contexts. For example, groups may develop a sense of heightened ethnicity in the context of specific historical events; likewise, their feeling of sharing a common group identity can result from being labeled as "outsiders" by dominant groups. Think, for example, of the heightened sense of group identity that has developed for Arab Americans in the aftermath of 9/11.

Patterns in the shifting meaning of ethnicity are being influenced by the changing contours of the U.S. population. Racial-ethnic groups comprise an increasing proportion of the U.S. population, and their numbers are expected to rise over the years ahead. One-quarter of the U.S. population is now Black, Hispanic, Asian American, Pacific Islander, or Native American; by 2050, non-Hispanic Whites are predicted to make up only slightly more than half the total population. At the same time that the nation is becoming more diverse, changes are occurring regarding which groups predominate. Latinos have recently exceeded African Americans as the largest minority group in the population. Hispanic, Native American, Asian American, and African American populations are also growing more rapidly than White American populations, with the greatest growth among Hispanics and Asian Americans.

Immigration into the United States is only one facet of the increasing importance that ethnicity and migration play in contemporary social life. Changes in ethnicity and migration are being enacted on a world stage—a global context that increasingly links nations together in a global social order where diversity is more typical than not. You need only consult an older world atlas to get a sense of the widespread changes that characterize the current period. *World cities* have been formed that are linked through international systems of commerce; within these cities, migrants play an important role in providing the labor that is needed to keep pace with the global economy. Migrant (or "guest") workers may provide the labor for multinational corporations, or they may provide the service work that increasingly characterizes postindustrial society. Some workers may provide agricultural labor for multinational producers of food; others perform domestic labor for middle- and upper-class families whose own lives are being transformed by changes in the world economic system. Wherever you look, the world is being changed by the increasingly global basis of modern life.

Such changes indicate that globalization is not just about what is happening elsewhere in the world—as if global studies were just a matter of comparing societies. Globalization is everywhere. No nation, including the United States, can really be understood without seeing how globalization is affecting life even at home. Patterns of life in any one society are now increasingly shaped by the connections between societies, and this is evident when you look at how ethnicity and migration are changing life in the United States. Cultural features associated with different cultural traditions are increasingly evident, even in communities in the United States that have been thought of as "all-American" towns. Youth cultures embrace common features whether in Russia, Mexico, Japan, or the United States because of the penetration of world capitalist markets. Thus, hip-hop plays on the streets of Buenos Aires and world music is heard on radio stations in the United States.

Within the borders of the United States, changes associated with migration are raising new questions about national identity. How will the United States define itself as a nation-state in the changing global context? How will the increased visibility of people of color in the United States, in Latin America, in Africa, in Asia, and within the borders of former European colonial powers shape the future? How are cultural representations and institutions changed in a world context marked by such an ethnic mix—both within and between countries? In this context, which groups are defined as races, and how does that map onto concepts of citizenship and the division of labor in different nations? What will it mean to be American in a nation where racial-ethnic diversity is so much a part of the national fabric?

Such questions have taken on heightened meaning in this post 9/11 society. Following the terrorist attacks on the United States, most in the United States were catapulted into a strong sense of national identity. But accompanying this have been more restrictive immigration practices and policies and restrictions on many civil liberties. International and "ethnic" students are increasingly under surveillance. The nation's response to immigration has largely been one of criminalizing immigrants and questioning their motives for coming to the United States. Such tensions bring increased importance to addressing how the nation can maintain great diversity within a framework of social justice.

Several themes emerge in rethinking ethnicity and migration through the lenses of race, class, and gender. Despite the ideology of the "melting pot," national identity in the United States has been closely linked to a history of White privilege. Beginning with the conquest of Native Americans, national policies have sanctioned the domination of Native Americans. C. Matthew Snipp's essay "The First Americans: American Indians" details the historical role of the state in forcing Native Americans from their homes and subordinating them to a new system of state control. Genocide, forced relocation, and regulation by state law have shaped Native American experience; state processes, whether in the form of war, legislation, or social policy, have contributed to the current place of Native Americans in U.S. society. At the same time, as Snipp notes, groups are resilient as they face adversity and often turn to the state for assistance. As the nation's first "ethnic" groups, Native Americans know all too well how the dynamics of race, class, and gender inequalities have grossly harmed their nations.

Indeed, as Lillian Rubin points out in "Is This a White Country, or What?" the term *American* is usually assumed to mean White. Other types of Americans, such as African Americans and Asian Americans, become distinguished from the "real" Americans by virtue of their race. More importantly, certain benefits are reserved for those deemed to be "deserving Americans." You might think of her article as you listen to contemporary diatribes against so called "illegal aliens." What different connotation emerges when you talk about immigrants as "illegal aliens" rather than simply as "undocumented workers?" Which nomenclature makes people seem to be "other," "distant," and "un-American?"

Many people view ethnic and racial inequality as the failure of people of color to assimilate into the mainstream as White ethnic groups allegedly have done. But as Mary C. Waters points out in "Optional Ethnicities: For Whites Only?" this view seriously misreads the role of race and ethnicity in shaping American national identity. Waters suggests that White Americans of European ancestry have *symbolic ethnicity,* meaning that their ethnic identity does not influence their lives unless they want it to. Waters contrasts this symbolic ethnic identity among many Whites with the socially enforced and imposed racial identity among African Americans. Because race operates as a physical marker in the United States, intersections of race and ethnicity operate differently for Whites and for people of color. White ethnics can thus have "ethnicity" without cost, but people of color pay the price for their ethnic identity.

The politics of immigration have permeated American society in recent years. No longer are immigrants concentrated mostly in particular parts of the country. Economic restructuring has distributed immigrant groups throughout the nation and into new destinations, such as rural areas in the South and Midwest, as well as small towns throughout America. This brings increased visibility to immigration as more people see the evidence of an increasingly diverse society. Such dispersion enriches the nation's culture, but also can fan the flames of anti-immigrant prejudices. This change reflects changes in the global economy, not just in people's attitudes.

Often such anti-immigration sentiments fall on those who are completely innocent of any wrong-doing. Carolina Bank Muñoz shows this to be the case for undocumented students—students who may have spent most or all of their lifetimes in U.S. schools, but then been denied the same rights to higher education as other U.S.-educated students. She articulates the need for social policies, such as the Dream Act (denied by Congress in 2011), that would extend certain rights of citizenship to such students. In this section, we see how ethnicity intertwines with race and class and gender in ways that shape the life chances of different groups.

SEXUALITY

The linkage between race, class, and gender is revealed within studies of sexuality, just as sexuality is a dimension of each. For example, constructing images about Black sexuality is central to maintaining institutional racism. Similarly, beliefs about women's sexuality structure gender oppression. Thus, sexuality operates as a system of power and inequality comparable to and intersecting with the systems of race, class, and gender.

The connection between race, class, gender, and sexuality is explored in the opening essay here by Patricia Hill Collins ("Prisons for Our Bodies, Closets for Our Minds: Racism, Heterosexism, and Black Sexuality"). Collins shows how historically racism has been buttressed by beliefs about Black sexuality. Sexuality has been used as the vehicle to support racial fears and racial subordination. Racial subordination was built on the exploitation of Black bodies—both as labor and as sexual objects. Strictures against certain interracial, sexual relationships (but not those between White men and African American women) are a way to maintain White racism and patriarchy. Collins also shows how stigmatizing the sexuality of African Americans can in some ways be likened to the experiences of lesbian, gay, bisexual, and transgendered persons. This is not to say that racial and sexual oppression are the same, but it is to say that these systems of oppression operate together producing social beliefs and social actions that oppress and exploit African American men and women in particular ways, while also oppressing sexual minorities.

Underlying the analysis of sexuality here are the concepts of *homophobia* (the fear and hatred of homosexuality) and *heterosexism* (the institutionalized power and privilege accorded to heterosexual behavior and identification). If only heterosexual forms of gender identity are labeled "normal," then gays, lesbians, and bisexuals become ostracized, oppressed, and defined as "socially deviant." Homophobia affects heterosexuals as well because it is part of the gender ideology used to distinguish "normal" men and women from those deemed deviant. Thus, young boys learn a rigid view of masculinity—one often associated with violence, bullying, and degrading others—to avoid being perceived as a "fag." The oppression of lesbians and gay men is then linked to the structure of gender in everyone's lives.

The institutionalized structures and beliefs that define and enforce heterosexual behavior as the only natural and permissible form of sexual expression are what is meant by *heterosexism*. As Jonathan Ned Katz points out ("The Invention of Heterosexuality"), heterosexism is a specific historic construction; its meaning and presumed significance have evolved and changed at distinct historical times.

Understanding this rests on understanding that sexuality, like race, class, gender, and ethnicity, is a social construction. Sexuality has a biological context because bodies make sex pleasurable, but the real significance of sexuality lies in its social dimensions. Thus, ideas about gender in society influence how we experience sexuality and how sexuality is controlled. Homophobia is used as a mechanism to enforce the construction of masculinity. The hatred directed toward lesbians, gays, and bisexuals is thus part of the system by which gender is created and maintained. In this regard, sexuality and gender are deeply linked.

The linkage between gender and sexuality is especially acute in the popular media and popular culture. Many have argued that contemporary youth culture is highly sexualized, involving very public, highly sexualized displays of and by women. Indeed, many young women take sexual freedom as a sign of their independence and freedom, without criticizing the sexual objectification of women. Rosalind Gill ("Beyond the 'Sexualization of Culture' Thesis: An Intersectional Analysis of 'Sixpacks,' 'Midriffs,' and 'Hot Lesbians' in Advertising") challenges monolithic views about the "sexualization of culture," arguing that media images have specific grounding in race, class, gender, and heteronormative assumptions. When you think about sexuality within an intersectional framework of race, class, and gender, many new questions and new perspectives come to light. One of these is examined by Chung-suk Han ("Darker Shades of Queer: Race and Sexuality at the Margins"). Han discusses the situation for gay men of color in a context where gay men are generally depicted as White (and stereotypically rich). His essay reveals the importance of remembering how the workings of race, class, gender, and sexuality can play out even within different more progressive communities.

Denise Brennan ("Selling Sex for Visas: Sex Tourism as a Stepping-stone to International Migration") reveals another dimension to the discussion of sexuality—sex as work. Sex workers sell their bodies or images of their bodies for money. Brennan emphasizes how sex workers are part of an international system of sex tourism, thus bringing a global perspective to the discussion of sexuality. Sex work on a global scale is also linked to world politics about race and class, given the specific position of women of color in international sex work. Women's sexuality is used to promote tourism and where images of the "exotic other" are used to attract more affluent classes to various regions of the world, thus linking sexuality to processes of migration and more global analyses of race, class, and gender.

REFERENCES

Acker, Joan. 1992. "Gendered Institutions: From Sex Roles to Gendered Institutions." *Contemporary Sociology* 21 (September): 565–569.

Charles, Camille Z. 2006. *Won't You Be My Neighbor? Race, Class, and Residence in Los Angeles.* New York: Russell Sage Foundation.

Collins, Patricia Hill. 2004. *Black Sexual Politics: African Americans, Gender, and the New Racism.* New York: Routledge.

Collins, Patricia Hill. 2000. *Black Feminist Thought: Knowledge, Consciousness, and the Politics of Empowerment.* New York: Routledge.

DeNavas-Walt, Carmen, Bernadette D. Proctor, and Jessica C. Smith. 2010. *Income, Poverty, and Health Insurance: Coverage in the United States: 2009, P60-238.* Washington, DC: U.S. Census Bureau.

DeNavas-Walt, Carmen, Bernadette D. Proctor, and Jessica C. Smith. 2011. *Income, Poverty, and Health Insurance: Coverage in the United States: 2010, P60-239.* Washington, DC: U.S. Census Bureau.

Eitzen, D. Stanley. 1999. *Fair and Foul: Beyond the Myths and Paradoxes of Sport.* Lanham, MD: Rowman and Littlefield.

Feagin, Joe R. 2000. *Racist America: Roots, Current Realities, and Future Reparations.* New York: Routledge.

Frye, Marilyn. 1983. *The Politics of Reality.* Trumansburg, NY: The Crossing Press.

Glaze, Lauren E. 2010. *Correctional Populations in the United States 2009.* Washington, DC: Bureau of Justice Statistics.

Glenn, Evelyn Nakano. 2002. *Unequal Freedom: How Race and Gender Shaped American Citizenship and Labor.* Cambridge, MA: Harvard University Press.

Hannaford, Ivan. 1996. *Race: The History of an Idea in the West.* Baltimore: Johns Hopkins University Press.

Lacy, Karyn R. 2007. *Blue-Chip Black: Race, Class, and Status in the New Black Middle Class.* Berkeley, CA: University of California Press.

Le, C.N. 2008. "Socioeconomic Statistics & Demographics." *Asian-Nation: The Landscape of Asian America.* www.asian-nation.org/demographics.html (April 11).

National Collegiate Athletic Association. 2010. Trends in *Graduation Success Rates and Federal Graduation Rates at Division I Institutions. Report for NCAA—Division I Schools.* Indianapolis, IN: NCAA. www.ncaa.org.

Oliver, Melvin L., and Thomas M. Shapiro. 2006. *Black Wealth/White Wealth: A New Perspective on Racial Inequality*, 2nd ed. New York: Routledge.

Omi, Michael, and Howard Winant. 1994. *Racial Formation in the United States: From the 1960s to the 1990s*, 2nd ed. New York: Routledge.

Pattillo, Mary. 2007. *Black on the Block: The Politics of Race and Class in the City*. Chicago, IL: University of Chicago Press.

Rodriguez, Clara. 2000. *Changing Race: Latinos, the Census, and the History of Ethnicity*. New York: New York University Press.

8

Seeing More Than Black and White

ELIZABETH MARTINEZ

The racial and ethnic landscape has changed too much in recent years to view it with the same eyes as before. We are looking at a multi-dimensional reality in which race, ethnicity, nationality, culture and immigrant status come together with breathtakingly new results. We are also seeing global changes that have a massive impact on our domestic situation, especially the economy and labor force. For a group of Korean restaurant entrepreneurs to hire Mexican cooks to prepare Chinese dishes for mainly African-American customers, as happened in Houston, Texas, has ceased to be unusual.

The ever-changing demographic landscape compels those struggling against racism and for a transformed, non-capitalist society to resolve several strategic questions. Among them: doesn't the exclusively Black-white framework discourage the perception of common interests among people of color and thus sustain White Supremacy? Doesn't the view that only African Americans face serious institutionalized racism isolate them from potential allies? Doesn't the Black-white model encourage people of color to spend too much energy understanding our lives in relation to whiteness, obsessing about what white society will think and do?

That tendency is inevitable in some ways: the locus of power over our lives has long been white (although big shifts have recently taken place in the color of capital, as we see in Japan, Singapore and elsewhere). The oppressed have always survived by becoming experts on the oppressor's ways. But that can become a prison of sorts, a trap of compulsive vigilance. Let us liberate ourselves, then, from the tunnel vision of whiteness and behold the many colors around us! Let us summon the courage to reject outdated ideas and stretch our imaginations into the next century.

For a Latina to urge recognizing a variety of racist models is not, and should not be, yet another round in the Oppression Olympics. We don't need more competition among different social groups for the gold medal of "Most Oppressed." We don't need more comparisons of suffering between women and Blacks, the disabled and the gay, Latino teenagers and white seniors, or whatever. Pursuing some hierarchy of oppression leads us down dead-end streets where we will never find the linkage between different oppressions and how to

SOURCE: From Elizabeth Martinez, *De Colores Means All of Us: Latina Views for a MultiColored Century* (Cambridge, MA: South End Press, 1998). Reprinted by permission of South End Press.

overcome them. To criticize the exclusively Black-white framework, then, is not some resentful demand by other people of color for equal sympathy, equal funding, equal clout, equal patronage or other questionable crumbs. Above all, it is not a devious way of minimizing the centrality of the African-American experience in any analysis of racism.

The goal in re-examining the Black-white framework is to find an effective strategy for vanquishing an evil that has expanded rather than diminished. Racism has expanded partly as a result of the worldwide economic recession that followed the end of the post-war boom in the early 1970s, with the resulting capitalist restructuring and changes in the international division of labor. Those developments generated feelings of insecurity and a search for scapegoats. In the United States racism has also escalated as whites increasingly fear becoming a weakened, minority population in the next century. The stage is set for decades of ever more vicious divide-and-conquer tactics.

What has been the response from people of color to this ugly White Supremacist agenda? Instead of uniting, based on common experience and needs, we have often closed our doors in a defensive, isolationist mode, each community on its own. A fire of fear and distrust begins to crackle, threatening to consume us all. Building solidarity among people of color is more necessary than ever—but the exclusively Black-white definition of racism makes such solidarity more difficult than ever.

We urgently need twenty-first century thinking that will move us beyond the Black-white framework without negating its historical role in the construction of U.S. racism. We need a better understanding of how racism developed both similarly and differently for various peoples, according to whether they experienced genocide, enslavement, colonization or some other structure of oppression. At stake is the building of a united anti-racist force strong enough to resist White Supremacist strategies of divide-and-conquer and move forward toward social justice for all....

... African Americans have reason to be uneasy about where they, as a people, will find themselves politically, economically and socially with the rapid numerical growth of other folk of color. The issue is not just possible job loss, a real question that does need to be faced honestly. There is also a feeling that after centuries of fighting for simple recognition as human beings, Blacks will be shoved to the back of history again (like the back of the bus). Whether these fears are real or not, uneasiness exists and can lead to resentment when there's talk about a new model of race relations. So let me repeat: in speaking here of the need to move beyond the bipolar concept, the goal is to clear the way for stronger unity against White Supremacy. The goal is to identify our commonalities of experience and needs so we can build alliances.

The commonalities begin with history, which reveals that again and again peoples of color have had one experience in common: European colonization and/or neo-colonialism with its accompanying exploitation. This is true for all indigenous peoples, including Hawaiians. It is true for all Latino peoples, who were invaded and ruled by Spain or Portugal. It is true for people in Africa, Asia and the Pacific Islands, where European powers became the colonizers.

People of color were victimized by colonialism not only externally but also through internalized racism—the "colonized mentality."

Flowing from this shared history are our contemporary commonalities. On the poverty scale, African Americans and Native Americans have always been at the bottom, with Latinos nearby. In 1995 the U.S. Census found that Latinos have the highest poverty rate, 24 percent. Segregation may have been legally abolished in the 1960s, but now the United States is rapidly moving toward resegregation as a result of whites moving to the suburbs. This leaves people of color—especially Blacks and Latinos—with inner cities that lack an adequate tax base and thus have inadequate schools. Not surprisingly, Blacks and Latinos finish college at a far lower rate than whites. In other words, the victims of U.S. social ills come in more than one color. Doesn't that indicate the need for new, inclusive models for fighting racism? Doesn't that speak to the absolutely urgent need for alliances among peoples of color?

With greater solidarity, justice for people of color could be won. And an even bigger prize would be possible: a U.S. society that advances beyond "equality," beyond granting people of color a respect equal to that given to Euro-Americans. Too often "equality" leaves whites still at the center, still embodying the Americanness by which others are judged, still defining the national character....

... Innumerable statistics, reports and daily incidents should make it impossible to exclude Latinos and other non-Black populations of color when racism is discussed, but they don't. Police killings, hate crimes by racist individuals and murders with impunity by border officials should make it impossible, but they don't. With chilling regularity, ranch owners compel migrant workers, usually Mexican, to repay the cost of smuggling them into the United States by laboring the rest of their lives for free. The 45 Latino and Thai garment workers locked up in an El Monte, California, factory, working 18 hours a day seven days a week for $299 a month, can also be considered slaves (and one must ask why it took three years for the Immigration and Naturalization Service to act on its own reports about this horror) (*SanFrancisco Examiner,* August 8, 1995). Abusive treatment of migrant workers can be found all over the United States. In Jackson Hole, Wyoming, for example, police and federal agents rounded up 150 Latino workers in 1997, inked numbers on their arms and hauled them off to jail in patrol cars and a horse trailer full of manure (*Los Angeles Times,* September 6, 1997).

These experiences cannot be attributed to xenophobia, cultural prejudice or some other, less repellent term than racism. Take the case of two small Latino children in San Francisco who were found in 1997 covered from head to toe with flour. They explained they had hoped to make their skin white enough for school. There is no way to understand their action except as the result of fear in the racist climate that accompanied passage of Proposition 187, which denies schooling to the children of undocumented immigrants. Another example: Mexican and Chicana women working at a Nabisco plant in Oxnard, California, were not allowed to take bathroom breaks from the assembly line and were told to wear diapers instead. Can we really imagine white workers being treated that way? (The Nabisco women did file a suit and won, in 1997.)

No "model minority" myth protects Asians and Asian Americans from hate crimes, police brutality, immigrant-bashing, stereotyping and everyday racist prejudice. Scapegoating can even take their lives, as happened with the murder of Vincent Chin in Detroit some years ago....

WHY THE BLACK-WHITE MODEL?

A bipolar model of racism has never been really accurate for the United States. Early in this nation's history, Benjamin Franklin perceived a tri-racial society based on skin color—"the lovely white" (Franklin's words), the Black, and the "tawny," as Ron Takaki tells us in *Iron Cages*. But this concept changed as capital's need for labor intensified in the new nation and came to focus on African slave labor. The "tawny" were decimated or forcibly exiled to distant areas; Mexicans were not yet available to be the main labor force. As enslaved Africans became the crucial labor force for the primitive accumulation of capital, they also served as the foundation for the very idea of whiteness—based on the concept of blackness as inferior.

Three other reasons for the Black-white framework seem obvious: numbers, geography and history. African Americans have long been the largest population of color in the United States; only recently has this begun to change. Also, African Americans have long been found in sizable numbers in most parts of the United States, including major cities, which has not been true of Latinos until recent times. Historically, the Black-white relationship has been entrenched in the nation's collective memory for some 300 years—whereas it is only 150 years since the United States seized half of Mexico and incorporated those lands and their peoples. Slavery and the struggle to end it formed a central theme in this country's only civil war—a prolonged, momentous conflict. Above all, enslaved Africans in the United States and African Americans have created an unmatched heritage of massive, persistent, dramatic and infinitely courageous resistance, with individual leaders of worldwide note.

We also find sociological and psychological explanations of the Black-white model's persistence. From the days of Jefferson onward, Native Americans, Mexicans and later the Asian/Pacific Islanders did not seem as much a threat to racial purity or as capable of arousing white sexual anxieties as did Blacks. A major reason for this must have been Anglo ambiguity about who could be called white. Most of the Mexican *ranchero* elite in California had welcomed the U.S. takeover, and Mexicans were partly European—therefore "semi-civilized"; this allowed Anglos to see them as white, unlike lower-class Mexicans. For years Mexicans were legally white, and even today we hear the ambiguous U.S. Census term "Non-Hispanic whites."

Like Latinos, Asian Americans have also been officially counted as white in some historical periods. They have been defined as "colored" in others, with

"Chinese" being yet another category. Like Mexicans, they were often seen as not really white but not quite Black either. Such ambiguity tended to put Asian Americans along with Latinos outside the prevailing framework of racism.

Blacks, on the other hand, were not defined as white, could rarely become upper-class and maintained an almost constant rebelliousness. Contemporary Black rebellion has been urban: right in the Man's face, scary. Mexicans, by contrast, have lived primarily in rural areas until a few decades ago and "have no Mau-Mau image," as one Black friend said, even when protesting injustice energetically. Only the nineteenth-century resistance heroes labeled "bandits" stirred white fear, and that was along the border, a limited area. Latino stereotypes are mostly silly: snoozing next to a cactus, eating greasy food, always being late and disorganized, rolling big Carmen Miranda eyes, shrugging with self-deprecation "me no speek good eengleesh." In other words, *not serious.* This view may be altered today by stereotypes of the gangbanger, criminal or dirty immigrant, but the prevailing image of Latinos remains that of a debased white, at best....

Among other important reasons for the exclusively Black–white model, sheer ignorance leaps to mind. The oppression and exploitation of Latinos (like Asians) have historical roots unknown to most Americans. People who learn at least a little about Black slavery remain totally ignorant about how the United States seized half of Mexico or how it has colonized Puerto Rico....

One other important reason for the bipolar model of racism is the stubborn self-centeredness of U.S. political culture. It has meant that the nation lacks any global vision other than relations of domination. In particular, the United States refuses to see itself as one among some 20 countries in a hemisphere whose dominant languages are Spanish and Portuguese, not English. It has only a big yawn of contempt or at best indifference for the people, languages and issues of Latin America. It arrogantly took for itself alone the name of half the western hemisphere, America, as was its "Manifest Destiny," of course.

So Mexico may be nice for a vacation and lots of Yankees like tacos, but the political image of Latin America combines incompetence with absurdity, fat corrupt dictators with endless siestas. Similar attitudes extend to Latinos within the United States. My parents, both Spanish teachers, endured decades of being told that students were better off learning French or German. The mass media complain that "people can't relate to Hispanics (or Asians)." It takes mysterious masked rebels, a beautiful young murdered singer or salsa outselling ketchup for the Anglo world to take notice of Latinos. If there weren't a mushrooming, billion-dollar "Hispanic" market to be wooed, the Anglo world might still not know we exist. No wonder that racial paradigm sees only two poles.

The exclusively Black–white framework is also sustained by the "model minority" myth, because it distances Asian Americans from other victims of racism. Portraying Asian Americans as people who work hard, study hard, obey the established order and therefore prosper, the myth in effect admonishes Blacks and Latinos: "See, anyone can make it in this society if you try hard enough. The poverty and prejudice you face are all *your* fault."

The "model" label has been a wedge separating Asian Americans from others of color by denying their commonalities. It creates a sort of racial bourgeoisie, which White Supremacy uses to keep Asian Americans from joining forces with the poor, the homeless and criminalized youth. People then see Asian Americans as a special class of yuppie: young, single, college-educated, on the white-collar track—and they like to shop for fun. Here is a dandy minority group, ready to be used against others.

The stereotype of Asian Americans as whiz kids is also enraging because it hides so many harsh truths about the impoverishment, oppression and racist treatment they experience. Some do come from middle- or upper-class families in Asia, some do attain middle-class or higher status in the U.S., and their community must deal with the reality of class privilege where it exists. But the hidden truths include the poverty of many Asian/Pacific Islander groups, especially women, who often work under intolerable conditions, as in the sweatshops....

THE DEVILS OF DUALISM

Yet another cause of the persistent Black-white conception of racism is dualism, the philosophy that sees all life as consisting of two irreducible elements. Those elements are usually oppositional, like good and evil, mind and body, civilized and savage. Dualism allowed the invaders, colonizers and enslavers of today's United States to rationalize their actions by stratifying supposed opposites along race, color or gender lines. So mind is European, male and rational; body is colored, female and emotional. Dozens of other such pairs can be found, with their clear implications of superior-inferior. In the arena of race, this society's dualism has long maintained that if a person is not totally white (whatever that can mean biologically), he or she must be considered Black....

Racism evolves; our models must also evolve. Today's challenge is to move beyond the Black-white dualism that has served as the foundation of White Supremacy. In taking up this challenge, we have to proceed with both boldness and infinite care. Talking race in these United States is an intellectual minefield; for every observation, one can find three contradictions and four necessary qualifications from five different racial groups. Making your way through that complexity, you have to think: keep your eyes on the prize.

9

Color-Blind Privilege

The Social and Political Functions of Erasing the Color Line in Post Race America

CHARLES A. GALLAGHER

The young white male sporting a FUBU (African-American owned apparel company "For Us By Us") shirt and his white friend with the tightly set, perfectly braided cornrows blended seamlessly into the festivities at an all white bar mitzvah celebration. A black model dressed in yachting attire peddles a New England, yuppie boating look in Nautica advertisements. It is quite unremarkable to observe white, Asian or African Americans with dyed purple, blond or red hair. White, black and Asian students decorate their bodies with tattoos of Chinese characters and symbols. In cities and suburbs young adults across the color line wear hip-hop clothing and listen to white rapper Eminem and black rapper 50-cent. It went almost unnoticed when a north Georgia branch of the NAACP installed a white biology professor as its president. Subversive musical talents like Jimi Hendrix, Bob Marley and The Who are now used to sell Apple Computers, designer shoes and SUVs. Du-Rag kits, complete with bandana headscarf and elastic headband, are on sale for $2.95 at hip-hop clothing stores and family centered theme parks like Six Flags. Salsa has replaced ketchup as the best selling condiment in the United States. Companies as diverse as Polo, McDonalds, Tommy Hilfiger, Walt Disney World, Master Card, Skechers sneakers, IBM, Giorgio Armani and Neosporin antibiotic ointment have each crafted advertisements that show an integrated, multiracial cast of characters interacting and consuming their products in [a] post-race, color-blind world.

Americans are constantly bombarded by depictions of race relations in the media, which suggests that discriminatory racial barriers have been dismantled. Social and cultural indicators suggest that America is on the verge, or has already become, a truly color-blind nation. National polling data indicate that a majority of whites now believe discrimination against racial minorities no longer exists. A majority of whites believe that blacks have "as good a chance as whites" in procuring housing and employment or achieving middle class status while a 1995 survey of white adults found that a majority of whites (58%) believed that African Americans were "better off" finding jobs than whites (Gallup, 1997; Shipler, 1998). Much of white America now see[s] a level playing field, while a majority of black

SOURCE: From *Race, Gender & Class* 10 (2003): 22–37. Reprinted by permission of the author.

Americans sees a field [that] is still quite uneven.... The colorblind or race neutral perspective holds that in an environment where institutional racism and discrimination have been replaced by equal opportunity, one's qualifications, not one's color or ethnicity, should be the mechanism by which upward mobility is achieved. Color as a cultural style may be expressed and consumed through music, dress or vernacular but race as a system which confers privileges and shapes life chances is viewed as an atavistic and inaccurate accounting of U.S. race relations.

Not surprisingly, this view of society blind to color is not equally shared. Whites and blacks differ significantly, however, on their support for affirmative action, the perceived fairness of the criminal justice system, the ability to acquire the "American Dream," and the extent to which whites have benefited from past discrimination (Moore, 1995; Moore & Saad, 1995; Kaiser, 1995). This article examines the social and political functions colorblindness serves for whites in the United States. Drawing on interviews and focus groups with whites from around the country I argue that color-blind depictions of U.S. race relations serves [sic] to maintain white privilege by negating racial inequality. Embracing a color-blind perspective reinforces whites' belief that being white or black or brown has no bearing on an individual's or a group's relative place in the socio-economic hierarchy.

DATA AND METHOD

I use data from seventeen focus groups and thirty-individual interviews with whites from around the country. Thirteen of the seventeen focus groups were conducted in a college or university setting, five in a liberal arts college in the Rocky Mountains and the remaining eight at a large urban university in the Northeast. Respondents in these focus groups were selected randomly from the student population. Each focus group averaged six respondents ... equally divided between males and females. An overwhelming majority of these respondents were between the ages of eighteen and twenty-two years of age. The remaining four focus groups took place in two rural counties in Georgia and were obtained through contacts from educational and social service providers in each county. One county was almost entirely white (99.54%) and in the other county whites constituted a racial minority. These four focus groups allowed me to tap rural attitudes about race relations in environments where whites had little or consistent contact with racial minorities....

COLORBLINDNESS AS NORMATIVE IDEOLOGY

The perception among a majority of white Americans that the socio-economic playing field is now level, along with whites' belief that they have purged themselves of overt racist attitudes and behaviors, has made colorblindness the dominant lens through which whites understand contemporary race relations.

Colorblindness allows whites to believe that segregation and discrimination are no longer [an] issue because it is now illegal for individuals to be denied access to housing, public accommodations or jobs because of their race. Indeed, lawsuits alleging institutional racism against companies like Texaco, Denny's, Coke, and Cracker Barrel validate what many whites know at a visceral level is true: firms which deviate from the color-blind norms embedded in classic liberalism will be punished. As a political ideology, the commodification and mass marketing of products that signify color but are intended for consumption across the color line further legitimate colorblindness. Almost every household in the United States has a television that, according to the U.S. Census, is on for seven hours every day (Nielsen 1997). Individuals from any racial background can wear hip-hop clothing, listen to rap music (both purchased at Wal-Mart) and root for their favorite, majority black, professional sports team. Within the context of racial symbols that are bought and sold in the market, colorblindness means that one's race has no bearing on who can purchase a Jaguar, live in an exclusive neighborhood, attend private schools or own a Rolex.

The passive interaction whites have with people of color through the media creates the impression that little, if any, socio-economic difference exists between the races....

Highly visible and successful racial minorities like [former] Secretary of State Colin Powell and ... [Secretary of State] Condelleeza Rice are further proof to white America that the state's efforts to enforce and promote racial equality have been accomplished.

The new color-blind ideology does not, however, ignore race; it acknowledges race while disregarding racial hierarchy by taking racially coded styles and products and reducing these symbols to commodities or experiences that whites and racial minorities can purchase and share. It is through such acts of shared consumption that race becomes nothing more than an innocuous cultural signifier. Large corporations have made American culture more homogenous through the ubiquitousness of fast food, television, and shopping malls but this trend has also created the illusion that we are all the same through consumption. Most adults eat at national fast food chains like McDonalds, shop at mall anchor stores like Sears and J.C. Penney's and watch major league sports, situation comedies or television drama. Defining race only as cultural symbols that are for sale allows whites to experience and view race as nothing more than a benign cultural marker that has been stripped of all forms of institutional, discriminatory or coercive power. The post-race, color-blind perspective allows whites to imagine that depictions of racial minorities working in high status jobs and consuming the same products, or at least appearing in commercials for products whites desire or consume, is the same as living in a society where color is no longer used to allocate resources or shape group outcomes. By constructing a picture of society where racial harmony is the norm, the color-blind perspective functions to make white privilege invisible while removing from public discussion the need to maintain any social programs that are race-based.

How then, is colorblindness linked to privilege? Starting with the deeply held belief that America is now a meritocracy, whites are able to imagine that the socio-economic success they enjoy relative to racial minorities is a function of individual hard work, determination, thrift and investments in education. The color-blind perspective removes from personal thought and public discussion any taint or suggestion of white supremacy or white guilt while legitimating the existing social, political and economic arrangements which privilege whites. This perspective insinuates that class and culture, and not institutional racism, are responsible for social inequality. Colorblindness allows whites to define themselves as politically and racially tolerant as they proclaim their adherence to a belief system that does not see or judge individuals by the "color of their skin." This perspective ignores, as Ruth Frankenberg puts it, how whiteness is a "location of structural advantage" (2001 p. 1).... Colorblindness hides white privilege behind a mask of assumed meritocracy while rendering invisible the institutional arrangements that perpetuate racial inequality. The veneer of equality implied in colorblindness allows whites to present their place in the racialized social structure as one that was earned.

OPPORTUNITY HAS NO COLOR

Given this norm of colorblindness it was not surprising that respondents in this study believed that using race to promote group interests was a form of (reverse) racism....

Believing and acting as if America is now color-blind allows whites to imagine a society where institutional racism no longer exists and racial barriers to upward mobility have been removed. The use of group identity to challenge the existing racial order by making demands for the amelioration of racial inequities is viewed as racist because such claims violate the belief that we are a nation that recognizes the rights of individuals not rights demanded by groups....

The logic inherent in the color-blind approach is circular; since race no longer shapes life chances in a color-blind world there is no need to take race into account when discussing differences in outcomes between racial groups. This approach erases America's racial hierarchy by implying that social, economic and political power and mobility is [sic] equally shared among all racial groups. Ignoring the extent or ways in which race shapes life chances validates whites' social location in the existing racial hierarchy while legitimating the political and economic arrangements that perpetuate and reproduce racial inequality and privilege.

REFERENCES

Frankenberg, R. (2001). The mirage of an unmarked whiteness. In B. B. Rasmussen, E. Klineberg, I. J. Nexica & M. Wray (eds.) *The making and unmaking of whiteness*. Durham: Duke University Press.

Gallup Organization. (1997). *Black/white relations in the U.S.* June 10, pp. 1–5.

Kaiser Foundation. (1995). *The four Americas: Government and social policy* America's multi-racial and multi-ethnic society. Menlo Park, CA: Kaiser F Foundation.

Moore, D. (1995). "Americans' most important sources of information: Local n *The Gallup Poll Monthly*, September, pp. 2–5.

Moore, D. & Saad, L. (1995). No immediate signs that Simpson trial intensified racia animosity. *The Gallup Poll Monthly*, October, pp. 2–5.

Nielsen, A. C. (1997). *Information please almanac*. Boston: Houghton Mifflin.

Shipler, D. (1998). *A country of strangers: Blacks and whites in America*. New York: Vintage Books.

10

Supremacists Taught ...olar About Identity

ABBY L. FERBER

A few years ago, my work on white supremacy led me to the neo-Nazi tract *The New Order,* which proclaims: "The single serious enemy facing the white man is the Jew." I must have read that statement a dozen times. Until then, I hadn't thought of myself as the enemy.

When I began my research for a book on race, gender, and white supremacy, I could not understand why white supremacists so feared and hated Jews. But after being immersed in newsletters and periodicals for months, I learned that white supremacists imagine Jews as the masterminds behind a great plot to mix races and, thereby, to wipe the white race out of existence.

The identity of white supremacists, and the white racial purity they espouse, requires the maintenance of secure boundaries. For that reason, the literature I read described interracial sex as "the ultimate abomination." White supremacists see Jews as threats to racial purity, the villains responsible for desegregation, integration, the civil-rights movement, the women's movement, and affirmative action—each depicted as eventually leading white women into the beds of black men. Jews are believed to be in control everywhere, staging a multi-pronged attack against the white race. For *WAR,* the newsletter of White Aryan Resistance, the Jew "promotes a thousand social ills … [f]or which you'll have to foot the bills."

Reading white-supremacist literature is a profoundly disturbing experience, and even more difficult if you are one of those targeted for elimination. Yet, as a Jewish woman, I found my research to be unsettling in unexpected ways. I had not imagined that it would involve so much self-reflection. I knew white supremacists were vehemently anti-Semitic, but I was ambivalent about my Jewish identity and did not see it as essential to who I was. Having grown up in a large Jewish community, and then having attended a college with a large Jewish enrollment, my Jewishness was invisible to me—something I mostly ignored. As I soon learned, to white supremacists, that is irrelevant.

Contemporary white supremacists define Jews as non-white: "not a religion, they are an Asiatic *race,* locked in a mortal conflict with Aryan man," according to *The New Order.* In fact, throughout white-supremacist tracts, Jews are

SOURCE: From *The Chronicle of Higher Education,* May 7, 1999, pp. B6–B7. Reprinted with permission of the author.

described not merely as a separate race, but as an impure race, the product of mongrelization. Jews, who pose the ultimate threat to racial boundaries, are themselves imagined as the product of mixed-race unions.

Although self-examination was not my goal when I began, my research pushed me to explore the contradictions in my own racial identity. Intellectually, I knew that the meaning of race was not rooted in biology or genetics, but it was only through researching the white-supremacist movement that I gained a more personal understanding of the social construction of race. Reading white-supremacist literature, I moved between two worlds: one where I was white, another where I was the non-white seed of Satan; one where I was privileged, another where I was despised; one where I was safe and secure, the other where I was feared and thus marked for death.

According to white-supremacist ideology, I am so dangerous that I must be eliminated. Yet, when I put down the racist, anti-Semitic newsletters, leave my office, and walk outdoors, I am white.

Growing up white has meant growing up privileged. Sure, I learned about the historical persecutions of Jews, overheard the hushed references to distant relatives lost in the Holocaust. I knew of my grandmother's experiences with anti-Semitism as a child of the only Jewish family in a Catholic neighborhood. But those were just stories to me. Reading white supremacists finally made the history real.

While conducting my research, I was reminded of the first time I felt like an "other." Arriving in the late 1980s for the first day of graduate school in the Pacific Northwest, I was greeted by a senior graduate student with the welcome: "Oh, you're the Jewish one." It was a jarring remark, for it immediately set me apart. This must have been how my mother felt, I thought, when, a generation earlier, a college classmate had asked to see her horns. Having lived in predominantly Jewish communities, I had never experienced my Jewishness as "otherness." In fact, I did not even *feel* Jewish. Since moving out of my parents' home, I had not celebrated a Jewish holiday or set foot in a synagogue. So it felt particularly odd to be identified by this stranger as a Jew. At the time, I did not feel that the designation described who I was in any meaningful sense.

But whether or not I define myself as Jewish, I am constantly defined by others that way. Jewishness is not simply a religious designation that one may choose, as I once naïvely assumed. Whether or not I see myself as Jewish does not matter to white supremacists.

I've come to realize that my own experience with race reflects the larger historical picture for Jews. As whites, Jews today are certainly a privileged group in the United States. Yet the history of the Jewish experience demonstrates precisely what scholars mean when they say that race is a social construction.

At certain points in time, Jews have been defined as a non-white minority. Around the turn of the last century, they were considered a separate, inferior race, with a distinguishable biological identity justifying discrimination and even genocide. Today, Jews are generally considered white, and Jewishness is largely considered merely a religious or ethnic designation. Jews, along with other European ethnic groups, were welcomed into the category of "white" as beneficiaries of one of the largest affirmative-action programs in history—the 1944 GI Bill of Rights. Yet, when I read white-supremacist discourse, I am reminded that my ancestors were expelled from the dominant race, persecuted, and even killed.

Since conducting my research, having heard dozens of descriptions of the murders and mutilations of "race traitors" by white supremacists, I now carry with me the knowledge that there are many people out there who would still wish to see me dead. For a brief moment, I think that I can imagine what it must feel like to be a person of color in our society … but then I realize that, as a white person, I cannot begin to imagine that.

Jewishness has become both clearer and more ambiguous for me. And the questions I have encountered in thinking about Jewish identity highlight the central issues involved in studying race today. I teach a class on race and ethnicity, and usually, about midway through the course, students complain of confusion. They enter my course seeking answers to the most troubling and divisive questions of our time, and are disappointed when they discover only more questions. If race is not biological or genetic, what is it? Why, in some states, does it take just one black ancestor out of 32 to make a person legally black, yet those 31 white ancestors are not enough to make that person white? And, always, are Jews a race?

I have no simple answers. As Jewish history demonstrates, what is and is not a racial designation, and who is included within it, is unstable and changes over time—and that designation is always tied to power. We do not have to look far to find other examples: The Irish were also once considered non-white in the United States, and U.S. racial categories change with almost every census.

My prolonged encounter with the white-supremacist movement forced me to question not only my assumptions about Jewish identity, but also my assumptions about whiteness. Growing up "white," I felt raceless. As it is for most white people, my race was invisible to me. Reflecting the assumption of most research on race at the time, I saw race as something that shaped the lives of people of color—the victims of racism. We are not used to thinking about whiteness when we think about race. Consequently, white people like myself have failed to recognize the ways in which our own lives are shaped by race. It was not until others began identifying me as the Jew, the "other," that I began to explore race in my own life.

Ironically, that is the same phenomenon shaping the consciousness of white supremacists: They embrace their racial identity at the precise moment when they feel their privilege and power under attack. Whiteness historically has equaled power, and when that equation is threatened, their own whiteness becomes visible to many white people for the first time. Hence, white supremacists seek to make racial identity, racial hierarchies, and white power part of the natural order again. The notion that race is a social construct threatens that order. While it has become an academic commonplace to assert that race is socially constructed, the revelation is profoundly unsettling to many, especially those who benefit most from the constructs.

My research on hate groups not only opened the way for me to explore my own racial identity, but also provided insight into the question with which I began this essay: Why do white supremacists express such hatred and fear of Jews? This ambiguity in Jewish racial identity is precisely what white supremacists find so threatening. Jewish history reveals race as a social designation, rather than a God-given or genetic endowment. Jews blur the boundaries between whites and people of color, failing to fall securely on either side of the divide. And it is ambiguity that white supremacists fear most of all.

I find it especially ironic that, today, some strict Orthodox Jewish leaders also find that ambiguity threatening. Speaking out against the high rates of inter-marriage among Jews and non-Jews, they issue dire warnings. Like white supremacists, they fear assaults on the integrity of the community and fight to secure its racial boundaries, defining Jewishness as biological and restricting it only to those with Jewish mothers. For both white supremacists and such Orthodox Jews, intermarriage is tantamount to genocide.

For me, the task is no longer to resolve the ambiguity, but to embrace it. My exploration of white-supremacist ideology has revealed just how subversive doing so can be: Reading white-supremacist discourse through the lens of Jewish experience has helped me toward new interpretations. White supremacy is not a movement just about hatred, but even more about fear: fear of the vulnerability and instability of white identity and privilege. For white supremacists, the central goal is to naturalize racial identity and hierarchy, to establish boundaries.

Both my own experience and Jewish history reveal that to be an impossible task. Embracing Jewish identity and history, with all their contradictions, has given me an empowering alternative to white-supremacist conceptions of race. I have found that eliminating ambivalence does not require eliminating ambiguity.

11

The Contested Meanings of "Asian American"

Racial Dilemmas in the Contemporary U.S.

NAZLI KIBRIA

In the social and political discourse of the U.S. today, the idea of race is one that is prominent, and yet highly ambiguous in meaning. Among the developments that both contribute to and reflect the current uncertainties of race is the emergence of such racialized constructs and groups as Asian American, Latino/Hispanic and Native American. Bringing together distinct ethnic or national groups, these pan-ethnic collectivities constitute an increasingly important and visible aspect of racial politics in the U.S. Omi and Winant (1986) argue that these pan-ethnic formations are the basis for new forms of racialization or the development of new racial subjects.

In this article I explore the contests of racial meaning that are part of the contemporary pan-ethnic formation of "Asian American." There are many ways in which the concept of "Asian American" has become institutionalized in U.S. life. "Asian American" is evident, for example, in classifications of race in the census and other bureaucratic forms. Since the late 1960s, we have also seen the emergence of a vast array of groups and organizations under the banner of "Asian American." Central to these developments has been the definition of "Asian American" as signifying a racial minority group bound by common racial interests. This conceptualization is, however, an increasingly uncertain one, in ways that reflect such forces as the 1965 Immigration Act and its effects on the Asian American population. But also implicated are more general uncertainties about the meaning of race in the U.S. today. Embedded in the contestations of Asian American meaning nowadays are such larger questions as "What constitutes a racial group?" and "What is the character of racial disadvantage in the U.S. today?"

DILEMMAS OF RACIAL MEANING

Race is a system of power, one that draws on physical differences to construct and give meaning to racial groups and the hierarchy in which they are embedded

SOURCE: From *Ethnic and Racial Studies*, Vol. 21 No. 5, September 1998, pp. 939–958.
Copyright © 1998 Routledge. Reprinted by permission of the Taylor & Francis Group.

(Miles 1989; Sanjek 1994). The focus of this study is on the conditions and processes that produce racial categories and their meanings: the social construction of
race. Drawing on constructionist perspectives of race and ethnicity, I suggest that
the production of racial meaning is a dynamic and ongoing matter (Omi and
Winant 1986; Nagel 1994). By definition, given their embeddedness in a larger
system of power, racial categories reflect the externally imposed designations or
assignments of dominant groups upon others. But racial definition is also shaped
by the actions of the categorized group itself. That is, within the limits of prevailing structures of opportunity and constraint, racialized groups work to shape
their own identities. Cornell and Hartmann (1998) discuss this interaction of
structure and agency in the formation of identities:

> Identities are made, but by an interaction between circumstantial or
> human assignment on the one hand and assertion on the other. Con-
> struction involves both the passive experience of being "made" by
> external forces, including not only material circumstances but the claims
> that other persons or groups make about the group in question, and the
> active process by which the group "makes" itself.... This interaction is
> continuous, and it involves all those processes through which identities
> are made and remade, from the initial formation of a collective identity
> through its maintenance, reproduction, transformation, and even repu-
> diation over time. Construction refers not to a one-time event but an
> ongoing project (p. 80).

There are several thematic issues that are often apparent in the dynamics of
racial meaning. I suggest that it is useful to conceptualize these issues as ongoing
dilemmas or core sets of questions about racial categories. These dilemmas operate as points of focus or conceptual and thematic guideposts in the production of
racial meanings. The specific questions that flow from these dilemmas will vary
for racialized groups, in ways that reflect their particular circumstances. At the
same time, the questions that arise will also reflect larger ongoing debates on
race in the society in question. But whatever their specific content, I suggest
that these emergent questions become the lightning-rod for contests about the
racialized group and the meanings that surround it. Thus, for example, within
institutional forums of various sorts (for example, political groups, newspapers),
these questions may generate discussion and debate about the racialized group.
For individual members of the racialized group, the questions may provide a
focus for reflections on identity.

Among the dilemmas of racial meaning are issues of *boundary*. What are the
criteria for membership in this racial category? To put it simply, what are the
grounds for being identified by it? The distinguishing feature of racial categories
is that they involve what Miles (1989) describes as the signification of human
biological characteristics in such a way as to define and construct differentiated
social collectivities" (p. 75). The dilemma of racial boundaries involves questions
about this signification, in particular the interpretation of physical characteristics.

Racial boundaries reflect relations of power, in particular the ability of the
dominant group to construct and impose definitions upon others. At the same

time, those who are categorized cannot simply be understood as passive recipients of externally imposed definitions. For example, in *Who Is Black?* Davis (1991) describes the development of the "one-drop rule" or the rule of hypodescent for defining black identity in the U.S. Imposed by white people as a means of maintaining power and enforcing segregation, the "one-drop rule" has also been widely accepted and enforced by African Americans as a basis for defining membership. It is of note, however, that the definition of black identity has become a growing focus of debate in the U.S., reflecting such conditions as a rising rate of black outmarriage as well as the growing presence of immigrants from the Caribbean and Africa who sometimes resist incorporation into a black American identity (Waters 1994; Mathews 1996).

Closely related to issues of boundary are those of *community*. How do the racial boundaries suggest membership in a community? That is, how do the boundaries mark a community of persons who share more than simply an externally imposed designation? These issues are linked to processes of ethnicization or the making of an ethnic group. That is, questions of community are embedded in the dynamics by which a racialized group comes to define itself in terms of common ancestry, history and a set of cultural symbols that they see as capturing the essence of their peoplehood (Cornell and Hartmann 1998, p. 19). Reflecting their particular histories of incorporation and community-building in the U.S., racialized groups vary in the extent of their ethnicization. These differences can be expected to shape the specific questions of community that surround them....

Intrinsic to racial construction is the locating of groups within a racial hierarchy. The dilemmas of racial meaning thus include questions of *positioning*— what is the position of the group within the racial hierarchy? In short, where and how do they fit in? Stereotypes and images of racial groups, with their presumptions about characteristic racial behaviour and temperament, play an important part in this process of hierarchical location. These racial notions are developed and affirmed as they become embedded in social institutions such as the law, economy and education. Here they may take the form of systematic and cross-cutting institutional patterns or dynamics that shape the access of racial groups to societal resources. Because racial membership is widely *believed* to be a given, biological matter, the presumed traits of race, and the institutional conditions and inequalities with which they are intertwined, can also be seen as "natural," inherent.

As part of a system of power, racial categories and their meanings reflect the ability of dominant groups to impose their designations upon others. Thus, in the U.S. white people have exerted control over the production of racial images and stereotypes. Scholars further observe that white dominance has marked the very terms or terrain on which race definition has unfolded (Frankenberg 1993). That is, race definition in the U.S. has occurred in relation to "whiteness," which has been defined as the centre or norm—the standard against which the racial identities of others are defined and measured....

Thus in a variety of ways, racial groups can work to challenge and shape their racial location. For example, an important activity of contemporary racial minority movements has been the challenging and subverting of stereotypes. The movements have therefore been an important force in contesting such stereotypes and more generally, expanding and shaping the institutional and conceptual space within which they are discussed.

THE ASIAN AMERICAN CONSTRUCT AND
POST-1965 DEVELOPMENTS

"Asian American" is a contemporary, post–Civil Rights construct. While the Asian settlers of the late nineteenth and early twentieth centuries often found themselves categorized or lumped together as "Asiatics" or "Orientals" by the dominant U.S. society, they did not respond by banding together. Instead, they maintained separate institutions and identities and at certain times engaged in strategic acts of distancing or disidentification from one another (Hayano 1981). But in the 1960s, young U.S.-born Asian American activists on college campuses, inspired by the civil rights struggles of the time, organized the Asian American movement as a means of political empowerment and mobilization. Rejecting the then common term "Oriental," they coined "Asian American," a term that has since gained currency.

Like the other pan-ethnic collectivities promoted by the movements of the time (for example, Native American, Latino), the Asian American concept has become an institutionalized dimension of the contemporary U.S. racial system. Like the others, it has been a basis for monitoring antidiscrimination efforts and of organizing access to government resources. Further reinforcing its significance has been the multiculturalist movement, which often draws on the pan-ethnic categories to organize the promotion of cultural diversity (Hollinger 1995). Accompanying these developments has been the mushrooming of Asian American groups and organizations which aim at mobilizing and advancing pan-Asian interests. The institutionalization of the Asian American movement is evident not only in the presence of an extensive network of pan-Asian associations of various sorts, but also in a series of pan-Asian publications (for example, *Asian Week, A. Magazine, Amerasia Journal*) whose stated goal is to serve a pan-Asian audience. One of the most important of the institutional developments has been the growth of Asian American Studies programmes and classes on college campuses. Among other things, Asian American Studies has provided an important forum for reflection and debate on the meaning of "Asian American."...

Critical to understanding the questions that surround the concept of Asian American today is the 1965 Immigration Reform Act, which lifted race restrictions and so opened the doors to Asian immigration. Prior to the late 1960s, the Asian American population was small in number. The flow of Asian immigration, most notably of Chinese and Japanese immigrants to the U.S. that had commenced in the mid-1800s, had been effectively halted with a series of restrictive immigration laws passed in the late nineteenth and early twentieth centuries. The 1970 census showed Asian Pacific Americans to be 0.7 per cent of the total U.S. population. But in 1990, reflecting the 1965 Act, their numbers had risen to 2.9 per cent (Min 1997a, p. 15). The rise in numbers, along with other developments, has affected the collective character of the Asian American population in important ways. For one thing, the foreign-born comprise a higher percentage of the Asian American population in comparison to the period preceding the 1965 Act (Espiritu 1992, p. 26; Barringer, Gardner and Levin 1993, p. 43). In addition, the high proportion of immigrants, along with a global context that in many instances facilitates "host" and "homeland" linkages, suggests processes of transnationalism to be an important feature of Asian American life today.

Concurrent with the expansion in Asian American numbers has been its growing ethnic diversity. In contrast to the largely Japanese and Chinese origins of Asian Americans in the first half of the century, today Asian Americans are highly diverse in their national origins. Asian Indians, Koreans and Filipinos, for example, are among the fastest growing segments of the Asian American population. There has also been a general shift away from the largely working-class origins of the late nineteenth—early twentieth—century Asian immigrants. While there is considerable socio-economic diversity within the Asian American population today, it is also the case that many post-1965 Asian immigrants come from professional, white-collar and highly educated backgrounds (Min 1997a, p. 17). The middle-class background of many Asian immigrants, in conjunction with the phenomenal growth and success of some Asian economies, has lent support and credence to the popular image of Asians as a "model minority": a group that is culturally programmed for economic success.

The middle-class image does not, however, do justice to the realities of socio-economic diversity within the new Asian immigrant stream. In fact, many analysts characterize the Asian American population today as polarized, consisting of two sharply disparate socio-economic segments (Mar and Kim 1994; Ong and Hee 1994). In contrast to those admitted in the 1970s, recent Asian immigrants have been less select and more diverse in their socioeconomic origins. Further contributing to a movement away from a purely middle-class profile is the entry since the mid-1970s of refugees from Vietnam, Cambodia and Laos whose levels of education and training are, on average, lower than that of the other post-1965 Asian immigrant groups (Barringer, Gardner and Levin 1993, pp. 31–32). Thus, while economic polarization is apparent *within* Asian ethnic groups, it is also one that can coincide with ethnic boundaries. In other words, there are important differences in the socioeconomic profiles of Asian-origin groups. For example, the income levels and poverty rates for Vietnamese Americans contrast sharply and negatively with those of Japanese or Filipino Americans.

The growth of the Asian American population has occurred at a time of complex transformation in the racial environment of the U.S. In the aftermath of the civil rights struggles, institutionalized racial barriers have been challenged in the U.S. But there has also been racial polarization and backlash, especially apparent in the last two decades. Feeding resentment and retreat from progressive policy has been the structural transformation of the U.S. economy, most notably the decline suffered by the U.S. manufacturing sector (Mar and Kim 1994). In recent years affirmative action programmes that were instituted in the 1960s and 1970s have come under attack. There has also been a rising tide of anti-immigrant sentiment, as expressed in calls for various restrictive measures, such as the curtailing of government welfare benefits and social services to immigrants....

By increasing the size and visibility of the Asian American population, post–1965 Asian immigration has lent vigour and urgency to issues of the Asian American community. But the new immigration has also posed challenges to the very notion of "Asian American," especially to its meaning and viability as a basis of community. Here it is important to point out that, as during the movement's inception, pan–Asian organizations today continue to involve primarily U.S.-born Asians (Espiritu and Ong 1994). The relative absence in these forums of immigrants, who constitute a significant proportion of the Asian American population, gives rise to questions about the possibility of the pan–Asian community.

Central to these questions is the meaningfulness of the notion of shared racial interest and location that has guided pan–Asian activity.... There is, for one thing, the element of ambiguity that surrounds Asian racial boundaries in the U.S. today. That is, the fact that not all of those who are institutionally or officially Asian American are widely seen or labelled as racially "Asian" during informal social encounters generates doubts about racial interest and location as bases of unity. The important role of an Asian American historical past in creating a sense of shared racial fate is also made uncertain by the post–1965 transformation of the Asian American population into a largely immigrant one. The U.S. roots of many Asian Americans are recent, not extending back to the late nineteenth—early twentieth—century Asian immigrations that are emphasized in Asian American history.

Uncertainties about shared racial location and interest also reflect the complexities of class that have been part of the post–1965 transformation of the Asian American population. As I have described, many post–1965 Asian immigrants have been highly educated and held white-collar and professional occupations in their countries of origin. For such persons, the working-class origins of Asian immigration that is emphasized in Asian American history as well as the agenda of solidarity with other racially oppressed groups may seem distant and irrelevant. Also important are the divisions of class as well as the occupational niche within the new Asian immigrant stream. These divisions are likely to feed into different experiences and perceptions of racism and thus the understanding of racial location and interest....

It is important to point out that there are ways in which contemporary conditions *do* support the vision of Asian American community as a racial and political one. In the current tide of racial resentment and backlash, Asian Americans often find themselves to be the target of fears about an invasion by foreign economic interests and by immigrants. These attacks tend to be indiscriminate, without regard to national origin and involving a wide range of Asian groups. More generally, the political advantages that can be derived from banding together still work to support a racial interest vision of the Asian American community. At the same time, as I have described here, the meaning of Asian American community is a highly contested matter. In ways that reflect post-1965 forces, a specific focus of contest is the vision of race-centred political community as it has been defined by the Asian American movement. This contest reflects larger uncertainties about the unitary nature of racial interests in the face of class and other divisions....

Positioning Dilemmas: Asians as a "Model Minority"

In affirming the identity of Asian Americans as a racial minority, commonality and solidarity with other oppressed groups of colour has been a central aspect of the ideology of pan-Asianism. Post-1965 conditions have, however, worked to generate questions about what Asian Americans share with other racial minorities by challenging the notion of Asian racial disadvantage. Among the relevant developments is the growth of the East Asian and Southeast Asian economies in recent decades. By the early 1980s, talk of the "Asian tiger economies" and the "Asian miracle," often attributed to "Confucian" values, had become commonplace in the U.S. Also relevant, of course, are the middle-class and professional backgrounds and socio-economic achievements of a substantial segment of the post-1965 Asian immigrants as I have described earlier. Even prior to the 1965 Act, in the years after World War II, there are indications of socio-economic mobility into the middle class among U.S.-born Asian Americans (Mar and Kim 1994).

The stereotype of Asians as a model minority has, I suggest, provided an important ideological focus for interpretation and debate of the uncertainties of Asian racial disadvantage and position raised by the[se] developments. In other words, the model minority stereotype and the debates surrounding it have offered a forum for understanding and considering the Asian racial position today. On the one hand, the model minority stereotype offers a way of making sense of the general indicators of Asian achievement. In essence, it does so by suggesting that Asians are a minority group endowed with cultural values such as a strong work ethic and devotion to education that predispose them to economic and educational achievement. Worth noting here is the radical contrast of this image to earlier ones—of Asians as [u]nassimilable, inscrutable, tricky and immoral heathens (Hurh and Kim 1989). This late nineteenth—early twentieth—century imagery provided the ideological backdrop to the exclusion acts that were to bar Asian immigration to the U.S. But the mid-1960s seem to have marked a shift in the U.S. imagery of Asians. Since that time, media reports on the educational and economic achievements of Asian Americans have become commonplace....

The model minority stereotype appears to situate Asians in a privileged position within the racial hierarchy of the U.S. The stereotype suggests the transient nature of racial disadvantage for Asian Americans. Simultaneously, it evokes the culturally rooted and ultimately essentialist nature of the chronic disadvantage of other racial minorities. A critical dimension of the stereotype is the suggestion that Asians are different from other racial minorities. Their cultural values, in particular their commitment to work, self-sufficiency, family and education, allow them to overcome disadvantage, in contrast to other minorities whose predisposition causes them to remain permanently mired in poverty and disability. As an explanatory variable, "culture" here takes on an essential, biological quality. Through this comparison, Asians emerge as a group for whom racial disadvantage, if it exists, is a temporary condition. In this sense they are heirs to the European immigrant story of successful triumph over adversity and, eventually, assimilation into the "mainstream." It is not surprising that media reports on Asian successes often draw parallels between the Asian and Jewish American experiences.

The Asian American movement has been a major force in challenging the model minority stereotype. Like other minority movements, it has approached the exposing of group stereotypes as an important political activity. But a critique of the model minority stereotype has also been important to the movement because of the questions that the stereotype can raise about the ideology of pan-Asianism. When left unpacked, the model minority stereotype brings into question the movement's core ideas of Asian racial disadvantage and commonality with other groups of colour. Critique of the stereotype has also been driven by a recognition of the ways in which its connotation of self-sufficiency and achievement work to undercut the access of Asian American service organizations to funding that targets disadvantaged populations....

Critique of the model minority stereotype has also focused on its larger implications for race relations and inequality in the U.S. Here the stereotype is seen as an instrument of white supremacy, used to pit Asians and other minorities against each other and thus weaken minority solidarity and power. Also pointed out are the ways in which the stereotype legitimates the absence or failure of programmes and measures to address racial inequality (Osajima 1988). That is, since Asian Americans are able to pull themselves up with their cultural bootstraps, racial disadvantage is not a structural aspect of U.S. life that requires active intervention.

CONCLUSION

Like the other pan-ethnic movements that emerged in the 1960s, "Asian American" was conceived by its founders as a strategy of political empowerment. Shared racial identity and interests provided the central rationale for pan-Asian organizing. The need to sustain and preserve this strategy is perhaps more pressing than ever, given the anti-immigrant sentiment and fears of Asian economic competition that mark the racial environment of the U.S. today. But this task is complicated by the sharp contests of meaning that surround the concept of "Asian American" today.

Post-1965 conditions have exacerbated the inherent contradictions and ambiguities of the racial interests framework that has guided pan-Asian activity.

Among those who are potentially encompassed by the Asian American concept, divisions of class, ethnicity and ascribed racial identity have become prominent. These divisions complicate efforts to articulate and forge a pan-Asian agenda based on common racial location and interests. I suggest that what can help to clarify the potential bases of solidarity among Asian Americans today are studies that actively compare racial experiences across different groups of Asian Americans, instead of simply assuming a common one. At present we know little about how the rank and file of those who are encompassed by the Asian American umbrella actually see and relate to the Asian American concept. Involvement in the Asian American movement and its institutions has been the province of primarily U.S.-born and middle-class Asians. The question [of] how other Asian Americans relate to the political and strategic understanding of "Asian American" that have been advanced by the Asian American movement is one that needs investigation.

Contestations of Asian American meaning are embedded in the larger uncertainties that mark the racial environment of the U.S. today. As I have discussed, questions about the identity of Asian Americans as a racial minority group emerge in the context of increased concern and confusion regarding the nature of racial disadvantage; in particular, whether it is reflective of essential inferiority. Also apparent is heightened uncertainty about the fundamental substance of racial boundaries. In established and popular U.S. understanding, racial boundaries are pure, discrete, immutable and easily defined by physical characteristics. Among the conditions that now explicitly challenge these ideas is the growth of a self-identified "multiracial" population that refuses to be racially categorized in unidimensional terms, as reflected in their demands for a "multiracial" category in the census (Marriot 1996). Also relevant are the racial ambiguities of the increasingly prominent Hispanic population who are identified as a racial group in some contexts, even though individuals within it experience different racial labels, ranging from "white" to "black" (Oboler 1995). In ways that reflect the more fluid character of racial classifications in Latin America, this population is also more likely to resist and thus to challenge the singular and discrete categories of the U.S. racial system (Fernandez 1992). The ambiguities of "Asian American" as a signifier of a racial group are part of these emergent challenges.

REFERENCES

Barringer, Herbert, Gardner, Robert and Levin, Michael. 1993. *Asians and Pacific Islanders in the United States*, New York: Russell Sage Foundation.

Cornell, Stephen and Hartmann, Douglas. 1998. *Ethnicity and Race: Making Identities in a Changing World*, Thousand Oaks, CA: Pine Forge Press.

Davis, F. James. 1991. *Who Is Black? One Nation's Definition*, University Park, PA: Pennsylvania State University Press.

Espiritu, Yen L. 1992. *Asian American Panethnicity: Bridging Institutions and Identities*, Philadelphia, PA: Temple University Press.

Espiritu, Yen L. and Ong, Paul. 1994. "Class constraints on racial solidarity among Asian Americans," in Paul Ong, Edna Bonacich and Lucie Cheng (eds.), *The New Asian Immigration in Los Angeles and Global Restructuring*, Philadelphia, PA: Temple University Press, pp. 295–321.

Fernandez, Carlos. 1992. "La Raza and the melting pot: a comparative look at multi-ethnicity," in Maria Root (ed.) *Racially Mixed People in America*, Newbury Park, CA: Sage Publications, pp. 126–43.

Frankenberg, Ruth. 1993. *White Women, Race Matters: The Social Construction of Whiteness*, Minneapolis, MN: University of Minnesota Press.

Hayano, David. 1981. "Ethnic identification and disidentification: Japanese-Americans view of Chinese-Americans," *Ethnic Groups*, vol. 3, no. 2, pp. 157–71.

Hollinger, David. 1995. *Postethnic America: Beyond Multiculturalism*, New York: Basic Books.

Hurh, Won Moo and Kim, Kwang Chung. 1989. "The 'success' image of Asian Americans: its validity, and its practical and theoretical implications," *Ethnic and Racial Studies*, vol. 12, no. 4: 512–38.

Mar, Don and Kim, Marlene. 1994. "Historical trends" in *The State of Asian Pacific America: Economic Diversity, Issues and Policies*, Los Angeles, CA: Asian Pacific American Public Policy Institute, pp. 13–31.

Marriot, Michel. 1996. "Multiracial Americans ready to claim their own identity," *The New York Times*, 20 July.

Mathews, Linda. 1996. "More than identity ride on a new racial category," *The New York Times*, 6 July.

Miles, Robert. 1989. *Racism*, London: Routledge.

Min, Pyong Gap. 1997a. "Introduction," in Pyong Gap Min (ed.), *Asian Americans: Contemporary Trends and Issues*, Thousand Oaks, CA: Sage Publications, pp. 1–9.

Nagel, Joane. 1994. "Constructing ethnicity: creating and recreating ethnic identity and culture," *Social Problems*, vol. 41, no. 1, pp. 152–76.

Oboler, Suzanne. 1995. *Ethnic Labels, Latino Lives*, Minneapolis, MN: University of Minnesota Press.

Omi, Michael and Winant, Howard. 1986. "Contesting the meaning of race in the post-civil rights movement era," in Silvia Pedraza and Ruben G. Rumbaut (eds.), *Origins and Destinies: Immigration, Race, and Ethnicity in America*, Belmont, CA: Wadsworth Publishing Co, pp. 470–78.

Ong, Paul and Hee, Suzanne. 1994. "Economic diversity" in *The State of Asian Pacific America: Economic Diversity, Issues and Policies*, Los Angeles, CA: Asian Pacific American Public Policy Institute, pp. 31–56.

Osajima, Keith. 1988. "Asian Americans as the model minority: an analysis of the popular press image in the 1960s and 1980s," in Gary Okihiro, Shirley Hune, Arthur Hansen and John Liu (eds.), *Reflections on Shattered Windows*, Pullman, WA: Washington State University Press, pp. 165–74.

Sanjek, Roger. 1994. "The enduring inequalities of race," in Steven Gregory and Roger Sanjek (eds), *Race*, New Brunswick, NJ: Rutgers University Press, pp. 1–17.

Waters, Mary. 1994. "Ethnic and racial identities of second-generation black immigrants in New York City," *International Migration Review*, vol. 28, no. 4, pp. 795–820.

12

Race as Class

HERBERT J. GANS

Humans of all colors and shapes can make babies with each other. Conse-quently most biologists, who define races as subspecies that cannot inter-breed, argue that scientifically there can be no human races. Nonetheless, lay people still see and distinguish between races. Thus, it is worth asking again why the lay notion of race continues to exist and to exert so much influence in human affairs.

Lay persons are not biologists, nor are they sociologists who argue these days that race is a social construction arbitrary enough to be eliminated if "society" chose to do so. The laity operates with a very different definition of race. They see that humans vary, notably in skin color, the shape of the head, nose, and lips, and quality of hair, and they choose to define the variations as individual races.

More important, the lay public uses this definition of race to decide whether strangers (the so-called "other") are to be treated as superior, inferior, or equal. Race is even more useful for deciding quickly whether strangers might be threat-ening and thus should be excluded. Whites often consider dark-skinned strangers threatening until they prove otherwise, and none more than African Americans.

Scholars believe the color differences in human skins can be traced to cli-matic adaptation. They argue that the high levels of melanin in dark skin origi-nally protected people living outside in hot, sunny climates, notably in Africa and South Asia, from skin cancer. Conversely, in cold climates, the low amount of melanin in light skins enabled the early humans to soak up vitamin D from a sun often hidden behind clouds. These color differences were reinforced by millen-nia of inbreeding when humans lived in small groups that were geographically and socially isolated. This inbreeding also produced variations in head and nose shapes and other facial features so that Northern Europeans look different from people from the Mediterranean area, such as Italians and, long ago, Jews. Like-wise, East African faces differ from West African ones, and Chinese faces from Japanese ones. (Presumably the inbreeding and isolation also produced the DNA patterns that geneticists refer to in the latest scientific revival and redefini-tion of race.)

Geographic and social isolation ended long ago, however, and human pop-ulation movements, intermarriage, and other occasions for mixing are eroding physical differences in bodily features. Skin color stopped being adaptive too

SOURCE: From *Contexts*, Vol. 4, Issue 4, pp. 17–21. © 2005 by the American Socio-logical Association. Reprinted by permission of the University of California Press.

after people found ways to protect themselves from the sun and could get their vitamin D from the grocery or vitamin store. Even so, enough color variety persists to justify America's perception of white, yellow, red, brown, and black races.

Never mind for the moment that the skin of "whites," as well as many East Asians and Latinos is actually pink; that Native Americans are not red; that most African Americans come in various shades of brown; and that really black skin is rare. Never mind either that color differences within each of these populations are as great as the differences between them, and that, as DNA testing makes quite clear, most people are of racially mixed origins, even if they do not know it. But remember that this color palette was invented by whites. Nonwhite people would probably divide the range of skin colors quite differently.

Advocates of racial equality use these contradictions to fight against racism. However, the general public also has other priorities. As long as people can roughly agree about who looks "white," "yellow," or "black" and find that their notion of race works for their purposes, they ignore its inaccuracies, inconsistencies, and other deficiencies.

Note, however, that only some facial and bodily features are selected for the lay definition of race. Some, like the color of women's nipples or the shape of toes (and male navels), cannot serve because they are kept covered. Most other visible ones, like height, weight, hairlines, ear lobes, finger or hand sizes—and even skin texture—vary too randomly and frequently to be useful for categorizing and ranking people or judging strangers. After all, your own child is apt to have the same stubby fingers as a child of another skin color or, what is equally important, a child from a very different income level.

RACE, CLASS, AND STATUS

In fact, the skin colors and facial features commonly used to define race are selected precisely because, when arranged hierarchically, they resemble the country's class-and-status hierarchy. Thus, whites are on top of the socioeconomic pecking order as they are on top of the racial one, while variously shaded non-whites are below them in socioeconomic position (class) and prestige (status).

The darkest people are for the most part at the bottom of the class-status hierarchy. This is no accident, and Americans have therefore always used race as a marker or indicator of both class and status. Sometimes they also use it to enforce class position, to keep some people "in their place." Indeed, these uses are a major reason for its persistence.

Of course, race functions as more than a class marker, and the correlation between race and the socioeconomic pecking order is far from statistically perfect: All races can be found at every level of that order. Still, the race-class correlation is strong enough to utilize race for the general ranking of others. It also becomes more useful for ranking dark-skinned people as white poverty declines so much that whiteness becomes equivalent to being middle or upper class.

The relation between race and class is unmistakable. For example, the 1998–2000 median household income of non-Hispanic whites was $45,500; of Hispanics (currently seen by many as a race) as well as Native Americans, $32,000; and of African Americans, $29,000. The poverty rates for these same groups were 7.8 percent among whites, 23.1 among Hispanics, 23.9 among blacks, and 25.9 among Native Americans. (Asians' median income was $52,600—which does much to explain why we see them as a model minority.)

True, race is not the only indicator used as a clue to socioeconomic status. Others exist and are useful because they can also be applied to ranking co-racials. They include language (itself a rough indicator of education), dress, and various kinds of taste, from given names to cultural preferences, among others.

American English has no widely known working-class dialect like the English Cockney, although "Brooklynese" is a rough equivalent, as is "black vernacular." Most blue-collar people dress differently at work from white-collar, professional, and managerial workers. Although contemporary American leisure-time dress no longer signifies the wearer's class, middle-income Americans do not usually wear Armani suits or French haute couture, and the people who do can spot the knockoffs bought by the less affluent.

Actually, the cultural differences in language, dress, and so forth that were socially most noticeable are declining. Consequently, race could become yet more useful as a status marker, since it is so easily noticed and so hard to hide or change. And in a society that likes to see itself as classless, race comes in very handy as a substitute.

THE HISTORICAL BACKGROUND

Race became a marker of class and status almost with the first settling of the United States. The country's initial holders of cultural and political power were mostly WASPs (with a smattering of Dutch and Spanish in some parts of what later became the United States). They thus automatically assumed that their kind of whiteness marked the top of the class hierarchy. The bottom was assigned to the most powerless, who at first were Native Americans and slaves. However, even before the former had been virtually eradicated or pushed to the country's edges, the skin color and related facial features of the majority of colonial America's slaves had become the markers for the lowest class in the colonies.

Although dislike and fear of the dark are as old as the hills and found all over the world, the distinction between black and white skin became important in America only with slavery and was actually established only some decades after the first importation of black slaves. Originally, slave owners justified their enslavement of black Africans by their being heathens, not by their skin color.

In fact, early Southern plantation owners could have relied on white indentured servants to pick tobacco and cotton or purchased the white slaves that were available then, including the Slavs from whom the term slave is derived. They also had access to enslaved Native Americans. Blacks, however, were cheaper, more plentiful, more easily controlled, and physically more able to survive the intense heat and brutal working conditions of Southern plantations.

After slavery ended, blacks became farm laborers and sharecroppers, de facto indentured servants, really, and thus they remained at the bottom of the class hierarchy. When the pace of industrialization quickened, the country needed new sources of cheap labor. Northern industrialists, unable and unwilling to recruit southern African Americans, brought in very poor European immigrants, mostly peasants. Because these people were near the bottom of the class hierarchy, they were considered nonwhite and classified into races. Irish and Italian newcomers were sometimes even described as black (Italians as "guineas"), and the eastern and southern European immigrants were deemed "swarthy."

However, because skin color is socially constructed, it can also be reconstructed. Thus, when the descendants of the European immigrants began to move up economically and socially, their skins apparently began to look lighter to the whites who had come to America before them. When enough of these descendants became visibly middle class, their skin was seen as fully white. The biological skin color of the second and third generations had not changed, but it was socially blanched or whitened. The process probably began in earnest just before the Great Depression and resumed after World War II. As the cultural and other differences of the original European immigrants disappeared, their descendants became known as white ethnics.

This pattern is now repeating itself among the peoples of the post-1965 immigration. Many of the new immigrants came with money and higher education, and descriptions of their skin color have been shaped by their class position. Unlike the poor Chinese who were imported in the 19th century to build the West and who were hated and feared by whites as a "yellow horde," today's affluent Asian newcomers do not seem to look yellow. In fact, they are already sometimes thought of as honorary whites, and later in the 21st century they may well turn into a new set of white ethnics. Poor East and Southeast Asians may not be so privileged, however, although they are too few to be called a "yellow horde."

Hispanics are today's equivalent of a "swarthy" race. However, the children and grandchildren of immigrants among them will probably undergo "whitening" as they become middle class. Poor Mexicans, particularly in the Southwest, are less likely to be whitened, however. (Recently a WASP Harvard professor came close to describing these Mexican immigrants as a brown horde.)

Meanwhile, black Hispanics from Puerto Rico, the Dominican Republic, and other Caribbean countries may continue to be perceived, treated, and mistreated as if they were African American. One result of that mistreatment is their low median household income of $35,000, which was just $1,000 more than that of non-Hispanic blacks but $4,000 below that of so-called white Hispanics.

Perhaps South Asians provide the best example of how race correlates with class and how it is affected by class position. Although the highly educated Indians and Sri Lankans who started coming to America after 1965 were often darker than African Americans, whites only noticed their economic success.

They have rarely been seen as nonwhites, and are also often praised as a model minority.

Of course, even favorable color perceptions have not ended racial discrimination against newcomers, including model minorities and other affluent ones. When they become competitors for valued resources such as highly paid jobs, top schools, housing, and the like, they also become a threat to whites. California's Japanese-Americans still suffer from discrimination and prejudice four generations after their ancestors arrived here.

AFRICAN-AMERICAN EXCEPTIONALISM

The only population whose racial features are not automatically perceived differently with upward mobility are African Americans: Those who are affluent and well educated remain as visibly black to whites as before. Although a significant number of African Americans have become middle class since the civil rights legislation of the 1960s, they still suffer from far harsher and more pervasive discrimination and segregation than nonwhite immigrants of equivalent class position. This not only keeps whites and blacks apart but prevents blacks from moving toward equality with whites. In their case, race is used both as a marker of class and, by keeping blacks "in their place," an enforcer of class position and a brake on upward mobility.

In the white South of the past, African Americans were lynched for being "uppity." Today, the enforcement of class position is less deadly but, for example, the glass ceiling for professional and managerial African Americans is set lower than for Asian Americans, and on-the-job harassment remains routine.

Why African-American upward economic mobility is either blocked or, if allowed, not followed by public blanching of skin color remains a mystery. Many explanations have been proposed for the white exceptionalism with which African Americans are treated. The most common is "racism," an almost innate prejudice against people of different skin color that takes both personal and institutional forms. But this does not tell us why such prejudice toward African Americans remains stronger than that toward other nonwhites.

A second explanation is the previously mentioned white antipathy to blackness, with an allegedly primeval fear of darkness extrapolated into a primordial fear of dark-skinned people. But according to this explanation, dark-skinned immigrants such as South Asians should be treated much like African Americans.

A better explanation might focus on "Negroid" features. African as well as Caribbean immigrants with such features—for example, West Indians and Haitians—seem to be treated somewhat better than African Americans. But this remains true only for new immigrants; their children are generally treated like African Americans.

Two additional explanations are class-related. For generations, a majority or plurality of all African Americans were poor, and about a quarter still remain so. In addition, African Americans continue to commit a proportionally greater share of the street crime, especially street drug sales—often because legitimate job opportunities are scarce. African Americans are apparently also more often arrested without cause. As one result, poor African Americans are more often considered undeserving than are other poor people, although in some parts of America, poor Hispanics, especially those who are black, are similarly stigmatized.

The second class-based explanation proposes that white exceptionalist treatment of African Americans is a continuing effect of slavery: They are still perceived as ex-slaves. Many hateful stereotypes with which today's African Americans are demonized have changed little from those used to dehumanize the slaves. (Black Hispanics seem to be equally demonized, but then they were also slaves, if not on the North American continent.) Although slavery ended officially in 1864, ever since the end of Reconstruction subtle efforts to discourage African-American upward mobility have not abated, although these efforts are today much less pervasive or effective than earlier.

Some African Americans are now millionaires, but the gap in wealth between average African Americans and whites is much greater than the gap between incomes. The African-American middle class continues to grow, but many of its members barely have a toehold in it, and some are only a few paychecks away from a return to poverty. And the African-American poor still face the most formidable obstacles to upward mobility. Close to a majority of working-age African-American men are jobless or out of the labor force. Many women, including single mothers, now work in the low-wage economy, but they must do without most of the support systems that help middle-class working mothers. Both federal and state governments have been punitive, even in recent Democratic administrations, and the Republicans have cut back nearly every antipoverty program they cannot abolish.

Daily life in a white-dominated society reminds many African Americans that they are perceived as inferiors, and these reminders are louder and more relentless for the poor, especially young men. Regularly suspected of being criminals, they must constantly prove that they are worthy of equal access to the American Dream. For generations, African Americans have watched immigrants pass them in the class hierarchy, and those who are poor must continue to compete with current immigrants for the lowest-paying jobs. If unskilled African Americans reject such jobs or fail to act as deferentially as immigrants, they justify the white belief that they are less deserving than immigrants. Blacks' resentment of such treatment gives whites additional evidence of their unworthiness, thereby justifying another cycle of efforts to keep them from moving up in class and status.

Such practices raise the suspicion that the white political economy and white Americans may, with the help of nonwhites who are not black, use African Americans to anchor the American class structure with a permanently lower-class population. In effect, America, or those making decisions in its name, could be seeking, not necessarily consciously, to establish an undercaste that cannot move out and up. Such undercastes exist in other societies: the gypsies of Eastern Europe, India's untouchables, "indigenous people," and "aborigines" in yet other places. But these are far poorer countries than the United States.

SOME IMPLICATIONS

The conventional wisdom and its accompanying morality treat racial prejudice, discrimination, and segregation as irrational social and individual evils that public policy can reduce but only changes in white behavior and values can eliminate. In fact, over the years, white prejudice as measured by attitude surveys has dramatically declined, far more dramatically than behavioral and institutional discrimination.

But what if discrimination and segregation are more than just a social evil? If they are used to keep African Americans down, then they also serve to eliminate or restrain competitors for valued or scarce resources, material and symbolic. Keeping African Americans from decent jobs and incomes as well as quality schools and housing makes more of these available to all the rest of the population. In that case, discrimination and segregation may decline significantly only if the rules of the competition change or if scarce resources, such as decent jobs, become plentiful enough to relax the competition, so that the African-American population can become as predominantly middle class as the white population. Then the stigmas, the stereotypes inherited from slavery, and the social and other arrangements that maintain segregation and discrimination could begin to lose their credibility. Perhaps "black" skin would eventually become as invisible as "yellow" skin is becoming.

THE MULTIRACIAL FUTURE

One trend that encourages upward mobility is the rapid increase in interracial marriage that began about a quarter century ago. As the children born to parents of different races also intermarry, more and more Americans will be multiracial, so that at some point far in the future the current quintet of skin colors will be irrelevant. About 40 percent of young Hispanics and two-thirds of young Asians now "marry out," but only about 10 percent of blacks now marry nonblacks—yet another instance of the exceptionalism that differentiates blacks.

Moreover, if race remains a class marker, new variations in skin color and in other visible bodily features will be taken to indicate class position. Thus, multiracials with "Negroid" characteristics could still find themselves disproportionately at the bottom of the class hierarchy. But what if at some point in the future everyone's skin color varied by only a few shades of brown? At that point, the dominant American classes might have to invent some new class markers.

If in some utopian future the class hierarchy disappears, people will probably stop judging differences in skin color and other features. Then lay Americans would probably agree with biologists that race does not exist. They might even insist that race does not need to exist.

13

Shadowy Lines That Still Divide

JANNY SCOTT AND DAVID LEONHARDT

There was a time when Americans thought they understood class. The upper crust vacationed in Europe and worshiped an Episcopal God. The middle class drove Ford Fairlanes, settled the San Fernando Valley, and enlisted as company men. The working class belonged to the AFL-CIO, voted Democratic, and did not take cruises to the Caribbean.

Today, the country has gone a long way toward an appearance of classlessness. Americans of all sorts are awash in luxuries that would have dazzled their grandparents. Social diversity has erased many of the old markers. It has become harder to read people's status in the clothes they wear, the cars they drive, the votes they cast, the god they worship, the color of their skin. The contours of class have blurred; some say they have disappeared.

But class is still a powerful force in American life. Over the past three decades it has come to play a greater, not lesser, role in important ways. At a time when education matters more than ever, success in school remains linked tightly to class. At a time when the country is increasingly integrated racially, the rich are isolating themselves more and more. At a time of extraordinary advances in medicine, class differences in health and life span are wide and appear to be widening.

And new research on mobility, the movement of families up and down the economic ladder, shows there is far less of it than economists once thought and less than most people believe. In fact, mobility, which once buoyed the working lives of Americans as it rose in the decades after World War II, has lately flattened out or possibly even declined, many researchers say.

Mobility is the promise that lies at the heart of the American dream. It is supposed to take the sting out of the widening gulf between the have-mores and the have-nots. There are poor and rich in the United States, of course, the argument goes; but as long as one can become the other, as long as there is something close to equality of opportunity, the differences between them do not add up to class barriers....

The trends are broad and seemingly contradictory: the blurring of the landscape of class and the simultaneous hardening of certain class lines; the rise in standards of living while most people remain moored in their relative places.

Even as mobility seems to have stagnated, the ranks of the elite are opening. Today, anyone may have a shot at becoming a United States Supreme Court justice or a CEO, and there are more and more self-made billionaires. Only

SOURCE: From *The New York Times*, May 15, 2005. Copyright © 2005 by The New York Times. All rights reserved. Reprinted by permission.

thirty-seven members of last year's Forbes 400, a list of the richest Americans, inherited their wealth, down from almost two hundred in the mid-1980s.

So it appears that while it is easier for a few high achievers to scale the summits of wealth, for many others it has become harder to move up from one economic class to another. Americans are arguably more likely than they were thirty years ago to end up in the class into which they were born.

A paradox lies at the heart of this new American meritocracy. Merit has replaced the old system of inherited privilege, in which parents to the manner born handed down the manor to their children. But merit, it turns out, is at least partly class-based. Parents with money, education, and connections cultivate in their children the habits that the meritocracy rewards. When their children then succeed, their success is seen as earned.

The scramble to scoop up a house in the best school district, channel a child into the right preschool program, or land the best medical specialist are all part of a quiet contest among social groups that the affluent and educated are winning in a rout.

"The old system of hereditary barriers and clubby barriers has pretty much vanished," said Eric Wanner, president of the Russell Sage Foundation, a social science research group in New York City that has published a series of studies on the social effects of economic inequality.

In place of the old system, Wanner said, have arisen "new ways of transmitting advantage that are beginning to assert themselves."

FAITH IN THE SYSTEM

Most Americans remain upbeat about their prospects for getting ahead. A recent *New York Times* poll on class found that 40 percent of Americans believed that the chance of moving up from one class to another had risen over the last thirty years, a period in which the new research shows that it has not. Thirty-five percent said it had not changed, and only 23 percent said it had dropped.

More Americans than twenty years ago believe it possible to start out poor, work hard, and become rich. They say hard work and a good education are more important to getting ahead than connections or a wealthy background.

"I think the system is as fair as you can make it," Ernie Frazier, a sixty-five-year-old real estate investor in Houston, said in an interview after participating in the poll. "I don't think life is necessarily fair. But if you persevere, you can overcome adversity. It has to do with a person's willingness to work hard, and I think it's always been that way."

Most say their standard of living is better than their parents' and imagine that their children will do better still. Even families making less than $30,000 a year subscribe to the American dream; more than half say they have achieved it or will do so.

But most do not see a level playing field. They say the very rich have too much power, and they favor the idea of class-based affirmative action to help those at the bottom. Even so, most say they oppose the government's taxing the assets a person leaves at death....

THE ATTRIBUTES OF CLASS

One difficulty in talking about class is that the word means different things to different people. Class is rank, it is tribe, it is culture and taste. It is attitudes and assumptions, a source of identity, a system of exclusion. To some, it is just money. It is an accident of birth that can influence the outcome of a life. Some Americans barely notice it; others feel its weight in powerful ways.

At its most basic, class is one way societies sort themselves out. Even societies built on the idea of eliminating class have had stark differences in rank. Classes are groups of people of similar economic and social position; people who, for that reason, may share political attitudes, lifestyles, consumption patterns, cultural interests, and opportunities to get ahead. Put ten people in a room and a pecking order soon emerges.

When societies were simpler, the class landscape was easier to read. Marx divided nineteenth-century societies into just two classes; Max Weber added a few more. As societies grew increasingly complex, the old classes became more heterogeneous. As some sociologists and marketing consultants see it, the commonly accepted big three—the upper, middle, and working classes—have broken down into dozens of microclasses, defined by occupations or lifestyles....

One way to think of a person's position in society is to imagine a hand of cards. Everyone is dealt four cards, one from each suit: education, income, occupation, and wealth, the four commonly used criteria for gauging class. Face cards in a few categories may land a player in the upper middle class. At first, a person's class is his parents' class. Later, he may pick up a new hand of his own; it is likely to resemble that of his parents, but not always.

Bill Clinton traded in a hand of low cards with the help of a college education and a Rhodes scholarship and emerged decades later with four face cards. Bill Gates, who started off squarely in the upper middle class, made a fortune without finishing college, drawing three aces.

Many Americans say that they too have moved up the nation's class ladder. In the *Times* poll, 45 percent of respondents said they were in a higher class than when they grew up, while just 16 percent said they were in a lower one. Overall, 1 percent described themselves as upper class, 15 percent as upper middle class, 42 percent as middle, 35 percent as working, and 7 percent as lower.

"I grew up very poor and so did my husband," said Wanda Brown, the fifty-eight-year-old wife of a retired planner for the Puget Sound Naval Shipyard who lives in Puyallup, Washington, near Tacoma. "We're not rich but we are comfortable and we are middle class and our son is better off than we are."

THE AMERICAN IDEAL

The original exemplar of American social mobility was almost certainly Benjamin Franklin, one of seventeen children of a candle maker. About twenty years ago, when researchers first began to study mobility in a rigorous way, Franklin seemed representative of a truly fluid society, in which the rags-to-riches trajectory was the readily achievable ideal, just as the nation's self-image promised.

In a 1987 speech, Gary S. Becker, a University of Chicago economist who would later win a Nobel Prize, summed up the research by saying that mobility in the United States was so high that very little advantage was passed down from one generation to the next. In fact, researchers seemed to agree that the grand-children of privilege and of poverty would be on nearly equal footing.

If that had been the case, the rise in income inequality beginning in the mid-1970s should not have been all that worrisome. The wealthy might have looked as if they were pulling way ahead, but if families were moving in and out of poverty and prosperity all the time, how much did the gap between the top and bottom matter?

But the initial mobility studies were flawed, economists now say. Some studies relied on children's fuzzy recollections of their parents' income. Others compared single years of income, which fluctuate considerably. Still others mis-read the normal progress people make as they advance in their careers, like from young lawyer to senior partner, as social mobility.

The new studies of mobility, which methodically track people's earnings over decades, have found far less movement. The economic advantage once believed to last only two or three generations is now believed to last closer to five. Mobility happens, just not as rapidly as was once thought....

One study, by the Federal Reserve Bank of Boston, found that fewer fami-lies moved from one quintile, or fifth, of the income ladder to another during the 1980s than during the 1970s and that still fewer moved in the 1990s than in the 1980s. A study by the Bureau of Labor Statistics also found that mobility declined from the 1980s to the 1990s.

The incomes of brothers born around 1960 have followed a more similar path than the incomes of brothers born in the late 1940s, researchers at the Chi-cago Federal Reserve and the University of California, Berkeley, have found. Whatever children inherit from their parents—habits, skills, genes, contacts, money—seems to matter more today.

Studies on mobility over generations are notoriously difficult, because they require researchers to match the earnings records of parents with those of their children. Some economists consider the findings of the new studies murky; it can-not be definitively shown that mobility has fallen during the last generation, they say, only that it has not risen. The data will probably not be conclusive for years.

Nor do people agree on the implications. Liberals say the findings are evi-dence of the need for better early-education and antipoverty programs to try to redress an imbalance in opportunities. Conservatives tend to assert that mobility remains quite high, even if it has tailed off a little.

But there is broad consensus about what an optimal range of mobility is. It should be high enough for fluid movement between economic levels but not so high that success is barely tied to achievement and seemingly random, econo-mists on both the right and left say....

One surprising finding about mobility is that it is not higher in the United States than in Britain or France. It is lower here than in Canada and some

Scandinavian countries but not as low as in developing countries like Brazil, where escape from poverty is so difficult that the lower class is all but frozen in place.

Those comparisons may seem hard to believe. Britain and France had hereditary nobilities; Britain still has a queen. The founding document of the United States proclaims all men to be created equal. The American economy has also grown more quickly than Europe's in recent decades, leaving an impression of boundless opportunity.

But the United States differs from Europe in ways that can gum up the mobility machine. Because income inequality is greater here, there is a wider disparity between what rich and poor parents can invest in their children. Perhaps as a result, a child's economic background is a better predictor of school performance in the United States than in Denmark, the Netherlands, or France, one study found.

"Being born in the elite in the U.S. gives you a constellation of privileges that very few people in the world have ever experienced," Levine said. "Being born poor in the U.S. gives you disadvantages unlike anything in Western Europe and Japan and Canada."

BLURRING THE LANDSCAPE

Why does it appear that class is fading as a force in American life?

For one thing, it is harder to read position in possessions. Factories in China and elsewhere churn out picture-taking cellphones and other luxuries that are now affordable to almost everyone. Federal deregulation has done the same for plane tickets and long-distance phone calls. Banks, more confident about measuring risk, now extend credit to low-income families, so that owning a home or driving a new car is no longer evidence that someone is middle class.

The economic changes making material goods cheaper have forced businesses to seek out new opportunities so that they now market to groups they once ignored. Cruise ships, years ago a symbol of the high life, have become the oceangoing equivalent of the Jersey Shore. BMW produces a cheaper model with the same insignia. Martha Stewart sells chenille jacquard drapery and scallop-embossed ceramic dinnerware at Kmart.

"The level of material comfort in this country is numbing," said Paul Bellew, executive director for market and industry analysis at General Motors. "You can make a case that the upper half lives as well as the upper 5 percent did fifty years ago."

Like consumption patterns, class alignments in politics have become jumbled. In the 1950s, professionals were reliably Republican; today they lean Democratic. Meanwhile, skilled labor has gone from being heavily Democratic to almost evenly split.

People in both parties have attributed the shift to the rise of social issues, like gun control and same-sex marriage, which have tilted many working-class voters rightward and upper-income voters toward the left. But increasing affluence plays an important role, too. When there is not only a chicken, but an organic, free-range chicken, in every pot, the traditional economic appeal to the working class can sound off-key.

Religious affiliation, too, is no longer the reliable class marker it once was. The growing economic power of the South has helped lift evangelical Christians into the middle and upper middle classes, just as earlier generations of Roman Catholics moved up in the mid-twentieth century. It is no longer necessary to switch one's church membership to Episcopal or Presbyterian as proof that one has arrived....

The once tight connection between race and class has weakened, too, as many African-Americans have moved into the middle and upper middle classes. Diversity of all sorts—racial, ethnic, and gender—has complicated the class picture. And high rates of immigration and immigrant success stories seem to hammer home the point: The rules of advancement have changed.

The American elite, too, is more diverse than it was. The number of corporate chief executives who went to Ivy League colleges has dropped over the past fifteen years. There are many more Catholics, Jews, and Mormons in the Senate than there were a generation or two ago. Because of the economic earthquakes of the last few decades, a small but growing number of people have shot to the top....

These success stories reinforce perceptions of mobility, as does cultural myth-making in the form of television programs like *American Idol* and *The Apprentice*.

But beneath all that murkiness and flux, some of the same forces have deepened the hidden divisions of class. Globalization and technological change have shuttered factories, killing jobs that were once stepping-stones to the middle class. Now that manual labor can be done in developing countries for two dollars a day, skills and education have become more essential than ever.

This has helped produce the extraordinary jump in income inequality. The after-tax income of the top 1 percent of American households jumped 139 percent, to more than $700,000, from 1979 to 2001, according to the Congressional Budget Office, which adjusted its numbers to account for inflation. The income of the middle fifth rose by just 17 percent, to $43,700, and the income of the poorest fifth rose only 9 percent.

For most workers, the only time in the last three decades when the rise in hourly pay beat inflation was during the speculative bubble of the 1990s. Reduced pensions have made retirement less secure.

Clearly, a degree from a four-year college makes even more difference than it once did. More people are getting those degrees than did a generation ago, but class still plays a big role in determining who does or does not. At 250 of the most selective colleges in the country, the proportion of students from upper-income families has grown, not shrunk....

Class differences in health, too, are widening, recent research shows. Life expectancy has increased overall; but upper-middle-class Americans live longer and in better health than middle-class Americans, who live longer and in better health than those at the bottom.

Class plays an increased role, too, in determining where and with whom affluent Americans live. More than in the past, they tend to live apart from everyone else, cocooned in their exurban châteaus. Researchers who have studied census data from 1980, 1990, and 2000 say the isolation of the affluent has increased.

Family structure, too, differs increasingly along class lines. The educated and affluent are more likely than others to have their children while married. They have fewer children and have them later, when their earning power is high. On average, according to one study, college-educated women have their first child at age thirty, up from twenty-five in the early 1970s. The average age among women who have never gone to college has stayed at about twenty-two.

Those widening differences have left the educated and affluent in a superior position when it comes to investing in their children....

The benefits of the new meritocracy do come at a price. It once seemed that people worked hard and got rich in order to relax, but a new class marker in upper-income families is having at least one parent who works extremely long hours (and often boasts about it). In 1973, one study found, the highest-paid tenth of the country worked fewer hours than the bottom tenth. Today, those at the top work more.

In downtown Manhattan, black cars line up outside Goldman Sachs's headquarters every weeknight around nine. Employees who work that late get a free ride home, and there are plenty of them. Until 1976, a limousine waited at 4:30 p.m. to ferry partners to Grand Central Terminal. But a new management team eliminated the late-afternoon limo to send a message: 4:30 is the middle of the workday, not the end.

A RAGS-TO-RICHES FAITH

Will the trends that have reinforced class lines while papering over the distinctions persist?

The economic forces that caused jobs to migrate to low-wage countries are still active. The gaps in pay, education, and health have not become a major political issue. The slicing of society's pie is more unequal than it used to be, but most Americans have a bigger piece than they or their parents once did. They appear to accept the trade-offs.

Faith in mobility, after all, has been consciously woven into the national self-image. Horatio Alger's books have made his name synonymous with rags-to-riches success, but that was not his personal story. He was a second-generation Harvard man, who became a writer only after losing his Unitarian ministry because of allegations of sexual misconduct. Ben Franklin's autobiography was punched up after his death to underscore his rise from obscurity.

The idea of fixed class positions, on the other hand, rubs many the wrong way. Americans have never been comfortable with the notion of a pecking order based on anything other than talent and hard work. Class contradicts their assumptions about the American dream, equal opportunity, and the reasons for their own successes and even failures. Americans, constitutionally optimistic, are disinclined to see themselves as stuck.

Blind optimism has its pitfalls. If opportunity is taken for granted, as something that will be there no matter what, then the country is less likely to do the hard work to make it happen. But defiant optimism has its strengths. Without confidence in the possibility of moving up, there would almost certainly be fewer success stories.

14

Is Capitalism Gendered and Racialized?

JOAN ACKER

Capitalism is racialized and gendered in two intersecting historical processes. ... First, industrial capitalism emerged in the United States dominated by white males, with a gender- and race-segregated labor force, laced with wage inequalities, and a society-wide gender division of caring labor. The processes of reproducing segregation and wage inequality changed over time, but segregation and inequality were not eliminated. A small group of white males still dominate the capitalist economy and its politics. The society–wide gendered division of caring labor still exists. Ideologies of white masculinity and related forms of consciousness help to justify capitalist practices. In short, conceptual and material practices that construct capitalist production and markets, as well as beliefs supporting those practices, are deeply shaped through gender and race divisions of labor and power and through constructions of white masculinity.

Second, these gendered and racialized practices are embedded in and replicated through the gendered substructures of capitalism. These gendered substructures exist in ongoing incompatible organizing of paid production activities and unpaid domestic and caring activities. Domestic and caring activities are devalued and seen as outside the "main business" (Smith 1999) of capitalism. The commodification of labor, the capitalist wage form, is an integral part of this process, as family provisioning and caring become dependent upon wage labor. The abstract language of bureaucratic organizing obscures the ongoing impact on families and daily life. At the same time, paid work is organized on the assumption that reproduction is of no concern. The separations between paid production and unpaid life-sustaining activities are maintained by corporate claims that they have no responsibility for anything but returns to shareholders. Such claims are more successful in the United States, in particular, than in countries with stronger labor movements and welfare states. These often successful claims contribute to the corporate processes of establishing their interests as more important than those of ordinary people.

SOURCE:. From *Class Questions, Feminist Answers*, ed. Joan Acker. Lanham, MD: Rowman and Littlefield, 2006, pp. 111–118. Reprinted by permission of Altamira Press, a member of the Rowman & Littlefield Publishing Group.

THE GENDERED AND RACIALIZED DEVELOPMENT
OF U.S. CAPITALISM

Segregations and Wage Inequalities

Industrial capitalism is historically, and in the main continues to be, a white male project, in the sense that white men were and are the innovators, owners, and holders of power. Capitalism developed in Britain and then in Europe and the United States in societies that were already dominated by white men and already contained a gender-based division of labor. The emerging waged labor force was sharply divided by gender, as well as by race and ethnicity with many variations by nation and regions within nations. At the same time, the gendered division of labor in domestic tasks was reconfigured and incorporated in a gendered division between paid market labor and unpaid domestic labor. In the United States, certain white men, unburdened by caring for children and households and already the major wielders of gendered power, buttressed at least indirectly by the profits from slavery and the exploitation of other minorities, were, in the nineteenth century, those who built the U.S. factories and railroads, and owned and managed the developing capitalist enterprise. As far as we know, they were also heterosexual and mostly of Northern European heritage. Their wives and daughters benefited from the wealth they amassed and contributed in symbolic and social ways to the perpetuation of their class, but they were not the architects of the new economy.

Recruitment of the labor force for the colonies and then the United States had always been transnational and often coercive. Slavery existed prior to the development of industrialism in the United States: Capitalism was built partly on profits from that source. Michael Omi and Howard Winant (1994, 265) contend that the United States was a racial dictatorship for 258 years, from 1607 to 1865. After the abolition of slavery in 1865, severe exploitation, exclusion, and domination of blacks by whites perpetuated racial divisions cutting across gender and some class divisions, consigning blacks to the most menial, low-paying work in agriculture, mining, and domestic service. Early industrial workers were immigrants. For example, except for the brief tenure (twenty-five years) of young, native-born white women workers in the Lowell, Massachusetts, mills, immigrant women and children were the workers in the first mass production industry in the United States, the textile mills of Massachusetts and Philadelphia, Pennsylvania (Perrow 2002). This was a gender and racial/ethnic division of labor that still exists, but now on a global basis. Waves of European immigrants continued to come to the United States to work in factories and on farms. Many of these European immigrants, such as impoverished Irish, Poles, and eastern European Jews were seen as non-white or not-quite-white by white Americans and were used in capitalist production as low-wage workers, although some of them were actually skilled workers (Brodkin 1998). The experiences of racial oppression built into industrial capitalism varied by gender within these racial/ethnic groups.

Capitalist expansion across the American continent created additional groups of Americans who were segregated by race and gender into racial and ethnic enclaves and into low-paid and highly exploited work. This expansion included the extermination and expropriation of native peoples, the subordination of Mexicans in areas taken in the war with Mexico in 1845, and the recruitment of Chinese and other Asians as low-wage workers, mostly on the west coast (Amott and Matthaei 1996; Glenn 2002).

Women from different racial and ethnic groups were incorporated differently than men and differently than each other into developing capitalism in the late nineteenth and early twentieth centuries. White Euro-American men moved from farms into factories or commercial, business, and administrative jobs. Women aspired to be housewives as the male breadwinner family became the ideal. Married white women, working class and middle class, were housewives unless unemployment, low wages, or death of their husbands made their paid work necessary (Goldin 1990, 133). Young white women with some secondary education moved into the expanding clerical jobs and into elementary school teaching when white men with sufficient education were unavailable (Cohn 1985). African Americans, both women and men, continued to be confined to menial work, although some were becoming factory workers, and even teachers and professionals as black schools and colleges were formed (Collins 2000). Young women from first- and second-generation European immigrant families worked in factories and offices. This is a very sketchy outline of a complex process (Kessler-Harris 1982), but the overall point is that the capitalist labor force in the United States emerged as deeply segregated horizontally by occupation and stratified vertically by positions of power and control on the basis of both gender and race.

Unequal pay patterns went along with sex and race segregation, stratification, and exclusion. Differences in the earnings and wealth (Keister 2000) of women and men existed before the development of the capitalist wage (Padavic and Reskin 2002). Slaves, of course, had no wages and earned little after abolition. These patterns continued as capitalist wage labor became the dominant form and wages became the primary avenue of distribution to ordinary people. Unequal wages were justified by beliefs about virtue and entitlement. A living wage or a just wage for white men was higher than a living wage or a just wage for white women or for women and men from minority racial and ethnic groups (Figart, Mutari, and Power 2002). African-American women were at the bottom of the wage hierarchy.

The earnings advantage that white men have had throughout the history of modern capitalism was created partly by their organization to increase their wages and improve their working conditions. They also sought to protect their wages against the competition of others, women and men from subordinate groups (for example, Cockburn 1983, 1991). This advantage also suggests a white male coalition across class lines (Connell 2000; Hartmann 1976), based at least partly in beliefs about gender and race differences and beliefs about the superior skills of white men. White masculine identity and self-respect were complexly involved in these divisions of labor and wages. This is another way in which capitalism is a gendered and racialized accumulation process (Connell 2000). Wage differences between white men and all other groups, as well as divisions of labor between these groups, contributed to profit and flexibility, by helping to maintain growing occupational areas, such as clerical work, as segregated and low paid. Where women worked in manufacturing or food processing, gender divisions of labor kept the often larger female work force in low-wage routine jobs, while males worked in other more highly paid, less routine, positions (Acker and Van Houten 1974). While white men might be paid more, capitalist organizations could benefit from this "gender/racial dividend." Thus, by maintaining divisions, employers could pay less for certain levels of skill, responsibility, and experience when the worker was not a white male.

This is not to say that getting a living wage was easy for white men, or that most white men achieved it. Labor-management battles, employers' violent tactics to prevent unionization, [and] massive unemployment during frequent economic depressions characterized the situation of white industrial workers as wage labor spread in the nineteenth and early twentieth centuries. During the same period, new white-collar jobs were created to manage, plan, and control the expanding industrial economy. This rapidly increasing middle class was also stratified by gender and race. The better-paid, more respected jobs went to white men; white women were secretaries and clerical workers; people of color were absent. Conditions and issues varied across industries and regions of the country. But, wherever you look, those variations contained underlying gendered and racialized divisions. Patterns of stratification and segregation were written into employment contracts in work content, positions in work hierarchies, and wage differences, as well as other forms of distribution.

These patterns persisted, although with many alterations, through extraordinary changes in production and social life. After World War II, white women, except for a brief period immediately after the war, went to work for pay in the expanding service sector, professional, and managerial fields. African Americans moved to the North in large numbers, entering industrial and service sector jobs. These processes accelerated after the 1960s, with the civil rights and women's movements, new civil rights laws, and affirmative action. Hispanics and Asian Americans, as well as other racial/ethnic groups, became larger proportions of the population, on the whole finding work in low-paid, segregated jobs. Employers continued, and still continue, to select and promote workers based on gender and racial identifications, although the processes are more subtle, and possibly less visible, than in the past (for example, Brown et al. 2003; Royster 2003). These processes continually recreate gender and racial inequities, not as cultural or ideological survivals from earlier times, but as essential elements in present capitalisms (Connell 1987, 103–106).

Segregating practices are a part of the history of white, masculine-dominated capitalism that establishes class as gendered and racialized. Images of masculinity support these practices, as they produce a taken-for-granted world in which certain men legitimately make employment and other economic decisions that affect the lives of most other people. Even though some white women and people from other-than-white groups now hold leadership positions, their actions are shaped within networks of practices sustained by images of masculinity (Wacjman 1998).

Masculinities and Capitalism

Masculinities are essential components of the ongoing male project, capitalism. While white men were and are the main publicly recognized actors in the history of capitalism, these are not just any white men. They have been, for example, aggressive entrepreneurs or strong leaders of industry and finance (Collinson and Hearn 1996). Some have been oppositional actors, such as self-respecting and tough workers earning a family wage, and militant labor leaders. They have been particular men whose locations within gendered and racialized social relations and practices can be partially captured by the concept of masculinity. "Masculinity" is a contested term. As Connell (1995, 2000), Hearn (1996), and others have pointed out, it should be pluralized as "masculinities," because in any society at any one time there are several ways of being a man. "Being a man" involves cultural images and practices. It always implies a contrast to an unidentified femininity.

Hegemonic masculinity can be defined as the taken-for-granted, generally accepted form, attributed to leaders and other influential figures at particular historical times. Hegemonic masculinity legitimates the power of those who embody it. More than one type of hegemonic masculinity may exist simultaneously, although they may share characteristics, as do the business leader and the sports star at the present time. Adjectives describing hegemonic masculinities closely follow those describing characteristics of successful business organizations, as Rosabeth Moss Kanter (1977) pointed out in the 1970s. The successful CEO and the successful organization are aggressive, decisive, competitive, focused on winning and defeating the enemy, taking territory from others. The ideology of capitalist markets is imbued with a masculine ethos. As R. W. Connell (2000, 35) observes, "The market is often seen as the antithesis of gender (marked by achieved versus ascribed status, etc.). But the market operates through forms of rationality that are historically masculine and involve a sharp split between instrumental reason on the one hand, emotion and human responsibility on the other" (Seidler 1989). Masculinities embedded in collective practices are part of the context within which certain men made and still make the decisions that drive and shape the ongoing development of capitalism. We can speculate that how these men see themselves, what actions and choices they feel compelled to make and they think are legitimate, how they and the world around them define desirable masculinity, enter into that decision making (Reed 1996). Decisions made at the very top reaches of (masculine) corporate power have consequences that are experienced as inevitable economic forces or disembodied social trends. At the same time, these decisions symbolize and enact varying hegemonic masculinities (Connell 1995). However, the embeddedness of masculinity within the ideologies of business and the market may become invisible, seen as just part of the way business is done. The relatively few women who reach the highest positions probably think and act within these strictures.

Hegemonic masculinities and violence are deeply connected within capitalist history: The violent acts of those who carried out the slave trade or organized colonial conquests are obvious examples. Of course, violence has been an essential component of power in many other socioeconomic systems, but it continues into the rational organization of capitalist economic activities. Violence is frequently a legitimate, if implicit, component of power exercised by bureaucrats as well as "robber barons." Metaphors of violence, frequently military violence, are often linked to notions of the masculinity of corporate leaders, as "defeating the enemy" suggests. In contemporary capitalism, violence and its links to masculinity are often masked by the seeming impersonality of objective conditions. For example, the masculinity of top managers, the ability to be tough, is involved in the implicit violence of many corporate decisions, such as those cutting jobs in order to raise profits and, as a result, producing unemployment. Armies and other organizations, such as the police, are specifically organized around violence. Some observers of recent history suggest that organized violence, such as the use of the military, is still mobilized at least partly to reach capitalist goals, such as controlling access to oil supplies. The masculinities of those making decisions to deploy violence in such a way are hegemonic, in the sense of powerful and exemplary. Nevertheless, the connections between masculinity, capitalism, and violence are complex and contradictory, as Jeff Hearn and Wendy Parkin (2001) make clear. Violence is always a possibility in mechanisms of control and domination, but it is not always evident, nor is it always used.

As corporate capitalism developed, Connell (1995) and others (for example, Burris 1996) argue that a hegemonic masculinity based on claims to expertise developed alongside masculinities organized around domination and control. Hegemonic masculinity relying on claims to expertise does not necessarily lead to economic organizations free of domination and violence, however (Hearn and Parkin 2001). Hearn and Parkin (2001) argue that controls relying on both explicit and implicit violence exist in a wide variety of organizations, including those devoted to developing new technology.

Different hegemonic masculinities in different countries may reflect different national histories, cultures, and change processes. For example, in Sweden in the mid-1980s, corporations were changing the ways in which they did business toward a greater participation in the international economy, fewer controls on currency and trade, and greater emphasis on competition. Existing images of dominant masculinity were changing, reflecting new business practices. This seemed to be happening in the banking sector, where I was doing research on women and their jobs (Acker 1994a). The old paternalistic leadership, in which primarily men entered as young clerks expecting to rise to managerial levels, was being replaced by young, aggressive men hired as experts and managers from outside the banks. These young, often technically trained, ambitious men pushed the idea that the staff was there to sell bank products to customers, not, in the first instance, to take care of the needs of clients. Productivity goals were put in place; nonprofitable customers, such as elderly pensioners, were to be encouraged not to come into the bank and occupy the staff's attention. The female clerks we interviewed were disturbed by these changes, seeing them as evidence that the men at the top were changing from paternal guardians of the people's interests to manipulators who only wanted riches for themselves. The confirmation of this came in a scandal in which the CEO of the largest bank had to step down because he had illegally taken money from the bank to pay for his housing. The amount of money was small; the disillusion among employees was huge. He had been seen as a benign father; now he was no better than the callous young men on the way up who were dominating the daily work in the banks. The hegemonic masculinity in Swedish banks was changing as the economy and society were changing.

Hegemonic masculinities are defined in contrast to subordinate masculinities. White working class masculinity, although clearly subordinate, mirrors in some of its more heroic forms the images of strength and responsibility of certain successful business leaders. The construction of working class masculinity around the obligations to work hard, earn a family wage, and be a good provider can be seen as providing an identity that both served as a social control and secured male advantage in the home. That is, the good provider had to have a wife and probably children for whom to provide. Glenn (2002) describes in some detail how this image of the white male worker also defined him as superior to and different from black workers.

Masculinities are not stable images and ideals, but [shift] with other societal changes. With the turn to neoliberal business thinking and globalization, there seem to be new forms. Connell (2000) identifies "global business masculinity," while Lourdes Beneria (1999) discusses the "Davos man," the global leader from business, politics, or academia who meets his peers once a year in the Swiss town of Davos to assess and plan the direction of globalization. Seeing masculinities as implicated in the ongoing production of global capitalism opens the possibility of seeing sexualities, bodies, pleasures, and identities as also implicated in economic relations.

In sum, gender and race are built into capitalism and its class processes through the long history of racial and gender segregation of paid labor and through the images and actions of white men who dominate and lead central capitalist endeavors. Underlying these processes is the subordination to production and the market of nurturing and caring for human beings, and the assignment of these responsibilities to women as unpaid work. Gender segregation that differentially affects women in all racial groups rests at least partially on the ideology and actuality of women as carers. Images of dominant masculinity enshrine particular male bodies and ways of being as different from the female and distanced from caring. ... I argue that industrial capitalism, including its present neoliberal form, is organized in ways that are, at the same time, antithetical and necessary to the organization of caring or reproduction and that the resulting tensions contribute to the perpetuation of gendered and racialized class inequalities. Large corporations are particularly important in this process as they increasingly control the resources for provisioning but deny responsibility for such social goals.

REFERENCES

Acker, Joan. 1994a. The Gender Regime of Swedish Banks. *Scandinavian Journal of Management* 10, no. 2: 117–30.

Acker, Joan, and Donald Van Houten. 1974. Differential Recruitment and Control: The Sex Structuring of Organizations. *Administrative Science Quarterly* 19 (June, 1974): 152–63.

Amott, Teresa, and Julie Matthaei. 1996. *Race, Gender, and Work: A Multi-cultural Economic History of Women in the United States*. Revised edition. Boston: South End Press.

Beneria, Lourdes. 1999. Globalization, Gender and the Davos Man. *Feminist Economics* 5, no. 3: 61–83.

Brodkin, Karen. 1998. Race, Class, and Gender: The Metaorganization of American Capitalism. *Transforming Anthropology* 7, no. 2: 46–57.

Brown, Michael K., Martin Carnoy, Elliott Currie, Troy Duster, David B. Oppenheimer, Marjorie M. Shultz, and David Wellman. 2003. *White-Washing Race The Myth of a Color-Blind Society*. Berkeley: University of California Press.

Burris, Beverly H. 1996. Technocracy, Patriarchy and Management. In *Men as Managers, Managers as Men*, ed. David L. Collinson and Jeff Hearn. London: Sage.

Cockburn, Cynthia. 1983. *Brothers*. London: Pluto Press.

———. 1991. *In the Way of Women: Men's Resistance to Sex Equality in Organization*. Ithaca, N.Y: ILR Press.

Cohn, Samuel. 1985. *The Process of Occupational Sex-Typing: The Femininization of Clerical Labor in Great Britain*. Philadelphia: Temple University Press.

Collins, Patricia Hill. 2000. *Black Feminist Thought*, second edition. New York and London: Routledge.

Collinson, David L., and Jeff Hearn. 1996. Breaking the Silence: On Men, Masculinities and Managements. In *Men as Managers, Managers as Men*, ed. David L. Collinson and Jeff Hearns. London: Sage.

Connell, R. W. 1987. *Gender & Power*. Stanford, Calif: Stanford University Press.

———. 1995. *Masculinities*. Berkeley: University of California Press.

———. 2000. *The Men and the Boys*. Berkeley: University of California Press.

Figart, Deborah M., Ellen Mutari, and Marilyn Power. 2002. *Living Wages, Equal Wages*. London and New York: Routledge.

Glenn, Evelyn Nakano. 2002. *Unequal Freedom: How Race and Gender Shaped American Citizenship and Labor*. Cambridge: Harvard University Press.

Goldin, Claudia. 1990. *Understanding the Gender Gap: An Economic History of American Women*. New York and Oxford: Oxford University Press.

Hearn, Jeff. 1996. Is Masculinity Dead? A Critique of the Concept of Masculinity/Masculinities. In *Understanding Masculinities: Social Relations and Cultural Arenas*, ed. M. Mac an Ghaill. Buckingham: Oxford University Press.

———. 2004. From Hegemonic Masculinity to the Hegemony of Men. *Feminist Theory* 5, no. 1: 49–72.

Hearn, Jeff, and Wendy Parkin. 2001. *Gender, Sexuality and Violence in Organizations*. London: Sage.

Kanter, Rosabeth Moss. 1977. *Men and Women of the Corporation*. New York: Basic Books.

Keister, Lisa. 2000. *Wealth in America: Trends in Wealth Inequality*. Cambridge: Cambridge University Press.

Kessler-Harris, Alice. 1982. *Out to Work: A History of Wage-Earning Women in the United States*. New York: Oxford University Press.

Omi, Michael, and Howard Winant. 1994. *Racial Formation in the United States*. New York: Routledge.

Padavic, Irene, and Barbara Reskin. 2002. *Women and Men at Work*, second edition. Thousand Oaks, Calif.: Pine Forge Press.

Perrow, Charles. 2002. *Organizing America*. Princeton and Oxford: Princeton University Press.

Reed, Rosslyn. 1996. Entrepreneurialism and Paternalism in Australian Management: A Gender Critique of the "Self-Made" Man. In *Men as Managers, Managers as Men*, ed. David L. Collinson and Jeff Hearn. London: Sage.

Royster, Deirdre A. 2003. *Race and the Invisible Hand: How White Networks Exclude Black Men from Blue-Collar Jobs*. Berkeley: University of California Press.

Seidler, Victor J. 1989. *Rediscovering Masculinity: Reason, Language, and Sexuality*. London and New York: Routledge.

Smith, Dorothy. 1999. *Writing the Social: Critique, Theory, and Investigation*. Toronto: University of Toronto Press.

Wacjman, Judy. 1998. *Managing Like a Man*. Cambridge: Polity Press.

15

Health and Wealth

Our Appalling Health Inequality Reflects and Reinforces Society's Other Gaps.

LAWRENCE R. JACOBS AND JAMES A. MORONE

A look at Americans' health reveals the astonishing inequalities in our society. American girls are born with a life expectancy that ranks 19th in the world (in another survey they fall to 28th). Male babies rank 31st—in a dead tie with Brunei. Among the 13 wealthiest countries, the United States ranks last or nearly so in almost every way we measure health: infant mortality, low birth weight, life expectancy at birth, life expectancy for infants. The average American boy lives three and a half fewer years than the average Japanese baby, despite higher rates of cigarette smoking in Japan. The American adolescent death rate is twice as high as, say, England's.

These dismal American averages mask vast differences across our population. A male born in some sections of Washington, D.C., for example, has a life expectancy 40 years lower than a woman born in many wealthy neighborhoods. In short, great differences in wealth match up to—indeed, they create—terrible differences in health.

Why do Americans come out so badly in the cross-national health statistics? Why are our infants more likely to die than those in, say, Croatia? Our health troubles have three interrelated causes: inequality, poverty, and the way we organize our health-care system.

Let's start with inequality. A famous study of the British civil service found that with each rung up the ladder of success, people suffered fewer fatal heart attacks—the clerks and messengers at the bottom were four times more likely to die than the executives at the top. Researchers following up this study reached a surprising conclusion that seems to hold up in one nation after another: The wider the inequality, the worse the nation's overall health.

Why should this be so? For one thing, falling behind in the race to make ends meet generates stress and physiological harm—the results are depression, hypertension, other illnesses, and high mortality rates. In addition, the middle-class scramble to get ahead erodes neighborly feelings, frays our communities,

SOURCE: Reprinted with permission from Lawrence R. Jacobs, and James A. Morone. "Health and Wealth: Our Appalling Health Inequality Reflects and Reinforces Society's Other Gaps." *The American Prospect* (May 17, 2004) Volume 15, Issue 6, http://www.prospect.org. The American Prospect, 1710 Rhode Island Avenue, NW, 12th Floor, Washington DC 20036. All rights reserved.

and lowers trust in institutions like churches and governments. All of these are factors in other countries. But most industrial nations buffer their citizens against economic uncertainty and lost jobs. In the United States, only the market winners get security.

Of course, American health problems go beyond inequality and are closely correlated with the poverty in which more than one in 10 Americans now live. Of our 34.6 million "poor" citizens, according to the U.S. Census Bureau, more than 14 million are "severely poor," meaning they don't even make it halfway to the federal poverty line. The numbers are worse for minorities, with nearly a quarter of blacks and more than a fifth of Hispanics living in poverty.

And poverty brings troubles like hunger (33 million Americans live with "food insecurity," as defined by the Department of Agriculture) and home-lessness (perhaps as many as 3.5 million a year), which disproportionately fall on kids. Poor neighborhoods face high crime, inferior schools, few good jobs, and inadequate health-care facilities. Instead, poverty attracts danger—too much alcohol and tobacco, illegal drugs, and fast foods. One observer after another has gone off to study poor communities and come back with the same report. The lives of the poor are full of stress and the struggle to get by.

People die younger in Harlem than in Bangladesh. Why? It is not what most people think—homicide, drug abuse, and AIDS are far down the list. Rather, as *The New England Journal of Medicine* reports, the leading causes of death in poor black neighborhoods are "unrelenting stress," "cardiovascular disease," "cancer," and "untreated medical conditions."

Finally, beyond the fundamentals—inequality and poverty—there is that stubborn American policy dilemma: No other industrial nation tolerates such yawning gaps in health insurance. According to the Congressional Budget Office, 43.6 million people were uninsured in 2002, with 19.9 million coming from the ranks of full-time workers; 74.7 million Americans under 65 were without health insurance for all or part of 2001 and 2002. Part of the problem is that workplace coverage is unraveling as more employers shift costs like pre-miums, co-payments, and coverage limitations onto their workers. Meanwhile, medical costs are rising faster than personal-income growth.

Simple medical care—annual check-ups, screenings, vaccinations, eyeglasses, dentistry—saves lives, improve well-being, and is shockingly uneven. Well-insured people get assigned hospital beds; the uninsured get patched up and sent back to the streets. From diagnostic procedures—prostate screenings, mam-mograms, and Pap smears—to treatment for asthma, the uninsured get less care, they get it later in their illnesses, and they are roughly three times more likely to have an adverse health outcome. The Institute of Medicine recently blamed gaps in insurance coverage for 17,000 preventable deaths a year.

Even middle-class parents worry about the next medical emergency or, in many cases, the routine trip to the doctor's office. Life without health insurance means constantly measuring aches and fevers against the next payday. Changing jobs brings a new set of anxieties about shifts in medical coverage. Health bills are the largest cause of personal bankruptcy in the United States.

Of course, no health-care system treats everyone the same way. But in America, our disparities are unusually wide and deep.

How can we reverse these trends and begin to build the good society? Recent experience counsels incremental reform that builds on past successes while pushing bold new proposals for the future.

As recent history shows, even half steps—like adding amendments to bipartisan legislation—can add up to something important. Back when the Reagan administration was attacking poverty programs while cutting taxes and running up enormous deficits, California Congressman Henry Waxman oversaw bipartisan support for a series of minor expansions in Medicaid eligibility. The result: In the late 1980s, the program grew to cover an additional 5 million children and 500,000 pregnant women.

While Bill Clinton's failure to pass national health insurance got most of the press, his administration quietly enacted the Children's Health Insurance Program for states in 1997. Using federal matching funds as a prod, the program pushed states to widen coverage to uninsured children, helping Medicaid reach 20 million kids by 2000 and funding non-Medicaid programs to cover an additional 2 million.

Even further below the national radar screen, the Robert Wood Johnson Foundation induced state governments to place health-care clinics directly in schools. Families in underserved neighborhoods suddenly—and usually for the first time—found it easy for their kids to get into a physician's office. Despite strong initial opposition from the cultural right over birth control, teachers, public-health advocates, parents, and community organizers have managed to open 1,498 school centers from Maine to California.

Reforms beyond medical care can also improve general living conditions and boost American health. The Earned Income Tax Credit, for example, has lifted millions of low-income workers and their children out of poverty. To be sure, making Americans healthy means addressing the economic insecurity that threatens these struggling families, forcing middle-class Americans to work double shifts and the poor to confront hunger and homelessness.

Making Americans healthy also means casting off the political torpor of this new Gilded Age and reclaiming a long-standing commitment to our neighbors and communities. Only great aspirations will galvanize a new populist politics and leverage our reluctant state.

There is not much mystery about what works. Other industrial countries rely on three familiar paths to good health. First, government plays an important role through such policies as family and housing allowances, universal health care, pensions, and tax credits. The generous welfare states of northern Europe and nations with more modest programs like France, Germany, and Canada all have poor, middle-class, and wealthy populations. However, all these nations achieve much narrower income gaps among groups than now exist in the United States.

A second type of policy fosters opportunity. Governments invest in education to expand the supply of skilled labor and help workers help themselves. Lowering the barriers to college education and worker retraining reduces the high premium for skilled labor. In addition, European governments collaborate with businesses by regularly adjusting the minimum wage and overseeing the negotiations between business and labor.

Finally, most wealthy nations maintain taxes. The new global economy was expected to spark dramatic tax cuts as governments competed with one another to create an attractive business climate and lure investment and skilled labor. In Europe and Canada, international pressures did not eviscerate the government's capacity to raise revenues. Instead, domestic support to maintain programs (and international pressure to limit deficits) barred governments from plunging into tax-cut wars.

In short, America's allies have tried to defend all their citizens from the worst effects of a global economy. The results across the industrial world are powerful: Policies that moderate income disparities turn out to be good for your health.

American public policy, has, on balance, gone the other way: Tax cuts, deregulation, and unmediated markets sabotage our incremental stabs at fostering real opportunity. Some individuals have grown fantastically wealthy; most struggle to make ends meet. The dirty policy secret lies in the health consequences: Our population suffers more illness and dies younger.

Our call to reform is simple: A civilized society should not accept gaping disparities in life and death, health and disability. Americans are too generous and fair-minded to tolerate so much preventable suffering. This moral vision undergirds a hardheaded analysis of the rapidly changing global economy that has reshuffled the distribution of money in American society and unsettled the life circumstances that nurture and protect the health of the country. The solutions are no mystery. Other nations successfully protect their people. So can we.

16

Sub-Prime as a Black Catastrophe

MELVIN L. OLIVER AND THOMAS M. SHAPIRO

No other recent economic crisis better illustrates the saying "when America catches a cold, African Americans get pneumonia" than the sub-prime mortgage meltdown. African Americans, along with other minorities and low-income populations, have been the targets of the sub-prime mortgage system. Blacks received a disproportionate share of these loans, leading to a "stripping" of their hard-won home-equity gains of the recent past and the near future. To fully understand how this has happened, we need to place this in the context of the continuing racial-wealth gap, the importance of home equity in the wealth portfolios of African Americans, and its intersection with the new financial markets of which sub-prime is but one manifestation.

Family financial assets play a key role in poverty reduction, social mobility, and securing middle-class status. Income helps families get along, but assets help them get and stay ahead. Those without the head start of family assets have a much steeper climb out of poverty. This generation of African Americans is the first one afforded the legal, educational, and job opportunities to accumulate financial assets essential to launching social mobility and sustaining well-being throughout the life course.

Despite legal gains in civil rights, however, asset inequality in America has actually been growing rapidly during the last 20 years. The assets that current generations own are heavily dependent on the legacies of their families of origin. Today's blacks still suffer from the fact that their parents and grandparents grew up in a rigidly segregated America, where opportunities to accumulate human and financial capital were strictly limited. So housing wealth is a disproportionate share of total black wealth.

Despite some income gains, African Americans own only 7 cents for every dollar of net worth that white Americans own; for Hispanics the figure is only slightly higher at 9 cents for every dollar. Even when middle-class accomplishments like income, job, and education are comparable, the racial-wealth gap is stuck stubbornly at about a quarter on the dollar.

HOUSING WEALTH: LAST IN, FIRST OUT

Prior to the sub-prime meltdown, the advances of African Americans in accumulating wealth depended heavily upon housing wealth. Home equity is the

SOURCE: Oliver, Melvin L., and Shapiro, Thomas M., "Sub-Prime as a Black Catastrophe." *The American Prospect* (9/20/08). Copyright © 2011 by The American Prospect, Inc. Reprinted by permission.

most important reservoir of wealth for average American families and disproportionately so for African Americans. For black households, home equity accounts for 63 percent of total average net worth. In sharp contrast, home equity represents only 38.5 percent of average white net worth. According to the Economic Policy Institute's *State of Working America 2008/2009*, black homeownership rates dropped a full percentage point between 2005 and 2007—the largest decrease for any racial or ethnic group.

Even though homeownership rates for African Americans are lower, and even though a "segregation tax" is in play because homes appreciate far more slowly in minority or even integrated neighborhoods, housing wealth is still a more prominent engine of wealth for African American families. Given this centrality of homeownership as a source of wealth accumulation for black families, and the racialized dynamics of housing markets, sub-prime has been a special disaster for black upward mobility.

Families use home equity to finance retirement, start small businesses, pay for college educations, and tide themselves over during hard times. Information from the Federal Reserve Flow of Funds on home equity cashed out during refinancing (loans refinanced above 105 percent of the balance on the original loan) tells a dramatic story. Refinancing resulting in cash pullouts skyrocketed from $15 billion in 1995 to $327 billion in 2006. This is housing wealth converted, potentially, into other investments, used to launch social mobility, pay down credit-card and store debt, or finance consumption.

Especially in minority and immigrant communities, home equity often is the collateral for small business start-ups. But, it appears that about half of this enormous recent cash pullout was not used for retirement or to invest in mobility but to pay down past debt. While the conversion of housing wealth into cash and then the infusion of it into the economy provided a boost to the economy, in reality it disguised a less-rosy economic picture. Often it was the only way for families to keep pace with rising essential costs and burgeoning debt.

Between 2003 and 2007, the amount of housing wealth extracted more than doubled from the previous period, as families pulled out $1.19 trillion—an incredible sum that allowed families to adjust to shrinking purchasing power and that significantly boosted gross national product. So, while homeownership reached historic highs, families today actually own a lesser share of their homes than at any previous time, because they have borrowed against their housing wealth.

Families typically spend more as house values increase and they can borrow against their equity. Then, as prices fall and credit is tightened, they spend less. For a time, up until the sub-prime meltdown, equity withdrawals acted as an engine of growth on the economy. The opposite is true now—the sharp drop in housing prices has become a drag on the economy. Real home equity fell 6.5 percent to $9.6 trillion in 2007. The 2008 State of the Nations Housing study reports that the switch from housing appreciation to depreciation, plus the 2007 slowdown in home equity withdrawals, trimmed about one-half of a percentage point from real consumer spending and more than one-third of a percentage point from total economic growth. Worse is still to come.

RACIAL TARGETING—AGAIN

Changes in the mortgage market, of which the current sub-prime meltdown is the most visible part of a larger pattern, were not racially neutral. Sub-prime loans were targeted at the African American community. With the recognition that average American families were accumulating trillions of dollars in housing wealth, "financial innovation" soon followed. New financial instruments, which relaxed (and sometimes ignored) rules and regulations, became the market's answer to broadening homeownership.

But the industry-promoted picture of sub-prime as an instrument of home-ownership opportunity for moderate income buyers is highly misleading. First, homeownership rates reached their historic highs before the zenith of sub-prime lending; and, second, increased access to credit brought homeownership opportunities within the reach of groups that had historically been denied access to credit. The issue became the terms of credit.

In hindsight, many critics now describe the sub-prime crisis as the conse-quence of bad loans to unqualified borrowers. In fact, the issue needs to be retrained to focus on the onerous terms of these loans. Data from the longest natural experiment in the field—the Community Advantage Program, a part-nership of Self-Help, Fannie Mae, and the Ford Foundation, where tens of thousand of loans were made beginning over a decade ago—show that home loans to apparently riskier populations, like lower-income, minority, and single-headed households, do not default at significantly higher rates than conven-tional loans to middle-class families do, as long as they are not the handiwork of predators.

The difference is that loans such as ones made through the Community Advantage Program had terms that were closer to conventional mortgages as opposed to the risky terms that have characterized sub-prime mortgages. The latter had high hidden costs, exploding adjustable rates, and prepayment penalties to preclude refinancing. When lower-income families have similar terms of credit as conventional buyers, and they are linked with a community-based social and organizational infrastructure that helps them become ready for home-ownership, they pay similar interest rates and default at similar rates.

PREDATORY LENDING AND WEALTH STRIPPING

Minority communities received a disproportionate share of sub-prime mortgages. As a result, they are suffering a disproportionate burden of the harm and losses. According to a Demos report, *Beyond the Mortgage Meltdown* (June 2008), in addition to being the target of mortgage companies specializing in sub-prime lending, minorities were steered away from safer, conventional loans by brokers who received incentives for jacking up the interest rate. Worst of all, African Americans who qualified for conventional mortgages were steered to riskier, and more profitable, sub-prime loans.

Households of color were more than three times as likely as white households to end up with riskier loans with features like exploding adjustable rates, deceptive teaser rates, and balloon payments. Good credit scores often made no difference, as profit incentives trumped sound policy. The line from redlining to sub-prime is direct, as is the culpability. Even many upper-income African Americans were steered into sub-prime mortgages.

The Center for Responsible Lending (and other groups) projects that 2.2 million borrowers who bought homes between 1998 and 2006 will lose their houses and up to $164 billion of wealth in the process. African American and Latino homeowners are twice as likely to suffer sub-prime-related home foreclosures as white homeowners are. Foreclosures are projected to affect one in 10 African American borrowers. In contrast, only about one in 25 white mortgage holders will be affected. African Americans and Latinos are not only more likely to have been caught in the sub-prime loan trap; they are also far more dependent, as a rule, on their homes as financial resources.

The Demos report finds that home equity, at its current total value of $20 trillion, represents the biggest source of wealth for most Americans, and, as we have noted, it is even more important for African Americans. The comparatively little bit of wealth accumulation in the African American community is concentrated largely in housing wealth.

One recent estimate places the total loss of wealth among African American households at between $72 billion and $93 billion for sub-prime loans taken out during the past eight years.

Forty years after the Fair Housing Act of 1968, housing markets are still segmented by class and race, what realtors politely call location, location, location. Homes appreciate most in value when they are situated in predominantly white communities, and they appreciate least in value when situated in low-income minority or integrated communities, except when those communities undergo gentrification (and often become predominantly white).

This perverse market logic is also reflected in the sub-prime crisis. Sub-prime loans and foreclosures are not randomly distributed but spatially concentrated in low-to-moderate income communities, especially minority communities. Thus, the wealth-stripping phenomenon, of which sub-prime lending schemes are the latest financial innovation to tap new sources of wealth, is even more devastating in African American and minority communities. In turn, foreclosures and the terms of credit in African American neighborhoods bring down home values in the entire community. The community impact adds an institutional level to the personal tragedies and downstream consequences.

This devastating impact is not confined to just those who have suffered foreclosures; there is a spillover effect in addition to the direct hit of 1.27 million foreclosures. An additional 40.6 million neighboring homes will experience devaluation because of sub-prime foreclosures that take place in their community.

The Center for Responsible Lending estimates that the total decline in house values and the tax base from nearby foreclosures will be $202 billion. The direct hit on housing wealth for homeowners living near foreclosed properties will cause property values to decrease by $5,000 on average.

It is not possible to analyze specifically the full spillover impact of sub-prime foreclosures on African Americans, largely because these data are not available yet. However, communities of color will be especially harmed, since these communities receive a disproportionate share of sub-prime home loans. We estimate that this lost home value translates into a decrease in the tax base, consumer expenditures, investment opportunities, and money circulating in communities of color. United for a Fair Economy estimates that borrowers or [*sic*] color have collectively lost between $164 billion and $213 billion in housing wealth as a result of sub-prime loans taken during the past eight years.

Whatever the exact figures, the bottom line is clear—after centuries of being denied any opportunity to accumulate wealth, after a few decades of having limited opportunities, and after a generation during which African American families accumulated significant wealth, the African American community now faces the greatest loss of financial wealth in its history. Institutional processes and racialized policy are trumping hard-earned educational, job, and income advances.

17

Lifting as We Climb: Women of Color, Wealth, and America's Future

Center for Community Economic Development

INTRODUCTION

The poverty gap, based on income, is among the most widely used barometers to measure the relative economic status of women of color. Twenty-nine percent of white women heads of households with children live in poverty, compared to 43 percent of African-American women and 46 percent of Latina women.[1] However, hidden beneath the poverty gap is an even more startling and consequential gap that often goes unnoticed—the wealth gap.... The wealth gap between women of color and the rest of the population [is] a gap that significantly limits the economic prospects of future generations and holds back the progress of the American economy.

The current economic crisis has revealed why wealth is so important to the stability of households. Wealth, or net worth, refers to the total value of one's assets minus debts. Without savings or wealth of some form, economic stability is built on a house of cards that quickly crumbles when income is cut or disrupted through job loss, reduced hours or pay, or if the family suffers an unexpected health emergency.

Using data from the 2007 Survey of Consumer Finances, the amounts and types of wealth owned by women of different races are analyzed. Explanation is given as to why women of color have less wealth than white women and men of their same race, locating the roots of the gap in past and present institutional factors. These include but are not limited to the ways in which government benefits, the tax code, and fringe benefits exclude many women of color from wealth-building opportunities that are provided to other segments of the American population.

Because gender and racial economic disparities have been studied separately, we have failed to recognize the daunting economic reality faced by women of color who experience the compounding negative economic effects of being both a woman and a person of color. Investing in programs and public policies that create asset building opportunities that fit the needs and build on the strengths of Native American, African-American, Latino, and Asian women, both native-born and immigrant, can bring economic security to a greater number

SOURCE: From Center for Community Economic Development. 2010. "Lifting As We Climb: Women of Color, Wealth, and America's Future." www.insightcced.org. Reprinted by permission.

of families, expand our economy, and increase our nation's capacity to compete in the global marketplace.

THE WEALTH GAP FOR WOMEN OF COLOR

Women of all races experience a gender wealth gap that is greater than the gender income gap (the median income for a women is three fourths of the median income for men, but the median wealth owned by women is less than half as much as men), but the disparities are greatest for women of color.[2] In fact, single black and Hispanic women have one penny of wealth for every dollar of wealth owned by their male counterparts and a fraction of a penny for every dollar of wealth owned by single white women.[3]

- Single black and Hispanic women have a median wealth of $100 and $120 respectively, which is less than 1 percent of the wealth of their same-race male counterparts. It is only a fraction of one-percent of the wealth of single white women.

- Nearly half of all single black and Hispanic women have zero, or negative wealth (negative wealth occurs when the value of debts is greater than the value of assets).

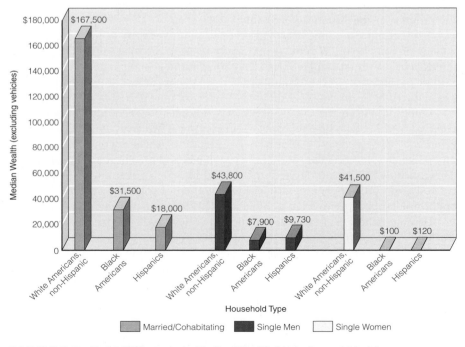

FIGURE 1 Racial Differences in Median Wealth (excluding vehicles) by Household Type, Ages 18–64, 2007.

- Never-married women of color have a median wealth of zero. In comparison, never-married white women have a median wealth of $2,600, never-married men of color $4,020, and never-married white men $16,310.

- Divorced women of color have a median wealth of $4,200, which is 26% of the wealth of divorced men of color ($16,100), 8% of the wealth of divorced white women ($52,120), and 5% of the wealth of divorced white men ($80,000).

- Black and Hispanic mothers with children under age 18 have a median wealth of zero. Black and Hispanic fathers have a median wealth of $10,960 and $2,400, respectively. White mothers have a median wealth of $7,970 and white fathers have $56,100.

- Prior to age 50, women of color have virtually no wealth at all.

WOMEN OF COLOR LESS LIKELY TO OWN ALL TYPES OF ASSETS

Asset building begins with cash savings, which form the base from which other assets can be built. Home ownership, retirement savings, and other types of assets such as businesses are important for helping families remain economically stable and for providing income during retirement. Women of color are less likely to have any of these types of assets, and when they do, the value of the assets is lower.

- About one-third of single Hispanic women and one-fourth of single black women are "unbanked," meaning they have no checking or savings account.

- While 57% of single white women own homes, only 33% of single black women and 28% of single Hispanic women are homeowners.

- Only 23% of single black women and 14% of single Hispanic women own stock. In contrast, 45% of single white women own stock and 51% of single white men own stock.

- Despite growth in minority-owned and women-owned businesses, women of color remain underrepresented among business owners. Only 1% of single Hispanic women and 4% of single black women own businesses compared to 8% of single white women.

WOMEN OF COLOR HAVE GREATER DEBT

Indebtedness is rising for all American households, but women of color—and especially black women—are more likely to have debt in the form of education loans, credit card debt, and installment debt.

- Twenty-five percent of women of color have education debt. Black women are most likely to have education debt, demonstrating their desire to prepare for better futures. While college loans are usually considered "good" debt, having too little income in relationship to debt payments can endanger financial security.

■   Almost half (48%) of all women of color have credit card debt. The cost of living is rising faster than wages, so many turn to credit cards simply to pay for the basic necessities of daily living.[4]

DISPARITIES IN RETIREMENT

Retirement experts speak of the "three-legged stool" of income from pensions, Social Security and personal savings that collectively support people during retirement. However, the data show that all three legs of women's retirement "stools" are shaky. In fact, women of color ages 65 and older are least likely to receive retirement income from pensions or from assets.

■   While 49% of white men and 30.5% of white women receive income from pensions, only 26% of black women, 17% of Asian women and 12.7% of Hispanic women receive any income from pensions.[5]

■   Whereas 66% of white men and 60.4% of white women receive income from assets, only 40% of Asian women, 25.4% of black women, and 23% of Hispanic women receive any income from assets.

WEALTH OF ASIAN-AMERICAN AND NATIVE
AMERICAN WOMEN

Because Asian Americans and Native Americans comprise a much smaller proportion of the U.S. population than blacks and Hispanics and because most surveys that measure wealth do not oversample these groups, our knowledge about their wealth is less robust—particularly for Native Americans. Often both groups are combined into a single "other" category.

According to data from the 2004 Survey of Income and Program Participation, Asian Americans have a higher median net worth than white non-Hispanic households. This is largely explained by the fact that much of the wealth of Asian Americans is in the form of home equity. When interpreting the wealth status of Asian Americans, it is also important to take into consideration that the Asian population is concentrated within a few major cities which have high costs of living and higher home values than the rest of the nation. When data is adjusted for these and other factors, Asian wealth levels are lower than for whites.[6] There is a great deal of variation in levels of wealth within the Asian-American group, depending on country of origin, refugee versus immigrant status, documented or undocumented status, educational levels, and length of time in the U.S. Little data is available on gender disparities.

Little is known about the wealth of Native Americans because their presence in national surveys is extremely small. One rare study used the National Longitudinal Survey of Youth (NLSY79), which contains some wealth questions and information on Native American ancestry.[7]

Based on this data, in 2000 the median wealth for Native Americans in the survey was $5,700, whereas the median wealth for the sample overall was $65,500, a ratio of only 8.7%. Data were not provided for men and women separately.

WHY WOMEN OF COLOR HAVE LESS WEALTH

The earnings of women of color are not converted to wealth as quickly because they are not linked with the "wealth escalator"—fringe benefits, favorable tax codes, and valuable government benefits.

- Women of color are more likely to work in service occupations—28% of black and 31% of Latina women compared to 19% of white women and only 12% of white men.[8] These jobs are the least likely to provide wealth-enhancing benefits such as retirement plans, paid sick days, and health insurance.[9]

- Women of color benefit less from tax advantages such as the home mortgage interest deduction because they are less likely to own homes. Due to residential segregation, their homes typically have less value and appreciate less quickly.

- Women of color depend more on Social Security because they lack other sources of retirement income. In fact, Social Security is the only source of retirement income for more than 25% of black women. But women of color receive lower Social Security benefits because of their lower earnings and because they are less likely to receive benefits as wives of high-income beneficiaries.[10]

- Women of color are less likely to meet eligibility requirements for unemployment insurance since part-time workers (primarily women) are often ineligible for benefits.[11]

- Women of color have been hard hit by predatory lending practices. Of low- and moderate-income borrowers, Hispanic women were almost one and a half times more likely and black women more than twice as likely to receive high-cost home loans as white women.[12]

- Many women of color who received subprime home loans could have qualified for conventional lower-cost mortgages.[13] Subprime home loans cost a borrower between $50,000 and $100,000 more than a comparable prime loan over the life of the loan.[14]

While structural inequities are the primary cause of the wealth gap for women of color, cultural factors also play a role.

- Cultural expectations for some women of color emphasize the importance of giving to family and community, often in the form of sharing economic resources.[15]

- Many groups of color have a greater distrust of financial institutions such as banks. Some Hispanic groups, for instance, are less likely to trust banks because they have experienced widespread volatility and uncertainty in banks in their home countries.[16]

HELPING WOMEN OF COLOR BUILD WEALTH SPEEDS ECONOMIC GROWTH

Providing women of color with equal opportunities to build wealth is an imperative for our nation's economic and political future. As the racial demographics of the United States continue to shift and our nation becomes majority minority, letting a large group stagnate financially is not only irresponsible, but detrimental to the nation's economic prosperity over the long run.

Even in the near term, as we work to rise above the economic crisis at hand, we need to maintain ground in this increasingly competitive global marketplace. We need the talents and resources of women of color as workers, consumers, and entrepreneurs to help fuel economic growth and recovery.

CONCLUSION

The dream of economic mobility and security for themselves and their children remains a chimera on the far horizon for the vast majority of women of color. Their ability to build wealth has been negatively impacted by the cumulative effect of historical policies restricting people of color and women from asset building opportunities, and by current policies and practices that continue to exacerbate those gaps. At the intersection of institutional barriers based on race and gender, the impact is not simply additive, it is exponential.

In the face of such overwhelming odds, women of color are some of the most resilient, resourceful and relied-upon people in our society. They raise children and take care of elders. They earn incomes, start businesses, create jobs for others, and donate their time and money to improving their communities. Last but not least, they are leaders working for a more equitable society.

But their stories are often buried in the aggregated data of the population as a whole, of people of color, or of all women. Moreover, the significance of wealth has been overlooked. As a result, their particular situation has not been well documented or understood, including their potential for current and future contribution to economic recovery and expansion.

Shining a spotlight on the overlooked issue of women of color and wealth can lead to new strategies to bring all Americans into the economic mainstream. Targeted policies can create opportunities that will lift women of color as they continue their climb toward economic security. Their futures are inextricably linked with the economic future of the nation.

NOTES

1. U.S. Census Bureau, Historical Poverty Tables, 2008.

2. Unless otherwise noted, statistics on the wealth gap for women of color refer to unmarried (i.e., never-married, divorced, widowed) and non-cohabitating people ages 18–64.

3. The statistics used for this paper are the author's calculations from various sources. Most of the data on assets is from the Federal Reserve Board's 2007 Survey of Consumer Finances released in 2009. For further information see: Chang, Mariko Lin. In press. *Shortchanged: Why women have less wealth and what can be done about it.* NY: Oxford University Press; and Mariko Chang. March 2010. Lifting as We Climb: Women of Color, Wealth, and America's Future. Insight Center for Community Economic Development

4. Garcia, Jose A. 2007. "Borrowing to Make Ends Meet: The Rapid Growth of Credit Card Debt in America." Demos.

5. Institute for Women's Policy Research. November 2007. Briefing Paper (#D480). "The Economic Security of Older Women and Men in the United States." http://www.iwpr.org/store/Details.cfm?ProdID=196&category

6. Patraporn, R. Varisa, Paul M. Ong, and Douglas Houston. 2009. "Closing the Asian-White Wealth Gap?" UCLA Asian American Studies Center.

7. Native American ancestry was determined based on a combination of factors, including self-identification. Classification was not based on verified tribal enrollees. Zagorsky, Jay L. 2006. "Native Americans' Wealth." Chapter 5 in Nembhard and Chiteji (eds.) *Wealth Accumulation and Communities of Color in the United States: Current Issues.* MI: The University of Michigan Press.

8. Bureau of Labor Statistics Current Population Survey. www.bls.gov/ops/tables .htm#annual

9. Bureau of Labor Statistics. www.bls.gov/nos/ebs/benefits/2008/benefits. htm#health; U.S. Department of Labor. "Employee Benefits in the United States, March 2008;" U.S. Department of Labor. "National Compensation Survey: Employee Benefits in Private Industry in the United States, March 2007."

10. U.S. Social Security Administration. Income of the Population 55 or Older, 2006, Table 8.B5. http://www.socialsecurity.gov/policy/docs/statcomps/income_pop55/ 2006/sect08.pdf (accessed February 12, 2010).

11. Simms, Margaret C. July 2008. "Weathering Job Loss: Unemployment Insurance." New Safety Net Paper 6. The Urban Institute. www.urban.org/UploadedPDF/ 411730_job_loss.pdf (accessed February 11, 2010).

12. National Council of Negro Women (in partnership with the National Community Reinvestment Coalition). "Income is No Shield, Part III-Assessing the Double Burden: Examining Racial and Gender Disparities in Mortgage Lending." June 2009. www.ncnw.org/images/double_burden.pdf (accessed February 5, 2010).

13. Rivera, Amaad, Brenda Cotto-Escalera, Anisha Desai, Jeanette Huezo, and Dedrick Muhammad. 2008. "Foreclosed: State of the Dream 2008." United for a Fair Economy.

14. National Council of Negro Women (in partnership with the National Community Reinvestment Coalition). "Income is No Shield, Part III-Assessing the Double

Burden: Examining Racial and Gender Disparities in Mortgage Lending." June 2009. www.ncnw.org/images/double_burden.pdf (accessed February 5, 2010).

15. Chiteji, Ngina and Darrick Hamilton. 2005. *"Family Matters: Kin Networks and Asset Accumulation" in Inclusion in the American Dream: Assets, Poverty, and Public Policy*, ed by Michael Sherraden, 87-111. New York: Oxford University Press; Evans, Connie E. 2006. "The Intersection of Gender, Race and Culture as Influences on African American Women's Financial Fitness, Asset Accumulation, and Wealth Attainment." Center for the Education of Women, University of Michigan; Single-tary, Michelle. "Study on Savings By Blacks and Whites Reveals Shades of Gray." *The Washington Post*, October 21, 2007. Mission Asset Fund. 2009. Survey Brief #2, Immigrant Financial Integration Initiative: Understanding the Financial Attitudes and Behaviors of Latino(a) Immigrants. www.missionassetfund.org/files/MAF Survey Brief Number 2 - Latina Immigrants_0.pdf (accessed February 6, 2010).

16. Lui, Meizhu Barbara Robles, Betsy Leondar-Wright, Rose Brewer, and Rebecca Adamson. 2006. *The Color of Wealth: The Story Behind the U.S. Racial Wealth Divide.* NY: The Free Press.

18

Sex and Gender Through the Prism of Difference

MAXINE BACA ZINN, PIERRETTE HONDAGNEU-SOTELO, AND MICHAEL MESSNER

"Men can't cry." "Women are victims of patriarchal oppression." "After divorces, single mothers are downwardly mobile, often moving into poverty." "Men don't do their share of housework and child care." "Professional women face barriers such as sexual harassment and a 'glass ceiling' that prevent them from competing equally with men for high-status positions and high salaries." "Heterosexual intercourse is an expression of men's power over women." Sometimes, the students in our sociology and gender studies courses balk at these kinds of generalizations. And they are right to do so. After all, some men are more emotionally expressive than some women, some women have more power and success than some men, some men do their share—or more—of housework and child care, and some women experience sex with men as both pleasurable and empowering. Indeed, contemporary gender relations are complex and changing in various directions, and as such, we need to be wary of simplistic, if handy, slogans that seem to sum up the essence of relations between women and men.

On the other hand, we think it is a tremendous mistake to conclude that "all individuals are totally unique and different," and that therefore all generalizations about social groups are impossible or inherently oppressive. In fact, we are convinced that it is this very complexity, this multifaceted nature of contemporary gender relations, that fairly begs for a sociological analysis of gender. We use the image of "the prism of difference" to illustrate our approach to developing this sociological perspective on contemporary gender relations. The *American Heritage Dictionary* defines "prism," in part, as "a homogeneous transparent solid, usually with triangular bases and rectangular sides, used to produce or analyze a continuous spectrum." Imagine a ray of light—which to the naked eye appears to be only one color—refracted through a prism onto a white wall. To the eye, the result is not an infinite, disorganized scatter of individual colors. Rather, the refracted light displays an order, a structure of relationships among the different colors—a rainbow. Similarly, we propose to use the "prism of difference"... to analyze a continuous spectrum of people, in order to show how gender is organized and experienced differently when refracted through the prism of

SOURCE: From Maxine Baca Zinn, Pierrette Hondagneu-Sotelo, and Michael Messner, eds. *Gender through the Prism of Difference* 2nd ed., pp. 1–6. Copyright © 2005. Reprinted by permission of Oxford University Press, Inc.

sexual, racial/ethnic, social class, physical abilities, age, and national citizenship differences.

EARLY WOMEN'S STUDIES: CATEGORICAL VIEWS OF "WOMEN" AND "MEN"

… It is possible to make good generalizations about women and men. But these generalizations should be drawn carefully, by always asking the questions "*which* women?" and "*which* men?" Scholars of sex and gender have not always done this. In the 1960s and 1970s, women's studies focused on the differences *between* women and men rather than *among* women and men. The very concept of gender, women's studies scholars demonstrated, is based on socially defined difference between women and men. From the macro level of social institutions such as the economy, politics, and religion, to the micro level of interpersonal relations, distinctions between women and men structure social relations. Making men and women *different* from one another is the essence of gender. It is also the basis of men's power and domination. Understanding this was profoundly illuminating. Knowing that difference produced domination enabled women to name, analyze, and set about changing their victimization.

In the 1970s, riding the wave of a resurgent feminist movement, colleges and universities began to develop women's studies courses that aimed first and foremost to make women's lives visible. The texts that were developed for these courses tended to stress the things that women shared under patriarchy—having the responsibility for housework and child care, the experience or fear of men's sexual violence, a lack of formal or informal access to education, and exclusion from high-status professional and managerial jobs, political office, and religious leadership positions (Brownmiller, 1975; Kanter, 1977).

The study of women in society offered new ways of seeing the world. But the 1970s approach was limited in several ways. Thinking of gender primarily in terms of differences *between* women and men led scholars to overgeneralize about both. The concept of patriarchy led to a dualistic perspective of male privilege and female subordination. Women and men were cast as opposites. Each was treated as a homogeneous category with common characteristics and experiences. This approach *essentialized* women and men. Essentialism, simply put, is the notion that women's and men's attributes and indeed women and men themselves are categorically different. From this perspective, male control and coercion of women produced conflict between the sexes. The feminist insight originally introduced by Simone De Beauvoir in 1953—that women, as a group, had been socially defined as the "other" and that men had constructed themselves as the subjects of history, while constructing women as their objects—fueled an energizing sense of togetherness among many women. As college students read books such as *Sisterhood Is Powerful* (Morgan, 1970), many of them joined organizations that fought—with some success—for equality and justice for women.

THE VOICES OF "OTHER" WOMEN

Although this view of women as an oppressed "other" was empowering for certain groups of women, some women began to claim that the feminist view of universal sisterhood ignored and marginalized their major concerns. It soon became apparent that treating women as a group united in its victimization by patriarchy was biased by too narrow a focus on the experiences and perspectives of women from more privileged social groups. "Gender" was treated as a generic category, uncritically applied to women. Ironically, this analysis, which was meant to unify women, instead produced divisions between and among them. The concerns projected as "universal" were removed from the realities of many women's lives. For example, it became a matter of faith in second-wave feminism that women's liberation would be accomplished by breaking down the "gendered public-domestic split." Indeed, the feminist call for women to move out of the kitchen and into the workplace resonated in the experiences of many of the college-educated white women who were inspired by Betty Friedan's 1963 book, *The Feminine Mystique*. But the idea that women's movement into workplaces was itself empowering or liberating seemed absurd or irrelevant to many working-class women and women of color. They were already working for wages, as had many of their mothers and grandmothers, and did not consider access to jobs and public life "liberating." For many of these women, liberation had more to do with organizing in communities and workplaces—often alongside men—for better schools, better pay, decent benefits, and other policies to benefit their neighborhoods, jobs, and families. The feminism of the 1970s did not seem to address these issues.

As more and more women analyzed their own experiences, they began to address the power relations that created differences among women and the part that privileged women played in the oppression of others. For many women of color, working-class women, lesbians, and women in contexts outside the United States (especially women in non-Western societies), the focus on male domination was a distraction from other oppressions. Their lived experiences could support neither a unitary theory of gender nor an ideology of universal sisterhood. As a result, finding common ground in a universal female victimization was never a priority for many groups of women.

Challenges to gender stereotypes soon emerged. Women of varied races, classes, national origins, and sexualities insisted that the concept of gender be broadened to take their differences into account (Baca Zinn et al., 1986; Hartmann, 1976; Rich, 1980; Smith, 1977). Many women began to argue that their lives were affected by their location in a number of different hierarchies: as African Americans, Latinas, Native Americans, or Asian Americans in the race hierarchy; as young or old in the age hierarchy; as heterosexual, lesbian, or bisexual in the sexual orientation hierarchy; and as women outside the Western industrialized nations, in subordinated geopolitical contexts. These arguments made it clear that women were not victimized by gender alone but by the historical and systematic denial of rights and privileges based on other differences as well.

MEN AS GENDERED BEINGS

As the voices of "other" women in the mid- to late 1970s began to challenge and expand the parameters of women's studies, a new area of scholarly inquiry was beginning to stir—a critical examination of men and masculinity. To be sure, in those early years of gender studies, the major task was to conduct studies and develop courses about the lives of women in order to begin to correct centuries of scholarship that rendered invisible women's lives, problems, and accomplishments. But the core idea of feminism—that "femininity" and women's subordination is a social construction—logically led to an examination of the social construction of "masculinity" and men's power. Many of the first scholars to take on this task were psychologists who were concerned with looking at the social construction of "the male sex role" (e.g., Pleck, 1976). By the late 1980s, there was a growing interdisciplinary collection of studies of men and masculinity, much of it by social scientists (Brod, 1987; Kaufman, 1987; Kimmel, 1987; Kimmel & Messner, 1989).

Reflecting developments in women's studies, the scholarship on men's lives tended to develop three themes: First, what we think of as "masculinity" is not a fixed, biological essence of men, but rather is a social construction that shifts and changes over time as well as between and among various national and cultural contexts. Second, power is central to understanding gender as a relational construct, and the dominant definition of masculinity is largely about expressing difference from—and superiority over—anything considered "feminine." And third, there is no singular "male sex role." Rather, at any given time there are various masculinities. R. W. Connell (1987; 1995; 2002) has been among the most articulate advocates of this perspective. Connell argues that hegemonic masculinity (the dominant form of masculinity at any given moment) is constructed in relation to femininities *as well as* in relation to various subordinated or marginalized masculinities. For example, in the United States, various racialized masculinities (e.g., as represented by African American men, Latino immigrant men, etc.) have been central to the construction of hegemonic (white middle-class) masculinity. This "othering" of racialized masculinities helps to shore up the privileges that have been historically connected to hegemonic masculinity. When viewed this way, we can better understand hegemonic masculinity as part of a system that includes gender as well as racial, class, sexual, and other relations of power.

The new literature on men and masculinities also begins to move us beyond the simplistic, falsely categorical, and pessimistic view of men simply as a privileged sex class. When race, social class, sexual orientation, physical abilities, immigrant, or national status are taken into account, we can see that in some circumstances, "male privilege" is partly—sometimes substantially—muted (Kimmel & Messner, 2004). Although it is unlikely that we will soon see a "men's movement" that aims to undermine the power and privileges that are connected with hegemonic masculinity, when we begin to look at "masculinities" through the prism of difference, we can begin to see similarities and possible points of coalition between and among certain groups of women and men (Messner,

1998). Certain kinds of changes in gender relations—for instance, a national family leave policy for working parents—might serve as a means of uniting particular groups of women and men.

GENDER IN INTERNATIONAL CONTEXTS

It is an increasingly accepted truism that late twentieth-century increases in transnational trade, international migration, and global systems of production and communication have diminished both the power of nation-states and the significance of national borders. A much more ignored issue is the extent to which gender relations—in the United States and elsewhere in the world—are increasingly linked to patterns of global economic restructuring. Decisions made in corporate headquarters located in Los Angeles, Tokyo, or London may have immediate repercussions on how women and men thousands of miles away organize their work, community, and family lives (Sassen, 1991). It is no longer possible to study gender relations without giving attention to global processes and inequalities....

Around the world, women's paid and unpaid labor is key to global development strategies. Yet it would be a mistake to conclude that gender is molded from the "top down." What happens on a daily basis in families and workplaces simultaneously constitutes and is constrained by structural transnational institutions. For instance, in the second half of the twentieth century young, single women, many of them from poor rural areas, were (and continue to be) recruited for work in export assembly plants along the U.S.-Mexico border, in East and Southeast Asia, in Silicon Valley, in the Caribbean, and in Central America. While the profitability of these multinational factories depends, in part, on management's ability to manipulate the young women's ideologies of gender, the women ... do not respond passively or uniformly, but actively resist, challenge, and accommodate. At the same time, the global dispersion of the assembly line has concentrated corporate facilities in many U.S. cities, making available myriad managerial, administrative, and clerical jobs for college educated women. Women's paid labor is used at various points along this international system of production. Not only employment but also consumption embodies global interdependencies. There is a high probability that the clothing you are wearing and the computer you use originated in multinational corporate headquarters and in assembly plants scattered around third world nations. And if these items were actually manufactured in the United States, they were probably assembled by Latin American and Asian-born women.

Worldwide, international labor migration and refugee movements are creating new types of multiracial societies. While these developments are often discussed and analyzed with respect to racial differences, gender typically remains absent. As several commentators have noted, the white feminist movement in the United States has not addressed issues of immigration and nationality. Gender, however, has been fundamental in shaping immigration policies (Chang,

1994; Hondagneu-Sotelo, 1994). Direct labor recruitment programs generally solicit either male or female labor (e.g., Filipina nurses and Mexican male farm workers), national disenfranchisement has particular repercussions for women and men, and current immigrant laws are based on very gendered notions of what constitutes "family unification." As Chandra Mohanty suggests, "analytically these issues are the contemporary metropolitan counterpart of women's struggles against colonial occupation in the geographical third world" (1991:23). Moreover, immigrant and refugee women's daily lives often challenge familiar feminist paradigms. The occupations in which immigrant and refugee women concentrate—paid domestic work, informal sector street vending, assembly or industrial piece work performed in the home—often blur the ideological distinction between work and family and between public and private spheres (Hondagneu-Sotelo, 2001; Parrenas, 2001).

FROM PATCHWORK QUILT TO PRISM

All of these developments—the voices of "other" women, the study of men and masculinities, and the examination of gender in transnational contexts—have helped redefine the study of gender. By working to develop knowledge that is inclusive of the experiences of all groups, new insights about gender have begun to emerge. Examining gender in the context of other differences makes it clear that nobody experiences themselves as solely gendered. Instead, gender is configured through cross-cutting forms of difference that carry deep social and economic consequences.

By the mid-1980s, thinking about gender had entered a new stage, which was more carefully grounded in the experiences of diverse groups of women and men. This perspective is a general way of looking at women and men and understanding their relationships to the structure of society. Gender is no longer viewed simply as a matter of two opposite categories of people, males and females, but a range of social relations among differently situated people. Because centering on difference is a radical challenge to the conventional gender framework, it raises several concerns. If we think of all the systems that converge to simultaneously influence the lives of women and men, we can imagine an infinite number of effects these interconnected systems have on different women and men. Does the recognition that gender can be understood only contextually (meaning that there is no singular "gender" per se) make women's studies and men's studies newly vulnerable to critics in the academy? Does the immersion in difference throw us into a whirlwind of "spiraling diversity" (Hewitt, 1992:316) whereby multiple identities and locations shatter the categories "women" and "men"? ...

We take a position directly opposed to an empty pluralism. Although the categories "woman" and "man" have multiple meanings, this does not reduce gender to a "postmodern kaleidoscope of lifestyles. Rather, it points to the *relational* character of gender" (Connell, 1992:736). Not only are masculinity and femininity relational, but different *masculinities* and *femininities* are interconnected through other social structures such as race, class, and nation. The concept of relationality suggests that the lives of different groups are interconnected even without face-to-face relations (Glenn, 2002:14). The meaning of "woman" is defined by the existence of women of different races and classes. Being a white woman in the United States is meaningful only insofar as it is set apart from and in contradistinction to women of color.

Just as masculinity and femininity each depend on the definition of the other to produce domination, differences *among* women and *among* men are also created in the context of structured relationships. Some women derive benefits from their race and class position and from their location in the global economy, while they are simultaneously restricted by gender. In other words, such women are subordinated by patriarchy, yet their relatively privileged positions within hierarchies of race, class, and the global political economy intersect to create for them an expanded range of opportunities, choices, and ways of living. They may even use their race and class advantage to minimize some of the consequences of patriarchy and/or to oppose other women. Similarly, one can become a man in opposition to other men. For example, "the relation between heterosexual and homosexual men is central, carrying heavy symbolic freight. To many people, homosexuality is the *negation* of masculinity…. Given that assumption, antagonism toward homosexual men may be used to define masculinity" (Connell, 1992:736).

In the past decade, viewing gender through the prism of difference has profoundly reoriented the field (Acker, 1999; Glenn, 1999, 2002; Messner, 1996; West & Fenstermaker, 1995). Yet analyzing the multiple constructions of gender does not just mean studying groups of women and groups of men as different. It is clearly time to go beyond what we call the "patchwork quilt" phase in the study of women and men—that is, the phase in which we have acknowledged the importance of examining differences within constructions of gender, but do so largely by collecting together a study here on African American women, a study there on gay men, a study on working-class Chicanas, and so on. This patchwork quilt approach too often amounts to no more than "adding difference and stirring." The result may be a lovely mosaic, but like a patchwork quilt, it still tends to overemphasize boundaries rather than to highlight bridges of inter-dependency. In addition, this approach too often does not explore the ways that social constructions of femininities and masculinities are based on and reproduce relations of power. In short, we think that the substantial quantity of research that has now been done on various groups and subgroups needs to be analyzed within a framework that emphasizes differences and inequalities not as discrete areas of separation, but as interrelated bands of color that together make up a spectrum….

REFERENCES

Acker, Joan. 1999. "Rewriting Class, Race and Gender: Problems in Feminist Rethinking" Pp. 44–69 in Myra Marx Ferree, Judith Lorber, and Beth B. Hess (eds.), *Revisioning Gender*. Thousand Oaks, CA: Sage Publications.

Baca Zinn, M., L., Weber Cannon, E., Higgenbotham, & B., Thornton Dill. 1986. "The Costs of Exclusionary Practices in Women's Studies," *Signs: Journal of Women in Culture and Society* 11: 290–303.

Brod, Harry (ed.). 1987. *The Making of Masculinities: The New Men's Studies*. Boston: Allen & Unwin.

Brownmiller, Susan. 1975. *Against Our Will: Men, Women, and Rape*. New York: Simon & Schuster.

Chang, Grace. 1994. "Undocumented Latinas: The New 'Employable Mothers.'" Pp. 259–285 in Evelyn Nakano Glenn, Grace Chang, and Linda Rennie Forcey (eds.), *Mothering, Ideology, Experience, and Agency*. New York and London: Routledge.

Connell, R. W. 1987. *Gender and Power*. Stanford, CA: Stanford University Press.

Connell, R. W. 1992. "A Very Straight Gay: Masculinity, Homosexual Experience, and the Dynamics of Gender," *American Sociological Review* 57: 735–751.

Connell, R. W. 1995. *Masculinities*. Berkeley: University of California Press.

Connell, R. W. 2002. *Gender*. Cambridge: Polity.

De Beauvoir, Simone. 1953. *The Second Sex*. New York: Knopf.

Glenn, Evelyn Nakano. 1999. "The Social Construction and Institutionalization of Gender and Race: An Integrative Framework," Pp. 3–43 in Myra Marx Ferree, Judith Lorber, and Beth B. Hess (eds.), *Revisioning Gender*. Thousand Oaks. CA: Sage Publications.

Glenn, Evelyn Nakano. 2002. *Unequal Sisterhood: How Race and Gender Shaped American Citizenship and Labor*. Cambridge, MA: Harvard University Press.

Hartmann, Heidi. 1976. "Capitalism, Patriarchy, and Job Segregation by Sex," *Signs: Journal of Women in Culture and Society* 1(3), part 2, spring: 137–167.

Hewitt, Nancy A. 1992. "Compounding Differences," *Feminist Studies* 18: 313–326.

Hondagneu-Sotelo, Pierrette. 1994. *Gendered Transitions: Mexican Experiences of Immigration*. Berkeley: University of California Press.

Hondagneu-Sotelo, Pierrette. 2001. *Doméstica: Immigrant Workers Cleaning and Caring in the Shadows of Affluence*. Berkeley: University of California Press.

Kanter, Rosabeth Moss. 1977. *Men and Women of the Corporation*. New York: Basic Books.

Kaufman, Michael. 1987. *Beyond Patriarchy: Essays by Men on Pleasure, Power, and Change*. Toronto and New York: Oxford University Press.

Kimmel, Michael S. (ed.). 1987. *Changing Men: New Directions in Research on Men and Masculinity*. Newbury Park, CA: Sage.

Kimmel, Michael S. 1996. *Manhood in America: A Cultural History*. New York: Free Press.

Kimmel, Michael S. & Michael A. Messner (eds.). 1989. *Men's Lives*. New York: Macmillan.

Kimmel, Michael S. & Michael A. Messner (eds.). 2004. *Men's Lives*, 6th ed. Boston: Pearson.

Messner, Michael A. 1996. "Studying Up on Sex," *Sociology of Sport Journal* 13: 221–237.

Messner, Michael A. 1998. *Politics of Masculinities: Men in Movements*. Thousand Oaks, CA: Sage Publications.

Mohanty, Chandra Talpade. 1991. "Cartographies of Struggle: Third World Women and the Politics of Feminism." Pp. 51–80 in Chandra Talpade Mohanty, Ann Russo, and Lourdes Torres, (eds.), *Third World Women and the Politics of Feminism.* Bloomington: Indiana University Press.

Morgan, Robin. 1970. *Sisterhood Is Powerful: An Anthology of Writing from the Women's Liberation Movement.* New York: Vintage Books.

Parrenas, Rhacel Salazar. 2001. *Servants of Globalization: Women, Migration and Domestic Work.* Stanford: Stanford University Press.

Pleck, J. H. 1976. "The Male Sex Role: Definitions, Problems, and Sources of Change," *Journal of Social Issues* 32: 155–164.

Rich, Adrienne. 1980. "Compulsory Heterosexuality and the Lesbian Experience," *Signs: Journal of Women in Culture and Society* 5: 631–660.

Sassen, Saskia. 1991. *The Global City: New York, London, Tokyo.* Princeton: Princeton University Press.

Smith, Barbara. 1977. *Toward a Black Feminist Criticism.* Freedom, CA: Crossing Press.

West, Candace & Sarah Fenstermaker. 1995. "Doing Difference," *Gender & Society* 9: 8–37.

19

The Myth of the Latin Woman
I Just Met a Girl Named María

JUDITH ORTIZ COFER

On a bus trip to London from Oxford University, where I was earning some graduate credits one summer, a young man, obviously fresh from a pub, spotted me and as if struck by inspiration went down on his knees in the aisle. With both hands over his heart he broke into an Irish tenor's rendition of "María" from *West Side Story*. My politely amused fellow passengers gave his lovely voice the round of gentle applause it deserved. Though I was not quite as amused, I managed my version of an English smile: no show of teeth, no extreme contortions of the facial muscles—I was at this time of my life practicing reserve and cool. Oh, that British control, how I coveted it. But María had followed me to London, reminding me of a prime fact of my life: you can leave the Island, master the English language, and travel as far as you can, but if you are a Latina, especially one like me who so obviously belongs to Rita Moreno's gene pool, the Island travels with you.

This is sometimes a very good thing—it may win you that extra minute of someone's attention. But with some people, the same things can make *you* an island—not so much a tropical paradise as an Alcatraz, a place nobody wants to visit. As a Puerto Rican girl growing up in the United States and wanting like most children to "belong," I resented the stereotype that my Hispanic appearance called forth from many people I met.

Our family lived in a large urban center in New Jersey during the sixties, where life was designed as a microcosm of my parents' casas on the island. We spoke in Spanish, we ate Puerto Rican food bought at the bodega, and we practiced strict Catholicism complete with Saturday confession and Sunday mass at a church where our parents were accommodated into a one-hour Spanish mass slot, performed by a Chinese priest trained as a missionary for Latin America.

As a girl I was kept under strict surveillance, since virtue and modesty were, by cultural equation, the same as family honor. As a teenager I was instructed on how to behave as a proper señorita. But it was a conflicting message girls got, since the Puerto Rican mothers also encouraged their daughters to look and act like women and to dress in clothes our Anglo friends and their mothers found too "mature" for our age. It was, and is, cultural, yet I often felt humiliated when

SOURCE: From *The Latin Deli: Prose & Poetry* by Judith Ortiz Cofer, pp. 148–154. The University of Georgia Press. Copyright © 1993 by Judith Ortiz Cofer. Reprinted by permission of the publisher.

I appeared at an American friend's party wearing a dress more suitable to a semi-formal than to a playroom birthday celebration. At Puerto Rican festivities, neither the music nor the colors we wore could be too loud. I still experience a vague sense of letdown when I'm invited to a "party" and it turns out to be a marathon conversation in hushed tones rather than a fiesta with salsa, laughter, and dancing—the kind of celebration I remember from my childhood.

I remember Career Day in our high school, when teachers told us to come dressed as if for a job interview. It quickly became obvious that to the barrio girls, "dressing up" sometimes meant wearing ornate jewelry and clothing that would be more appropriate (by mainstream standards) for the company Christmas party than as daily office attire. That morning I had agonized in front of my closet, trying to figure out what a "career girl" would wear because, essentially, except for Marlo Thomas on TV, I had no models on which to base my decision. I knew how to dress for school: at the Catholic school I attended we all wore uniforms; I knew how to dress for Sunday mass, and I knew what dresses to wear for parties at my relatives' homes. Though I do not recall the precise details of my Career Day outfit, it must have been a composite of the above choices. But I remember a comment my friend (an Italian-American) made in later years that coalesced my impressions of that day. She said that at the business school she was attending the Puerto Rican girls always stood out for wearing "everything at once." She meant, of course, too much jewelry, too many accessories. On that day at school, we were simply made the negative models by the nuns who were themselves not credible fashion experts to any of us. But it was painfully obvious to me that to the others, in their tailored skirts and silk blouses, we must have seemed "hopeless" and "vulgar." Though I now know that most adolescents feel out of step much of the time, I also know that for the Puerto Rican girls of my generation that sense was intensified. The way our teachers and classmates looked at us that day in school was just a taste of the culture clash that awaited us in the real world, where prospective employers and men on the street would often misinterpret our tight skirts and jingling bracelets as a come-on.

Mixed cultural signals have perpetuated certain stereotypes—for example, that of the Hispanic woman as the "Hot Tamale" or sexual firebrand. It is a one-dimensional view that the media have found easy to promote. In their special vocabulary, advertisers have designated "sizzling" and "smoldering" as the adjectives of choice for describing not only the foods but also the women of Latin America. From conversations in my house I recall hearing about the harassment that Puerto Rican women endured in factories where the "boss men" talked to them as if sexual innuendo was all they understood and, worse, often gave them the choice of submitting to advances or being fired.

It is custom, however, not chromosomes, that leads us to choose scarlet over pale pink. As young girls, we were influenced in our decisions about clothes and colors by the women—older sisters and mothers who had grown up on a tropical island where the natural environment was a riot of primary colors, where showing your skin was one way to keep cool as well as to look sexy. Most important of all, on the island, women perhaps felt freer to dress and move

Macro constitutions, policies

more provocatively, since, in most cases, they were protected by the traditions, mores, and laws of a Spanish/Catholic system of morality and machismo whose main rule was: *You may look at my sister, but if you touch her I will kill you.* The extended family and church structure could provide a young woman with a circle of safety in her small pueblo on the Island; if a man "wronged" a girl, everyone would close in to save her family honor.

This is what I have gleaned from my discussions as an adult with older Puerto Rican women. They have told me about dressing in their best party clothes on Saturday nights and going to the town's plaza to promenade with their girlfriends in front of the boys they liked. The males were thus given an opportunity to admire the women and to express their admiration in the form of *piropos:* erotically charged street poems they composed on the spot. I have been subjected to a few piropos while visiting the Island, and they can be outrageous, although custom dictates that they must never cross into obscenity. This ritual, as I understand it, also entails a show of studied indifference on the woman's part; if she is "decent," she must not acknowledge the man's impassioned words. So I do understand how things can be lost in translation. When a Puerto Rican girl dressed in her idea of what is attractive meets a man from the mainstream culture who has been trained to react to certain types of clothing as a sexual signal, a clash is likely to take place. The line I first heard based on this aspect of the myth happened when the boy who took me to my first formal dance leaned over to plant a sloppy overeager kiss painfully on my mouth, and when I didn't respond with sufficient passion said in a resentful tone: "I thought you Latin girls were supposed to mature early"—my first instance of being thought of as a fruit or vegetable—I was supposed to *ripen,* not just grow into womanhood like other girls.

It is surprising to some of my professional friends that some people, including those who should know better, still put others "in their place." Though rarer, these incidents are still commonplace in my life. It happened to me most recently during a stay at a very classy metropolitan hotel favored by young professional couples for their weddings. Late one evening after the theater, as I walked toward my room with my new colleague (a woman with whom I was coordinating an arts program), a middle-aged man in a tuxedo, a young girl in satin and lace on his arm, stepped directly into our path. With his champagne glass extended toward me, he exclaimed, "Evita!"

Our way blocked, my companion and I listened as the man half-recited, half-bellowed "Don't Cry for Me, Argentina." When he finished, the young girl said: "How about a round of applause for my daddy?" We complied, hoping this would bring the silly spectacle to a close. I was becoming aware that our little group was attracting the attention of the other guests. "Daddy" must have perceived this too, and he once more barred the way as we tried to walk past him. He began to shout-sing a ditty to the tune of Bamba"—except the lyrics were about a girl named María whose exploits all rhymed with her name and gonorrhea. The girl kept saying "Oh, Daddy" and looking at me with pleading eyes. She wanted me to laugh along with the others. My companion and I stood silently waiting for the man to end his offensive song. When he finished,

I looked not at him but at his daughter. I advised her calmly never to ask her father what he had done in the army. Then I walked between them and to my room. My friend complimented me on my cool handling of the situation. I confessed to her that I really had wanted to push the jerk into the swimming pool. I knew that this same man—probably a corporate executive, well educated, even worldly by most standards—would not have been likely to regale a white woman with a dirty song in public. He would perhaps have checked his impulse by assuming that she could be somebody's wife or mother, or at least *somebody* who might take offense. But to him, I was just an Evita or a María: merely a character in his cartoon-populated universe.

Because of my education and my proficiency with the English language, I have acquired many mechanisms for dealing with the anger I experience. This was not true for my parents, nor is it true for the many Latin women working at menial jobs who must put up with stereotypes about our ethnic group such as: "They make good domestics." This is another facet of the myth of the Latin women in the United States. Its origin is simple to deduce. Work as domestics, waitressing, and factory jobs are all that's available to women with little English and few skills. The myth of the Hispanic menial has been sustained by the same media phenomenon that made "Mammy" from *Gone with the Wind* America's idea of the black woman for generations; María, the housemaid or counter girl, is now indelibly etched into the national psyche. The big and the little screens have presented us with the picture of the funny Hispanic maid, mispronouncing words and cooking up a spicy storm in a shiny California kitchen.

This media-engendered image of the Latina in the United States has been documented by feminist Hispanic scholars, who claim that such portrayals are partially responsible for the denial of opportunities for upward mobility among Latinas in the professions. I have a Chicana friend working on a Ph.D. in philosophy at a major university. She says her doctor still shakes his head in puzzled amazement at all the "big words" she uses. Since I do not wear my diplomas around my neck for all to see, I too have on occasion been sent to that "kitchen," where some think I obviously belong.

One such incident that has stayed with me, though I recognize it as a minor offense, happened on the day of my first public poetry reading. It took place in Miami in a boat-restaurant where we were having lunch before the event. I was nervous and excited as I walked in with my notebook in hand. An older woman motioned me to her table. Thinking (foolish me) that she wanted me to autograph a copy of my brand new slender volume of verse, I went over. She ordered a cup of coffee from me, assuming that I was the waitress. Easy enough to mistake my poems for menus, I suppose. I know that it wasn't an intentional act of cruelty, yet with all the good things that happened that day, I remember that scene most clearly, because it reminded me of what I had to overcome before anyone would take me seriously. In retrospect I understand that my anger gave my reading fire, that I have almost always taken doubts in my abilities as a challenge—and that the result is, most times, a feeling of satisfaction at having won a convert when I see the cold, appraising eyes warm to my words, the body language change, the smile that indicates that I have opened some avenue for communication. That day I read to that woman and her lowered eyes told me that she was embarrassed at her little faux pas, and when I willed her to look up to me, it was my victory, and she graciously allowed me to punish her with my full attention. We shook hands at the end of the reading, and I never saw her again. She has probably forgotten the whole thing but maybe not.

Yet I am one of the lucky ones. My parents made it possible for me to acquire a stronger footing in the mainstream culture by giving me the chance at an education. And books and art have saved me from the harsher forms of ethnic and racial prejudice that many of my Hispanic *compañeras* have had to endure. I travel a lot around the United States, reading from my books of poetry and my novel, and the reception I most often receive is one of positive interest by people who want to know more about my culture. There are, however, thousands of Latinas without the privilege of an education or the entrée into society that I have. For them life is a struggle against the misconceptions perpetuated by the myth of the Latina as whore, domestic, or criminal. We cannot change this by legislating the way people look at us. The transformation, as I see it, has to occur at a much more individual level. My personal goal in my public life is to try to replace the old pervasive stereotypes and myths about Latinas with a much more interesting set of realities. Every time I give a reading, I hope the stories I tell, the dreams and fears I examine in my work, can achieve some universal truth which will get my audience past the particulars of my skin color, my accent, or my clothes.

I once wrote a poem in which I called us Latinas "God's brown daughters." This poem is really a prayer of sorts, offered upward, but also, through the human-to-human channel of art, outward. It is a prayer for communication, and for respect. In it, Latin women pray "in Spanish to an Anglo God/with a Jewish heritage," and they are "fervently hoping/that if not omnipotent,/at least He be bilingual."

20

Becoming Entrepreneurs

Intersections of Race, Class, and Gender at the Black Beauty Salon

ADIA M. HARVEY

A great deal of research has applied the concept of intersectionality to understand the ways race, class, and gender affect individuals. Feminist researchers have been particularly effective in promoting scholarship that examines race, gender, and class as interlocking categories of oppression, wherein the experiences of some minority groups (e.g., Black women) must be understood as a consequence of racial, gendered, and sometimes class-based inequality (Chafetz 1997; Collins 1990; King 1988). For these women, understanding their experiences involves more than just knowing how race and gender lead to inequality but mandates an understanding of how race is gendered and gender is racialized (Browne and Misra 2003). It is this totality that researchers seek to explore.

Accepting the premise that the intersections of race, gender, and class affect virtually all aspects of life, feminist researchers examine these intersections in social arenas such as the workplace, community organizations, media, and others (Collins 1990; Jewell 1993; Rollins 1985). Studies that address interacting oppressions in the workplace often focus on the ways that race, gender, and class shape minority women's access to certain jobs, compensation for work, and the perception of these women's suitability for various occupations (Browne 1999; Higginbotham 1997). These intersecting oppressions often relegate minority women into the bottom of the labor queue, where economic stability is precarious and they are over-represented in low-skill, low-status work (Browne and Kennelly 1999; Reskin and Roos 1990).

This study builds on existing studies of intersectionality and work by exploring the ways the intersections of race, gender, and class shape working-class Black women's experiences with entrepreneurship. While research has demonstrated that intersecting oppressions are likely to channel working-class Black women into occupations like janitorial, food, or health service work, little is known about how these interacting inequalities affect Black women's entrepreneurial activity. This study shifts the center to focus on a group whose entrepreneurial activity may differ structurally and experientially from the entrepreneurs who are normally the subject of this type of research....

SOURCE: From *Gender & Society*, Vol. 19 No. 6, December 2005, pp. 789–808.
© 2005 by Sociologists for Women in Society. Reprinted by permission of Sage Publications.

THEORETICAL FRAMEWORK

... The legal and social gains of the 1960s have had an enormous impact on Black women's occupational opportunities; nonetheless, structural, entrenched problems still remain for African American women in the workplace. Institutional and individual discrimination, the glass ceiling, and hostile workplace cultures are still realities that constrain occupational opportunities for many Black women. For working Black women, the relocation of jobs to suburbs, rise of the technology industry, and lack of educational opportunities also leave limited work options, and those that exist tend to be low paying with low prestige and little authority, job security, or benefits.

Faced with constrained opportunities in the labor market, business ownership can become an appealing pathway to economic support for some Black women (Smith 1992). Indeed, the increase in Black women's entrepreneurial activity speaks to its growing popularity as an alternative to the paid labor force. Between the years 1997 and 2004, the numbers of Black women entrepreneurs grew 33 percent, with recent estimates suggesting that there are now approximately 414,472 firms owned by African American women (Center for Women's Business Research 2004). This study focuses on an often understudied segment of business owners—Black working-class women—to explore how their entrepreneurial activity is influenced by race, gender, and class.

Black Women's Work in the Hair Industry

Structural disadvantages in the labor market have made entry into many fields virtually impossible for Black women, but some fields have been easier to access than others. For instance, Black women have a long history of work within the hair industry dating back to the early twentieth century. Black women in the late 1800s and early 1900s sold hair products in Black communities. Probably the most notable and successful of these early entrepreneurs was Madame C. J. Walker, the inventor of the straightening comb and first African American millionaire. Unlike other mostly white entrepreneurs at the time who promoted Black hair care products as ways to eliminate "kinky, snarly, ugly, curly hair," Madame Walker established her fortune by marketing her products as ones that would promote healthy hair growth and also employed other African American women as sales associates (Rooks 1996, 35).

Madame Walker's success in the hair industry reflects structural barriers that heightened the appeal of entrepreneurship in the hair industry for many African American women. With racial segregation a common practice, and women's occupational opportunities severely constrained, the intersection of race and gender meant that Black women were effectively excluded from most jobs. In addition, in keeping with the custom of racial segregation, whites in the beauty industry often refused to service Black customers. As such, the beauty industry was distinctive as a niche that Black women could enter relatively easily, pursue entrepreneurship, and maintain access to a relatively untapped consumer base. Their entrepreneurial work in this area reflects the assertion that minority and women's entrepreneurship often has elements of middleman minority character-istics. In this particular case, women in the beauty industry did not follow the typical middleman minority route of selling goods and services such as hair pro-ducts to the masses. They did, however, function as middleman minorities in that these entrepreneurs provided a service to groups with whom elites were loath to interact.

Boyd (2000) argues that the combination of racial oppression and labor mar-ket disadvantage during the great depression in fact led Black women to pursue entrepreneurship. As the economy declined and employment became increas-ingly scarce, many Black women were dismissed from positions they had held and increasingly relied on domestic service for necessary wages. However, some women gravitated to the beauty industry to supplement their earnings because it was one of few available options for additional income and because a demand for these services came from other Black women, who perceived that maintaining a well-groomed appearance was a necessity for securing employment. Labor mar-ket disadvantage, economic depression, and racial discrimination made entre-preneurship in the hair industry highly appealing for African American women during this time.

Today, African American women have more occupational opportunities than ever before, and many are moving into the private sector as well as into economically rewarding fields of entrepreneurship that were unavailable in the past, such as entertainment and consulting. Nonetheless, Black women's con-temporary entrepreneurship in the hair industry remains contextualized by per-vasive structural issues that they face in the occupational sphere, including discrimination, wage inequality, glass ceilings, and occupational sex segregation.

However, these studies address only how intersectionality shapes Black women's entrepreneurship at structural levels. While it is important to address how institutionalized racism, sexism, and classism channel Black women into certain occupational niches, equally important are the ways these interlocking factors shape the more micro-level, interactional aspects of entrepreneurship. This study explores, at the level of social interaction, how intersectionality affects Black women's work practices, patterns of business ownership, and experiences as entrepreneurs.

DATA COLLECTION

To assess how race, gender, and class shaped the entrepreneurial activity of working-class Black women, I conducted interviews with 11 Black female hair salon owners in a major city in the mid-Atlantic United States. The women I interviewed owned their businesses for time periods ranging from 2 to 15 years. All interviews took place in the women's shops and generally lasted one hour to an hour and a half. The majority of interviews were tape-recorded, with the exception of three women who were uncomfortable being recorded. In these cases I took copious notes.

The women ranged in age from 25 to 50, and all had worked as stylists before eventually becoming salon owners. I contacted respondents through a

snowball method. Four of the respondents had worked as stylists for other owners included in this study before eventually becoming owners themselves. Two
women were sole operators, meaning they hired no employees. Of the other
nine women, the number of employees ranged from 2 to 13 stylists. Nine owners had children; of these nine, their children ranged in age from toddlers to
adulthood. The locations of the salons covered a wide range, including
working-class neighborhoods, middle-class areas on the outskirts of the city,
and urban, lower-class neighborhoods. Similarly, clienteles ranged in income
but did not vary widely with regard to race and gender. Some owners had a
handful of white or Asian customers, but at each shop, the majority of the clientele were Black women. The location of the shop tended to mirror the social
class of the clientele, for example, shops in middle-class neighborhoods usually
had mostly middle-class women as a customer base.

All of the owners interviewed here hailed from working-class backgrounds.... As members of the working class, these women had been previously
employed as sales associates, fast food workers, or telemarketers, positions that
generated annual incomes of about $25,000 or less annually. They did not own
their homes, although six had owned cars, and none had amassed any other
forms of wealth or educational attainment past high school (although all owners
completed cosmetology school). Furthermore, none of these women were raised
in families where the combined household incomes reached the middle-class or
upper-middle-class levels Sernau describes of $40,000 to $75,000 per year....

FINDINGS

Rather than describing the ways certain aspects of entrepreneurship were shaped
by race or gender or class, I have chosen instead to pinpoint key aspects of the
women's entrepreneurial activity to highlight how these various facets were
shaped by some combination of these social constructs. In this way, a much
clearer picture is visible of how one or more of these issues determine the contours of working-class Black women's entrepreneurship. The intersections of
race, class, and gender (or some combination of these) are most clearly revealed
in two areas of working-class Black women's entrepreneurship: the process of
becoming entrepreneurs and relationships with stylists.

Becoming Entrepreneurs

In some ways, the women's decision to become entrepreneurs reflects the findings of existing studies of gender and entrepreneurial activity. Specifically, for
many women in this study, entrepreneurship is a gendered choice in that it
allows them to balance the demands of work and family. The desire to balance
work and family is an important factor that often distinguishes women's motivations for entrepreneurial activity from [those of] men. In this regard, then, Black
women's initial choice to become entrepreneurs is shaped by gendered factors
that appeal to many other women regardless of race or class. Many women stated

that as salon owners, it was much easier for them to meet the challenges of working full-time and raising children. They were able to set their own hours and could structure their time to devote adequate attention to children and work.

Chandra, for example, is 25 with a one-year-old daughter. Chandra's salon is located in what she acknowledges is a "bad location" because of the rampant drug trade and related criminal activity nearby. She has owned her salon for nearly two years and employs two other stylists, one of whom is her best friend, Sarah. In response to a question about why she decided to open her own salon, Chandra stated, "I can honestly tell you, the most rewarding part [of owning a salon] is knowing that [because] I do have a small baby, I can come and go as I please, and I don't have to answer to anybody." In contrast to Chandra, Danette, 35, owns a salon located on the outskirts of the city in a quiet, middle-class, tree-lined neighborhood. Like many of the women interviewed for this study, Danette's salon is spacious, with a bulletin board announcing upcoming community events and African American art lining the walls. Danette also indicates that part of the lure of entrepreneurship was the opportunity it offered her to balance work and family. She too spoke of bringing her son to the salon when he was a toddler and allowing him to entertain himself in the shop while she worked: "Everybody loved [my son], he was cute…. Everybody loved him so much that he used to be in the hair, have hair everywhere, and they would tell me, 'Oh, just leave him alone…. And then when he got older he would come in Saturday night and clean up for me, things like that." The decision to become a business owner because it offers the advantage of balancing work and family thus reflects one particular way in which entrepreneurial decisions are influenced by gender. However, aspects of intersectionality are present here too vis-à-vis the unspoken fact that as working-class women, professional child care may be too costly for them to consider it a financially viable option….

The decision to become entrepreneurs in the beauty industry represents a choice that clearly reflects gendered norms, given that the hair salon capitalizes on and profits from societal messages about gender and appearance. Perhaps more important, the appreciation for "making women beautiful" was frequently cited as one of the motivations that led them to this line of work. A number of owners stated that they loved their work: they loved working with hair and enjoyed the ability to "make women look good." Maxine is 34 and owns a small, sparsely furnished salon on the ground level of a four-story walk-up building. Her shop is located in a middle-class neighborhood adjacent to a major university. Prior to opening her salon, she was a stylist at a shop located a few doors down from the one she now owns. She was compelled to enter the hair industry because "I've always loved working with hair, and I had been doing it since I was a teenager. The best part of this work is fixing people up, making them look nice; they smile, say thank you. It makes me feel good." The decision to become entrepreneurs in the hair industry itself indicates how gender shapes entrepreneurial activity. It is clear that these women's motivations for entrepreneurship in this field reflect the gendered notion that it is appropriate for women to devote attention, time, and money to meeting dominant standards of beauty.

However, Black women's entrepreneurial work is in some ways shaped by gendered factors that might be applicable to other women entrepreneurs; in other ways, their initial pathway to entrepreneurship is specifically delineated by the intersection of race and gender. Gender ideology demands that women should be attractive, but overlapping racial messages often insist that Black women do not meet beauty ideals. Interestingly, the social significance attached to women's hair exposes this contradiction. Hair has long been considered a symbol of feminine beauty and attractiveness, but long, straight, blonde hair has been epitomized as the female ideal. Instead, Black women's hair in its natural state is more likely to be short, tightly curled, and dark—the opposite of the ideal.

For Black women, the social significance of hair reflects a "gendered racism" that has very particular and specific implications. The racialized standard of beauty can produce a conundrum for African American women in a society where hair is unequivocally connected to beauty and adherence to beauty ideals is linked to "access to rewards such as employment, income, and self-esteem" (St. Jean and Feagin 1998, 74). While many women experience contradictions and tensions that stem from the symbolic significance of hair, race and gender intersect to shape these issues in a manner that reinforces mainstream culture's messages of African American women's inferiority (hooks 1992; Weitz 2001; Wilson and Russell 1996).

Gendered racism labels Black women's hair unattractive and unappealing; consequently, a large consumer base exists for Black women who work in the hair industry. As such, many of the women interviewed became salon owners with full confidence that there was a market for their services. Their awareness of this market and their ability to tap into it stemmed both from prior experience as stylists and from their first-hand familiarity with the ways in which gendered racism shapes Black women's feelings about their hair. Greta, a tall, confident woman in her thirties, runs a small but cozy salon in the same neighborhood as Maxine's. She worked alone for a year before hiring two stylists to join her in her shop. Greta states that she never worries about losing business because "[Black women] will go get our hair done. Even when I had to work alone, I had customers, because we will get our hair done. We'll take the rent money to go get our hair done! We don't believe in looking crazy." Greta's remarks indicate her familiarity with her customers' attitudes about hair and the importance, for some Black women, of making sure hair is professionally styled even if it comes at the expense of other necessities. Interestingly, she also likens not having professionally done hair to "looking crazy," further emphasizing the perception that Black women's hair needs professional styling to be considered appealing....

Coming from working-class backgrounds also shapes these women's entrepreneurial activity in a way that differs from middle- and upper-class women's decisions to enter into self-employment. Budig (2001) finds that many upper-class women engage in entrepreneurship as a response to the glass ceiling they experience in the workplace. Entrepreneurship thus becomes a way for them to avoid or minimize the gender discrimination they face as paid workers. Similarly, Davies-Netzley (2000) cites this as one of the primary factors that leads white and Latina women into entrepreneurship. While these women are able to attain middle- or upper-class status (e.g., cars, homes, benefits), they are often motivated to become entrepreneurs as they recognize that institutionalized and individual workplace discrimination will eventually limit (or has already limited) their continued upward mobility.

The working-class Black women interviewed for this study, however, face different circumstances. Prior to becoming salon owners, these women worked primarily in the low-skill, low-wage service jobs in which Black women are disproportionately represented: fast food, factory work, telemarketing, filing clerks, and sales associates. Entrepreneurship, then, did not provide a way to continue upward mobility as much as it was a step toward upward mobility. Unlike middle- and upper-class women who became entrepreneurs to avoid a glass ceiling, the women interviewed here became entrepreneurs as a route to securing basic financial stability that middle- and upper-class women already enjoy. While class plays a role in both groups of women's entrepreneurial work, it shapes working-class women's advent into entrepreneurship differently than middle- and upper-class women's.

Furthermore, working-class salon owners' economic needs intersected with gender-based ones in that many sought economic stability to meet familial responsibilities. Chandra has owned her salon for nearly two years and explicitly stated that financial considerations were why she made the move from working as a stylist at a chain salon to opening her own establishment, but she concedes that familial concerns were important as well: "I have a baby, and at Supercuts, you have to split like 50 percent with them, and they take out for the shampoo, they take out for the conditioner, so I said, I'm not going to work for them, because I was barely making $500 every two weeks. Being here, I can make $500 in a day, or in two days, and it's mine." Chandra's statement underscores how becoming an entrepreneur was shaped by her class position as well as by racialized, gendered factors. As an African American woman, Chandra's hardships in the labor market likely reflect the complex interplay of gender and race. As a working-class mother, economic stability is absolutely essential, to the point where the deduction of shampoo and other supplies from her wages meant that she was not earning enough to support her family. Salon ownership, therefore, offers Chandra a way to sidestep some forms of racial and gendered discrimination and to achieve the economic stability necessary to provide for her family....

Relationships with Stylists: The Helping Ideology

The relationship between Black women salon owners and the stylists they hire is the other key area of entrepreneurship that is shaped by the interplay of race, gender, and class. Salon owners interviewed for this study typically hired 1 to 3 stylists to work in their shops. Miranda and Lola, both in their mid-thirties, are the only two owners interviewed who work alone. In contrast, Lana employed 14 stylists, more than any other owner interviewed for this study. All the owners operated salons in which they charged stylists booth rent. Under this system, stylists pay a monthly rent to owners for the use of space at the salon. Stylists are then permitted to keep all money from business transactions but are not guaranteed any benefits such as medical insurance, dental insurance, retirement plans, or sick leave. Although stylists are technically independent contractors, they must follow the rules and guidelines established by the salon owner, and either party can terminate the working relationship.

The relationship between owners and stylists is unique in a number of respects. Possibly most striking is the ideology of help and support that shapes, to varying degrees, owners' relationships with stylists. Owners generally profess a willingness to assist stylists in their professional development and growth as well as a readiness to help them succeed in the business. In its most extreme cases, this helping ideology motivates owners to encourage stylists to pursue entrepreneurship and to open their own salons. Lana, an owner in her mid-forties who has owned her salon for nearly 15 years, is perhaps the most interesting embodiment of this philosophy in practice. Her salon is large and upscale, with two floors, and is usually bustling with her clientele of mostly middle-class African American women. Lana is opinionated and outspoken and tells her stylists frequently that she expects them to eventually go into business for themselves. When asked the most rewarding part of owning a salon, Lana's reply was similar to that of many other owners: "Having people leave here and become salon owners, and being successful at it. This is really a field where Black women can do well, ... and it's important that people understand that and respect this industry." Lana also requires that her stylists attend mandatory workshops she sponsors on money management, customer service, and business ethics so that they will have exposure to the knowledge she deems necessary for successful entrepreneurship.

Interestingly, Tanisha and Greta, two other owners interviewed for this study, worked as stylists in Lana's salon before opening their own shops. Lana's relationship with Tanisha in particular exemplifies how the nature of the relationship between owners and stylists is sometimes shaped by a willingness to help other Black women prosper in the industry. Tanisha and Lana first met at church, and after learning that Tanisha styled hair for several of the women in the congregation, Lana decided to take Tanisha under her wing. Tanisha began working full-time for Lana immediately after high school and remained at this salon for three years before moving on to open her own salon....

This helping ideology—owners' desire to help stylists successfully shift to entrepreneurship—stems from a sense of gender/racial solidarity. Yet this gender/racial solidarity also has a dimension of class solidarity. Other research describes gender/racial solidarity among middle- and upper-class Black women as the manifestation of a desire to help their lower-class counterparts (Giddings 1984; McDonald 1997; White 1999). This willingness to help is based on shared experience of racial and gender discrimination but is also characterized by the willingness to reach across class lines to help those less economically privileged. However, in the case of Black female salon owners, their sense of gender/racial solidarity extended to other working-class women in the same class position. Owners did not seek to uplift less fortunate Black women; instead, their sense of gender/racial solidarity compelled them to help other Black women of the same class position. As such, the helping ideology that often underlies the relationship between owner and stylist is very clearly influenced by a sense of race, gender, and class solidarity that facilitates a level of support that Black women in other workplaces rarely experience.

Contradictions, Tensions, and Problems

Predictably, the intersections of race, gender, and class can cause working–class Black women to face certain specific obstacles with entrepreneurship. One such obstacle is the issue of access to start-up capital. Many women interviewed stated unhesitatingly that getting access to start-up funding was the most difficult part of becoming entrepreneurs.... Denise, for instance, is 37 and owns a salon that was passed down to her by her mother on retirement. The salon is located outside the city limits in a suburban area that draws a mostly middle–class clientele.... Denise owns two other salons in different parts of the city as well, both of which service mostly working-class women. Like most of the other women in this study, Denise relied on unconventional strategies to procure necessary start-up capital. She stated, "The government will give money to[,] seems like every-body else. But when you go to get money for small businesses, *we* [she makes a gesture to indicate the two of us, as Black women] have to go through a lot. So the hard part was trying to get help from the government. That was the hard part. The easy part was having faith in the Lord's will." Like most of the other women interviewed, Denise secured her start-up capital for buying the other salons through meticulous savings as well as loans from family and friends. She interprets her difficulty securing funding as a common pitfall that Black women face at the onset of entrepreneurship. Significantly, she attributes this obstacle to race and gender.

These women's status as working-class Black women also shaped their access to social networks in ways that likely undermined their potential for economic returns as entrepreneurs.... In the process of becoming owners, many women relied on those in their social networks for information about available property that they could purchase or rent out to open their salons. Consequently, many women opened salons near the ones where they had previously worked as stylists. This was beneficial in that it helped them remain accessible to former clients, but it also reflected minimal knowledge of real estate in other parts of the city that might have offered other benefits such as better parking, an untapped customer base, or lower property taxes. Maxine states, "I moved right up the street when a place became available. In hindsight, that was not such a good idea. The parking around here is bad. I should not have moved so fast, and if I had to do it again, I wouldn't have rushed. I actually hope to move somewhere else pretty soon." Race and gender play a role here in constraining some owners' business decisions. Social networks that comprise primarily other African American women can limit the degree and breadth of information and knowledge available to these owners....

Finally, the success these women experience as business owners must be examined in the context of entrepreneurial ventures at large. Although race, gender, and class shape their entry into salon ownership and their relationships with stylists, it is important to note that these women still experience ghettoization within the entrepreneurial sector. Like the majority of Black-owned businesses, these women's entrepreneurial ventures are located in the personal service sector, which tends to generate fewer economic returns than emerging (and male-dominated) fields of minority entrepreneurship such as construction and manufacturing. Furthermore, women's businesses in general tend to be "smaller than men's" and involved in a "narrow range of activity"—characteristics that are applicable to these women's hair salons (Grasmuck and Espinal 2000, 233). For many of these women, owning a hair salon is an economically sound business decision but one that does not exempt them from the pitfalls and challenges that are common to Black female entrepreneurs. Gender, race, and class shape these women's entrepreneurial choices to channel them into a field that, relative to other avenues of entrepreneurship, offers fewer returns and less prestige.

CONCLUSION

The results of this study demonstrate that intersectionality is a key factor in working-class Black women's entrepreneurial activity in two important regards: the transition to entrepreneurship and the nature of relationships with stylists. Through the use of intersectionality as a theoretical lens, it becomes increasingly clear that race, gender, and class have a definitive impact on the market these entrepreneurs try to reach as well as these women's actions, decisions, and experiences as entrepreneurs. The use of an intersectional framework therefore provides greater insight into the social processes of entrepreneurship.

REFERENCES

Boyd, Robert, 2000. Race, labor market disadvantage, and survivalist entrepreneurship: Black women in the urban North during the great depression. *Sociological Forum* 15 (4): 647–70.

Browne, Irene, 1999. Latinas and African American women in the labor market. In *Latinas and African American women at work*, edited by Irene Browne. New York: Russell Sage.

Browne, Irene, and Ivy Kennelly, 1999. Stereotypes and realities: Images of African American women in the labor market. In *Latinas and African American women at work*, edited by Irene Browne. New York: Russell Sage.

Browne, Irene, and Joya Misra, 2003. The intersection of gender and race in the labor market. *Annual Review of Sociology* 29: 487–513.

Budig. Michelle J, 2001. Gender, family status, and job characteristics: Determinants of self-employment participation for a young cohort, 1979–1998. Paper presented at annual meetings of the American Sociological Association, Anaheim, CA, August.

Center for Women's Business Research. 2004. *African American women-owned businesses in the United States, 2004: A fact sheet*, Washington, DC: Center for Women's Business Research.

Chafetz, Janet Salzman, 1997: Feminist theory and sociology: Underutilized contributions for mainstream theory. *Annual Review of Sociology* 23: 97–120.

Collins, Patricia Hill, 1990. *Black feminist thought: Knowledge, consciousness, and the politics of empowerment.* New York: Routledge.

Davies-Netzley, Sally, 2000. *Gendered capital: Entrepreneurial women in American society.* New York: Garland Press.

Giddings, Paula, 1984. *When and where I enter: The impact of Black women on race and sex in America.* New York: Bantam.

Grasmuck, Sherri, and Rosario Espinal, 2000. Market success or female autonomy? Income, ideology, and empowerment among microentrepreneurs in the Dominican Republic. *Gender & Society* 14 (2): 231–55.

Higginbotham, Elizabeth, 1997. Black professional women: Job ceilings and employment sectors. In *Workplace/women's place*, edited by Dana Dunn. Los Angeles: Roxbury.

hooks, bell, 1992. *Black looks: Race and representation.* Boston: South End.

Jewell, K. Sue, 1993. *From mammy to Miss America and beyond.* New York: Routledge.

King, Deborah, 1988. Multiple jeopardy, multiple consciousness. *Signs: Journal of Women in Culture and Society* 14 (1): 42–72.

McDonald, Katrina Bell, 1997. Black activist mothering. *Gender & Society* 11 (6): 773–95.

Reskin, Barbara, and Patricia Roos, 1990. *Job queues, gender queues.* Philadelphia: Temple University Press.

Rollins, Judith, 1985. *Between women: Domestics and their employers.* Philadelphia: Temple University Press.

Rooks, Noliwe, 1996. *Hair raising: Beauty, culture, and African-American women.* New Brunswick, NJ: Rutgers University Press.

Sernau, Scott, 2001. *World's apart: Social inequalities in a new century.* Thousand Oaks, CA: Pine Forge.

Smith, Wade A, 1992. Race, gender, and entrepreneurial orientation. *National Journal of Sociology* 6 (2): 141–55.

St. Jean, Yanick, and Joe Feagin, 1998. *Double burden.* Armonk, NY: M. E. Sharpe.

Weitz, Rose, 2001. Women and their hair: Seeking power through resistance and accommodation. *Gender & Society* 15 (5): 667–86.

White, Deborah Gray, 1999. *Too heavy a load: Black women in defense of themselves, 1894–1994.* New York: W. W. Norton.

Wilson, Midge, and Kathy Russell, 1996. *Divided sisters: Bridging the gap between Black and white women.* New York: Doubleday.

21

The Well-Coiffed Man

Class, Race, and Heterosexual Masculinity in the Hair Salon

KRISTEN BARBER

Few people know what exactly to make of the metrosexual, a man who turns himself into a project in the seeming pursuit of the body beautiful. Traditionally associated with women and with gay men, the body beautiful has been tightly linked to the concept of femininity. In her book on *The Male Body*, Bordo (1999) suggests that the media now positions men as sexualized objects of the gaze, just as it has done for women. She claims that women, for the first time in recent history, are now encouraged to consume the beautified male bodily form. As a result, Bordo and others contend that the sexualization of men in the media and their participation in appearance-enhancing practices destabilize traditional gender dichotomies. This seeming subversion leads Bordo to exclaim, "I never dreamed that 'equality' would move in the direction of men worrying *more* about their looks rather than women worrying less" (1999, 217).

Scholars who study the meaning of *women's* body work help us understand that participation in beauty culture is not rooted solely in gendered relationships, it is tied up with interlocking systems of race, class, gender, sexuality, and age. That is, for women beauty work is often about more than beauty, it is about appropriating and expressing a particular social status by grooming the body in a particular way. Few scholars have examined the meaning of men's participation in beauty culture, however.

In this article, I use a case study of Shear Style,[1] a small hair salon in a Southern California suburb, to explore how men hair salon clients make sense of their participation in beauty work. I find that men at Shear Style empty beauty work of its association with feminized aesthetics and instead construct it as a practice necessary for them to embody a class-based masculinity. But not all men are able to participate in the beauty industry. Rather, it is men with enough disposable income, "all the money," who are able to purchase beauty work and beauty products—or "grooming products" as they are often called for men—that promise to deliver aesthetics compatible with social standards. For the men at Shear Style, preference for "stylish" and "superior" hair embodies expectations of white professional-class masculinity. These men use beauty work in a way that distinguishes themselves

SOURCE: Barber, Kristen. 2008. "The Well-Coiffed Man: Class, Race, and Heterosexual Masculinity in the Hair Salon." *Gender & Society* 22 (August) pp. 455–476. Reprinted by permission of Sage Publications, Inc.

from white working-class men, while at the same time distancing themselves from the feminizing character of the "women's" hair salon.

LITERATURE REVIEW

The beauty industry generates services and products with which women groom their bodies according to social expectations of raced, classed, and sexualized femininities. The hair salon is a key space in the perpetuation of women's body projects. Hair is a social symbol that allows people to associate themselves with others along the lines of race, class, gender, sexuality, and age. Women cut, shape, and dye their hair in ways that express social location and talk about their hair in ways that define their relationships with other women (Gimlin 1996; Jacobs Huey 2006).

The Hair Salon as a Gendered Space for Women

Feminist scholars interested in women, beauty, and the body find that women's participation in beauty work is about relationships, pleasure, and "achiev[ing] a look, and sometimes also a feeling, which is regarded as 'appropriate' in relation to categories of gender, age, sexuality, class and ethnicity" (Black 2004, 11). The hair salon is one space in which women shape their bodies and relationships with others in ways that mark them as members of particular social groups. Since we have long associated beauty work and the hair salon with "women's culture," the hair salon is a space in which men do not venture, or do so infrequently. In this way, "the salon both reflects and reinforces divisions along gender, [as well as] ethnicity and class lines" (Black 2004, 11). Hair salons are for women, barbershops are for men. Under this ideological regime, a heterosexual man in a hair salon is an anomaly who transgresses gender boundaries by moving into a women's space and by participating in a beauty practice traditionally associated with women.

Men: From the Laboring Body to
the Flannel Suit to the Well-coiffed Man

Postwar America between 1945 and 1960 saw rapid economic growth, and with it came the corporation and an exponential increase in white-collar jobs. While men's identities continued to be bound up with their jobs, opportunities for intellectual work grew. Many middle-class white men flooded into corporations, trading in their denim work-jumpers for grey flannel suits. As it became necessary for the corporate man to interact with customers and clients, interpersonal skills, personality, and appearance became essential hiring and firing criteria (Luciano 2001).

Luciano (2001) describes how the emphasis on the appearance of middle-class white men emerged from capitalist notions of who a successful professional-class man is. Corporations encouraged these men to package their bodies and personalities for success. Employers correlated softness with Communism, fatness with laziness, and saw baldness as a detriment to sales. Marketers quickly produced sales gimmicks that attached the accomplishment of "professional success" to products such as toupees and pomade, products that had long been considered symbols of vanity, narcissism, and, thus, femininity.

It is no longer enough for some men to work hard, they must also look good.

Thus far, most research on men and hair care has focused on the Black barbershop. This research reveals Black barbershops as spaces for community and the socialization of Black boys into Black men (Alexander 2003; Williams 1993; Wright 1998). In a society that marginalizes Black masculinity, the Black barbershop acts as a safe place for men to congregate, socialize, and reject oppressive stereotyped notions of masculinity. In the Black barbershop, appearance and the cutting of hair are often secondary to the conversations that are important in perpetuating culture and community (Alexander 2003). Research on the Black barbershop and the women's hair salon shows us how race, class, and gender are constructed differently in different spaces.

Unlike the barbershop, the hair salon is not an obvious place in which men participate in the reproduction of masculinity. Certainly it is not a space in which men could create community with other men. Hence, for those men who choose the salon, it is their participation in feminized beauty work that becomes salient.

THE STUDY

In this study, I employed both ethnographic methods and in-depth interviews to explore the roles of class, gender, race, and heterosexuality in the purchase of beauty work by men in a small Southern California hair salon, Shear Style. From October 2006 through February 2007, I conducted 40 hours of observations, 15 formal in-depth interviews with men salon clients, and a group interview with three of the four salon stylists.[2] The small size of the salon, with only four work stations in close proximity to each other, made it an ideal space for me to observe client/stylist interaction, talk with men patrons, and recruit interview respondents. I had a number of informal conversations with men as they waited for their appointments or exited the salon.

Of the 15 men I formally interviewed, 14 are white and one is Mexican. Over the course of the study, I saw only one Asian man and two Latino men with appointments for haircuts. Thirteen of the men identify as heterosexual, while two did not disclose their sexual orientation or marital status. Of the heterosexual men, 10 are married, two are single, and one is divorced. Their ages range from 30 to 63, with the average being 49. While one of the men I formally interviewed is a stay-at-home father, the rest are professional white-collar workers.

The salon's services are priced as follows: A woman's haircut costs $65, dye costs $65 per color, and a man's haircut costs $45. This price for a man's haircut is more than three times as much as the cost at the Supercuts and the barbershop which are located right across the street from the salon. While women often purchase a cut and a color, which can take several hours to complete, men's haircuts usually take half an hour. I observed only one man have his hair colored; he declined to be interviewed for this study.³

THE SALON

The context of Shear Style is important to an understanding of what it means for men to enter the gendered space of the "women's" hair salon. Shear Style is a feminized space marked by pink walls, fresh flowers, and regular cookie samplers. The men at the salon transgress gender boundaries and risk feminization to enter the salon and get their hair cut there. Eleven of the 15 men I interviewed had followed their hairstylists from a prior salon to Shear Style. With its white walls and clean lines, the atmosphere of the previous salon is much more androgynous and, in contrast, Shear Style is understood by the men as "feminine."

The men are keenly aware that they are outnumbered in this gendered space, reporting that they rarely see other men in the salon. Hamilton, a 57-year-old white investment manager, points to the women around him as he describes his aversion to the salon, "It's jammed full of housewives … and there are bimbos walking around chatting." Hamilton trivializes the salon and the practice of beauty work in the salon by associating it with "bimbos" and "housewives." His privileged position as a white upper-middle-class man likely allows him to feel he has the authority to make such disparaging remarks. Many of my respondents recommended steps the salon could take to better serve its men clients. For example, two men proposed that flat screen televisions be affixed to the walls and tuned to the sports and the news channels. Also, many of the men would like to see available reading material other than "gossip" and bridal magazines. Evan, a 37-year-old white stay-at-home-father, says that men's magazines would be "kind of nice to see … Maybe I'll suggest that next time I see her [his stylist]." He laughs, "Where's my *GQ*? Where's my *Esquire*?" Evan's laugh implies that asking for "men's magazines" is not appropriate since the salon mainly belongs to women. By recommending magazines such as *GQ* and *Esquire* (magazines targeted to middle- and upper-middle-class white men), instead of *Ebony* or *JET* (which are aimed at Black audiences), Evan suggests that he and the other men at the salon are interested in appropriating a particular kind of professional-class whiteness.

The pink walls, throw pillows on the waiting bench, fresh weekly flowers, and cookie samplers further feminize the salon in the eyes of the men. Neil, a 43-year-old white engineer, describes the salon as a "fairly feminine atmosphere." When I asked him what it is about the salon that makes it "feminine," he told me that you have to "perch" on a cushioned bench while waiting, "and men don't perch" He then gestured toward the maroon, purple, and gold tasseled throw pillows that decorated the bench on which we sat and exclaimed that the "the pillows, food, [and] décor" make him feel he is in a "woman's space." Relying on my assumed reading of him as a man, this description allowed Neil to situate himself in contrast to the feminized character of the salon.

The gender composition of the salon's clientele, the lack of amenities for men, and the décor of the salon provides clues to the men that they are in a "women's" space. By articulating an understanding of the space as feminine the men set themselves against and create distance from the potentially contaminating "feminine" character of the salon. They contend the salon is not an appropriate space for *them*, as men, despite their regular visits. While the men enter the salon to purchase beauty work from the stylists, they simultaneously maintain a sense of masculinity by distancing themselves from the "feminine" salon and by situating themselves as anomalies in this space.

MEN'S MOTIVATIONS FOR GOING TO THE HAIR SALON

Three themes emerge from this study that help us to understand why the men at Shear Style purchase beauty work in a "women's" hair salon: (1) Because they enjoy the salon as a place of leisure, luxury, and pampering; (2) For the personalized relationships they feel they form with their women stylists; and (3) To obtain a stylish haircut they conflate with white professional-class aesthetics. As I discuss these themes in turn, what emerges is a picture of how the men at Shear Style understand the salon as a place important to the appropriation of a white professional-class embodiment, how they contrast the salon and the barbershop to differentiate themselves from white working-class masculinity, and how they resist feminization while transgressing gender boundaries.

The Pampered Heterosexual Man: Leisure, Luxury, and Touch in the Hair Salon

Bodily pampering, which includes attention to appearance and which takes place within a traditionally feminized space, could be considered a feminized form of leisure. However, the class and race privilege of the men at Shear Style allow them to access this leisure and pleasure without marginalization. The men also maintain a sense of masculinity by marking the salon services as *less* feminine than those offered by nail salons and spas. While they can afford to pay for stylish hair, they would not pay for nail care, for example. Finally, by heterosexualizing the touch that accompanies this pampering the men resist feminization and instead position themselves as heterosexually masculine.

Many of my respondents described the salon as a place of leisure, where they go to relax and pause from their hectic daily lives. The men at Shear Style are professionals who live in a speeded-up metropolis where work and family life collide and compete, and where people's social lives often lose out to the occupational expectations of white-collar "success" The 45 minutes they spend at the salon, waiting and getting their hair cut, is time for themselves, and time to relax and enjoy the services the salon provides. For example, Coby, a 30-year-old Mexican realtor, says,

> You know, I don't have a lot of time; I'm really busy during the week. I have a son. I have a five-year old, [and] so I don't have a lot of time to do things for myself. So this is cool. I like to just come out and get some coffee or something … It's kind of relaxing; it's something I do for myself.

While Coby is pleased with the haircut he receives at the salon, it is the wait for the haircut he seems to enjoy most. As work becomes increasingly demanding, the salon serves an important function in the lives of the men at Shear Style; they pay for space in which they can escape from their busy lives. This is despite the feminized décor of the salon and the "girly" pillows which decorate the waiting bench.

While many of the men described the salon as a place of leisure, there are any number of places they could go to find time for themselves. So, why do these men choose the gendered space of the "woman's" hair salon? Shear Style provides the men with services that are unique to the salon experience, services that make many of the men feel "pampered" and taken care of. Sharma and Black's interviews with women beauty workers show that " 'pampering' was seen as a service which the stressed and hardworking (female) client deserved and needed" (2001, 918). This pampering comes in the form of paid touch, which is a key aspect of the hair salon experience. As I observed,

> You know, there's nothing like putting your head back and getting a little head massage while you're getting your hair washed, it feels good. Sal's [a hypothetical barber] not going to do that … That more than anything, it's one of those things where it just feels good. You get a little pampered; you get detail.

Mack notes that it is in the hair salon where beauty services include physical pleasure. He suggests that the men's barbershop does not include touch that can be described as pleasurable or pampering. In his work on the Black barbershop, Alexander (2003) describes being touched by the barber as *secretly* pleasurable. That is, while he enjoys having the barber touch his scalp, this touch could never be discussed openly as pleasurable since doing so might be interpreted by other men as homoerotic. Comparatively, the heterosocial interactions the men have with their women salon stylists allow the men to access paid touch without their presumed heterosexuality falling under suspicion.

By couching paid touch in heterosexuality, the men position themselves as heterosexual and resist the potential feminization of pampering in a hair salon. They clearly receive pleasure from the scalp massages and shampooing; but state that they enjoy this aspect of pampering solely because they are being touched by a woman. "It's like when you go to get a massage [and] they want to know if you want a man or a woman [masseuse]; I'm like, 'a woman sounds nice,' " Don told me. Sam, a 53-year-old white architect, agrees, "I would say I prefer women. It's kind of like a massage too. They ask, 'do you want a man or a woman?' I always prefer a woman." Don and Sam suggest that a massage from a man would be uncomfortable for them, presumably because it does not fit with their sense of themselves as heterosexual. Therefore, touch in the salon is possible and pleasurable for the men because it is done within the context of a hetero-sexual interaction. If the men were shampooed by another man, such as a barber, they might not enjoy, or at least discuss, the pampering aspect of paid touch because it could compromise their association with privileged heterosexuality.

Although some of the men classed their hair salon experience by describing it as akin to a "mini spa day," they had not, and declared that they would not, enter a spa. These men define the spa as a place where people (presumably women) purchase services such as pedicures, manicures, facials, massages, body-wraps, and mud-baths. They counterpose their more utilitarian haircut to spa and nail services, which they describe as not a "priority" for them. As one man notes, "I think it's one of those things where I value my spare time so much that it's like when I look at my priorities and things I want to do in my time, [the] spa isn't there." I briefly spoke with a man who was leaving the salon after his haircut; he said that the salon is a nice way to "sneak pampering in while doing something you would have to do anyway." He elaborated by telling me that for professional-class men, paying $45 for a haircut is more acceptable than spending money in a spa or a nail salon because it can be veiled as a necessity. Many of the men admitted that it would probably feel good to get a pedicure, for example, but said such services do not serve a utili-tarian purpose and, therefore, cannot be adequately disguised as masculine.

Personalized Relationships and Gendered Care Work

The relationships the men have with their stylists involve both touch and the exchange of personal information. Many of the men at Shear Style have grown attached and loyal to their stylists partially because of relationships they believe they share.

Building relationships with clients is an integral part of the workers' tasks in many body service occupations. Sharma and Black (2001) show that when "beauty therapists" draw customers into a conversation, they often ask questions regarding the customer's family and recent events in the customer's life. The men perceive their stylists' interest in their families as sincere and mutual. Patrick, a 61-year-old white corporate official, reports that he and his stylist have "become kind of friends … She tells me about [her daughter] and stuff, and I tell her about my sons and daughter." This personalized relationship motivates the men to come back to get their hair cut by the same stylist time and time again. For example, I asked one man why he comes back to Rosa for his haircuts and he said, "She's very friendly and she knows my kid's, my son's name. We keep kind of going on what happened last time, if something happened in my personal life, we talk about it." Another man said that he patronizes the salon because, "You know my daughters have come here, my wife has come here; so we [my stylist and I] can talk about *our* families."

Unlike the men, the stylists perceive their relationships with their clients as simply part of the job. They confided that they are not always interested in their clients' lives or families. This demonstrates the way in which beauty work involves the physical labor of cutting and styling hair as well as emotional labor. The stylists recognize the care work they perform, as well as the fact that it is not valued as such, claiming that their men clients come to them for a hair-cut because "it's cheaper than a psychiatrist." The men clients, however, do not see their relationships with their stylists as one-sided; nor do they understand their stylists as paid informal therapists. This is because emotional labor is often taken for granted and naturalized as part of women's essential character. Further-more, the men believe their relationships with their stylists are genuine, not "marred" by economic exchange. By personalizing these relationships, they make invisible the fact that they are paying for *body labor*: both the physical and emotional labor of women beauty workers.

In the men's discussions of their relationships with their stylists, they not only appropriate women's body work, they also establish themselves as members of a particular class. The men position themselves as "classy" by comparing "salon talk" to barbershop talk and by describing the barbershop as a place for the expression of working-class masculinity in which men talk about "beer and pussy," sports and cars. While the men explicitly class the barbershop and the hair salon, they also implicitly racialize these settings. Though these men do not discuss race, by setting up the barbershop as a place for the expression of a masculinity that differs from their own, they define the barbershop as a space for a *white* working-class masculinity. One man describes the difference between the salon and the barbershop as "night and day" and explains that, "You have garage talk [read as barbershop talk] and you have salon talk." By describing the conver-sation in the barbershop as "garage talk," the man genders the two spaces and suggests that traditionally "masculine" topics, which do not include feeling and family, are the only things discussed at the barbershop.

While the men at Shear Style have the option of getting their hair cut in a men's barbershop, they describe the barbershop as a place that does not provide them with *caring* relationships. The value the men place on women as sincere care workers is not surprising given the expectation that women perform emo-tional labor while providing services. However, by comparing the salon to the barbershop, the men indicate that the barbershop acts to uphold "traditional" notions of masculinity by informally discouraging the sharing of intimate infor-mation. Consequently, they differentiate themselves from what they describe as the white working-class masculinity of barbershop men, whom they say prefer to talk about "beer and pussy," to successfully situate themselves as progressive, professional-class white men.

A Stylish and Classed Haircut

The men at Shear Style also set themselves apart as members of a particular class by describing salon hair care as important to the accomplishment of a "stylish" white professional-class embodiment. They conflate salon hair care with "stylish," customized, and contemporary haircuts. For example, Mack notes, "[If I] want something a little more stylish, I'll come to the salon because the salon develops more current styles [and] different techniques [that are] more relevant." The men understand the salon as a space in which they are able to purchase current trends in hair, and they attribute the ability to deliver this style to the women hairstylists whom they suggest have a "high taste level" and are highly skilled. The men trust their stylists and take comfort in knowing they will get a "good" haircut each and every time they come to the salon. "I know that I'm going to have a consistently good haircut every time I go [to the salon]," one man told me. This consistency gives the men "peace of mind," and alleviates the worry and the stress they feel when they have their hair cut elsewhere.

The men's desires for aesthetic enhancement are potentially threatening to their masculinity since, as men, their sense of self-worth is not supposed to be tied to how they look. To counteract this potential threat, the men claim they do not *want* to look stylish for themselves; rather they *need* to look good to succeed professionally. They construct their purchase of beauty work in the salon as a practice that helps them to compete in the workplace and to persuade their clients that they are professional, responsible, and will do the job well. Tom says,

> I mean, you know, I have clients. That means before they become clients, I have to win them over. Now who are they going to go with? The person who has ... this great appearance package [pointing to himself] including grooming, style, professionalism, mannerism ... Who are they going to go with, that person, or are they going to go with somebody who looks like they came in and dressed by accident or [that they are] indifferent about their hair?

The men at Shear Style embed the meaning of their beauty work in professionalism. That is, their purchase of beauty services in the hair salon becomes about fulfilling what they interpret as expectations of white professional-class masculinity.

Few of these clients directly acknowledge the role class and occupation play in both their desire for and ability to purchase beauty work and "style" in the hair salon. Kerry, a 50-year-old white marketer, first describes the typical male client in the salon as "somebody who has more money than somebody who can only afford 10 dollars." However, he quickly reassesses his answer and decides, "It doesn't matter how rich or how poor you are, you have budgets and you have allocations. And some people who make less money will spend more on entertainment than people who make a lot of money. What's your priority[?]" In this way, Kerry marks himself, and the other men at Shear Style, as distinct in "priority," not privilege. This works to erase class privilege, although the purchase of salon hair care is made possible by the men's income and is rooted in the ability to succeed in a professional white-collar occupation.

The men also solidify their class status by again distancing themselves from the masculinity they associate with the "old school" barbershop. They reject the barbershop as a place where men purchase mass produced hairstyles by an out-of-date barber. For example, one man told me, "I think there is a difference; I think she [his stylist] cuts hair a little bit better [than a barber]." A good, stylish haircut is one that is current and modern, and is in contrast with the haircut the men feel they would receive from "old barber[s]." "The male barber is just bad," Hamilton exclaims, "80-year-old barbers who can't see just chop your hair." These men believe that white professional-class men do not need to get their hair cut at the barbershop since they can afford the "superior" and "customized" work of a salon stylist. Rather, they claim that working-class men purchase what they consider the inferior haircuts of the barbershop. As Evan says, "I can't see a mechanic working at, or a grease-monkey working at a Jiffy-Lube, or something like that, going to a shop that charges 65 bucks for a haircut." Evan differentiates the clients of the salon and barbershop in terms of class, and also sets the men salon clients up as superior by derogatorily referring to white working-class men as "grease-monkeys."

The men clients at Shear Style contrast the salon with the barbershop to justify their presence in a "women's" space. They refer to the barbershop as "old" and out of date, allowing them to position themselves in contrast as contemporary stylish men who rightfully seek the beauty work of women. Both the haircuts and the space of the barbershop are associated with an out-of-date style.

By contrasting themselves with the traditional "machismo" barbershop which they say does not deliver its supposed working-class customers with "style," the men at Shear Style construct themselves as a class of "new men": progressive, stylish, and professional.

★★★

The men at Shear Style—through the purchase and consumption of beauty work—appropriate embodied symbols of educational and cultural capital that distinguish them as raced, classed, sexualized, and gendered. While within a space defined as feminine, the men maintain a sense of masculinity by situating themselves as anomalies in a "women's" salon and by heterosexualizing their interactions with the women hairstylists. This research allows us to see how race and especially class privilege are reproduced by men through the consumption of beauty work, and how this privilege both allows men a "pass" to enter into a "women's" space while protecting them from the powerless aspects associated with the feminine culture of beauty.

NOTES

1. The names of the salon, the stylists, and the clients have been changed to protect the privacy of those involved in the study.
2. The fourth stylist did not cut men's hair.
3. A number of men declined to be formally interviewed because of family and work engagements.

REFERENCES

Alexander, Bryant Keith. 2003. Fading, twisting, and weaving: An interpretive ethnography of the Black barbershop as cultural space. *Qualitative Inquiry* 9 (1): 105–28.

Black, Paula. 2004. *The beauty industry: Gender, culture, pleasure.* New York: Routledge.

Bordo, Susan. 1999. *The male body: A new look at men in public and in private.* New York: Farrar, Straus and Giroux.

Jacobs-Huey, Lanita. 2006. *From the kitchen to the parlor: Language and becoming in African American women's hair care.* New York: Oxford University Press.

Luciano, Lynne. 2001. *Looking good: Male body image in modern America.* New York: Hill and Wang.

Sharma, Ursula, and Paula Black. 2001. Look good, feel better: Beauty therapy as emotional labor. *Sociology* 35 (4): 913–31.

Williams, Louis. 1993. The relationship between a Black barbershop and the community that supports it. *Human Mosaic* 27: 29–33.

Wright, Earl II. 1998. More than just a haircut: Sociability within the urban African American barbershop. *Challenge: A Journal of Research on African American Men* 9: 1–13.

22

The Culture of Black Femininity and School Success

CARLA O'CONNOR, R. L'HEUREUX LEWIS, AND JENNIFER MUELLER

Researchers who study conventions of black femininity repeatedly document the finding that black people raise their girls to be assertive and independent (Lewis, 1975; Slevin & Wingrove, 1998). Additionally, black families socialize their girls not only to take the roles of wife and mother, but to assume the role of worker. The heightened attention to developing voice, independence, and the worker identity has been linked to black men's historic marginalization in the workforce relative to white men. Black families necessarily had to raise their daughters in ways that would facilitate their girls' ability to assume partial, if not full, responsibility for the financial survival of their families upon becoming women (Lewis, 1975; Ward, 1996).

On the one hand, researchers have argued that this orientation toward voice and independence is a positive adaptation. Unlike white women—particularly upper-middle-class, white women—black women are not necessarily expected to silence their experiences, thoughts, and desires in relations with others. Researchers have subsequently suggested that these differential expectations might explain why black girls generally have higher self-esteem than their white and Latina counterparts and why they are able to maintain this esteem throughout adolescence (AAUW, 1991). Fordham (1993) also showed how this orientation toward voice enables the "loud black girl" to "retrieve a safe cultural space" for herself in school—one in which she is not rendered invisible. Instead, she can creatively subvert the expectations that she be female (rather than male) in a black (and not a white) sense.

Finally, Holland and Eisenhart (1990) showed that as a consequence of how black femininity is articulated, black *women*, compared to their white counterparts, were less preoccupied with romance and were less likely to be manipulated by men. Researchers, however, associate these same expressions of black women's agency with negative educational outcomes.

The same voice and power that are said to protect women against the loss of esteem and the loss of a culturally specific gendered self are also said to place them at academic risk when they produce psychological isolation and are realized in conflict with school officials (Fordham, 1993; Taylor, Gilligan, & Sullivan,

SOURCE: Reprinted with permission from *Beyond Silenced Voices: Class, Race, and Gender in United States Schools, Revised Edition*, edited by Lois Weis and Michelle Fine, the State University of New York Press. Copyright © 2005, State University of New York. All rights reserved.

1995). In the case of Fordham's (1993) work, she identified a high-achieving African American female who used her voice to affirm her existence. This expression of voice "propel[led] this young woman to the margins of [']good behavior' but never "actually forc[ed] her into the realm of 'bad behavior.'" Fordham, however, maintained that most "loud black girls" are not as strategic in their use of voice, and such "loudness" eventually "mutilates the academic achievement of large numbers of female African American students" (p. 5). Researchers also find that it is often the most assertive low-income and minority girls who leave or are pushed out of school (Fine, 1991; Fine & Zane, 1991). Those more likely to stay in school mute their own voices and express high conformity and limited political awareness.

In a similarly contradictory fashion, Holland and Eisenhart (1990) showed that black women focused less of their attention on men and were less manipulated by them. However, this independence did not translate into strong and consistent efforts in school. Having determined that grades of "C" would be sufficient for completing college, the black women were more likely than the white women to accept this grade. Holland and Eisenhart attributed this less than competitive performance to the black women's recognition that blacks were not equitably rewarded for their efforts in school. In the case of this text, there was nothing about the culture of black femininity that mitigated the negative effects associated with the perception of a limited opportunity structure. The culture of black femininity may have been cast in agentic terms, but such agency did not translate into competitive academic performance.

In reporting on the liberatory potential and practices associated with the culture of black femininity, this [paper] recognizes that conventions of femininity are not static phenomena. Rather, these conventions change over time in response to shifting social and economic opportunities. Importantly, then, our 19 women are distributed across 3 age cohorts that will be elaborated upon below. This cohort distribution affords us the opportunity to show how cultural conventions of black femininity were differently articulated across time to affect the women's experience with school success. In total then, this [paper] will not only discuss the liberatory potential and practices of the culture of black femininity in relation to how well and how far these women went in school but also will show that elements of this culture were differentially reflected over time and thus differentially taken up (or not) by the women to facilitate their educational success.

DOCUMENTING THE WOMEN'S VOICES
AND EXPERIENCES

The data for this chapter derive from a larger investigation of the life histories of 19 black women who were first-generation college graduates and grew up in low-income and working-class households. The women were born and then first attended a postsecondary institution within the following age cohorts. Cohort I (the precivil rights era cohort) includes those women who were born between 1926 and 1931 and first attended a postsecondary school before the 1950s. Cohort II (the post-civil rights era cohort) includes those women who were born between 1946 and 1955 and first attended a postsecondary school during the mid-1960s to mid-1970s. Cohort III (the post-Reagan era cohort) represents those women who were born between 1966 and 1970 and first attended a postsecondary school after 1984. All of the respondents were first-generation college graduates, and in all but two cases they were the first in their families to receive a baccalaureate degree. In every case, they were the first females in their immediate families to earn this degree. All of the respondents also attended what I will refer to as "Midwest University" (MU).

We captured the women's life stories via in-depth individual interviews that were audiotaped and transcribed verbatim. For the purposes of this paper, data analysis was directed toward answering the following questions: (1) What gender-related narratives were communicated in these women's life stories? (2) How did these gender-related narratives reveal constraints to and/or opportunities for educational achievement and mobility? (3) What were the processes by which the women developed, maintained, and exercised commitment to school to attain high levels of education given the representation of gender-based constraints and opportunities that were particular to their space and time? More specifically, we focused on how the women articulated the ways that gender (including how it may have intersected with race and/or social class) operated in their "everyday," as well as within the greater social context in which they grew up.

At the time in which the women shared their life stories, they were employed in or retired from a wide variety of professions that signaled their evident upward mobility in light of their lower social class origins. In our efforts to document those origins, we learned that for the most part these women grew up in homes that were solidly working class. With few exceptions, they grew up in two-parent households. And with only one breadwinner in the home, their families in most cases (even when extremely large) generated the financial resources necessary to avoid public or even charitable assistance. Most of the women additionally reported that for much of their childhood their family owned their home (though for the oldest cohort the homes were in most cases built rather than bought by their parents or guardians). In the majority of the cases, such homeownership coincided with the upward mobility of the family. That is, more than half of the women indicated that over time (particularly in relation to buying a first or a larger house) their families moved to "better," safer, and more economically stable communities.

Most of the women also grew up with fairly well educated caretakers. Although none of their primary caretakers was a graduate of a four-year college, seven of the women had at least one parent or guardian who had received some postsecondary school training, and thirteen had at least one caretaker who had graduated from high school.

THE CULTURE OF BLACK FEMININITY:
DEVELOPING VOICE, INDEPENDENCE,
AND POSSIBILITY

Across the age cohorts the women discussed the messages they received within the community with regards to how black women should look, how they should behave, and for what roles they should assume or hope. In many instances, these messages seemed oriented toward circumscribing these women's bodies and, to a lesser degree, their life chances. And as might be expected, Cohort I was most apt to discuss how these conventions were directed toward such circumscription. Only in this cohort did women report that it was not only schooling agents but also their own family members who sought to suppress their educational ambitions precisely because they were women.

Across the age cohorts, the women's stories also revealed that physical indices of femininity required women to be light in complexion, have fine and long hair, and be "thick" (neither fat nor skinny) in body. However, these physical requirements of femininity were more salient in the experiences of Cohort I.

The youngest cohort (Cohort III) made the least reference to the influence of these requirements. While these findings are consistent with other accounts of how skin color and hair length, in particular, operated in the black community over time (e.g., Russell, 1992), these women's life stories suggest that these physical requirements of beauty had important educational effects. This was particularly revealed in the case of Cohort I members who, in the absence of school-sanctioned support for their educational ambitions, had to rely heavily on informal social networks to access information and encouragement for their college attendance. Their stories revealed that their physical appearance determined whether they would be welcomed into more elite social circles that had accordant access to knowledge and resources upon which they could draw to facilitate their educational mobility.

With regards to how the women's lives were otherwise circumscribed, they reported (without evident distinctions between the cohorts) that while growing up, restrictions were placed upon their bodies. Sometimes these restrictions were articulated via their inability to move as freely about public space as compared to their brothers and other male counterparts.

At other times, the women found that these bodily restrictions were a function of how "etiquette" prescribed how, where, when, and with whom they could or could not use or move their bodies. Sometimes indices of etiquette referred to dating etiquette or if and when they could begin dating and in accordance with what rules. The women indicated that compared to the males in their family, they encountered severe or at least more severe dating rules and regulations. Otherwise, etiquette lessons focused on defining the women's physical comportment.

… Repeatedly, the women conveyed that they received communications that they should not be constrained in their ambitions … [*because*] they were women. Sometimes these communications were conveyed in light of what was not said, as when Tia reported that she was "never told that women can't do anything." Other times, these communications were modeled on the bodies of other women. Sometimes this modeling was reflected in the actions of seemingly ordinary women pursuing more routine activities. For example, when we asked Desiree Strong whether anyone had ever taught her what it meant to be a woman, in general, or a black woman, in particular, she responded:

> Well, first of all, my mother … taught me in her own way that *you could do anything*. My mother didn't start driving till I think she was in her 40s.… Got tired of waiting on my father to teach her how to drive. Saved her little change, bought her own car, took driving lessons, was driving. (Original emphasis)

[The data we found] demonstrate the multiple ways in which women in this study came to know and/or emulate the voice and power of black women. Candace's aunt taught her, if only indirectly, that black women not only can wield financial power but also can move things and benefit from a no-nonsense stance. Desiree's mother showed her that women need not wait on a man but can act efficaciously in their "everyday" efforts to pursue their own interests. Angela Davis and the extraordinary women in Kingsley's life provided evidence of black women's political will and agency. These different representations of black women's voice and power were weaved throughout the participants' life stories. Importantly, when the women further reported on their familiarity with black women's voice and power and the processes by which they became acculturated to these expressions of womanhood, they often discussed how family members often established an explicit link between these expressions and the receipt of higher education. For example, Leona Holmes' mother told her, "Honey, get your degree so that you don't have to be tied down to no Negro.… You can always take care of yourself." Tia Richardson explained that her parents wanted her sister and her "to be able to go to college so that we could, you know, provide for ourselves and, you know, be self-sufficient so where we would have to necessarily rely on someone else to provide … for us." Nora Bentley explained that her father provided her with the most "predominate" message regarding dating or the role that romantic relationships should play in her life. His message was, "Get a degree. Don't depend upon any man."

Nearly to a person, the members of Cohort III indicated that they received explicit exhortations that they needed to get an education so that they could enact their agency by being independent of men.

COPING WITH CONSTRAINTS

Having been socialized toward voice and power as a consequence of explicit exhortation, as well as via intimate and publicly available models of black women's individual, economic, social, and political agency, it is not surprising to learn that the women in this study described themselves as "pushy" "strong," "loud," "aggressive," "assertive," "demanding," "determined." Having asked one participant, L'Nette Farnsworth, why she used one of the adjectives above to describe herself (in her case, she was explaining why she would describe herself as strong), we were not surprised to receive the following response:

> I feel that I can take a stand even though I am female.... I don't have to succumb to certain issues because, you know, men are supposed to be the superior sex and, you know, wear the pants in the family.... I'm not afraid to speak up for what I believe, and ... if I feel that someone is stepping on my toes or because I am a woman or whatever, ... I'm not afraid to challenge someone as it relates to being a woman.

Cohort III members (along with some members of Cohort II) explained that it was this very socialization that explained, in part, their pursuit of and persistence in higher education. Members of Cohort III indicated that much of their motivation to go far in school was fostered in light of the explicit exhortation that they achieve high levels of education in order to wield their power in relationship with or independent of men. Sometimes they would hold onto these exhortations when they encountered educational obstacles.

In contrast to Cohort III, Cohorts I and II specifically indicated how they actually employed their voice and power in their effort to have others respond to, support, or not hinder their own school-related pursuits.

SUMMARY AND CONCLUSION

These women's stories require us to reassess previous research, which suggests that black women's socialization toward voice and power is likely to put them at risk for school failure. Though little research has been conducted on this topic, what work is available suggests that when black girls express their voice and power in school, they are placed at an academic disadvantage. They experience psychological isolation, find themselves in conflict with school officials, and are more likely to be "pushed" out of school. The findings of this study, however, require us to generate a more complicated picture regarding how the culture of black femininity can shape the educational experiences and outcomes of black girls. More precisely, the findings reported herein reveal how the culture of black femininity can be productive in relation to schooling.

The life stories featured in this paper show how the culture of black femininity impacted black women's ability to imagine that black women in general, and they in particular, could act efficaciously despite gender-based barriers that operate in the "everyday" and in society at large. Just as important, these stories convey how this culture (articulated via exhortation or feminine models) informed the cognitive and behavioral processes by which the women actually resisted those constraints that threatened their academic success and persistence.

The women in Cohorts II and III indicated that, in part, their positive valuation of education, their college aspirations, and their persistence through school were fostered by their family members' efforts in socializing them to become self-sufficient (i.e., independent of men). Additionally, the women were able to invoke their voice and power in ways that enabled them to resist those efforts, cultures, and practices directed at circumscribing their educational experiences and outcomes. In most instances, these women experienced such circumscription in academic settings and at the hands of schooling agents. However, they would draw on black feminine agency to protest inequitable treatment and to claim school spaces that had been previously denied to blacks. Often the women's protests and their willingness to be in conflict with school officials produced (more) positive educational outcomes. Conflict in these instances was productive. In the absence of engaging in conflict, these women would have been further marginalized in academic settings. Additionally, they would have resigned themselves to inequitable grading practices and lower grades.

Although these findings reveal the productive character of the women's socialization towards black feminine agency, we must be mindful of evidence that the voice and power of black women is at best suppressed and often negatively sanctioned in schools. What is at stake when schools and educators seek to suppress, sanction, or push out those traits that are central to cultivation of worldviews and strategies that can expand black people's life chances? If we have any chance of working toward a more just society, schools and educators must take up the challenge to explore how they might build upon, rather than work against, the socially productive nature of black femininity.

REFERENCES

American Association of University Women & Greenberg Lake the Analysis Group. (1991). *Shortchanging girls, shortchanging America. A nationwide poll to assess self esteem, educational experiences, interest in math and science, and career aspirations of girls and boys ages 9–15*, Washington, D.C.: American Association of University Women.

Fine, M. (1991). *Framing dropouts: Notes on the politics of an urban public high school*. Albany, NY: State University of New York Press.

Fine, M. & Zane, N. (1991). Bein' wrapped too tight: When low-income women drop out of high school. *Women's Studies Quarterly, 1 & 2*, 77–99.

Fordham, S. (1996). *Blacked out: Dilemmas of race, identity, and success at Capital High*. Chicago, IL: University of Chicago Press.

Holland, D. C. & Eisenhart, M. A. (1990). *Educated in romance; Women, achievement, and college culture*. Chicago, IL: University of Chicago Press.

Lewis, D. K. (1975). The black family: Socialization and sex roles. *Phylon, 36*, 221–238.

Russell, K. (1992). *The color complex: The politics of skin color among African Americans*. New York: Harcourt Brace Jovanovich.

Slevin, K. F., & Wingrove, C. R. (1998). *From stumbling blocks to stepping stones: The life experiences of fifty professional African American women*. New York: New York University Press.

Taylor, J. M., Gilligan, C., & Sullivan, A. M. (1995). *Between voice and silence: Women and girls, race and relationship*. Cambridge, MA: Harvard University Press.

Ward, J. V. (1996). Raising resisters: The role of truth telling in the psychological development of African American girls. In B. J. Ross Leadbeater and N. Way (Eds.), *Urban girls: Resisting stereotypes, creating identities*. New York: New York University Press.

23

The First Americans
American Indians

C. MATTHEW SNIPP

By the end of the nineteenth century, many observers predicted that American Indians were destined for extinction. Within a few generations, disease, warfare, famine, and outright genocide had reduced their numbers from millions to less than 250,000 in 1890. Once a self-governing, self-sufficient people, American Indians were forced to give up their homes and their land, and to subordinate themselves to an alien culture. The forced resettlement to reservation lands or the Indian Territory (now Oklahoma) frequently meant a life of destitution, hunger, and complete dependency on the federal government for material needs.

Today, American Indians are more numerous than they have been for several centuries. While still one of the most destitute groups in American society, tribes have more autonomy and are now more self-sufficient than at any time since the last century. In cities, modern pan-Indian organizations have been successful in making the presence of American Indians known to the larger community, and have mobilized to meet the needs of their people (Cornell 1988; Nagel 1986; Weibel-Orlando 1991). In many rural areas, American Indians and especially tribal governments have become increasingly more important and increasingly more visible by virtue of their growing political and economic power. The balance of this [reading] is devoted to explaining their unique place in American society.

THE INCORPORATION OF AMERICAN INDIANS

The current political and economic status of American Indians is the result of the process by which they were incorporated into Euro-American society (Hall 1989). This amounts to a long history of efforts aimed at subordinating an otherwise self-governing and self-sufficient people that eventually culminated in widespread economic dependency. The role of the U.S. government in this process can be seen in the five major historical periods of federal Indian relations:

SOURCE: From Silvia Pedraza and Rubén G. Rumbaut, eds., *Origins and Destinies: Immigration, Race, and Ethnicity in America.* (Belmont, CA: Wadsworth, 1996), pp. 390–403. Reprinted by permission.

removal, assimilation, the Indian New Deal, termination and relocation, and self-determination.

Removal

In the early nineteenth century, the population of the United States expanded rapidly at the same time that the federal government increased its political and military capabilities. The character of Indian-American relations changed after the War of 1812. The federal government increasingly pressured tribes settled east of the Appalachian Mountains to move west to the territory acquired in the Louisiana Purchase. Numerous treaties were negotiated by which the tribes relinquished most of their land and eventually were forced to move west.

Initially the federal government used bargaining and negotiation to accomplish removal, but many tribes resisted (Prucha 1984). However, the election of Andrew Jackson by a frontier constituency signaled the beginning of more forceful measures to accomplish removal. In 1830 Congress passed the Indian Removal Act, which mandated the eventual removal of the eastern tribes to points west of the Mississippi River, in an area which was to become the Indian Territory and is now the state of Oklahoma. Dozens of tribes were forcibly removed from the eastern half of the United States to the Indian Territory and newly created reservations in the west, a long process ridden with conflict and bloodshed.

As the nation expanded beyond the Mississippi River, tribes of the plains, southwest, and west coast were forcibly settled and quarantined on isolated reservations. This was accompanied by the so-called Indian Wars—a bloody chapter in the history of Indian-White relations (Prucha 1984; Utley 1963). This period in American history is especially remarkable because the U.S. government was responsible for what is unquestionably one of the largest forced migrations in history.

The actual process of removal spanned more than a half-century and affected nearly every tribe east of the Mississippi River. Removal often meant extreme hardships for American Indians, and in some cases this hardship reached legendary proportions. For example, the Cherokee removal has become known as the "Trail of Tears." In 1838, nearly 17,000 Cherokees were ordered to leave their homes and assemble in military stockades (Thornton 1987, p. 117). The march to the Indian Territory began in October and continued through the winter months. As many as 8,000 Cherokees died from cold weather and diseases such as influenza (Thornton 1987, p. 118).

According to William Hagan (1979), removal also caused the Creeks to suffer dearly as their society underwent a profound disintegration. The contractors who forcibly removed them from their homes refused to do anything for "the large number who had nothing but a cotton garment to protect them from the sleet storms and no shoes between them and the frozen ground of the last stages of their hegira. About half of the Creek nation did not survive the migration and the difficult early years in the West" (Hagan 1979, pp. 77–81). In the West, a band of Nez Perce men, women, and children, under the leadership of Chief

Joseph, resisted resettlement in 1877. Heavily outnumbered, they were pursued by cavalry troops from the Wallowa valley in eastern Oregon and finally captured in Montana near the Canadian border. Although the Nez Perce were eventually captured and moved to the Indian Territory, and later to Idaho, their resistance to resettlement has been described by one historian as "one of the great military movements in history" (Prucha 1984, p. 541).

Assimilation

Near the end of the nineteenth century, the goal of isolating American Indians on reservations and the Indian Territory was finally achieved. The Indian population also was near extinction. Their numbers had declined steadily throughout the nineteenth century, leading most observers to predict their disappearance (Hoxie 1984). Reformers urged the federal government to adopt measures that would humanely ease American Indians into extinction. The federal government responded by creating boarding schools and the allotment acts—both were intended to "civilize" and assimilate American Indians into American society by Christianizing them, educating them, introducing them to private property, and making them into farmers. American Indian boarding schools sought to accomplish this task by indoctrinating Indian children with the belief that tribal culture was an inferior relic of the past and that Euro-American culture was vastly superior and preferable. Indian children were forbidden to wear their native attire, to eat their native foods, to speak their native language, or to practice their traditional religion. Instead, they were issued Euro-American clothes, and expected to speak English and become Christians. Indian children who did not relinquish their culture were punished by school authorities. The curriculum of these schools taught vocational arts along with "civilization" courses.

The impact of allotment policies is still evident today. The 1887 General Allotment Act (the Dawes Severalty Act) and subsequent legislation mandated that tribal lands were to be allotted to individual American Indians ... and the surplus lands left over from allotment were to be sold on the open market. Indians who received allotted tribal lands also received citizenship, farm implements, and encouragement from Indian agents to adopt farming as a livelihood (Hoxie 1984, Prucha 1984).

For a variety of reasons, Indian lands were not completely liquidated by allotment, many Indians did not receive allotments, and relatively few changed their lifestyles to become farmers. Nonetheless, the allotment era was a disaster because a significant number of allottees eventually lost their land. Through tax foreclosures, real estate fraud, and their own need for cash, many American Indians lost what for most of them was their last remaining asset (Hoxie 1984).

Allotment took a heavy toll on Indian lands. It caused about 90 million acres of Indian land to be lost, approximately two-thirds of the land that had belonged to tribes in 1887 (O'Brien 1990). This created another problem that continues to vex many reservations: "checkerboarding." Reservations that were subjected to allotment are typically a crazy quilt composed of tribal lands, privately owned "fee" land, and trust land belonging to individual Indian families.

Checkerboarding presents reservation officials with enormous administrative problems when trying to develop land use management plans, zoning ordinances, or economic development projects that require the construction of physical infrastructure such as roads or bridges.

The Indian New Deal

The Indian New Deal was short-lived but profoundly important. Implemented in the early 1930s along with the other New Deal programs of the Roosevelt administration, the Indian New Deal was important for at least three reasons. First, signaling the end of the disastrous allotment era as well as a new respect for American Indian tribal culture, the Indian New Deal repudiated allotment as a policy. Instead of continuing its futile efforts to detribalize American Indians, the federal government acknowledged that tribal culture was worthy of respect. Much of this change was due to John Collier, a long-time Indian rights advocate appointed by Franklin Roosevelt to serve as Commissioner of Indian Affairs (Prucha 1984).

Like other New Deal policies, the Indian New Deal also offered some relief from the Great Depression and brought essential infrastructure development to many reservations, such as projects to control soil erosion and to build hydroelectric dams, roads, and other public facilities. These projects created jobs in New Deal programs such as the Civilian Conservation Corps and the Works Progress Administration.

An especially important and enduring legacy of the Indian New Deal was the passage of the Indian Reorganization Act (IRA) of 1934. Until then, Indian self-government had been forbidden by law. This act allowed tribal governments, for the first time in decades, to reconstitute themselves for the purpose of overseeing their own affairs on the reservation. Critics charge that this law imposed an alien form of government, representative democracy, on traditional tribal authority. On some reservations, this has been an on-going source of conflict (O'Brien 1990). Some reservations rejected the IRA for this reason, but now have tribal governments authorized under different legislation.

Termination and Relocation

After World War II, the federal government moved to terminate its long-standing relationship with Indian tribes by settling the tribes' outstanding legal claims, by terminating the special status of reservations, and by helping reservation Indians relocate to urban areas (Fixico 1986). The Indian Claims Commission was a special tribunal created in 1946 to hasten the settlement of legal claims that tribes had brought against the federal government. In fact, the Indian Claims Commission became bogged down with prolonged cases, and in 1978 the commission was dissolved by Congress. At that time, there were 133 claims still unresolved out of an original 617 that were first heard by the commission three decades earlier (Fixico 1986, p. 186). The unresolved claims that were still pending were transferred to the Federal Court of Claims.

Congress also moved to terminate the federal government's relationship with Indian tribes. House Concurrent Resolution (HCR) 108, passed in 1953, called for steps that eventually would abolish all reservations and abolish all special programs serving American Indians. It also established a priority list of reservations slated for immediate termination. However, this bill and subsequent attempts to abolish reservations were vigorously opposed by Indian advocacy groups such as the National Congress of American Indians. Only two reservations were actually terminated, the Klamath in Oregon and the Menominee in Wisconsin. The Menominee reservation regained its trust status in 1975 and the Klamath reservation was restored in 1986.

The Bureau of Indian Affairs (BIA) also encouraged reservation Indians to relocate and seek work in urban job markets. This was prompted partly by the desperate economic prospects on most reservations, and partly because of the federal government's desire to "get out of the Indian business." The BIA's relocation programs aided reservation Indians in moving to designated cities, such as Los Angeles and Chicago, where they also assisted them in finding housing and employment. Between 1952 and 1972, the BIA relocated more than 100,000 American Indians (Sorkin 1978). However, many Indians returned to their reservations (Fixico 1986). For some American Indians, the return to the reservation was only temporary; for example, during periods when seasonal employment such as construction work was hard to find.

Self-Determination

Many of the policies enacted during the termination and relocation era were steadfastly opposed by American Indian leaders and their supporters. As these programs became stalled, critics attacked them for being harmful, ineffective, or both. By the mid-1960s, these policies had very little serious support. Perhaps inspired by the gains of the Civil Rights movement, American Indian leaders and their supporters made "self-determination" the first priority on their political agendas. For these activists, self-determination meant that Indian people would have the autonomy to control their own affairs, free from the paternalism of the federal government.

The idea of self-determination was well received by members of Congress sympathetic to American Indians. It also was consistent with the "New Federalism" of the Nixon administration. Thus, the policies of termination and relocation were repudiated in a process that culminated in 1975 with the passage of the American Indian Self-Determination and Education Assistance Act, a profound shift in federal Indian policy. For the first time since this nation's founding, American Indians were authorized to oversee the affairs of their own communities, free of federal intervention. In practice, the Self-Determination Act established measures that would allow tribal governments to assume a larger role in reservation administration of programs for welfare assistance, housing, job training, education, natural resource conservation, and the maintenance of reservation roads and bridges (Snipp and Summers 1991). Some reservations also have their own police forces and game wardens, and can issue licenses and levy taxes. The Onondaga tribe in upstate New York have taken their sovereignty one step further by issuing passports that are internationally recognized. Yet there is a great deal of variability in terms of how much autonomy tribes have over reservation affairs. Some tribes, especially those on large and well-organized reservations have nearly complete control over their reservations, while smaller reservations with limited resources often depend heavily on BIA services....

CONCLUSION

Though small in number, American Indians have an enduring place in American society. Growing numbers of American Indians occupy reservation and other trust lands, and equally important has been the revitalization of tribal governments. Tribal governments now have a larger role in reservation affairs than ever in the past. Another significant development has been the urbanization of American Indians. Since 1950, the proportion of American Indians in cities has grown rapidly. These American Indians have in common with reservation Indians many of the same problems and disadvantages, but they also face other challenges unique to city life.

The challenges facing tribal governments are daunting. American Indians are among the poorest groups in the nation. Reservation Indians have substantial needs for improved housing, adequate health care, educational opportunities, and employment, as well as developing and maintaining reservation infrastructure. In the face of declining federal assistance, tribal governments are assuming an ever-larger burden. On a handful of reservations, tribal governments have assumed completely the tasks once performed by the BIA.

As tribes have taken greater responsibility for their communities, they also have struggled with the problems of raising revenues and providing economic opportunities for their people. Reservation land bases provide many reservations with resources for development. However, these resources are not always abundant, much less unlimited, and they have not always been well managed. It will be yet another challenge for tribes to explore ways of efficiently managing their existing resources. Legal challenges also face tribes seeking to exploit unconventional resources such as gambling revenues. Their success depends on many complicated legal and political contingencies.

Urban American Indians have few of the resources found on reservations, and they face other difficult problems. Preserving their culture and identity is an especially pressing concern. However, urban Indians have successfully adapted to city environments in ways that preserve valued customs and activities—powwows, for example, are an important event in all cities where there is a large Indian community. In addition, pan-Indianism has helped urban Indians set aside tribal differences and forge alliances for the betterment of urban Indian communities.

These alliances are essential, because unlike reservation Indians, urban American Indians do not have their own form of self-government. Tribal governments do not have jurisdiction over urban Indians. For this reason, urban Indians must depend on other strategies for ensuring that the needs of their community are met, especially for those new to city life. Coping with the transition to urban life poses a multitude of difficult challenges for many American Indians. Some succumb to these problems, especially the hardships of unemployment, economic deprivation, and related maladies such as substance abuse, crime, and violence. But most successfully overcome these difficulties, often with help from other members of the urban Indian community.

Perhaps the greatest strength of American Indians has been their ability to find creative ways for dealing with adversity, whether in cities or on reservations. In the past, this quality enabled them to survive centuries of oppression and persecution. Today this is reflected in the practice of cultural traditions that Indian people are proud to embrace. The resilience of American Indians is an abiding quality that will no doubt ensure that they will remain part of the ethnic mosaic of American society throughout the twenty-first century and beyond.

REFERENCES

Cornell, Stephen. 1988. *The Return of the Native: American Indian Political Resurgence.* New York: Oxford University Press.

Fixico, Donald L. 1986. *Termination and Relocation: Federal Indian Policy, 1945–1960.* Albuquerque, NM: University of New Mexico Press.

Hagan, William T. 1979. *American Indians.* Chicago, IL: University of Chicago Press.

Hall, Thomas D. 1989. *Social Change in the Southwest, 1350–1880.* Lawrence, KS: University Press of Kansas.

Hoxie, Frederick E. 1984. *A Final Promise: The Campaign to Assimilate the Indians, 1880–1920.* Lincoln, NE: University of Nebraska Press.

Nagel, Joanne. 1986. "American Indian Repertoires of Contention." Paper presented at the annual meeting of the American Sociological Association, San Francisco, CA.

O'Brien, Sharon. 1990. *American Indian Tribal Governments.* Norman, OK: University of Oklahoma Press.

Prucha, Francis Paul. 1984. *The Great Father.* Lincoln, NE: University of Nebraska Press.

Snipp, C. Matthew and Gene F. Summers. 1991. "American Indian Development Policies," pp. 166–180 in *Rural Policies for the 1990s*, edited by Cornelia Flora and James A. Christenson. Boulder, CO: Westview Press.

Sorkin, Alan L. 1978. *The Urban American Indian.* Lexington, MA: Lexington Books.

Thornton, Russell. 1987. *American Indian Holocaust and Survival: A Population History since 1942.* Norman, OK: University of Oklahoma Press.

Utley, Robert M. 1963. *The Last Days of the Sioux Nation.* New Haven: Yale University Press.

Weibel-Orlando, Joan. 1991. *Indian Country, L.A.* Urbana, IL: University of Illinois Press.

24

"Is This a White Country, or What?"

LILLIAN B. RUBIN

"They're letting all these coloreds come in and soon there won't be any place left for white people," broods Tim Walsh, a thirty-three-year-old white construction worker. "It makes you wonder: Is this a white country, or what?"

It's a question that nags at white America, one perhaps that's articulated most often and most clearly by the men and women of the working class. For it's they who feel most vulnerable, who have suffered the economic contractions of recent decades most keenly, who see the new immigrants most clearly as direct competitors for their jobs.

It's not whites alone who stew about immigrants. Native-born blacks, too, fear the newcomers nearly as much as whites—and for the same economic reasons. But for whites the issue is compounded by race, by the fact that the newcomers are primarily people of color. For them, therefore, their economic anxieties have combined with the changing face of America to create a profound uneasiness about immigration—a theme that was sounded by nearly 90 percent of the whites I met, even by those who are themselves first-generation, albeit well-assimilated, immigrants.

Sometimes they spoke about this in response to my questions; equally often the subject of immigration arose spontaneously as people gave voice to their concerns. But because the new immigrants are dominantly people of color, the discourse was almost always cast in terms of race as well as immigration, with the talk slipping from immigration to race and back again as if these are not two separate phenomena. If we keep letting all them foreigners in, pretty soon there'll be more of them than us and then what will this country be like?" Tim's wife, Mary Anne, frets. I mean, this is *our* country, but the way things are going, white people will be the minority in our own country. Now does that make any sense?"

Such fears are not new. Americans have always worried about the strangers who came to our shores, fearing that they would corrupt our society, dilute our culture, debase our values. So I remind Mary Anne, "When your ancestors came here, people also thought we were allowing too many foreigners into the country. Yet those earlier immigrants were successfully integrated into the American society. What's different now?"

SOURCE: From *Families on the Fault Line* by Lillian B. Rubin, pp. 172–196, HarperCollins Publishers. Copyright © 1994 by Lilian B. Rubin. Reprinted by permission of the author.

"Oh, it's different, all right," she replies without hesitation. "When my people came, the immigrants were all white. That makes a big difference."...

Listening to Mary Anne's words I was reminded again how little we Americans look to history for its lessons, how impoverished is our historical memory.

For, in fact, being white didn't make "a big difference" for many of those earlier immigrants. The dark-skinned Italians and the eastern European Jews who came in the late nineteenth and early twentieth centuries didn't look very white to the fair-skinned Americans who were here then. Indeed, the same people we now call white—Italians, Jews, Irish—were seen as another race at that time. Not black or Asian, it's true, but an alien other, a race apart, although one that didn't have a clearly defined name. Moreover, the racist fears and fantasies of native-born Americans were far less contained then than they are now, largely because there were few social constraints on their expression.

When, during the nineteenth century, for example, some Italians were taken for blacks and lynched in the South, the incidents passed virtually unnoticed. And if Mary Anne and Tim Walsh, both of Irish ancestry, had come to this country during the great Irish immigration of that period, they would have found themselves defined as an inferior race and described with the same language that was used to characterize blacks: "low-browed and savage, grovelling and bestial, lazy and wild, simian and sensual."[1] Not only during that period but for a long time afterward as well, the U.S. Census Bureau counted the Irish as a distinct and separate group, much as it does today with the category it labels "Hispanic."

But there are two important differences between then and now, differences that can be summed up in a few words: the economy and race. Then, a growing industrial economy meant that there were plenty of jobs for both immigrant and native workers, something that can't be said for the contracting economy in which we live today. True, the arrival of the immigrants, who were more readily exploitable than native workers, put Americans at a disadvantage and created discord between the two groups. Nevertheless, work was available for both.

Then, too, the immigrants—no matter how they were labeled, no matter how reviled they may have been—were ultimately assimilable, if for no other reason than that they were white. As they began to lose their alien ways, it became possible for native Americans to see in the white ethnics of yesteryear a reflection of themselves. Once this shift in perception occurred, it was possible for the nation to incorporate them, to take them in, chew them up, digest them, and spit them out as Americans—with subcultural variations not always to the liking of those who hoped to control the manners and mores of the day, to be sure, but still recognizably white Americans.

Today's immigrants, however, are the racial other in a deep and profound way.... And integrating masses of people of color into a society where race conciousness lies at the very heart of our central nervous system raises a whole new set of anxieties and tensions....

The increased visibility of other racial groups has focused whites more self-consciously than ever on their own racial identification. Until the new immigration shifted the complexion of the land so perceptibly, whites didn't think of themselves as white in the same way that Chinese know they're Chinese and African-Americans know they're black. Being white was simply a fact of life, one that didn't require any public statement, since it was the definitive social value against which all others were measured. "It's like everything's changed and I don't know what happened," complains Marianne Bardolino. "All of a sudden you have to be thinking all the time about these race things. I don't remember growing up thinking about being white like I think about it now. I'm not saying I didn't know there was coloreds and whites; it's just that I didn't go along thinking, *Gee, I'm a white person.* I never thought about it at all. But now with all the different colored people around, you have to think about it because they're thinking about it all the time."

"You say you feel pushed now to think about being white, but I'm not sure I understand why. What's changed?" I ask.

"I told you," she replies quickly, a small smile covering her impatience with my question. "It's because they think about what they are, and they want things their way, so now I have to think about what I am and what's good for me and my kids." She pauses briefly to let her thoughts catch up with her tongue, then continues. "I mean, if somebody's always yelling at you about being black or Asian or something, then it makes you think about being white. Like, they want the kids in school to learn about their culture, so then I think about being white and being Italian and say: What about my culture? If they're going to teach about theirs, what about mine?"

To which America's racial minorities respond with bewilderment. "I don't understand what white people want," says Gwen Tomalson. "They say if black kids are going to learn about black culture in school, then white people want their kids to learn about white culture. I don't get it. What do they think kids have been learning about all these years? It's all about white people and how they live and what they accomplished. When I was in school you wouldn't have thought black people existed for all our books ever said about us."

As for the charge that they're "thinking about race all the time," as Marianne Bardolino complains, people of color insist that they're forced into it by a white world that never lets them forget. "If you're Chinese, you can't forget it, even if you want to, because there's always something that reminds you," Carol Kwan's husband, Andrew, remarks tartly. "I mean, if Chinese kids get good grades and get into the university, everybody's worried and you read about it in the papers."

While there's little doubt that racial anxieties are at the center of white concerns, our historic nativism also plays a part in escalating white alarm. The new immigrants bring with them a language and an ethnic culture that's vividly expressed wherever they congregate. And it's this also, the constant reminder of an alien presence from which whites are excluded, that's so troublesome to them.

The nativist impulse isn't, of course, given to the white working class alone. But for those in the upper reaches of the class and status hierarchy—those whose children go to private schools, whose closest contact with public transportation is the taxi cab—the immigrant population supplies a source of cheap labor, whether as nannies for their children, maids in their households, or workers in their businesses. They may grouse and complain that "nobody speaks English anymore," just as working-class people do. But for the people who use immigrant labor, legal or illegal, there's a payoff for the inconvenience—a payoff that doesn't exist for the families in this study but that sometimes costs them dearly. For while it may be true that American workers aren't eager for many of the jobs immigrants are willing to take, it's also true that the presence of a large immigrant population—especially those who come from developing countries where living standards are far below our own—helps to make these jobs undesirable by keeping wages depressed well below what most American workers are willing to accept....

It's not surprising, therefore, that working-class women and men speak so angrily about the recent influx of immigrants. They not only see their jobs and their way of life threatened, they feel bruised and assaulted by an environment that seems suddenly to have turned color and in which they feel like strangers in their own land. So they chafe and complain: "They come here to take advantage of us, but they don't really want to learn our ways," Beverly Sowell, a thirty-three-year old white electronics assembler, grumbles irritably. "They live different than us; it's like another world how they live. And they're so clannish. They keep to themselves, and they don't even *try* to learn English. You go on the bus these days and you might as well be in a foreign country; everybody's talking some other language, you know, Chinese or Spanish or something. Lots of them have been here a long time, too, but they don't care; they just want to take what they can get."

But their complaints reveal an interesting paradox, an illuminating glimpse into the contradictions that beset native-born Americans in their relations with those who seek refuge here. On the one hand, they scorn the immigrants; on the other, they protest because they "keep to themselves." It's the same contradiction that dominates black-white relations. Whites refuse to integrate blacks but are outraged when they stop knocking at the door, when they move to sustain the separation on their own terms' [sic]—in black theme houses on campuses, for example, or in the newly developing black middle-class suburbs.

I wondered, as I listened to Beverly Sowell and others like her, why the same people who find the life ways and languages of our foreign-born population offensive also care whether they "keep to themselves."

"Because like I said, they just shouldn't, that's all," Beverly says stubbornly. "If they're going to come here, they should be willing to learn our ways—you know what I mean, be real Americans. That's what my grandparents did, and that's what they should do."

"But your grandparents probably lived in an immigrant neighborhood when they first came here, too," I remind her.

"It was different," she insists. "I don't know why; it was. They wanted to be Americans; these here people now, I don't think they do. They just want to take advantage of this country["]....

"Everything's changed, and it doesn't make sense. Maybe you get it, but I don't. We can't take care of our own people and we keep bringing more and more foreigners in. Look at all the homeless. Why do we need more people here when our own people haven't got a place to sleep?"

"Why do we need more people here?"—a question Americans have asked for two centuries now. Historically, efforts to curb immigration have come during economic downturns, which suggests that when times are good, when American workers feel confident about their future, they're likely to be more generous in sharing their good fortune with foreigners. But when the economy falters, as it did in the 1990s, and workers worry about having to compete for jobs with people whose standard of living is well below their own, resistance to immigration rises. "Don't get me wrong; I've got nothing against these people," Tim Walsh demurs. "But they don't talk English, and they're used to a lot less, so they can work for less money than guys like me can. I see it all the time; they get hired and some white guy gets left out."

It's this confluence of forces—the racial and cultural diversity of our new immigrant population; the claims on the resources of the nation now being made by those minorities who, for generations, have called America their home; the failure of some of our basic institutions to serve the needs of our people; the contracting economy, which threatens the mobility aspirations of working-class families—all these have come together to leave white workers feeling as if everyone else is getting a piece of the action while they get nothing. "I feel like white people are left out in the cold," protests Diane Johnson, a twenty-eight-year-old white single mother who believes she lost a job as a bus driver to a black woman. "First it's the blacks; now it's all those other colored people, and it's like everything always goes their way. It seems like a white person doesn't have a chance anymore. It's like the squeaky wheel gets the grease, and they've been squeaking and we haven't," she concludes angrily.

Until recently, whites didn't need to think about having to "squeak"—at least not specifically as whites. They have, of course, organized and squeaked at various times in the past—sometimes as ethnic groups, sometimes as workers. But not as whites. As whites they have been the dominant group, the favored ones, the ones who could count on getting the job when people of color could not. Now suddenly there are others—not just individual others but identifiable groups, people who share a history, a language, a culture, even a color—who lay claim to some of the rights and privileges that formerly had been labeled for whites only." And whites react as if they've been betrayed, as if a sacred promise has been broken. They're white, aren't they? They're *real* Americans, aren't they? This is their country, isn't it?

The answers to these questions used to be relatively unambiguous. But not anymore. Being white no longer automatically assures dominance in the politics of a multiracial society. Ethnic group politics, however, has a long and fruitful history. As whites sought a social and political base on which to stand, therefore, it was natural and logical to reach back to their ethnic past. Then they, too, could be "something"; they also would belong to a group; they would have a name, a history, a culture, and a voice. "Why is it only the blacks or Mexicans or Jews that are 'something'?" asks Tim Walsh. "I'm Irish, isn't that something, too? Why doesn't that count?"

In reclaiming their ethnic roots, whites can recount with pride the tribulations and transcendence of their ancestors and insist that others take their place in the line from which they have only recently come. "My people had a rough time, too. But nobody gave us anything, so why do we owe them something? Let them pull their share like the rest of us had to do," says Al Riccardi, a twenty-nine-year-old white taxi driver.

From there it's only a short step to the conviction that those who don't progress up that line are hampered by nothing more than their own inadequacies or, worse yet, by their unwillingness to take advantage of the opportunities offered them. "Those people, they're hollering all the time about discrimination," Al continues, without defining who "those people" are. "Maybe once a long time ago that was true, but not now. The problem is that a lot of those people are lazy. There's plenty of opportunities, but you've got to be willing to work hard."

He stops a moment, as if listening to his own words, then continues, "Yeah, yeah, I know there's a recession on and lots of people don't have jobs. But it's different with some of those people. They don't really want to work, because if they did, there wouldn't be so many of them selling drugs and getting in all kinds of trouble."

"You keep talking about 'those people' without saying who you mean," I remark.

"Aw c'mon, you know who I'm talking about," he says, his body shifting uneasily in his chair. "It's mostly the black people, but the Spanish ones, too."

In reality, however, it's a no-win situation for America's people of color, whether immigrant or native born. For the industriousness of the Asians comes in for nearly as much criticism as the alleged laziness of other groups. When blacks don't make it, it's because, whites like Al Riccardi insist, their culture doesn't teach respect for family; because they're hedonistic, lazy, stupid, and/or criminally inclined. But when Asians demonstrate their ability to overcome the obstacles of an alien language and culture, when the Asian family seems to be the repository of our most highly regarded traditional values, white hostility doesn't disappear. It just changes its form. Then the accomplishments of Asians, the speed with which they move up the economic ladder, aren't credited to their superior culture, diligence, or intelligence—even when these are granted—but to the fact that they're "single minded," "untrustworthy," "clannish drones," "narrow people" who raise children who are insufficiently "well rounded."[2]...

Not surprisingly, as competition increases, the various minority groups often are at war among themselves as they press their own particular claims, fight over turf, and compete for an ever-shrinking piece of the pie. In several African-American communities, where Korean shopkeepers have taken the place once held by Jews, the confrontations have been both wrenching and tragic. A Korean grocer in Los Angeles shoots and kills a fifteen-year-old black girl for allegedly trying to steal some trivial item from the store.[3] From New York City to Berkeley, California, African-Americans boycott Korean shop owners who, they charge, invade their neighborhoods, take their money, and treat them disrespectfully.[4] But painful as these incidents are for those involved, they are only symptoms of a deeper malaise in both communities—the contempt and distrust in which the Koreans hold their African-American neighbors, and the rage of blacks as they watch these new immigrants surpass them.

Latino-black conflict also makes headlines when, in the aftermath of the riots in South Central Los Angeles, the two groups fight over who will get the lion's share of the jobs to rebuild the neighborhood. Blacks, insisting that they're being discriminated against, shut down building projects that don't include them in satisfactory numbers. And indeed, many of the jobs that formerly went to African-Americans are now being taken by Latino workers. In an article entitled "Black vs. Brown," Jack Miles, an editorial writer for the *Los Angeles Times,* reports that janitorial firms serving downtown Los Angeles have almost entirely replaced their unionized black work force with non-unionized immigrants."[5]...

But the disagreements among America's racial minorities are of little interest or concern to most white working-class families. Instead of conflicting groups, they see one large mass of people of color, all of them making claims that endanger their own precarious place in the world. It's this perception that has led some white ethnics to believe that reclaiming their ethnicity alone is not enough, that so long as they remain in their separate and distinct groups, their power will be limited. United, however, they can become a formidable countervailing force, one that can stand fast against the threat posed by minority demands. But to come together solely as whites would diminish their impact and leave them open to the charge that their real purpose is simply to retain the privileges of whiteness. A dilemma that has been resolved, at least for some, by the birth of a new entity in the history of American ethnic groups—the "European-Americans."[6]...

At the University of California at Berkeley, for example, white students and their faculty supporters insisted that the recently adopted multicultural curriculum include a unit of study of European-Americans. At Queens College in New York City, where white ethnic groups retain a more distinct presence, Italian-American students launched a successful suit to win recognition as a disadvantaged minority and gain the entitlements accompanying that status, including special units of Italian-American studies.

White high school students, too, talk of feeling isolated and, being less sophisticated and wary than their older sisters and brothers, complain quite openly that there's no acceptable and legitimate way for them to acknowledge a white identity. "There's all these things for all the different ethnicities, you know, like clubs for black kids and Hispanic kids, but there's nothing for me and my friends to join," Lisa Marshall, a sixteen-year-old white high school student, explains with exasperation. "They won't let us have a white club because that's supposed to be racist. So we figured we'd just have to call it something else, you know, some ethnic thing, like Euro-Americans. Why not? They have African-American clubs."

Ethnicity, then, often becomes a cover for "white," not necessarily because these students are racist but because racial identity is now such a prominent feature of the discourse in our social world. In a society where racial consciousness is so high, how else can whites define themselves in ways that connect them to a community and, at the same time, allow them to deny their racial antagonisms?

Ethnicity and race—separate phenomena that are now inextricably entwined. Incorporating newcomers has never been easy, as our history of controversy and violence over immigration tells us.[7] But for the first time, the new immigrants are also people of color, which means that they tap both the nativist and racist impulses that are so deeply a part of American life. As in the past, however, the fear of foreigners, the revulsion against their strange customs and seemingly unruly ways, is only part of the reason for the anti-immigrant attitudes that are increasingly being expressed today. For whatever xenophobic suspicions may arise in modern America, economic issues play a critical role in stirring them up.

REFERENCES

Alba, Richard D. *Ethnic Identity*. New Haven: Yale University Press, 1990.

Roediger, David R. *The Wages of Whiteness*. New York: Verso, 1991.

NOTES

1. David R. Roediger, *The Wages of Whiteness* (New York: Verso, 1991), p. 133.

2. These were, and often still are, the commonly held stereotypes about Jews. Indeed, the Asian immigrants are often referred to as "the new Jews."

3. Soon Ja Du, the Korean grocer who killed fifteen-year-old Latasha Harlins, was found guilty of voluntary manslaughter and sentenced to four hundred hours of community service, a $500 fine, reimbursement of funeral costs to the Harlins family, and five years' probation.

4. The incident in Berkeley didn't happen in the black ghetto, as most of the others did. There, the Korean grocery store is near the University of California campus, and the woman involved in the incident is an African-American university student who was maced by the grocer after an argument over a penny.

5. Jack Miles, "Blacks vs. Browns," *Atlantic Monthly* (October 1992), pp. 41–68.

6. For an interesting analysis of what he calls "the transformation of ethnicity," see Richard D. Alba, *Ethnic Identity* (New Haven, CT: Yale University Press, 1990).

7. In the past, many of those who agitated for a halt to immigration were immigrants or native-born children of immigrants. The same often is true today. As antiimmigrant sentiment grows, at least some of those joining the fray are relatively recent arrivals. One man in this study, for example—a fifty-two-year-old immigrant from Hungary—is one of the leaders of an anti-immigration group in the city where he lives.

25

Optional Ethnicities

For Whites Only?

MARY C. WATERS

What does it mean to talk about ethnicity as an option for an individual? To argue that an individual has some degree of choice in their ethnic identity flies in the face of the commonsense notion of ethnicity many of us believe in— that one's ethnic identity is a fixed characteristic, reflective of blood ties and given at birth. However, social scientists who study ethnicity have long concluded that while ethnicity is based on a *belief* in a common ancestry, ethnicity is primarily a *social* phenomenon, not a biological one (Alba 1985, 1990; Barth 1969; Weber [1921] 1968, p. 389). The belief that members of an ethnic group have that they share a common ancestry may not be a fact. There is a great deal of change in ethnic identities across generations through intermarriage, changing allegiances, and changing social categories. There is also a much larger amount of change in the identities of individuals over their lives than is commonly believed. While most people are aware of the phenomenon known as "passing"—people raised as one race who change at some point and claim a different race as their identity—there are similar life course changes in ethnicity that happen all the time and are not given the same degree of attention as racial "passing."

White Americans of European ancestry can be described as having a great deal of choice in terms of their ethnic identities. The two major types of options White Americans can exercise are (1) the option of whether to claim any specific ancestry, or to just be "White" or American, [Lieberson (1985) called these people "unhyphenated Whites"] and (2) the choice of which of their European ancestries to choose to include in their description of their own identities. In both cases, the option of choosing how to present yourself on surveys and in everyday social interactions exists for Whites because of social changes and societal conditions that have created a great deal of social mobility, immigrant assimilation, and political and economic power for Whites in the United States. Specifically, the option of being able to not claim any ethnic identity exists for Whites of European background in the United States because they are the majority group—in terms of holding political and social power, as well as being a numerical majority. The option of choosing among different ethnicities in their

SOURCE: From Silvia Pedraza and Rubén G. Rumbaut, eds., *Origins and Destinies: Immigration, Race and Ethnicity in America* (Belmont, CA: Wadsworth, 1996), pp. 444–54. Reprinted by permission.

family backgrounds exists because the degree of discrimination and social distance attached to specific European backgrounds has diminished over time....

SYMBOLIC ETHNICITIES FOR WHITE AMERICANS

What do these ethnic identities mean to people and why do they cling to them rather than just abandoning the tie and calling themselves American? My own field research with suburban Whites in California and Pennsylvania found that later-generation descendants of European origin maintain what are called "symbolic ethnicities. Symbolic ethnicity is a term coined by Herbert Gans (1979) to refer to ethnicity that is individualistic in nature and without real social cost for the individual. These symbolic identifications are essentially leisure-time activities, rooted in nuclear family traditions and reinforced by the voluntary enjoyable aspects of being ethnic (Waters 1990). Richard Alba (1990) also found later-generation Whites in Albany, New York, who chose to keep a tie with an ethnic identity because of the enjoyable and voluntary aspects to those identities, along with the feelings of specialness they entailed. An example of symbolic ethnicity is individuals who identify as Irish, for example, on occasions such as Saint Patrick's Day, on family holidays, or for vacations. They do not usually belong to Irish American organizations, live in Irish neighborhoods, work in Irish jobs, or marry other Irish people. The symbolic meaning of being Irish American can be constructed by individuals from mass media images, family traditions, or other intermittent social activities. In other words, for later-generation White ethnics, ethnicity is not something that influences their lives unless they want it to. In the world of work and school and neighborhood, individuals do not have to admit to being ethnic unless they choose to. And for an increasing number of European-origin individuals whose parents and grandparents have intermarried, the ethnicity they claim is largely a matter of personal choice as they sort through all of the possible combinations of groups in their genealogies....

RACE RELATIONS AND SYMBOLIC ETHNICITY

However much symbolic ethnicity is without cost for the individual, there is a cost associated with symbolic ethnicity for the society. That is because symbolic ethnicities of the type described here are confined to White Americans of European origin. Black Americans, Hispanic Americans, Asian Americans, and American Indians do not have the option of a symbolic ethnicity at present in the United States. For all of the ways in which ethnicity does not matter for White Americans, it does matter for non-Whites. Who your ancestors are does affect your choice of spouse, where you live, what job you have, who your friends are, and what your chances are for success in American society, if those ancestors happen not to be from Europe. The reality is that White ethnics have a lot more choice and room for maneuver than they themselves think they do.

The situation is very different for members of racial minorities, whose lives are strongly influenced by their race or national origin regardless of how much they may choose not to identify themselves in terms of their ancestries.

When White Americans learn the stories of how their grandparents and great-grandparents triumphed in the United States over adversity, they are usually told in terms of their individual efforts and triumphs. The important role of labor unions and other organized political and economic actors in their social and economic successes are left out of the story in favor of a generational story of individual Americans rising up against communitarian, Old World intolerance, and New World resistance. As a result, the "individualized" voluntary, cultural view of ethnicity for Whites is what is remembered.

One important implication of these identities is that they tend to be very individualistic. There is a tendency to view valuing diversity in a pluralist environment as equating all groups. The symbolic ethnic tends to think that all groups are equal; everyone has a background that is their right to celebrate and pass on to their children. This leads to the conclusion that all identities are equal and all identities in some sense are interchangeable—"I'm Italian American, you're Polish American. I'm Irish American, you're African American." The important thing is to treat people as individuals and all equally. However, this assumption ignores the very big difference between an individualistic symbolic ethnic identity and a socially enforced and imposed racial identity.

My favorite example of how this type of thinking can lead to some severe misunderstandings between people of different backgrounds is from the *Dear Abby* advice column. A few years back a person wrote in who had asked an acquaintance of Asian background where his family was from. His acquaintance answered that this was a rude question and he would not reply. The bewildered White asked Abby why it was rude, since he thought it was a sign of respect to wonder where people were from, and he certainly would not mind anyone asking HIM about where his family was from. Abby asked her readers to write in to say whether it was rude to ask about a person's ethnic background. She reported that she got a large response, that most non-Whites thought it was a sign of disrespect, and Whites thought it was flattering:

> Dear Abby,
> I am 100 percent American and because I am of Asian ancestry I am often asked "What are you?" It's not the personal nature of this question that bothers me, it's the question itself. This query seems to question my very humanity. "What am I? Why I am a person like everyone else!"
> Signed, *A REAL AMERICAN*

> Dear Abby,
> Why do people resent being asked what they are? The Irish are so proud of being Irish, they tell you before you even ask. Tip O'Neill has never tried to hide his Irish ancestry.
> Signed, *JIMMY*.

(Reprinted by permission of Universal Press Syndicate)

In this exchange Jimmy cannot understand why Asians are not as happy to be asked about their ethnicity as he is, because he understands his ethnicity and theirs to be separate but equal. Everyone has to come from somewhere—his family from Ireland, another's family from Asia—each has a history and each should be proud of it. But the reason he cannot understand the perspective of the Asian American is that all ethnicities are not equal; all are not symbolic, cost-less, and voluntary. When White Americans equate their own symbolic ethnici-ties with the socially enforced identities of non-White Americans, they obscure the fact that the experiences of Whites and non-Whites have been qualitatively different in the United States and that the current identities of individuals partly reflect that unequal history.

In the next section I describe how relations between Black and White students on college campuses reflect some of these asymmetries in the understanding of what a racial or ethnic identity means. While I focus on Black and White students in the following discussion, you should be aware that the myriad other groups in the United States—Mexican Americans, American Indians, Japanese Americans—all have some degree of social and individual influences on their identities, which reflect the group's social and economic history and present circumstance.

RELATIONS ON COLLEGE CAMPUSES

Both Black and White students face the task of developing their race and ethnic identities. Sociologists and psychologists note that at the time people leave home and begin to live independently from their parents, often ages eighteen to twenty-two, they report a heightened sense of racial and ethnic identity as they sort through how much of their beliefs and behaviors are idiosyncratic to their families and how much are shared with other people. It is not until one comes in close contact with many people who are different from oneself that individuals realize the ways in which their backgrounds may influence their individual personality. This involves coming into contact with people who are different in terms of their ethnicity, class, religion, region, and race. For White students, the ethnicity they claim is more often than not a symbolic one—with all of the voluntary, enjoyable, and intermittent characteristics I have described above.

Black students at the university are also developing identities through interactions with others who are different from them. Their identity development is more complicated than that of Whites because of the added element of racial discrimination and racism, along with the "ethnic" developments of finding others who share their background. Thus Black students have the positive attraction of being around other Black students who share some cultural elements, as well as the need to band together with other students in a reactive and oppositional way in the face of racist incidents on campus.

Colleges and universities across the country have been increasing diversity among their student bodies in the last few decades. This has led in many cases to strained relations among students from different racial and ethnic backgrounds. The 1980s and 1990s produced a great number of racial incidents and high racial tensions on campuses. While there were a number of racial incidents that were due to bigotry, unlawful behavior, and violent or vicious attacks, much of what happens among students on campuses involves a low level of tension and awkwardness in social interactions.

Many Black students experience racism personally for the first time on campus. The upper-middle-class students from White suburbs were often isolated enough that their presence was not threatening to racists in their high schools. Also, their class background was known by their residence and this may have prevented attacks being directed at them. Often Black students at the university who begin talking with other students and recognizing racial slights will remember incidents that happened to them earlier that they might not have thought were related to race.

Black college students across the country experience a sizeable number of incidents that are clearly the result of racism. Many of the most blatant ones that occur between students are the result of drinking. Sometimes late at night, drunken groups of White students coming home from parties will yell slurs at single Black students on the street. The other types of incidents that happen include being singled out for special treatment by employees, such as being followed when shopping at the campus bookstore, or going to the art museum with your class and the guard stops you and asks for your I.D. Others involve impersonal encounters on the street—being called a nigger by a truck driver while crossing the street, or seeing old ladies clutch their pocketbooks and shake in terror as you pass them on the street. For the most part these incidents are not specific to the university environment, they are the types of incidents middle-class Blacks face every day throughout American society, and they have been documented by sociologists (Feagin 1991).

In such a climate, however, with students experiencing these types of incidents and talking with each other about them, Black students do experience a tension and a feeling of being singled out. It is unfair that this is part of their college experience and not that of White students. Dealing with incidents like this, or the ever-present threat of such incidents, is an ongoing developmental task for Black students that takes energy, attention, and strength of character. It should be clearly understood that this is an asymmetry in the "college experience" for Black and White students. It is one of the unfair aspects of life that results from living in a society with ongoing racial prejudice and discrimination. It is also very understandable that it makes some students angry at the unfairness of it all, even if there is no one to blame specifically. It is also very troubling because, while most Whites do not create these incidents, some do, and it is never clear until you know someone well whether they are the type of person who could do something like this. So one of the reactions of Black students to these incidents is to band together.

In some sense then, as Blauner (1992) has argued, you can see Black students coming together on campus as both an "ethnic" pull of wanting to be together to share common experiences and community, and a "racial" push of banding together defensively because of perceived rejection and tension from Whites. In this way the ethnic identities of Black students are in some sense similar to, say, Korean students wanting to be together to share experiences. And it is an ethnicity that is generally much stronger than, say, Italian Americans. But for Koreans who come together there is generally a definition of themselves as "different from" Whites. For Blacks reacting to exclusion, there is a tendency for the coming together to involve both being "different from" but also "opposed to" Whites.

The anthropologist John Ogbu (1990) has documented the tendency of minorities in a variety of societies around the world, who have experienced severe blocked mobility for long periods of time, to develop such oppositional identities. An important component of having such an identity is to describe others of your group who do not join in the group solidarity as devaluing and denying their very core identity. This is why it is not common for successful Asians to be accused by others of acting "White" in the United States, but it is quite common for such a term to be used by Blacks and Latinos. The oppositional component of a Black identity also explains how Black people can question whether others are acting "Black enough." On campus, it explains some of the intense pressures felt by Black students who do not make their racial identity central and who choose to hang out primarily with non-Blacks. This pressure from the group, which is partly defining itself by not being White, is exacerbated by the fact that race is a physical marker in American society. No one immediately notices the Jewish students sitting together in the dining hall, or the one Jewish student sitting surrounded by non-Jews, or the Texan sitting with the Californians, but everyone notices the Black student who is or is not at the "Black table" in the cafeteria.

An example of the kinds of misunderstandings that can arise because of different understandings of the meanings and implications of symbolic versus oppositional identities concerns questions students ask one another in the dorms about personal appearances and customs. A very common type of interaction in the dorm concerns questions Whites ask Blacks about their hair. Because Whites tend to know little about Blacks, and Blacks know a lot about Whites, there is a general asymmetry in the level of curiosity people have about one another. Whites, as the numerical majority, have had little contact with Black culture; Blacks, especially those who are in college, have had to develop bicultural skills—knowledge about the social worlds of both Whites and Blacks. Miscommunication and hurt feelings about White students' questions about Black students' hair illustrate this point. One of the things that happens freshman year is that White students are around Black students as they fix their hair. White students are generally quite curious about Black students' hair—they have basic questions such as how often Blacks wash their hair, how they get it straightened or curled, what products they use on their hair, how they comb it, etc. Whites often wonder to themselves whether they should ask these questions. One thought experiment Whites perform is to ask themselves whether a particular question would upset them. Adopting the "do unto others" rule, they ask themselves, "If a Black person was curious about my hair would I get upset?" The answer usually is "No, I would be happy to tell them." Another example is an Italian American student wondering to herself, "Would I be upset if someone asked me about calamari?" The answer is no, so she asks her Black roommate about collard greens, and the roommate explodes with an angry response such as, "Do you think all Black people eat watermelon too?" Note that if this Italian American knew her friend was Trinidadian American and asked about peas and rice the situation would be more similar and would not necessarily ignite underlying tensions.

Like the debate in *Dear Abby,* these innocent questions are likely to lead to resentment. The issue of stereotypes about Black Americans and the assumption that all Blacks are alike and have the same stereotypical cultural traits has more power to hurt or offend a Black person than vice versa. The innocent questions about Black hair also bring up a number of asymmetries between the Black and White experience. Because Blacks tend to have more knowledge about Whites than vice versa, there is not an even exchange going on, the Black freshman is likely to have fewer basic questions about his White roommate than his White roommate has about him. Because of the differences historically in the group experiences of Blacks and Whites there are some connotations to Black hair that don't exist about White hair. (For instance, is straightening your hair a form of assimilation, do some people distinguish between women having "good hair" and "bad hair" in terms of beauty and how is that related to looking "White"?) Finally, even a Black freshman who cheerfully disregards or is unaware that there are these asymmetries will soon slam into another asymmetry if she willingly answers every innocent question asked of her. In a situation where Blacks make up only 10 percent of the student body, if every non-Black needs to be educated about hair, she will have to explain it to nine other students. As one Black student explained to me, after you've been asked a couple of times about something so personal you begin to feel like you are an attraction in a zoo, that you are at the university for the education of the White students.

INSTITUTIONAL RESPONSES

Our society asks a lot of young people. We ask young people to do something that no one else does as successfully on such a wide scale—that is to live together with people from very different backgrounds, to respect one another, to appreciate one another, and to enjoy and learn from one another. The successes that occur every day in this endeavor are many, and they are too often overlooked. However, the problems and tensions are also real, and they will not vanish on their own. We tend to see pluralism working in the United States in much the same way some people expect capitalism to work. If you put together people with various interests and abilities and resources, the "invisible hand" of capitalism is supposed to make all the parts work together in an economy for the common good.

There is much to be said for such a model—the invisible hand of the market can solve complicated problems of production and distribution better than any "visible hand" of a state plan. However, we have learned that unequal power relations among the actors in the capitalist marketplace, as well as "externalities" that the market cannot account for, such as long-term pollution, or collusion between corporations, or the exploitation of child labor, means that state regulation is often needed. Pluralism and the relations between groups are very similar. There is a lot to be said for the idea that bringing people who belong to different ethnic or racial groups together in institutions with no interference will have good consequences. Students from different backgrounds will make friends if they share a dorm room or corridor, and there is no need for the institution to do any more than provide the locale. But like capitalism, the invisible hand of pluralism does not do well when power relations and externalities are ignored. When you bring together individuals from groups that are differentially valued in the wider society and provide no guidance, there will be problems. In these cases the "invisible hand" of pluralist relations does not work, and tensions and disagreements can arise without any particular individual or group of individuals being "to blame." On college campuses in the 1990s some of the tensions between students are of this sort. They arise from honest misunderstandings, lack of a common background, and very different experiences of what race and ethnicity mean to the individual.

The implications of symbolic ethnicities for thinking about race relations are subtle but consequential. If your understanding of your own ethnicity and its relationship to society and politics is one of individual choice, it becomes harder to understand the need for programs like affirmative action, which recognize the ongoing need for group struggle and group recognition, in order to bring about social change. It also is hard for a White college student to understand the need that minority students feel to band together against discrimination. It also is easy, on the individual level, to expect everyone else to be able to turn their ethnicity on and off at will, the way you are able to, without understanding that ongoing discrimination and societal attention to minority status makes that impossible for individuals from minority groups to do. The paradox of symbolic ethnicity is that it depends upon the ultimate goal of a pluralist society, and at the same time makes it more difficult to achieve that ultimate goal. It is dependent upon the concept that all ethnicities mean the same thing, that enjoying the traditions of one's heritage is an option available to a group or an individual, but that such a heritage should not have any social costs associated with it.

As the Asian Americans who wrote to *Dear Abby* make clear, there are many societal issues and involuntary ascriptions associated with non-White identities. The developments necessary for this to change are not individual but societal in nature. Social mobility and declining racial and ethnic sensitivity are closely associated. The legacy and the present reality of discrimination on the basis of race or ethnicity must be overcome before the ideal of a pluralist society, where all heritages are treated equally and are equally available for individuals to choose or discard at will, is realized.

REFERENCES

Alba, Richard D. 1985. *Italian Americans: Into the of Twilight Ethnicity*. Englewood Cliffs, NJ: Prentice-Hall.

Alba, Richard D. 1990. *Ethnic Identity: The Transformation of White America*. New Haven: Yale University Press.

Barth, Frederick. 1969. *Ethnic Groups and Boundaries*. Boston: Little, Brown.

Blauner, Robert. 1992. "Talking Past Each Other: Black and White Languages of Race." *American Prospect* (Summer): 55–64.

Feagin, Joe R. 1991. "The Continuing Significance of Race: Anti-Black Discrimination in Public Places." *American Sociological Review* 56: 101–17.

Gans, Herbert. 1979. "Symbolic Ethnicity: The Future of Ethnic Groups and Cultures in America." *Ethnic and Racial Studies* 2: 1–20.

Lieberson, Stanley. 1985. *Making It Count: The Improvement of Social Research and Theory*. Berkeley: University of California Press.

Ogbu, John. 1990. "Minority Status and Literacy in Comparative Perspective." *Daedalus* 119: 141–69.

Waters, Mary C. 1990. *Ethnic Options: Choosing Identities in America*. Berkeley: University of California Press.

Weber, Max. [1921]/1968. *Economy and Society: An Outline of Interpretive Sociology*. Eds. Guenther Roth and Claus Wittich, trans. Ephraim Fischoff. New York: Bedminister Press.

26

A Dream Deferred: Undocumented Students at CUNY

CAROLINA BANK MUÑOZ

I first became aware of the difficulties for undocumented students at the City University of New York (CUNY) when I started teaching a course at Brooklyn College, a CUNY campus, on the sociology of immigration. On the first day of class, five students requested appointments to speak with me in private. This was extremely unusual to say the least. All five students were undocumented and had family members who were undocumented. They were hoping I could help. As one student put it, "I'm hoping you can teach me how to get my papers." I had to explain that I was not a lawyer, nor was the class about how to immigrate "legally," but about the social process of immigration. Needless to say, the students were deeply disappointed, but nevertheless stayed enrolled in the course. One student in particular made a tremendous impression on me.

Luisa came to the United States when she was five years old.[1] Her father was diagnosed with a rare and serious illness and they initially migrated so that he could be treated. Like many other immigrants, they obtained a visa to visit the United States. Once Luisa's father was treated and recovering, they decided to remain in the United States. They overstayed their visa, and from one night to the next became undocumented. Luisa attended public school while both of her parents worked in the garment industry. After Proposition 187 passed in California, Luisa's parents decided that it was time to leave California and move to New York, where the anti-immigrant climate was less intense.

During high school, Luisa worked after school as a seamstress in the factory where her mother worked. Her father was now a union janitor and their financial situation had stabilized substantially. In her last year of high school, Luisa started researching colleges. At that point she realized that there were very few opportunities for undocumented immigrants. She had been in this country for twelve years. She had learned English, worked hard, and made good grades. Yet, she was not going to be able to simply apply to college like many of her classmates. Despite her 3.8 GPA, Luisa would have to attend a community college because she simply could not afford to pay full tuition at a 4-year college and she was not eligible for any federal loans. After working full-time and attending school part-time for three years, Luisa had finally saved enough

SOURCE: Munoz, Carolina Bank, "A Dream Deferred: Undocumented Students at CUNY," *Radical Teacher* 84 (Spring) pp. 8–17. Copyright © 2008 by City University of New York. Reprinted by permission.

money to enroll at Brooklyn College. During her first year at BC, Luisa's brother was deported. She used her entire savings to bring her brother back to the United States across the U.S.–Mexico border and was forced to drop out. She was devastated to have to delay her education, but her family was the priority. After a two year hiatus from school, Luisa was able to re-enroll at Brooklyn College. That very semester she enrolled in my immigration course. While Luisa's story is incredible, it is not exceptional.

Hundreds of undocumented students across the country have similar stories.[2]

In fact, over 60,000 undocumented students, the vast majority of whom are people of color, graduate from high school every year (UCLA Labor Center 2007). Most of these students migrated to the United States at a young age along with parents or other family members. Yet they are subject to the same harsh immigration policies as their parents who predominantly work in the low wage sector. The United States is the only home that most of these students know, but they are forced to live in the shadows of American society, living in fear of Immigration and Customs Enforcement (ICE) with marginal access to good jobs or a college education.

For these students, a college education is usually only a dream. In fact, only five to ten percent of these students make it to college (UCLA Labor Center 2007, NILC 2006). Undocumented college students have no access to federal and state student aid, work study programs, or many scholarships. Furthermore, since the 1990s but especially since September 11th, 2001, access to higher education for undocumented students has been severely curtailed. Many states passed laws that required colleges and universities to charge non-resident tuition to undocumented students (Gonzales 2007). Nonresident tuition is often 2 to 3 times more expensive than in-state tuition, making it nearly impossible for undocumented students to attend college.

The half-dozen undocumented students in my class and the more than two thousand undocumented students at CUNY (according to the CUNY Immigration and Citizenship Project) have had to overcome tremendous adversity to be at the university. Over the course of the semester, several of my students saw their family members deported, one student successfully evaded a workplace raid by Immigration and Customs Enforcement, and two students had to drop out because they simply could not afford to stay in school. Since my first experience with these five students, I have run into dozens of undocumented college students across CUNY who have had to drop out of college, find work, save money, and return to college a few years later. Many never return to school because they simply do not earn enough in the low wage sector or underground economy to afford a college education.

Undocumented students are systematically denied access to a college education by a flawed immigration system that has roots in institutionalized racism. At its root, contemporary immigration policy is inherently flawed because it seeks an individual solution to a structural problem. In the latest round of immigration reform, legislators have focused on either blocking the flow of migration through "solutions" such as a border fence or severely limiting it through guest worker programs and other means. These policies treat immigration as a faucet that can be turned on and off. In fact, immigration is far more complex. There are structural conditions and policies that *force* people to migrate. The North American Free Trade Agreement (NAFTA), structural adjustment policies, and war all impact migration. NAFTA, in particular, has been instrumental in increased forced migration from Mexico. NAFTA resulted in removing tariffs that were protecting Mexican farmers without removing U.S. subsidies to U.S. producers. As a result, the Mexican market was flooded with underpriced agricultural goods from the United States, especially corn. Unable to compete with these underpriced goods, Mexican farmers had no choice but to leave their land and seek employment in other parts of Mexico. Many displaced farmers migrate to large cities in Mexico to work in factories. As those factory jobs disappear or are exported to other countries, they have nowhere to go but the United States (Bank Muñoz 2008). Ironically, then, U.S. economic and trade policies are significantly responsible for the increase in immigration from Mexico and other Latin American countries.

The ongoing backlash against immigration disproportionately affects immigrants of color (militias and vigilante groups such as the Minutemen for the most part are not violently attacking "illegal" Germans). As Ngai (2004) aptly puts it "restrictive immigration laws produced new categories of racial difference.... The legal racialization of these ethnic groups' national origin cast them as permanently foreign and unassimilable to the nation" (7–8). A perfect example is that in the contemporary immigration debate, Latina/o immigrants are racialized as "illegal" even if they were born in the United States or otherwise hold "legal" status (Bank Muñoz 2008). This blanket racialization falls on undocumented students in very particular ways.

The crisis for access to higher education for undocumented students affects not only the students who are in college or trying to get into college now, but also younger undocumented students who drop out of high school because they see that they have no opportunities for upward mobility. We are facing the possibility of a lost generation of extraordinarily bright and talented students....

TEACHING AND WORKING WITH
IMMIGRANT STUDENTS

Working with immigrant students, and particularly undocumented students of color, offers various opportunities and challenges. On the one hand, their life experience provides them with an intuitive sense of the global economy and racial and class disparities. Many of them come from the Global South and have experienced poverty, racism, and exploitation. These students also tend to have a greater understanding of world politics. Needless to say, their knowledge and experiences contribute tremendously to a vibrant classroom environment. I recall a particularly intense classroom discussion over the idea of reparations. Native born Blacks were arguing that only Black people who can prove a link to slavery in the United States should benefit from reparations while Caribbean

Blacks argued that they were also entitled to reparations because they were forced to migrate to the United States due to the devastations of globalization and colonization. All students in the class, immigrants and native born alike, benefited tremendously from this exchange.

✳   In my experience, undocumented students are among the most self-motivated and focused students I have had, perhaps because they have the most to gain or lose. There are very few paths to obtain permanent residence and a green card. One can either acquire it through a family member (spouse, parent, etc.) who is a U.S. citizen or through employment. Employment is often the best option for undocumented students. In this case, employers have to make a case for why a foreign national (instead of a U.S. citizen) is better suited for the position (USCIS 2008). A college degree gives undocumented students, especially in high demand fields, some hope of finding a job in which an employer will be able to help secure their immigration status. On the other hand, having to drop out of school minimizes the chances that these students will find good jobs and a road to citizenship. Therefore, college recruitment and retention of undocumented students is imperative to enhancing their life chances.

Unfortunately, undocumented students face extreme barriers to succeeding in school and completing their college education even when they overcome the barriers to getting into college. Immigrant students in general and undocumented students in particular often live in poor neighborhoods with underfunded public schools. This is also true for other students of color. However, immigrant students have the additional barriers of having had to transition from schools in their native countries to a new method of U.S. education and learn U.S. English.[3] As a result they often have weaker writing and public speaking skills than other students. Additionally, as I have already mentioned, many of these students have significant barriers outside the classroom, which limit their prospects for campus based [sic] activism.

WHAT CAN WE DO? *SUPPORT THE DREAM ACT*

It is not good enough to rely on states to change their policies regarding undocumented students. We need a federal policy that would affect all states so that all undocumented students have access to higher education in the United States. To this end, lawmakers have been trying to pass the Dream Act, which would give undocumented students a road to citizenship. Several variations of the Dream Act have been introduced in Congress since 2001. While no form of the legislation has passed, it has gained significant momentum. The Dream Act would make two major changes to current law. It would "permit certain immigrant students who have grown up in the U.S. to apply for temporary legal status and eventually obtain permanent status and become eligible for citizenship if they go to college or serve in the U.S. military" (NILC, 2007). It would also "eliminate a federal provision that penalizes states that provide in-state tuition without regard to immigration status" (NILC 2007).

Under the Dream Act, students of "good moral character" who came to the United States at age fifteen or younger would obtain conditional permanent resident status (six years) upon acceptance to college, graduation from a U.S. high school, or receiving a GED. Students would also be able to qualify for the federal work study program and for student loans. At the end of the six-year conditional period, students would be granted unrestricted lawful permanent resident status if "during the conditional period the immigrant has maintained good moral character, avoided lengthy trips abroad, and either 1) graduated from a 2 year college or studied for at least 2 years towards a B.A. or higher degree, or 2) served in the U.S. Military for 2 years" (NILC, 2007).

The Dream Act is far from perfect. The condition of "good moral character," for example, is troubling. How is moral character defined? How would gay students, activist students, students who have been arrested for acts of civil disobedience, and students in left organizations fare under this conditon? What kind of invasive investigations into their moral character would they be subjected to? These are all important questions, and as a result of the vagueness of the concept, a significant layer of students would not be eligible to reap the benefits of the Dream Act. Furthermore, the option of participating in military service to obtain permanent resident status is deeply problematic. It gives military recruiters who already prey on communities of color further ammunition to convince these students to participate in military service instead of going to college. States would have an incentive to encourage young undocumented immigrants to go into the military since it would save them the costs of granting instate tuition and save federal government Pell grant money. Moreover, given that recruiters use free college tuition as a carrot for potential recruits, joining the military might look especially attractive to young undocumented immigrants.

Moreover, the Dream Act would not require (or prohibit) states to provide instate tuition, nor would students qualify for federal Pell grants. They would, however, be eligible for federal work study and student loans. In short, the Dream Act has many problems. However, it would offer undocumented students a path to citizenship. Most importantly, states would not be restricted from providing their own financial aid to students. Given that we are unlikely to see progressive immigration reform in the near future, it is important to support the Dream Act as a first step towards change.

CONCLUSION

Why is higher education for undocumented immigrant students so important? It's a fundamental issue of rights. [Today], we would be hard pressed to find individuals who do not believe that women and native born minorities should be given access to higher education. What is so different about someone who at the age of five was brought to this country by their parents? The United States is the only country that a majority of these students know. They are going to stay, work, pay taxes, and possibly raise families in this country. As I have mentioned, earning a B.A. degree significantly improves the life chances of all citizens.

An education means access to better jobs, which means access to savings, which means access to the accumulation of wealth that is passed on from one generation to the next. A majority of these students will be contributing to our economy, culture, and collective conscience. As a nation, we should want undocumented students to be empowered through education.

Currently, our immigration policy is sending the message that undocumented students who have been raised in the United States are disposable. Undocumented students' dreams and aspirations are shattered every year, as they realize that they have few possibilities for obtaining a college degree. Many lose hope for the future and begin the slow decline into accepting their fate. Others turn to crime and violence as a method of releasing their frustrations with inequality. Undocumented students who migrated to the country with their parents should not be expected to pay the price of a flawed immigration system....

NOTES

1. All names have been changed to protect students' identities.
2. For an excellent resource on this issue see *Underground Undergrads: UCLA Undocumented Immigrant Students Speak Out*, reviewed in this issue.
3. I use the term U.S. English, because many immigrants from the Caribbean already speak English, but the writing norms and vocabulary for U.S. English are different. So while these students speak and write English, they often have to relearn it to reflect U.S. norms.

REFERENCES

Bank Muñoz, Carolina. 2008. *Transnational Tortillas: Race, Gender and Shop Floor Politics in Mexico and the United States*. Ithaca: Cornell University Press.

Gonzales, Roberto. 2007. "Wasted Talent and Broken Dreams: The Lost Potential of Undocumented Students." *Immigration Policy in Focus*, v5 (13). Immigration Policy Center. www.immigrationpolicy.org

National Immigration Law Center. 2006. "Basic Facts About In-State Tuition for Undocumented Immigrant Students." www.nilc.org

National Immigration Law Center. 2007. "The Dream Act: Basic Facts." www.nilc.org

Ngai, Mae M. 2005. *Impossible Subjects: Illegal Aliens and the Making of Modern America*. Princeton: Princeton University Press.

UCLA Labor Center. 2007. "Undocumented Students, Unfulfilled Dreams...." Report. www.labor.ucla.edu

United States Citizenship and Immigration Services. www.uscis.gov.

27

Prisons for Our Bodies, Closets for Our Minds
Racism, Heterosexism, and Black Sexuality

PATRICIA HILL COLLINS

White fear of black sexuality is a basic ingredient of white racism.

Cornel West

or African Americans, exploring how sexuality has been manipulated in defense of racism is not new. Scholars have long examined the ways in which "white fear of black sexuality" has been a basic ingredient of racism. For example, colonial regimes routinely manipulated ideas about sexuality in order to maintain unjust power relations. Tracing the history of contact between English explorers and colonists and West African societies, historian Winthrop Jordan contends that English perceptions of sexual practices among African people reflected preexisting English beliefs about Blackness, religion, and animals. American historians point to the significance of sexuality to chattel slavery. In the United States, for example, slaveowners relied upon an ideology of Black sexual deviance to regulate and exploit enslaved Africans. Because Black feminist analyses pay more attention to women's sexuality, they too identify how the sexual exploitation of women has been a basic ingredient of racism. For example, studies of African American slave women routinely point to sexual victimization as a defining feature of American slavery. Despite the important contributions of this extensive literature on race and sexuality, because much of the literature assumes that sexuality means heterosexuality, it ignores how racism and heterosexism influence one another.

In the United States, the assumption that racism and heterosexism constitute two separate systems of oppression masks how each relies upon the other for meaning. Because neither system of oppression makes sense without the other, racism and heterosexism might be better viewed as sharing one history with similar yet disparate effects on all Americans differentiated by race, gender, sexuality, class, and nationality. People who are positioned at the margins of both systems

SOURCE: From Patricia Hill Collins, *Black Sexual Politics: African Americans and the New Racism.* pp. 87–88, 95–105, 114–116. New York: Routledge, 2004. Reprinted by permission of the Taylor & Francis Group.

and who are harmed by both typically raise questions about the intersections of racism and heterosexism much earlier and/or more forcefully than those people who are in positions of privilege. In the case of intersections of racism and heterosexism, Black lesbian, gay, bisexual, and transgendered (LGBT) people were among the first to question how racism and heterosexism are interconnected. As African American LGBT people point out, assuming that all Black people are heterosexual and that all LGBT people are White distorts the experiences of LGBT Black people. Moreover, such comparisons misread the significance of ideas about sexuality to racism and race to heterosexism.

Until recently, questions of sexuality in general, and homosexuality in particular, have been treated as crosscutting, divisive issues within antiracist African American politics. The consensus issue of ensuring racial unity subordinated the allegedly crosscutting issue of analyzing sexuality, both straight and gay alike. This suppression has been challenged from two directions. Black women, both heterosexual and lesbian, have criticized the sexual politics of African American communities that leave women vulnerable to single motherhood and sexual assault. Black feminist and womanist projects have challenged Black community norms of a sexual double standard that punishes women for behaviors in which men are equally culpable. Black gays and lesbians have also criticized these same sexual politics that deny their right to be fully accepted within churches, families, and other Black community organizations. Both groups of critics argue that ignoring the heterosexism that underpins Black patriarchy hinders the development of a progressive Black sexual politics....

Developing a progressive Black sexual politics requires examining how racism and heterosexism mutually construct one another.

MAPPING RACISM AND HETEROSEXISM:
THE PRISON AND THE CLOSET

... Racism and heterosexism, the prison and the closet, appear to be separate systems, but LGBT African Americans point out that *both* systems affect their everyday lives. If racism and heterosexism affect Black LGBT people, then these systems affect *all* people, including heterosexual African Americans. Racism and heterosexism certainly converge on certain key points. For one, both use similar state-sanctioned institutional mechanisms to maintain racial and sexual hierarchies. For example, in the United States, racism and heterosexism both rely on segregating people as a mechanism of social control. For racism, segregation operates by using race as a visible marker of group membership that enables the state to relegate Black people to inferior schools, housing, and jobs. Racial segregation relies on enforced membership in a visible community in which racial discrimination is tolerated. For heterosexism, segregation is enforced by pressuring LGBT individuals to remain closeted and thus segregated from one another. Before social movements for gay and lesbian liberation, sexual segregation meant that refusing to claim homosexual identities virtually eliminated any group-based political action to resist heterosexism. For another, the state has played a very important role in sanctioning both forms of oppression. In support of racism, the state sanctioned laws that regulated where Black people could live, work, and attend school. In support of heterosexism, the state maintained laws that refused to punish hate crimes against LGBT people, that failed to offer protection when LGBT people were stripped of jobs and children, and that generally sent a message that LGBT people who came out of the closet did so at their own risk.

Racism and heterosexism also share a common set of practices that are designed to discipline the population into accepting the status quo. These disciplinary practices can best be seen in the enormous amount of attention paid both by the state and organized religion to the institution of marriage. If marriage were in fact a natural and normal occurrence between heterosexual couples and if it occurred naturally within racial categories, there would be no need to regulate it. People would naturally choose partners of the opposite sex and the same race. Instead, a series of laws have been passed, all designed to regulate marriage. For example, for many years, the tax system has rewarded married couples with tax breaks that have been denied to single taxpayers or unmarried couples. The message is clear—it makes good financial sense to get married. Similarly, to encourage people to marry within their assigned race, numerous states passed laws banning interracial marriage. These restrictions lasted until the landmark Supreme Court decision in 1967 that overturned state laws. The state has also passed laws designed to keep LGBT people from marrying. In 1996, the U.S. Congress passed the Federal Defense of Marriage Act that defined marriage as a "legal union between one man and one woman." In all of these cases, the state perceives that it has a compelling interest in disciplining the population to marry and to marry the correct partners.

Racism and heterosexism also manufacture ideologies that defend the status quo. When ideologies that defend racism and heterosexism become taken-for-granted and appear to be natural and inevitable, they become hegemonic. Few question them and the social hierarchies they defend. Racism and heterosexism both share a common cognitive framework that uses binary thinking to produce hegemonic ideologies. Such thinking relies on oppositional categories. It views race through two oppositional categories of Whites and Blacks, gender through two categories of men and women, and sexuality through two oppositional categories of heterosexuals and homosexuals. A master binary of normal and deviant overlays and bundles together these and other lesser binaries. In this context, ideas about "normal" race (Whiteness, which ironically, masquerades as racelessness), "normal" gender (using male experiences as the norm), and "normal" sexuality (heterosexuality, which operates in a similar hegemonic fashion) are tightly bundled together. In essence, to be completely "normal," one must be White, masculine, and heterosexual, the core hegemonic White masculinity. This mythical norm is hard to see because it is so taken-for-granted. Its antithesis, its Other, would be Black, female, and lesbian, a fact that Black lesbian feminist Audre Lorde pointed out some time ago.

Within this oppositional logic, the core binary of normal/deviant becomes ground zero for justifying racism and heterosexism. The deviancy assigned to race and that assigned to sexuality becomes an important point of contact between the two systems. Racism and heterosexism both require a concept of sexual deviancy for meaning, yet the form that deviance takes within each system differs. For racism, the point of deviance is created by a *normalized White heterosexuality* that depends on a *deviant Black heterosexuality* to give it meaning. For heterosexism, the point of deviance is created by this very same *normalized White heterosexuality* that now depends on a *deviant White homosexuality*. Just as racial normality requires the stigmatization of the sexual practices of Black people, heterosexual normality relies upon the stigmatization of the sexual practices of homosexuals. In both cases, installing White heterosexuality as normal, natural, and ideal requires stigmatizing alternate sexualities as abnormal, unnatural, and sinful.

The purpose of stigmatizing the sexual practices of Black people and those of LGBT people may be similar, but the content of the sexual deviance assigned to each differs. Black people carry the stigma of *promiscuity* or excessive or unrestrained heterosexual desire. This is the sexual deviancy that has both been assigned to Black people and been used to construct racism. In contrast, LGBT people carry the stigma of *rejecting* heterosexuality by engaging in unrestrained homosexual desire. Whereas the deviancy associated with promiscuity (and, by implication, with Black people as a race) is thought to lie in an *excess* of heterosexual desire, the pathology of homosexuality (the invisible, closeted sexuality that becomes impossible within heterosexual space) seemingly resides in the *absence* of it.

While analytically distinct, in practice, these two sites of constructed deviancy work together and both help create the "sexually repressive culture" in America.... Despite their significance for American society overall, here I confine my argument to the challenges that confront Black people. Both sets of ideas frame a hegemonic discourse of *Black sexuality* that has at its core ideas about an assumed promiscuity among heterosexual African American men and women and the impossibility of homosexuality among Black gays and lesbians. How have African Americans been affected by and reacted to this racialized system of heterosexism (or this sexualized system of racism)?

AFRICAN AMERICANS AND THE RACIALIZATION
OF PROMISCUITY

Ideas about Black promiscuity that produce contemporary sexualized spectacles such as Jennifer Lopez, Destiny's Child, Ja Rule, and the many young Black men on the U.S. talk show circuit have a long history. Historically, Western science, medicine, law, and popular culture reduced an African-derived aesthetic concerning the use of the body, sensuality, expressiveness, and spirituality to an ideology about *Black sexuality*. The distinguishing feature of this ideology was its reliance on the idea of Black promiscuity. The possibility of distinctive and worthwhile African-influenced worldviews on anything, including sexuality, as well as the heterogeneity of African societies expressing such views, was collapsed into an imagined, pathologized Western discourse of what was thought to be essentially African. To varying degrees, observers from England, France, Germany, Belgium, and other colonial powers perceived African sensuality, eroticism, spirituality, and/or sexuality as deviant, out of control, sinful, and as an essential feature of racial difference....

With all living creatures classified in this way, Western scientists perceived African people as being more natural and less civilized, primarily because African people were deemed to be closer to animals and nature, especially the apes and monkeys whose appearance most closely resembled humans. Like African people, animals also served as objects of study for Western science because understanding the animal kingdom might reveal important insights about civilization, culture, and what distinguished the human "race" from its animal counterparts as well as the human "races" from one another....

Those most proximate to animals, those most lacking civilization, also were those humans who came closest to having the sexual lives of animals. Lacking the benefits of Western civilization, people of African descent were perceived as having a biological nature that was inherently more sexual than that of Europeans. The primitivist discourse thus created the category of "beast" and the sexuality of such beasts as "wild." The legal classification of enslaved African people as chattel (animal-like) under American slavery that produced controlling images of bucks, jezebels, and breeder women drew meaning from this broader interpretive framework.

Historically, this ideology of Black sexuality that pivoted on a Black heterosexual promiscuity not only upheld racism but it did so in gender-specific ways. In the context of U.S. society, beliefs in Black male promiscuity took diverse forms during distinctive historical periods. For example, defenders of chattel slavery believed that slavery safely domesticated allegedly dangerous Black men because it regulated their promiscuity by placing it in the service of slave owners. Strategies of control were harsh and enslaved African men who were born in Africa or who had access to their African past were deemed to be the most dangerous. In contrast, the controlling image of the rapist appeared after emancipation because Southern Whites feared that the unfettered promiscuity of Black freedmen constituted a threat to the Southern way of life....

The events themselves may be over, but their effects persist under the new racism. This belief in an inherent Black promiscuity reappears today. For example, depicting poor and working-class African American inner-city neighborhoods as dangerous urban jungles where SUV-driving White suburbanites come to score drugs or locate prostitutes also invokes a history of racial and sexual conquest. Here sexuality is linked with danger, and understandings of both draw upon historical imagery of Africa as a continent replete with danger and peril to the White explorers and hunters who penetrated it. Just as contemporary safari tours in Africa create an imagined Africa as the "White man's playground" and mask its economic exploitation, jungle language masks social relations of hyper-segregation that leave working-class Black communities isolated, impoverished, and dependent on a punitive welfare state and an illegal international drug trade. Under this logic, just as wild animals (and the proximate African natives) belong in nature preserves (for their own protection), unassimilated, undomesticated poor and working-class African Americans belong in racially segregated neighborhoods....

African American women also live with ideas about Black women's promiscuity and lack of sexual restraint. Reminiscent of concerns with Black women's fertility under slavery and in the rural South, contemporary social welfare policies also remain preoccupied with Black women's fertility. In prior eras, Black women were encouraged to have many children. Under slavery, having many children enhanced slave owners' wealth and a good "breeder woman" was less likely to be sold. In rural agriculture after emancipation, having many children ensured a sufficient supply of workers. But in the global economy of today, large families are expensive because children must be educated. Now Black women are seen as producing too many children who contribute less to society than they take. Because Black women on welfare have long been seen as undeserving, long-standing ideas about Black women's promiscuity become recycled and redefined as a problem for the state....

RACISM AND HETEROSEXISM REVISITED

On May 11, 2003, a stranger killed fifteen-year-old Sakia Gunn who, with four friends, was on her way home from New York's Greenwich Village. Sakia and her friends were waiting for the bus in Newark, New Jersey, when two men got out of a car, made sexual advances, and physically attacked them. The women fought back, and when Gunn told the men that she was a lesbian, one of them stabbed her in the chest.

Sakia Gunn's murder illustrates the connections among class, race, gender, sexuality, and age. Sakia lacked the protection of social class privilege. She and her friends were waiting for the bus in the first place because none had access to private automobiles that offer protection for those who are more affluent. In Gunn's case, because her family initially did not have the money for her funeral, she was scheduled to be buried in a potter's grave. Community activists took up a collection to pay for her funeral. She lacked the gendered protection provided by masculinity. Women who are perceived to be in the wrong place at the wrong time are routinely approached by men who feel entitled to harass and proposition them. Thus, Sakia and her friends share with all women the vulnerabilities that accrue to women who negotiate public space. She lacked the protection of age—had Sakia and her friends been middle-aged, they may not have been seen as sexually available. Like African American girls and women, regardless of sexual orientation, they were seen as approachable. Race was a factor, but not in a framework of interracial race relations. Sakia and her friends were African American, as were their attackers. In a context where Black men are encouraged to express a hyper-heterosexuality as the badge of Black masculinity, women like Sakia and her friends can become important players in supporting patriarchy. They challenged Black male authority, and they paid for the transgression of refusing to participate in scripts of Black promiscuity. But the immediate precipitating catalyst for the violence that took Sakia's life was her openness about her lesbianism. Here, homophobic violence was the prime factor. Her death illustrates how deeply entrenched homophobia can be among many African American men and women, in this case, beliefs that resulted in an attack on a teenaged girl.

How do we separate out and weigh the various influences of class, gender, age, race, and sexuality in this particular incident? Sadly, violence against Black girls is an everyday event. What made this one so special? Which, if any, of the dimensions of her identity got Sakia Gunn killed? There is no easy answer to this question, because *all* of them did. More important, how can any Black political agenda that does not take *all* of these systems into account, including sexuality, ever hope adequately to address the needs of Black people as a collectivity? One expects racism in the press to shape the reports of this incident. In contrast to the 1998 murder of Matthew Shepard, a young, White, gay man in Wyoming, no massive protests, nationwide vigils, and renewed calls for federal hate crimes legislation followed Sakia's death. But what about the response of elected and appointed officials? The African American mayor of Newark decried the crime, but he could not find the time to meet with community activists who wanted programmatic changes to retard crimes like Sakia's murder. The principal of her high school became part of the problem. As one activist described it, "students at Sakia's high school weren't allowed to hold a vigil. And the kids wearing the rainbow flag were being punished like they had on gang colors."

Other Black leaders and national organizations spoke volumes through their silence. The same leaders and organizations that spoke out against the police beating of Rodney King by Los Angeles area police, the rape of immigrant Abner Louima by New York City police, and the murder of Timothy Thomas by Cincinnati police said nothing about Sakia Gunn's death. Apparently, she was just another unimportant little Black girl to them. But to others, her death revealed the need for a new politics that takes the intersections of racism and heterosexism as well as class exploitation, age discrimination, and sexism into account. Sakia was buried on May 16 and a crowd of approximately 2,500 people attended her funeral. The turnout was unprecedented: predominantly Black, largely high school students, and mostly lesbians. Their presence says that as long as African American lesbians like high school student Sakia Gunn are vulnerable, then every African American woman is in danger; and if all Black women are at risk, then there is no way that any Black person will ever be truly safe or free.

28

The Invention of Heterosexuality

JONATHAN NED KATZ

Heterosexuality is old as procreation, ancient as the lust of Eve and Adam. That first lady and gentleman, we assume, perceived themselves, behaved, and felt just like today's heterosexuals. We suppose that heterosexuality is unchanging, universal, essential: ahistorical.

Contrary to that common sense conjecture, the concept of heterosexuality is only one particular historical way of perceiving, categorizing, and imagining the social relations of the sexes. Not ancient at all, the idea of heterosexuality is a modern invention, dating to the late nineteenth century. The heterosexual belief, with its metaphysical claim to eternity, has a particular, pivotal place in the social universe of the late nineteenth and twentieth centuries that it did not inhabit earlier. This essay traces the historical process by which the heterosexual idea was created as a historical and taken-for-granted....

By not studying the heterosexual idea in history, analysts of sex, gay and straight, have continued to privilege the "normal" and "natural" at the expense of the "abnormal" and "unnatural." Such privileging of the norm accedes to its domination, protecting it from questions. By making the normal the object of a thoroughgoing historical study we simultaneously pursue a pure truth and a sex-radical and subversive goal: we upset basic preconceptions. We discover that the heterosexual, the normal, and the natural have a history of changing definitions. Studying the history of the term challenges its power.

Contrary to our usual assumption, past Americans and other peoples named, perceived, and socially organized the bodies, lusts, and intercourse of the sexes in ways radically different from the way we do. If we care to understand this vast past sexual diversity, we need to stop promiscuously projecting our own hetero and homo arrangement. Though lip service is often paid to the distorting, ethnocentric effect of such conceptual imperialism, the category heterosexuality continues to be applied uncritically as a universal analytical tool. Recognizing the time-bound and culturally specific character of the heterosexual category can help us begin to work toward a thoroughly historical view of sex....

SOURCE: "The Invention of Heterosexuality" by Jonathan Ned Katz from *Socialist Review* 20 (January - March 1990) pp. 7–34. Reprinted by permission of the author.

BEFORE HETEROSEXUALITY: EARLY VICTORIAN TRUE LOVE, 1820–1860

In the early nineteenth-century United States, from about 1820 to 1860, the heterosexual did not exist. Middle-class white Americans idealized a True Womanhood, True Manhood, and True Love, all characterized by "purity"—the freedom from sensuality.[1] Presented mainly in literary and religious texts, this True Love was a fine romance with no lascivious kisses. This ideal contrasts strikingly with late nineteenth- and twentieth-century American incitements to a hetero sex.[2]

Early Victorian True Love was only realized within the mode of proper procreation, marriage, the legal organization for producing a new set of correctly gendered women and men. Proper womanhood, manhood, and progeny—not a normal male-female eros—was the main product of this mode of engendering and of human reproduction.

The actors in this sexual economy were identified as manly men and womanly women and as procreators, not specifically as erotic beings or heterosexuals. Eros did not constitute the core of a heterosexual identity that inhered, democratically, in both men and women. True Women were defined by their distance from lust. True Men, though thought to live closer to carnality, and in less control of it, aspired to the same freedom from concupiscence.

Legitimate natural desire was for procreation and a proper manhood or womanhood; no heteroerotic desire was thought to be directed exclusively and naturally toward the other sex; lust in men was roving. The human body was thought of as a means towards procreation and production; penis and vagina were instruments of reproduction, not of pleasure. Human energy, thought of as a closed and severely limited system, was to be used in producing children and in work, not wasted in libidinous pleasures.

The location of all this engendering and procreative labor was the sacred sanctum of early Victorian True Love, the home of the True Woman and True Man—a temple of purity threatened from within by the monster masturbator, an archetypal early Victorian cult figure of illicit lust. The home of True Love was a castle far removed from the erotic exotic ghetto inhabited most notoriously then by the prostitute, another archetypal Victorian erotic monster....

LATE VICTORIAN SEX-LOVE: 1860–1892

"Heterosexuality" and "homosexuality" did not appear out of the blue in the 1890s. These two eroticisms were in the making from the 1860s on. In late Victorian America and in Germany, from about 1860 to 1892, our modern idea of an eroticized universe began to develop, and the experience of a hetero-lust began to be widely documented and named....

In the late nineteenth-century United States, several social factors converged to cause the eroticizing of consciousness, behavior, emotion, and identity that

became typical of the twentieth-century Western middle class. The transformation of the family from producer to consumer unit resulted in a change in family
members' relation to their own bodies; from being an instrument primarily of
work, the human body was integrated into a new economy, and began more
commonly to be perceived as a means of consumption and pleasure. Historical
work has recently begun on how the biological human body is differently integrated into changing modes of production, procreation, engendering, and pleasure so as to alter radically the identity, activity, and experience of that body.[3]

The growth of a consumer economy also fostered a new pleasure ethic. This
imperative challenged the early Victorian work ethic, finally helping to usher in
a major transformation of values. While the early Victorian work ethic had touted the value of economic production, that era's procreation ethic had extolled
the virtues of human reproduction. In contrast, the late Victorian economic ethic
hawked the pleasures of consuming, while its sex ethic praised an erotic pleasure
principle for men and even for women.

In the late nineteenth century, the erotic became the raw material for a new
consumer culture. Newspapers, books, plays, and films touching on sex, "normal"
and "abnormal," became available for a price. Restaurants, bars, and baths opened,
catering to sexual consumers with cash. Late Victorian entrepreneurs of desire
incited the proliferation of a new eroticism, a commoditized culture of pleasure.

In these same years, the rise in power and prestige of medical doctors
allowed these upwardly mobile professionals to prescribe a healthy new sexuality.
Medical men, in the name of science, defined a new ideal of male-female
relationships that included, in women as well as men, an essential, necessary, normal eroticism. Doctors, who had earlier named and judged the sex-enjoying
woman a "nymphomaniac," now began to label women's *lack* of sexual pleasure
a mental disturbance, speaking critically, for example, of female "frigidity" and
"anesthesia."[4]

By the 1880s, the rise of doctors as a professional group fostered the rise of a
new medical model of Normal Love, replete with sexuality. The new Normal
Woman and Man were endowed with a healthy libido. The new theory of Normal Love was the modern medical alternative to the old Cult of True Love. The
doctors prescribed a new sexual ethic as if it were a morally neutral, medical
description of health. The creation of the new Normal Sexual had its counterpart
in the invention of the late Victorian Sexual Pervert. The attention paid the
sexual abnormal created a need to name the sexual normal, the better to distinguish the average him and her from the deviant it.

HETEROSEXUALITY: THE FIRST YEARS, 1892–1900

In the periodization of heterosexual American history suggested here, the years
1892 to 1900 represent "The First Years" of the heterosexual epoch, eight key
years in which the idea of the heterosexual and homosexual were initially and

tentatively formulated by U.S. doctors. The earliest-known American use of the word "heterosexual" occurs in a medical journal article by Dr. James G. Kiernan of Chicago, read before the city's medical society on March 7, 1892, and published that May—portentous dates in sexual history.[5] But Dr. Kiernan's heterosexuals were definitely not exemplars of normality. Heterosexuals, said Kiernan, were defined by a mental condition, "psychical hermaphroditism." Its symptoms were "inclinations to both sexes." These heterodox sexuals also betrayed inclinations "to abnormal methods of gratification," that is, techniques to insure pleasure without procreation. Dr. Kiernan's heterogeneous sexuals did demonstrate "traces of the normal sexual appetite" (a touch of procreative desire). Kiernan's normal sexuals were implicitly defined by a monolithic other-sex inclination and procreative aim. Significantly, they still lacked a name.

Dr. Kiernan's article of 1892 also included one of the earliest-known uses of the word "homosexual" in American English. Kiernan defined "Pure homosexuals" as persons whose "general mental state is that of the opposite sex." Kiernan thus defined homosexuals by their deviance from a gender norm. His heterosexuals displayed a double deviance from both gender and procreative norms.

Though Kiernan used the new words heterosexual and homosexual, an old procreative standard and a new gender norm coexisted uneasily in his thought. His word heterosexual defined a mixed person and compound urge, abnormal because they wantonly included procreative and non-procreative objectives, as well as same-sex and different-sex attractions.

That same year, 1892, Dr. Krafft-Ebing's influential *Psychopathia Sexualis* was first translated and published in the United States.[6] But Kiernan and Krafft-Ebing by no means agreed on the definition of the heterosexual. In Krafft-Ebing's book, "hetero-sexual" was used unambiguously in the modern sense to refer to an erotic feeling for a different sex. "Homosexual" referred unambiguously to an erotic feeling for a "same sex." In Krafft-Ebing's volume, unlike Kiernan's article, heterosexual and homosexual were clearly distinguished from a third category, a "psycho-sexual hermaphroditism," defined by impulses toward both sexes.

Krafft-Ebing hypothesized an inborn "sexual instinct" for relations with the "opposite sex," the inherent "purpose" of which was to foster procreation. Krafft-Ebing's erotic drive was still a reproductive instinct. But the doctor's clear focus on a different-sex versus same-sex sexuality constituted a historic, epochal move from an absolute procreative standard of normality toward a new norm. His definition of heterosexuality as other-sex attraction provided the basis for a revolutionary, modern break with a centuries-old procreative standard.

It is difficult to overstress the importance of that new way of categorizing. The German's mode of labeling was radical in referring to the biological sex, masculinity or femininity, and the pleasure of actors (along with the procreant purpose of acts). Krafft-Ebing's heterosexual offered the modern world a new norm that came to dominate our idea of the sexual universe, helping to change it from a mode of human reproduction and engendering to a mode of pleasure. The heterosexual category provided the basis for a move from a production-oriented, procreative imperative to a consumerist pleasure principle—an institutionalized pursuit of happiness....

Only gradually did doctors agree that heterosexual referred to a normal, "other-sex" eros. This new standard-model heterosex provided the pivotal term for the modern regularization of eros that paralleled similar attempts to standardize masculinity and femininity, intelligence, and manufacturing.[7] The idea of heterosexuality as the master sex from which all others deviated was (like the idea of the master race) deeply authoritarian. The doctors' normalization of a sex that was hetero proclaimed a new heterosexual separatism—an erotic apartheid that forcefully segregated the sex normals from the sex perverts. The new, strict boundaries made the emerging erotic world less polymorphous—safer for sex normals. However, the idea of such creatures as heterosexuals and homosexuals emerged from the narrow world of medicine to become a commonly accepted notion only in the early twentieth century. In 1901, in the comprehensive *Oxford English Dictionary,* "heterosexual" and "homosexual" had not yet made it.

THE DISTRIBUTION OF THE HETEROSEXUAL MYSTIQUE: 1900–1930

In the early years of this heterosexual century the tentative hetero hypothesis was stabilized, fixed, and widely distributed as the ruling sexual orthodoxy: The Heterosexual Mystique. Starting among pleasure-affirming urban working-class youths, southern blacks, and Greenwich Village bohemians as defensive subculture, heterosex soon triumphed as dominant culture.[8]

In its earliest version, the twentieth-century heterosexual imperative usually continued to associate heterosexuality with a supposed human "need," "drive," or "instinct" for propagation, a procreant urge linked inexorably with carnal lust as it had not been earlier. In the early twentieth century, the falling birth rate, rising divorce rate, and "war of the sexes" of the middle class were matters of increasing public concern. Giving vent to heteroerotic emotions was thus praised as enhancing baby-making capacity, marital intimacy, and family stability. (Only many years later, in the mid-1960s, would heteroeroticism be distinguished completely, in practice and theory, from procreativity and male-female pleasure sex justified in its own name.)

The first part of the new sex norm—hetero—referred to a basic gender divergence. The "oppositeness" of the sexes was alleged to be the basis for a universal, normal, erotic attraction between males and females. The stress on the sexes' "oppositeness," which harked back to the early nineteenth century, by no means simply registered biological differences of females and males. The early twentieth-century focus on physiological and gender dimorphism reflected the deep anxieties of men about the shifting work, social roles, and power of men over women, and about the ideals of womanhood and manhood. That gender anxiety is documented, for example, in 1897, in *The New York Times'* publication of the Reverend Charles Parkhurst's diatribe against female "andro-maniacs," the preacher's derogatory, scientific-sounding name for women who tried to "minimize distinctions by which manhood and womanhood are differentiated."[9] The stress on gender difference was a conservative response to the changing social-sexual division of activity and feeling which gave rise to the independent "New Woman" of the 1880s and eroticized "Flapper" of the 1920s.

The second part of the new hetero norm referred positively to sexuality. That novel upbeat focus on the hedonistic possibilities of male-female conjunctions also reflected a social transformation—a revaluing of pleasure and procreation, consumption and work in commercial, capitalist society. The democratic attribution of a normal lust to human females (as well as males) served to authorize women's enjoyment of their own bodies and began to undermine the early Victorian idea of the pure True Woman—a sex-affirmative action still part of women's struggle. The twentieth-century Erotic Woman also undercut the nineteenth-century feminist assertion of women's moral superiority, cast suspicions of lust on women's passionate romantic friendships with women, and asserted the presence of a menacing female monster, "the lesbian."[10]...

In the perspective of heterosexual history, this early twentieth-century struggle for the more explicit depiction of an "opposite-sex" eros appears in a curious new light. Ironically, we find sex-conservatives, the social purity advocates of censorship and repression, fighting against the depiction not just of sexual perversity but also of the new normal hetero-sexuality. That a more open depiction of normal sex had to be defended against forces of propriety confirms the claim that heterosexuality's predecessor, Victorian True Love, had included no legitimate eros....

THE HETEROSEXUAL STEPS OUT: 1930–1945

In 1930, in *The New York Times,* heterosexuality first became a love that dared to speak its name. On April 20th of that year, the word "heterosexual" is first known to have appeared in *The New York Times Book Review.* There, a critic described the subject of André Gide's *The Immoralist* proceeding "from a hetero-sexual liaison to a homosexual one." The ability to slip between sexual categories was referred to casually as a rather unremarkable aspect of human possibility. This is also the first known reference by *The Times* to the new hetero/homo duo.[11]

In September the second reference to the hetero/homo dyad appeared in *The New York Times Book Review,* in a comment on Floyd Dell's *Love in the Machine Age.* This work revealed a prominent antipuritan of the 1930s using the dire threat of homosexuality as his rationale for greater heterosexual freedom. *The Times* quoted Dell's warning that current abnormal social conditions kept the young dependent on their parents, causing "infantilism, prostitution and homosexuality." Also quoted was Dell's attack on the inculcation of purity" that "breeds distrust of the opposite sex." Young people, Dell said, should be "permitted to develop normally to heterosexual adulthood." "But," *The Times* reviewer emphasized, "such a state already exists, here and now." And so it did. Heterosexuality, a new gender-sex category, had been distributed from the narrow, rarified realm of a few doctors to become a nationally, even internationally, cited aspect of middle-class life.[12]...

HETEROSEXUAL HEGEMONY: 1945–1965

The "cult of domesticity" following World War II—the reassociation of women with the home, motherhood, and child-care; men with fatherhood and wage work outside the home—was a period in which the predominance of the hetero norm went almost unchallenged, an era of heterosexual hegemony. This was an age in which conservative mental-health professionals reasserted the old link between heterosexuality and procreation. In contrast, sex-liberals of the day strove, ultimately with success, to expand the heterosexual ideal to include within the boundaries of normality a wider-than-ever range of nonprocreative, premarital, and extramarital behaviors. But sex-liberal reform actually helped to extend and secure the dominance of the heterosexual idea, as we shall see when we get to Kinsey.

The postwar sex-conservative tendency was illustrated in 1947, in Ferdinand Lundberg and Dr. Marynia Farnham's books, *Modern Woman: The Lost Sex*. Improper masculinity and femininity was exemplified, the authors decreed, by "engagement in heterosexual relations ... with the complete intent to see to it that they do not eventuate in reproduction."[13] Their procreatively defined heterosex was one expression of a postwar ideology of fecundity that, internalized and enacted dutifully by a large part of the population, gave rise to the postwar baby boom.

The idea of the feminine female and masculine male as prolific breeders was also reflected in the stress, specific to the late 1940s, on the homosexual as sad symbol of "sterility"—that particular loaded term appears incessantly in comments on homosex dating to the fecund forties.

In 1948, in *The New York Times Book Review,* sex liberalism was in ascendancy. Dr. Howard A. Rusk declared that Alfred Kinsey's just published report on *Sexual Behavior in the Human Male* had found "wide variations in sex concepts and behavior." This raised the question: "What is 'normal' and 'abnormal'?" In particular, the report had found that "homosexual experience is much more common than previously thought," and "there is often a mixture of both homo and hetero experience."[14]

Kinsey's counting of orgasms indeed stressed the wide range of behaviors and feelings that fell within the boundaries of a quantitative, statistically accounted heterosexuality. Kinsey's liberal reform of the hetero/homo dualism widened the narrow, old hetero category to accord better with the varieties of social experience. He thereby contradicted the older idea of a monolithic, qualitatively defined, natural procreative act, experience, and person.[15]

Though Kinsey explicitly questioned "whether the terms 'normal' and 'abnormal' belong in a scientific vocabulary," his counting of climaxes was generally understood to define normal sex as majority sex. This quantified norm constituted a final, society-wide break with the old qualitatively defined reproductive standard. Though conceived of as purely scientific, the statistical definition of the normal as the-sex-most-people-are-having substituted a new, quantitative moral standard for the old, qualitative sex ethic—another triumph for the spirit of capitalism.

Kinsey also explicitly contested the idea of an absolute, either/or antithesis between hetero and homo persons. He denied that human beings "represent two discrete populations, heterosexual and homosexual." The world, he ordered, "is not to be divided into sheep and goats." The hetero/homo division was not nature's doing: "Only the human mind invents categories and tries to force facts into separated pigeon-holes. The living world is a continuum."[16]

With a wave of the taxonomist's hand, Kinsey dismissed the social and historical division of people into heteros and homos. His denial of heterosexual and homosexual personhood rejected the social reality and profound subjective force of a historically constructed tradition which, since 1892 in the United States, had cut the sexual populaton in two and helped to establish the social reality of a heterosexual and homosexual identity.

On the one hand, the social construction of homosexual persons has led to the development of a powerful gay liberation identity politics based on an ethnic group model. This has freed generations of women and men from a deep, painful, socially induced sense of shame, and helped to bring about a society-wide liberalization of attitudes and responses to homosexuals.[17] On the other hand, contesting the notion of homosexual and heterosexual persons was one early, partial resistance to the limits of the hetero/homo construction. Gore Vidal, rebel son of Kinsey, has for years been joyfully proclaiming:

> ...there is no such thing as a homosexual or a heterosexual person. There are only homo- or heterosexual acts. Most people are a mixture of impulses if not practices, and what anyone does with a willing partner is of no social or cosmic significance.
>
> So why all the fuss? In order for a ruling class to rule, there must be arbitrary prohibitions. Of all prohibitions, sexual taboo is the most useful because sex involves everyone.... We have allowed our governors to divide the population into two teams. One team is good, godly, straight; the other is evil, sick, vicious.[18]

HETEROSEXUALITY QUESTIONED: 1965–1982

By the late 1960s, anti-establishment counter culturalists, fledgling feminists, and homosexual-rights activists had begun to produce an unprecedented critique of sexual repression in general, of women's sexual repression in particular, of marriage and the family—and of some forms of heterosexuality....

Heterosexual History: Out of The Shadows

Our brief survey of the heterosexual idea suggests a new hypothesis. Rather than naming a conjunction old as Eve and Adam, heterosexual designates a word and concept, a norm and role, an individual and group identity, a behavior and feeling, and a peculiar sexual-political institution particular to the late nineteenth and twentieth centuries.

Because much stress has been placed here on heterosexuality as word and concept, it seems important to affirm that heterosexuality (and homosexuality) came into existence before it was named and thought about. The formulation of the heterosexual idea did not create a heterosexual experience or behavior; to suggest otherwise would be to ascribe determining power to labels and concepts. But the titling and envisioning of heterosexuality did play an important role in consolidating the construction of the heterosexual's social existence. Before the wide use of the word "heterosexual," I suggest, women and men did not mutually lust with the same profound, sure sense of normalcy that followed the distribution of "heterosexual" as universal sanctifier.

According to this proposal, women and men make their own sexual histories. But they do not produce their sex lives just as they please. They make their sexualities within a particular mode of organization given by the past and altered by their changing desire, their present power and activity, and their vision of a better world. That hypothesis suggests a number of good reasons for the immediate inauguration of research on a historically specific heterosexuality.

The study of the history of the heterosexual experience will forward a great intellectual struggle still in its early stages. This is the fight to pull heterosexuality, homosexuality, and all the sexualities out of the realm of nature and biology [and] into the realm of the social and historical. Feminists have explained to us that anatomy does not determine our gender destinies (our masculinities and femininities). But we've only recently begun to consider that *biology does not settle our erotic fates*. The common notion that biology determines the object of sexual desire, or that physiology and society together cause sexual orientation, are determinisms that deny the break existing between our bodies and situations and our desiring. Just as the biology of our hearing organs will never tell us why we take pleasure in Bach or delight in Dixieland, our female or male anatomies, hormones, and genes will never tell us why we yearn for women, men, both, other, or none. That is because desiring is a self-generated project of individuals within particular historical cultures. Heterosexual history can help us see the place of values and judgments in the construction of our own and others' pleasures, and to see how our erotic tastes—our aesthetics of the flesh—are socially institutionalized through the struggle of individuals and classes.

The study of heterosexuality in time will also help us to recognize the *vast historical diversity of sexual emotions and behaviors*—a variety that challenges the monolithic heterosexual hypothesis. John D'Emilio and Estelle Freedman's *Intimate Matters: A History of Sexuality in America* refers in passing to numerous substantial changes in sexual activity and feeling: for example, the widespread use of contraceptives in the nineteenth century, the twentieth-century incitement of the female orgasm, and the recent sexual conduct changes by gay men in response to the AIDS epidemic. It's now a commonplace of family history that people in particular classes feel and behave in substantially different ways under different, historical conditions. Only when we stop assuming an invariable essence of heterosexuality will we begin the research to reveal the full variety of sexual emotions and behaviors.

The historical study of the heterosexual experience can help us *understand the erotic relationships of women and men in terms of their changing modes of social organization*. Such model analysis actually characterizes a sex history well underway. This suggests that the eros-gender-procreation system (the social ordering of lust, femininity and masculinity, and baby-making) has been linked closely to a society's particular organization of power and production. To understand the subtle history of heterosexuality we need to look carefully at correlations between (1) society's organization of eros and pleasure; (2) its mode of engendering persons as feminine or masculine (its making of women and men); (3) its ordering of human reproduction; and (4) its dominant political economy. This General Theory of Sexual Relativity proposes that substantial historical changes in the social organization of eros, gender, and procreation have basically altered the activity and experience of human beings within those modes.

A historical view locates heterosexuality and homosexuality in time, helping us distance ourselves from them. This distancing can help us formulate new questions that clarify our long-range sexual-political goals: What has been and is the social function of sexual categorizing? Whose interests have been served by the division of the world into heterosexual and homosexual? Do we dare not draw a line between those two erotic species? Is some sexual naming socially necessary? Would human freedom be enhanced if the sex-biology of our partners in lust was of no particular concern, and had no name? In what kind of society could we all more freely explore our desire and our flesh?

As we move [into the present], a new sense of the historical making of the heterosexual and homosexual suggests that these are ways of feeling, acting, and being with each other that we can together unmake and radically remake according to our present desire, power, and our vision of a future political-economy of pleasure.

NOTES

1. Barbara Welter, "The Cult of True Womanhood: 1820–1860," *American Quarterly*, vol. 18 (Summer 1966); Welter's analysis is extended here to include True Men and True Love.

2. Some historians have recently told us to revise our idea of sexless Victorians: their experience and even their ideology, it is said, were more erotic than we previously

thought. Despite the revisionists, I argue that "purity" was indeed the dominant, early Victorian, white middle-class standard. For the debate on Victorian sexuality see John D'Emilio and Estelle Freedman, *Intimate Matters: A History of Sexuality in America* (New York: Harper & Row, 1988), p. xii.

3. See, for example, Catherine Gallagher and Thomas Laqueur, eds., "The Making of the Modern Body: Sexuality and Society in the Nineteenth Century," *Representations,* no. 14 (Spring 1986) (republished, Berkeley: University of California Press, 1987).

4. This reference to females reminds us that the invention of heterosexuality had vastly different impacts on the histories of women and men. It also differed in its impact on lesbians and heterosexual women, homosexual and heterosexual men, the middle class and working class, and on different religious, racial, national, and geographic groups.

5. Dr. James G. Kieman, "Responsibility in Sexual Perversion," *Chicago Medical Recorder,* vol. 3 (May 1892), pp. 185–210.

6. R. von Krafft-Ebing, *Psychopathia Sexualis, with Especial Reference to Contrary Sexual Instinct: A Medico-Legal Study,* trans. Charles Gilbert Chaddock (Philadelphia: F. A. Davis, 1892), from the 7th and revised German ed. Preface, November 1892.

7. For the standardization of gender see Lewis Terman and C. C. Miles, *Sex and Personality, Studies in Femininity and Masculinity* (New York: McGraw Hill, 1936). For the standardization of intelligence see Lewis Terman, *Stanford-Binet Intelligence Scale* (Boston: Houghton Mifflin, 1916). For the standardization of work, see "scientific management" and "Taylorism" in Harry Braverman, *Labor and Monopoly Capital: The Degradation of Work in the Twentieth Century* (New York: Monthly Review Press, 1974).

8. See D'Emilio and Freedman, *Intimate Matters,* pp. 194–201, 231, 241, 295–96; Ellen Kay Trimberger, "Feminism, Men, and Modern Love: Greenwich Village, 1900–1925," in *Powers of Desire: The Politics of Sexuality,* ed. Ann Snitow, Christine Stansell, and Sharon Thompson (New York: Monthly Review Press, 1983), pp. 131–52; Kathy Peiss, " 'Charity Girls' and City Pleasures: Historical Notes on Working Class Sexuality, 1880–1920," in *Powers of Desire,* pp. 74–87; and Mary P. Ryan, "The Sexy Saleslady: Psychology, Heterosexuality, and Consumption in the Twentieth Century," in her *Womanhood in America,* 2nd ed. (New York: Franklin Watts, 1979), pp. 151–82.

9. [Rev. Charles Parkhurst], "Woman. Calls Them Andromaniacs. Dr. Parkhurst So Characterizes Certain Women Who Passionately Ape Everything That Is Mannish. Woman Divinely Preferred. Her Supremacy Lies in Her Womanliness, and She Should Make the Most of It—Her Sphere of Best Usefulness the Home," *The New York Times,* May 23, 1897, p. 16:1.

10. See Lisa Duggan, "The Social Enforcement of Heterosexuality and Lesbian Resistance in the 1920s," in *Class, Race, and Sex: The Dynamics of Control,* ed. Amy Swerdlow and Hanah Lessinger (Boston: G. K. Hall, 1983), pp. 75–92; Rayna Rapp and Ellen Ross, "The Twenties Backlash: Compulsory Heterosexuality, the Consumer Family, and the Waning of Feminism," in *Class, Race, and Sex;* Christina Simmons, "Companionate Marriage and the Lesbian Threat," *Frontiers,* vol. 4, no. 3 (Fall 1979), pp. 54–59; and Lillian Faderman, *Surpassing the Love of Men* (New York: William Morrow, 1981).

11. Louis Kronenberger, review of André Gide, *The Immoralist, New York Times Book Review,* April 20, 1930, p. 9.

12. Henry James Forman, review of Floyd Dell, *Love in the Machine Age* (New York: Farrar & Rinehart), *New York Times Book Review,* September 14, 1930, p. 9.

13. Ferdinand Lundberg and Dr. Marynia F. Farnham, *Modern Woman: The Lost Sex* (New York: Harper, 1947).

14. Dr. Howard A. Rusk, *New York Times Book Review,* January 4, 1948, p. 3.

15. Alfred Kinsey, Wardell B. Pomeroy, and Clyde E. Martin, *Sexual Behavior in the Human Male* (Philadelphia: W. B. Saunders, 1948), pp. 199–200.

16. Kinsey, *Sexual Behavior,* pp. 637, 639.

17. See Steven Epstein, "Gay Politics, Ethnic Identity: The Limits of Social Constructionism," *Socialist Review* 93/94 (1987), pp. 9–54.

18. Gore Vidal, "Someone to Laugh at the Squares With" [Tennessee Williams], *New York Review of Books,* June 13, 1985; reprinted in his *At Home: Essays, 1982–1988* (New York: Random House, 1988), p. 48.

29

An Intersectional Analysis of 'Sixpacks', 'Midriffs' and 'Hot Lesbians' in Advertising

ROSALIND GILL

In the last few years there have been a growing number of discussions of the 'sexualization of culture'. This phrase speaks to a range of different things.

In the analysis presented here I will seek to provide a (feminist) inter-[sectional] analysis that pays attention to gender, class, age, sexuality and racialization within practices of 'sexualization' in advertising.

This approach is a material-discursive one that understands representations as not merely representing the world but as constitutive and generative. It focuses on the repetition of figures across different media sites in such a way that they seem to take on a life of their own.

In this analysis I will be examining the figures of the 'sixpack', the 'midriff' and the 'hot lesbian'. Rather than tracking them across different media or genres, I will be focusing on their repetition and materialization in multiple types of 'sexualized' advertising, examining the forms of in/visibility they make possible and using an intersectional approach to go beyond the tendency to speak of sexualization as if it were a singular, unmarked process.

MEN AS SEX OBJECTS: THE 'SIXPACKS'

One of the most profound shifts in visual culture in the last two decades has been the proliferation of representations of the male body. However, it is not simply that there are more images of men circulating, but that a specific kind of representational practice has emerged for depicting the male body: namely an idealized and eroticized aesthetic showing a toned, young body. What is significant about this type of representation is that it codes men's bodies in ways that give permission for them to be looked at and desired.

This transformation has prompted much discussion, with claims that 'we are all objectified now' and that idealized-sexualized representational strategies are

SOURCE: Gill, Rosalind, "Beyond the 'Sexualization of Culture: Thesis: An Intersectional analysis of 'Sixpacks,' 'Midriffs,' and 'Hot Lesbians' in Advertising." *Sexualities* 12: 137–160. Copyright © 2009 by Sage Publications, Inc. Reprinted by permission.

no longer limited to women's bodies. There is a growing sense in much writing that visual culture has become *equalized*, and that we are *all* today subject to relentless sexualization.

I want to suggest that, despite the apparent similarities, there are in fact profound differences in the ways in which men's and women's bodies are represented sexually. Moreover, these patterns of 'sexualization' have different determinants, employ different modes of representation, and are likely to be read in radically different ways because of long, distinct histories of gender representations and the politics of looking.

At a general level the representations can be understood as part of the shift away from the 'male as norm' in which masculinity lost its unmarked status and became visible as gendered. Sally Robinson (2000) argues that white masculinity was rendered visible through pressure from black and women's liberation movements, which were highly critical of its hegemony. A variety of new social movements galvanized the creation of the 'new man', the reinvention of masculinity along more gentle, emotional and communicative lines.

Moreover, the shift had significant economic determinants: retailers, marketers and magazine publishers were keen to develop new markets and had affluent men in their sights as the biggest untapped source of high spending consumers.

As Rowena Chapman (1988) argues, 'new man' was a contradictory formation, representing both a response to critique from progressive social movements, and a gleam in the eyes of advertisers, marketers and companies aspiring to target young and affluent men. Perhaps the figure of the metrosexual that has come to prominence more recently symbolizes the extent to which marketing-driven constructions won out over more explicitly political articulations of 'new' masculinity.

The radical transformation in the portrayal of men in mainstream visual culture began more than 20 years ago. By the early 1990s the eroticized representation of male bodies was well established, particularly in fashion and fragrance advertising and the emerging market for male grooming products. But rather than a diversity of different representations of the male body, most advertisements belong to a very specific type. The models are generally white, they are young, they are muscular and slim, they are usually clean-shaven (with perhaps the exception of a little designer stubble), and they have particular facial features which connote a combination of softness and strength – strong jaw, large lips and eyes, and soft looking, clear skin, this combination of muscularity/hardness and softness in the particular 'look' of the models allows them to manage contradictory expectations of men and masculinity as strong and powerful but also gentle and tender – they embody, in a sense, a cultural contradiction about what a man is 'meant to be'.

Older bodies are strikingly absent and there are strong and persistent patterns of racialization to be found in the corpus of eroticized images. White bodies are over-represented, but they are frequently not Anglo-American or northern European bodies, but bodies that are coded as 'Latin', with dark hair and olive skin, referencing long histories of sexual Othering and exoticism. Black, African American and African Caribbean bodies are also regularly represented in a highly eroticized manner, but these bodies are usually reserved for products associated with sport, drawing on cultural myths about black male sexuality and physical prowess. It is also worth noting that adverts depicting black men frequently use black male celebrities (e.g. Tiger Woods, Thierry Henri), in contrast to the unknown models who are used when the sexy body is white.

In advertising, a number of strategies were developed to deal with the anxieties and threats produced by this shift. On the one hand, many advertisements used models with an almost 'phallic muscularity'—the size and hardness of the muscles 'standing in for' male power.

The use of photographic conventions mise-en-scène from 'high art' also served as a distancing device to diffuse some of the potential threats engendered by 'sexualizing' the male body. Giving the representations an 'arthouse' look and feel through the use of black and white photography or 'sculpted' models that made reference to classical iconography, offered the safety of distance, as well as connoting affluence, sophistication and 'class'.

The organization of gazes within advertisements also works to diminish the transgressive threat. Men tend not to smile or pout, nor to deploy any of the bodily gestures or postures of the 'ritualised subordination' of women in advertising, and nor are they depicted in mirror shots – so long a favoured mode for conveying women's narcissism. In contrast, in what we might call 'sixpack advertising,' men are generally portrayed standing or involved in some physical activity, and they look back at the viewer in ways reminiscent of street gazes to assert dominance or look up or off, indicating that their interest is elsewhere. They are mostly pictured alone in ways that reference the significance of independence as a value marking hegemonic masculinity (Connell, 1995), or they are pictured with a beautiful woman – to 'reassure' viewers of their heterosexuality.

In multiple ways, then, advertising images of the last two decades have been designed to offset or diffuse some of the anxieties and threats generated by presenting men as objects of an 'undifferentiated' sexual gaze. Neither hegemonic masculinity nor the institution of heterosexuality have been destroyed – and 'sexualized' representation of the male body has not proved incommensurable with male dominance. Rather it appears that a highly specific set of modes of representing the male body have emerged – which are quite different from sexualized representations of women's bodies.

SEXUAL OBJECTIFICATION POSTFEMINIST STYLE: THE 'MIDRIFFS'

If advertising's representation of men has undergone a dramatic shift, then so too have 'sexualized' representations of women been reinvented. In the last decade or so a new figure has been constructed to sell to women: a young, attractive, heterosexual woman who knowingly and deliberately plays with her sexual power and is always 'up for' sex. Following Douglas Rushkoff (n.d.) I will call her the 'midriff'.

Where once sexualized representations of women in advertising presented them as passive, mute *objects* of an assumed male gaze, today women are presented as active, desiring, sexual *subjects* who choose to present themselves in a seemingly objectified manner because it suits their (implicitly liberated) interests to do so.

A crucial aspect of the shift from objectification to sexual subjectification is that this is framed in advertising through a discourse of playfulness, freedom and, above all, choice. Women are presented as not seeking men's approval but as pleasing themselves, and, in so doing, they just happen to win men's admiration.

✳ A discourse of empowerment is also central. Contemporary advertising targeted at the midriffs suggests, above all, that buying the product will empower you. 'I pull the strings' asserts a beautiful woman in a black Wonderbra; 'Empower your eyes,' says an advert for Shiseido mascara; 'Discover the power of femininity. Defy conventions and take the lead' reads an advert for Elizabeth Arden beauty products. What is on offer in all these adverts is a specific kind of power – the sexual power to bring men to their knees. Empowerment is tied to possession of a slim and alluring young body, whose power is the ability to attract male attention and (sometimes) female envy.

As with representations of the male body, this change in 'sexualized' depictions of young women had a number of different determinants. Women's increasing financial independence throughout the 1980s and 1990s meant that they became targets for new products, and also forced a reconsideration of earlier modes of representation: showing a woman draped over a car – to take an emblematic image of sexism from the 1970s – may not be the best strategy if the aim is to sell that car to women. Moreover by the late 1980s and early 1990s advertisers had begun to recognize the significance of many women's anger at being objectified and bombarded with unattainable, idealized images of femininity, and also sought to appropriate some of the cultural power and energy of feminism.

The construction of the midriff represented one of a number of different ways in which advertising responded to feminism, and picked up and amplified other selective trends. This figure is notable for opening up a new vocabulary for the 'sexualized' representation of women in advertising, which aimed to banish the emphasis on passivity and objectification in favour of a modernized version of heterosexual femininity as feisty, sassy, and sexually agentic. This new set of meanings was produced through the combination of sexualized representations of women's bodies (focusing in particular on breasts, bottoms and flowing hair, made-up lips and eyes), juxtaposed with written or verbal texts that purported to speak of women's new sexual agency.

Like the erotic presentation of men's bodies, the 'sexualized' representation of women's is highly patterned. It is clear that only *some* women can be sexual subjects: women who are young, white, heterosexual and conventionally attractive. Sexual subjectification operates within a resolutely heteronormative economy. It is not that lesbians are not presented as sexy (see next section) but that their sex appeal is not constructed in lying in the agentic, knowing playful 'I' of the midriff. Moreover, older women, fat women, and any women who do not live up to the increasingly narrow normative judgements of female attractiveness are excluded from the pleasurable, empowering world of midriff advertising.

It is also worth considering the patterns of classing and racialization that are evident in this form of 'sexualization'. As a phenomenon it is striking to note how *white* the figure of the midriffs is. Black women's bodies are sexualized in advertising, to be sure, but mostly in ways that differ sharply from the figure of the active, knowing, desiring sexual subject examined here. Rather, black women's bodies still seem more likely to be portrayed as objects, signalling sexual promise, soul or authenticity. In a 2007 advertisement for Revlon cosmetics, for example, the black model's face is cropped, her eyes, wet-look lipstick and half-open mouth encoding mystery, pleasure and seductive sexual promise. Unlike the figure of the 'sexualized' midriff, this woman does not address us, does not make reference to her own pleasure or desire, and is not presented as a sexual subject. In a racist visual economy, the meaning of these types of sexualization is different depending if the bodies are black or white.

In ways that are less evident I would also suggest that this form of representational practice is profoundly classed. It defines itself against both an outdated sexual puritanism associated with the 'old' bourgeoisie *and*, simultaneously, against the 'looseness' or 'sluttishness' of contemporary constructions of (racialized) lower-class ('white trash') female sexuality. The female sexuality constructed in midriff advertising occupies a new space constructed around a middle-class 'respectable' sexiness.

QUEER CHIC AND SEXUALIZATION:
'HOT LESBIANS'

The final figure I want to consider is that of the 'hot lesbian' who is seen increasingly in contemporary advertising. Again, as with the previous two figures examined, the focus is not on 'representations of lesbians' in general but on a specific and relatively stable representational practice for the 'sexualized' depiction of woman–woman relations.

The last 10 years have witnessed an increasing number of representations of lesbians in media and culture.

Advertising is no exception. In June 2007 Commercial Closet, a web-based organization that monitors gay-themed advertisements, identified no fewer than 3500 adverts from 33 countries. This proliferation is partly the result of flourishing LGBTQ creativity in the wake of HIV and AIDS, the growing confidence of queer media and a recognition by companies of the significance of the pink economy. It is also a result of the cultural coolness currently accruing to queer sexualities; 'queering' an advertisement or deploying lesbian and gay themes has come to be regarded within the industry as an easy way of adding desirable 'edginess' to a product's image, and instantly giving it a more trendy, contemporary feel. The same process can be seen in music, as evidenced by the phenomenal success of Katy Perry's single 'I Kissed a Girl'.

The figure of the 'luscious lesbian' within advertising is notable for her extraordinarily attractive, conventionally feminine appearance. Women depicted in this way are almost always slim yet curvaceous, flawlessly made up and beautiful. Whilst this marks a rupture with earlier negative portrayals of lesbians as 'manly' or 'ugly', such representations have been criticized for packaging lesbianism within heterosexual norms of female attractiveness (Ciasullo, 2001). Ciasullo argues that such portrayals seem to work to annihilate the butch. Like the midriff, then, the 'hot lesbian' seems to rest on multiple exclusions, and in this case those excluded are precisely those with visibility in establishing lesbianism as a political identity: women who reject a traditionally feminine presentation.

The packaging of 'lesbians' within conventional norms of heterosexual feminine attractiveness is one way in which the figure appears to be constructed primarily for a straight male gaze. The figure never appears alone (unlike the midriff or the sixpack) but is almost always depicted kissing, touching or locked in an embrace with another woman. Two main strategies appear to dominate this kind of representation: either each woman will be shown with her 'other' e.g. a black woman with a blonde light-skinned woman, in ways reminiscent of many soft porn scenarios in which men choose their 'type' or 'flava'. Or, alternatively, they will be shown with another woman whom they resemble closely. This 'doubling' is, of course, another common male sexual fantasy which plays out in porn and is implicitly alluded to in many advertisements. Other scenarios also draw on the codes of heterosexual male porn: in an online advert for FCUK clothing, Fashion versus Style, two scantily clad women are seen wrestling, until the fight inevitably becomes sexual play and the pair tumble and writhe together erotically. Not only is this notable for being a stock scene from soft porn, but it is also markedly different from the way in which gay *men* are presented in advertisements. Whilst lesbian women rarely appear in mainstream advertisements *except* in this highly sexualized manner, gay men are rarely portrayed kissing or even touching – and the kind of erotic contact displayed between women in the FCUK advert would be unimaginable between two men, even in cinema advertising which is often more liberally regulated than that of terrestrial TV. Indeed, notwithstanding Calvin Klein's Guitar Kiss and a few other celebrated adverts in which two men embrace, albeit rather chastely, for the most part gay men are signified through stylish and attractive appearance or through a series of negative stereotypes rather than intimate conduct. The figure of the hot lesbian is therefore marked out from both representations of heterosexual women and from representations of gay men.

The sexualization of 'lesbian' bodies, then, seems to be constructed in relation to heterosexuality not as an autonomous or independent sexual identity. An example should make this critique clear. In this advertisement, Kiss Cool, a chewing gum is shown having electric and erotic effects. A young woman chews the gum and suddenly zooms to a haystack where she is kissing a man. The scene then cuts to a car where the man is kissing a different woman. After this kiss the new lover is suddenly transported onto a sofa, and is kissing a different man.

And so it goes on until finally the young woman is in a nightclub kissing another man, before proceeding to kiss a woman – to the man's intense surprise and then apparent amusement.

It would be hard to sustain the idea that the woman featured in this advert is a lesbian or even that she is bisexual. Her kiss with another woman is clearly marked as transgressive in a way that the other kisses were not, and the camera's focus on her boyfriend's shock and then amusement reinforces the heteronormative economy of gazes in this advert. We as (presumed heterosexual) viewers are invited to look to him to provide a guide to how to react to this kiss: it is sexy, to be sure, has produced a frisson, but is ultimately not to be taken seriously.

This is an example of what Diamond (2005) has called 'hetero flexibility' to denote heterosexual women 'experimenting' sexually with other women. It presents girl-on-girl action as exciting, fun, but, crucially, as entirely unthreatening to heterosexuality. Arguably, one of the pernicious aspects of this is that it allows advertisers to buy into the 'hot', 'now' social currency of queer whilst erasing lesbianism as such. In a truly queer world in which sexual identities no longer mattered this might be welcomed, but in a context in which heteronormativity remains powerful, and non-normative sexualities are marginalized, it appears entirely cynical.

CONCLUSION

I have argued that general claims about the 'sexualization of culture' have paid insufficient attention to the different ways in which different bodies are represented erotically. Through an analysis of three key figures from contemporary advertising – the sexy male 'sixpack', the active heterosexually desiring 'midriff' and the 'hot lesbian' – I have sought to demonstrate that power relations of gender and sexuality remain pivotal to processes of 'sexualization'.

The representations of the 'sexy' men and women we see in advertising are not equivalent: they have different meanings, different histories and are constructed in radically different ways. I have argued that eroticized representations of the male body are designed both to produce and disavow homoerotic desire, as well as to deal with the manifold anxieties occasioned by making the male body the object of multiple gazes. In contrast, in this postfeminist moment, the 'sexualization' of women's bodies requires an emphasis on women's pleasure and empowerment. However traditional or objectified the visual representation of women's bodies, it must be framed within a discourse of fun, freedom and female agency. Indeed, there often seems to be an inverse relationship between the visual and verbal texts of midriff advertising such that the more closely an image borrows from the vocabulary of heterosexual soft porn, the more the advertisement's written or verbal text will stress women's empowerment. Gay men and lesbian women too are sexualized in very different ways in mainstream advertising. Queer chic as it operates within advertising is not an undifferentiated practice, but one that is structured both by heteronormativity and by gender.

'Sexualization' remains an ongoing process in advertising, but in order to understand it, we need to move beyond a generic, undifferentiated notion, to look at the ways in which advertising's commodified 'sexiness' links to gender, sexuality, class, race and age.

REFERENCES

Chapman, R. (1988) 'The Great Pretender: Variations on a New Man Theme', in R. Chapman and J. Rutherford (eds) *Male Order: Unwrapping Masculinity*, pp. 225–48. London: Lawrence and Wishart.

Ciasullo, A. (2001) 'Making Her (In)visible: Cultural Representations of Lesbianism and the Lesbian Body in the 1990s', *Feminist Studies* 27(3): 477–508.

Connell, R. W. (1995) *Masculinities*. Cambridge: Polity.

Diamond, L. (2005) ' "I'm Straight but I Kissed a Girl": the Trouble with American Media Representations of Female-Female Sexuality', *Feminism and Psychology* 15(1): 140–10.

Rushkoff, D. (n.d.) *The Merchants of Cool* (PBS documentary), URL (accessed 1 December 2008): www.pbs.org/wgbh/pages/frontline/shows/cool/interviews/rushkoff.html

30

Darker Shades of Queer: Race and Sexuality at the Margins

CHONG-SUK HAN

By now, I've listened with a mild sense of amusement as countless gay leaders, an overwhelmingly white group, attempt to explain that their oppressed status as sexual minorities provides them with an enlightened sense of social justice that enables them to understand the plight of those who are racially oppressed. As sexual minorities, they say they share a history of oppression. This "shared history of oppression," they explain, provides them with exceptional insight and somehow absolves them of blame when it comes to the racial social hierarchy. Certainly, this view is not unique. People of color also believe that they possess special insights when it comes to social justice. Yet, two things are bitterly clear about our "shared" American experiences. One, a shared history of oppression rarely leads to coalition building among those who have been systematically denied their rights. More devastatingly, such shared experiences of oppression rarely lead to sympathy for others who are also marginalized, traumatized, and minimized by the dominant society. Rather, all too miserably, those who should naturally join in fighting discrimination find it more comforting to join their oppressors in oppressing others. In fact, they trade in one oppressed status for the other. Doing so, many gay white folks become more racist than non-gay folks while many people of color become more homophobic. As a gay man of color, I see this on a routine basis, whether it be racism in the gay community or homophobia in communities of color. And it pisses me off.

Psychologists have theories, I'm sure, about why such things happen. Perhaps some of us feel some comfort in the ability to claim at least a small share of the privileges and benefits of belonging to the so-called majority group. Maybe gay white folks use their race to buy into the mainstream while straight people of color use their sexuality for the same purpose. If they have something in common with those with power, perhaps some of that power will trickle down or rub off on them. I doubt it, but it's easy to believe. It's even comforting. What easier way to make oneself feel better than to marginalize others. But for now, it doesn't really matter why they do what they do. I'm not interested in why it happens. Rather, I'm interested in exposing it, condemning it, shaming it, and stopping it. Many gay activists want to believe that there aren't

SOURCE: From Shira Tarrant, *Men Speak Out: Views on Gender Sex and Power*, pp. 86–93. New York: Routledge, 2008. Reprinted by permission of the Taylor & Francis Group.

issues of racism within the gay community. As members of an oppressed group, they like to think that they are above oppressing others. Yet, looking around any gayborhood, something becomes blatantly clear to those of us on the outside looking in. Within the queer spaces that have sprung up in once neglected and forgotten neighborhoods, inside the slick new storefronts, in trendy restaurants, and on magazine covers, gay America has given a whole new meaning to the term "whitewash."

Whiteness in the gay community is everywhere, from what we see, what we experience, and more importantly, what we desire. Media images now popular in television and film such as *Queer as Folk, Queer Eye for the Straight Guy, The L-Word,* and the like promote a monolithic image of the gay community as being overwhelmingly upper-middle class—if not simply rich—and white. These images aren't new. Flipping through late night television, I'm often struck by how white and rich gay characters are whether they are on reruns of the *Golden Girls* or rebroadcasts of *Making Love.* The only difference now is that there are more of them, and they aren't so angst ridden about their sexuality. We now live in a country where gay TV characters are out and proud instead of languishing in perpetual shame, hiding in the shadows. They revel in their sexuality, at least on the screen. I can't help but think, though, that their revelation comes largely from their privilege of whiteness, the privilege to "be like everyone else," with only one minor difference.

Even the most perfunctory glance through gay publications exposes the paucity of non-white gay images. It's almost as if no gay men or women of color exist outside of fantasy cruises to Jamaica, Puerto Rico, or the "Orient." To the larger gay community, our existence, as gay men and women of color, is merely a footnote, an inconvenient fact that is addressed in the most insignificant and patronizing way. Sometime between Stonewall and *Will and Grace,* gay leaders decided that the best way to be accepted was to mimic upper-middle-class white America. As a sexually marginalized group, the idea is to take what they can from the dominant group and claim it as their own. Rarely a day goes by when I don't read something, by some "gay leader," about the purchasing power of the gay "community."

Sometimes, racism in the gay community takes on a more explicit form aimed at excluding men and women of color from gay institutions. All over the country, gay people of color are routinely asked for multiple forms of identification to enter the most basic of gay premises, the gay bar. Some might argue that this is a minor irritant. But in a country where gays and lesbians are expected to hide their sexuality while in public spaces, gay bars and other gay businesses provide us with the few opportunities to freely be ourselves. Denied access to the bars, gay men and women of color often lose the ability to see and socialize with others like us who also turn to these allegedly safe places for not only their social aspects but for their affirming aspects, as well. Isolated incidents might be easily forgotten, but news reports and buzz on various online forums expose such practices as endemic in gay communities. From New York to Los Angeles, from Seattle to Miami, the borders of gayness are patrolled by those who deny the existence of gay men and women of color.

And much like a neighborhood with more than six black families is quickly labeled "black" and, hence, avoided by many white buyers, a bar with more than six non-white folks is quickly labeled "ethnic" and avoided by the white clientele. Except, of course, those on the prowl for something "different" to wet their sexual appetite. In so many ways, their actions are nothing more than a cheap version of sexual tourism.

More importantly, gay men and women of color are routinely denied leadership roles in gay organizations that purport to speak for all of us. In effect, it is the needs and concerns of a largely middle-class gay white community that come to the forefront of what is thought to be a gay cause. Interjecting race in these community organizations is no easy task. On too many occasions, gay men and women of color have been told not to muddy the waters of the primary goal by bringing in concerns that might be addressed elsewhere. When mainstream gay organizations actually address issues of race, gay white men and women continue to set the agenda for what is and is not considered appropriate for discussion. During one community forum on race, the organizers, again an overwhelmingly white bunch, informed the audience that we would not actually talk about racism, as "everyone is capable of racism." With one sweeping generalization, this group of white men trivialized the everyday experiences of the men, and a few women, of color in the audience by denying our personal experiences and by turning the accusation back on us. It's a funny feeling to sit in an audience as an Asian gay man and watch as middle-class white men claim victimhood in racist America. Ultimately, isn't that what their claim that "everyone is capable of racism" boils down to? Isn't what they're really saying that they are also victims of racism?

Sadly, racism isn't just about how we are treated. It's also about how we treat ourselves. The primacy of whiteness in the gay community often manifests as internalized racism. In his 2002 online essay, "No Blacks Allowed," Keith Boykin argues that "in a culture that devalues black males and elevates white males," black men deal with issues of self-hatred that white men do not. Boykin argues that this racial self-hatred makes gay black men see other gay black men as unsuitable sexual partners and white males as the ultimate sexual partners.

This desire for white male companionship is not limited just to black men, and neither is racial self-hatred. Rather, it seems to be pandemic among many gay men of color. Even the briefest visit to a gay bar betrays the dirty secret that gay men of color don't see each other as potential life partners. Rather, we see each other as competitors for the few white men who might be willing to date someone considered lower on the racial hierarchy. We spend our energy and time contributing to the dominance of whiteness by putting white men on the pedestal while ignoring those who would otherwise be our natural allies.

The primacy of white masculinity in the gay community is no accident. It is a carefully choreographed racialization of men of color that mimics the masculine hierarchy in straight communities. Related to the "middle-classing" and "whitening" of gay America, images of gay men have mirrored the mainstream. No longer the Nellie queens of the 1960s, stereotypical images of gay men have changed from swivel-hipped sissies to muscle-bound he-men.

So ingrained is the image of the "ideal" man within the gay community, "straight-acting" is a selling point in gay personal ads.

But masculinity doesn't exist in a vacuum. Masculinity is built upon the femininity of Others. Lacking female femininity, the gay community thrusts this role upon Asian men. Given the way that Asian men are gendered in the mainstream, this isn't a difficult task. If gay white men are masculine, they are masculine compared to gay Asian men. If masculinity is desirable and femininity is not, then clearly white men are desirable but Asian men are not.

While black men have escaped the feminization inflicted on Asian men, their masculinity is also heavily gendered. Rather than lacking masculinity, they are hyper-masculine, outside of the norm and, thus, to be feared. Gay Latino men and gay Native American men fare no better in the gay imagination. In the gay white mind, they are exotic beings who exist for the pleasure of white male consumption. Much like heterosexual male claims to the right of sexual consumption of women, many gay white men have claimed the right to pick and choose what they want from their sexual partners by positioning whiteness as a bargaining tool in sexual encounters.

Ironically, we strive for the attention of the very same white men who view us as nothing more than an inconvenience. "No femmes, no fats, and no Asians" is a common quote found in many gay personal ads, both in print and in cyberspace. Gay white men routinely tell us that we are lumped with the very least of desirable men within the larger gay community. In this way, we are reduced to no more than one of many characteristics that are considered undesirable. Rather than confronting this racism, many of my gay Asian brothers have become apologists for this outlandish racist behavior. We damage ourselves by not only allowing it, but actively participating in it. We excuse their racist behavior because we engage in the same types of behavior. When seeking sexual partners for ourselves, we also exclude "femmes, fats, and Asians."

The rationale we use, largely to fool ourselves, to justify the inability of seeing each other as potential partners and allies, is laughable at best. Many Asian guys have told me that dating other Asians would be like "dating [their] brother, father, uncle, etc." Yet, we never hear white men argue that dating other white men would be like dating their brothers or fathers. This type of logic grants individuality to white men while feeding into the racist stereotype that all of "us" are indistinguishable from one another and therefore easily interchangeable.

Some of us rely on tired stereotypes. Boykin writes about the professional gay black man who degrades other black men as being of a lower social class while thinking nothing of dating blue-collar white men. The Asian version is that they are "Americanized" and looking for "American" men. In effect, we help white men oppress other men of color by buying into the racial social structure that the former group has created.

This self-hatred and willingness to join our oppressors in oppressing others is clearly evident in an April 4, 2006, column posted on Advocate.com. Jasmyne Cannick, a self-described black lesbian, argues that we should not extend any rights to "illegal" immigrants until gays are granted full rights. When did equality become a zero-sum game? Does extending human rights to immigrants somehow limit the amount of equal rights left for us? This is as absurd as claiming that extending marriage rights to gays and lesbians would somehow weaken heterosexual marriages. Extending rights to immigrants does not limit our rights in any way. To the contrary, it reinforces the commitment to equality and fairness for everyone, including gays and lesbians. As for Cannick's argument that these immigrants are "illegal," I might remind her that until the Supreme Court ruled on *Lawrence v. Texas* in 2003 millions of us were engaging in the horribly illegal activity of loving someone of the same sex. Thousands of us also got married "illegally." Why does Cannick believe that we have the right to challenge laws that brand us or our actions illegal, while others do not share this right? Clearly, those with the real power at the *Advocate* wanted to make a statement and found a pawn in Cannick. Her willingness to stand so firmly against granting rights to "illegal" immigrants simply reinforces the erroneous belief that it is acceptable for some members of society not to have the exact same rights enjoyed by others. Shamefully and unapologetically, Cannick quotes Audre Lorde at the very same time she uses the master's tools to build a fence around another oppressed group.

If we are invisible in the dominant gay community, perhaps we are doubly so in our own communities of color. If we are a footnote in the gay community, we are an endnote in communities of color—an inconvenient fact that is buried in the back out of view. We are told by family and friends that being gay is a white problem. We are told, early in life, that we must avoid such stigma at all costs. When we try to interject issues of sexuality, we are told that there is precious little time to waste on trivial needs while we pursue racial justice. Cannick writes that "lesbians and gays should not be second-class citizens. Our issues should not get bumped to the back of the line in favor of extending rights to people who have entered this country illegally." I've heard that exact sentence elsewhere. In fact, nearly verbatim, except "entered this country illegally" was replaced with "chosen an immoral lifestyle." See how that works? Master's tools.

I've seen those who are marginalized use the master's tools in numerous instances, now too legion to list. Citing Leviticus, some people of color who are also members of the clergy have vehemently attacked homosexuality as an abomination. This is the same Leviticus that tells us that wearing cloth woven of two fabrics and eating pork or shrimp is an abomination punishable by death. Yet not surprisingly, rarely do Christian fundamentalists picket outside of a Gap or a Red Lobster. If hypocrisy has a border, those wielding Leviticus as their weapon of choice must have crossed it by now. It must be convenient to practice a religion with such disdain that the word of God need only be obeyed when it reinforces one's own hatred and bigotry. How else do we explain those who condemn *Brokeback Mountain* based on their religious views while, in the same breath, praise *Walk the Line,* a movie about two adulterous country singers?

More problematic is that we choose to practice historic amnesia by ignoring the fact that Leviticus was used by slave owners to justify slavery by arguing that

God allowed the owning of slaves and selling of daughters. Anti–miscegenation laws, too, were justified using the Bible. In 1965, Virginia trial court judge Leon Bazile sentenced an interethnic couple who were married in Washington, D.C., to a jail term using the Bible as his justification. In his ruling, he wrote, "Almighty God created the races white, black, yellow, malay and red, and he placed them on separate continents. The fact that he separated the races shows that he did not intend for the races to mix." Scores of others also used the story of Phinehas, who distinguished himself in the eyes of God by murdering an interracial couple, thereby preventing a plague to justify their own bigotry. Have we forgotten that the genocide and removal of Native Americans was also largely justified on biblical grounds?

Have we simply decided to pick and choose the parts of the Bible that reinforce our own prejudices and use it against others in the exact same way that it has been used against us? Have we really gotten so adept at using the master's tools that he no longer needs to use them himself to keep us all in our place?

Given the prevalence of negative racial attitudes in the larger gay community and the homophobia in communities of color, gay people of color have to begin building our own identities. For gay people of color to be truly accepted by both the gay community and communities of color, we must form connections with each other first and build strong and lasting coalitions with each other rather than see each other as competitors for the attention of potential white partners. We must begin confronting whiteness where it stands while simultaneously confronting homophobia. More importantly, we must begin doing this within our own small circle of gay people of color. We must confront our own internalized racism that continues to put gay white people on a pedestal while devaluing other gays and lesbians of color. Certainly, this is easier said than done. The task at hand seems insurmountable. In Seattle, a group of gay, lesbian, and transgendered social activists from various communities of color have launched the Queer People of Color Liberation Project. Through a series of live performances, they plan on telling their own stories to counter the master narratives found within the larger gay community and within communities of color.

Certainly, gay people of color have allies both in the mainstream gay community and in our communities of color. Recently, Khalil Hassam, a high school student in Seattle, won a national ACLU scholarship for opposing prejudice. Hassam, the only Muslim student at University Prep High School, decided to fight for justice after a Muslim speaker made derogatory comments about homosexuals. Despite his own marginalized status as a Muslim American, Hassam confronted the homophobia found within his own community. Examples such as these are scattered throughout the country. Nonetheless, there is much more that allies, both straight and gay, can do to promote social justice. We must see gay rights and civil rights not as exclusive, but as complementary. All too often, even those on the left support "other" causes out of a Niemoellerian fear of having no one left to speak up for us if the time should come. I propose that the motivation to join in political efforts should come not from such fears, but from the belief that there are no such "other" causes. Rather, as Martin Luther King, Jr., reminded us, "an injustice anywhere is a threat to justice everywhere." We must remind ourselves, contrary to what Cannick may want us to believe, that social justice is not a zero-sum game. Granting rights to others does not diminish our rights. It is the exact opposite. Ensuring that rights are guaranteed to others ensures that they are guaranteed to us.

Ultimately, the crisis for gay men of color is one of masculinity. The centrality of masculinity within the gay community leads to rejecting all those outside of the masculine norm. Likewise, the centrality of masculinity in communities of color leads to stripping away masculinity based on stereotypical perceptions of homosexuality. Gay men of color are told, by both communities, that we are somehow not masculine enough to be full members in either community. Some of us have tried to attain this mythical masculine norm. We spend hours at the gym toning our bodies and building our muscles to fit with the gay masculine norm. Some of us disguise our speech, alter our style, and watch our steps in an effort to appear more straight to the untrained eye. And while a few of us may escape the cage of the masculine abnormality, we leave our brothers behind. "You're pretty masculine for an Asian guy," we are told. "You don't act like a black guy," they say. Ironically, both of these betray the cage of acceptable masculinity that binds us to the mythical norm. Asian men are not masculine enough, black men are too masculine. The narrow range of acceptable masculinity is reserved for white men—gay or straight—because, ultimately, it benefits them. Rather than see the explicit racist statement embedded in this compliment, we secretly blush and giggle at the attention. I can't speak eloquently or competently about lesbian experiences with the expectations of femininity they experience. Yet, whenever I listen to my gay Asian sisters speak of needing to be more lipstick to attract the butch white woman, I wonder what's happening over there in the lesbian community. It sounds vaguely familiar.

The real solution lies not in mimicking the masculinity found in the larger society, but in abolishing it. We need to think about re-envisioning what it means to be masculine and, for that matter, what it means to be feminine and the social value we place on each of these categories. Sadly, I don't have the answers, at least not yet. But I can't help yet feel that feminist scholars who attack the gendered hierarchy have a point. Perhaps one of them will find the answer someday. Until then, I can continue to make sure that concerns regarding race are brought up in the gay community and homophobia is confronted in communities of color. Perhaps one day, I won't need to do either.

31

Selling Sex for Visas

Sex Tourism as a Stepping-stone to International Migration

DENISE BRENNAN

On the eve of her departure for Germany to marry her German client–turned-boyfriend, Andrea, a Dominican sex worker, spent the night with her Dominican boyfriend. When I dropped by the next morning to wish her well, her Dominican boyfriend was still asleep. She stepped outside, onto her porch. She could not lie about her feelings for her soon-to-be husband. "No," she said, "it's not love." But images of an easier life for herself and her two daughters compelled her to migrate off the island and out of poverty. She put love aside—at least temporarily.

Andrea, like many Dominican sex workers in Sosúa, a small town on the north coast of the Dominican Republic, makes a distinction between marriage *por amor* (for love) and marriage *por residencia* (for visas). After all, why waste a marriage certificate on romantic love when it can be transformed into a visa to a new land and economic security?

Since the early 1990s, Sosúa has been a popular vacation spot for male European sex tourists, especially Germans. Poor women migrate from throughout the Dominican Republic to work in Sosúa's sex trade; there, they hope to meet and marry foreign men who will sponsor their migration to Europe. By migrating to Sosúa, these women are engaged in an economic strategy that is both familiar and altogether new: they are attempting to capitalize on the very global linkages that exploit them. These poor single mothers are not simply using sex work in a tourist town with European clients as a survival strategy; they are using it as an *advancement* strategy.

The key aims of this strategy are marriage and migration off the island. But even short of these goals, Sosúa holds out special promise to its sex workers, who can establish ongoing transnational relationships with the aid of technologies such as fax machines at the phone company in town (the foreign clients and the women communicate about the men's return visits in this manner) and international money wires from clients overseas. Sosúa's sex trade also stands apart from that of many other sex-tourist destinations in the developing world in

SOURCE: From *Global Woman: Nannies, Maids, and Sex Workers in the New Economy*, by Barbara Ehrenreich and Arlie Russell Hochschild, eds., 2003, pp. 154–161, 168. Reprinted by permission of the author.

that it does not operate through pimps, nor is it tied to the drug trade; young women are not trafficked to Sosúa, and as a result they maintain a good deal of control over their working conditions.

Certainly, these women still risk rape, beatings, and arrest; the sex trade is dangerous, and Sosúa's is no exception. Nonetheless, Dominican women are not coerced into Sosúa's trade but rather end up there through networks of female family members and friends who have worked there. Without pimps, sex workers keep all their earnings; they are essentially working freelance. They can choose the bars and nightclubs in which to hang out, the number of hours they work, the clients with whom they will work, and the amount of money to charge.

There has been considerable debate over whether sex work can be anything but exploitative. The stories of Dominican women in Sosúa help demonstrate that there is a wide range of experiences within the sex trade, some of them beneficial, others tragic.... I have been particularly alarmed at the media's monolithic portrayal of sex workers in sex-tourist destinations, such as Cuba, as passive victims easily lured by the glitter of consumer goods. These overly simplistic and implicitly moralizing stories deny that poor women are capable of making their own labor choices. The women I encountered in Sosúa had something else to say.

SEX WORKERS AND SEX TOURISTS

Sex workers in Sosúa are at once independent and dependent, resourceful and exploited. They are local agents caught in a web of global economic relations. To the extent that they can, they try to take advantage of the men who are in Sosúa to take advantage of them. The European men who frequent Sosúa's bars might see Dominican sex workers as exotic and erotic because of their dark skin color; they might pick one woman over another in the crowd, viewing them all as commodities for their pleasure and control. But Dominican sex workers often see the men, too, as readily exploitable—potential dupes, walking visas, means by which the women might leave the island, and poverty, behind.

Even though only a handful of women have actually married European men and migrated off the island, the possibility of doing so inspires women to move to Sosúa from throughout the island and to take up sex work. Once there, however, Dominican sex workers are beholden to their European clients to deliver visa sponsorships, marriage proposals, and airplane tickets. Because of the differential between sex workers and their clients in terms of mobility, citizenship, and socioeconomic status, these Dominican sex workers might seem to occupy situations parallel to those that prevail among sex workers throughout the developing world. Indeed, I will recount stories here of disappointment, lies, and unfulfilled dreams. Yet some women make modest financial gains through Sosúa's sex trade—gains that exceed what they could achieve working in export-processing zones or domestic service, two common occupations among poor Dominican women. These jobs, on average, yield fewer than 1,000 pesos ($100) a month, whereas sex workers in Sosúa charge approximately 500 pesos for each encounter with a foreign client.

Sex tourism, it is commonly noted, is fueled by the fantasies of white, First World men who exoticize dark-skinned "native" bodies in the developing world, where they can buy sex for cut-rate prices. These two components—racial stereotypes and the economic disparity between the developed and the developing worlds—characterize sex-tourist destinations everywhere. But male sex tourists are not the only ones who travel to places like Sosúa to fulfill their fantasies. Many Dominican sex workers look to their clients as sources not only of money, marriage, and visas, but also of greater gender equity than they can hope for in the households they keep with Dominican men. Some might hope for romance and love, but most tend to fantasize about greater resources and easier lives.

Yet even for the women with the most pragmatic expectations, there are few happy endings. During the time I spent with sex workers in Sosúa, I, too, became invested in the fantasies that sustained them through their struggles. Although I learned to anticipate their return from Europe, disillusioned and divorced, I continued to hope that they would find financial security and loving relationships. Similarly, Sosúa's sex workers built their fantasies around the stories of their few peers who managed to migrate as the girlfriends or wives of European tourists—even though nearly all of these women returned, facing downward mobility when they did so. Though only a handful of women regularly receive money wires from clients in Europe, the stories of those who do circulate among sex workers like Dominicanized versions of Hollywood's *Pretty Woman*.

The women who pursue these fantasies in Sosúa tend to be pushed by poverty and single motherhood. Of the fifty women I interviewed and the scores of others I met, only two were not mothers. The practice of consensual unions (of not marrying but living together), common among the poor in the Dominican Republic, often leads to single motherhood, which then puts women under significant financial pressure. Typically, these women receive no financial assistance from their children's fathers. I met very few sex workers who had sold sex before migrating to Sosúa, and I believe that the most decisive factor propelling these women into the sex trade is their status as single mothers. Many women migrated to Sosúa within days of their partners' departure from the household and their abandonment of their financial obligations to their children.

Most women migrated from rural settings with meager job opportunities, among them sporadic agricultural work, low-wage hairstyling out of one's home, and waitressing. The women from Santo Domingo, the nation's capital, had also held low-paying jobs, working in domestic service or in *zonas francas* (export-processing zones). Women who sell sex in Sosúa earn more money, more quickly than they can in any other legal job available to poor women with limited educations (most have not finished school past their early teens) and skill bases. These women come from *los pobres,* the poorest class in the

Dominican Republic, and they simply do not have the social networks that would enable them to land work, such as office jobs, that offer security or mobility. Rather, their female-based social networks can help them find factory jobs, domestic work, restaurant jobs, or sex work.

Sex work offers women the possibility of making enough money to start a savings account while covering their own expenses in Sosúa and their children's expenses back home. These women tend to leave their children in the care of female family members, but they try to visit and to bring money at least once a month. If their home communities are far away and expensive to get to, they return less frequently. Those who manage to save money use it to buy or build homes back in their home communities. Alternatively, they might try to start small businesses, such as *colmados* (small grocery stores), out of their homes.

While saving money is not possible in factory or domestic work, sex workers, in theory at least, make enough money to build up modest savings. In practice, however, it is costly to live in Sosúa. Rooms in boardinghouses rent for 30 to 50 pesos a day, while apartments range from 1,500 to 3,000 pesos a month, and also incur start-up costs that most women cannot afford (such as money for a bed and cooking facilities). Since none of the boarding-houses have kitchens, women must spend more for take-out or restaurant meals. On top of these costs, they must budget for bribes to police officers (for release from jail), since sex workers usually are arrested two to five times a month. To make matters worse, the competition for clients is so fierce, particularly during the low-volume tourist seasons, that days can go by before a woman finds a client. Many sex workers earn just enough to cover their daily expenses in Sosúa while sending home modest remittances for their children. Realizing this, and missing their children, most women return to their home communities in less than a year, just as poor as when they first arrived....

The sex workers I interviewed, who generally have no immediate family members abroad, have never had reliable transnational resources available to them. Not only do they not receive remittances but they cannot migrate legally through family sponsorship. Sex workers' transnational romantic ties act as surrogate family-migration networks. Consequently, migration to Sosúa from other parts of the Dominican Republic can be seen as both internal and international, since Sosúa is a stepping-stone to migration to other countries. For some poor young women, hanging out in the tourist bars of Sosúa is a better use of their time than waiting in line at the United States embassy in Santo Domingo. Carla, a first-time sex worker, explained why Sosúa draws women from throughout the country: "We come here because we dream of a ticket," she said, referring to an airline ticket. But without a visa—which they can obtain through marriage—that airline ticket is of little use.

If sex workers build their fantasies around their communities' experiences of migration, the fantasies sex tourists hope to enact in Sosúa are often first suggested through informal networks of other sex tourists. Sosúa first became known among European tourists by word of mouth. Most of the sex tourists I met in Sosúa had been to other sex-tourist destinations as well. These seasoned sex tourists, many of whom told me that they were "bored" with other

destinations, decided to try Sosúa and Dominican women based on the recommendations of friends. This was the case for a group of German sex tourists who
were drinking at a bar on the beach. They nodded when the German bar owner
explained, "Dominican girls like to fuck." One customer chimed in, "With
German women it's over quickly. But Dominican women have fiery blood....
When the sun is shining it gives you more hormones."

The Internet is likely to increase the traffic of both veteran and first-time
sex tourists to previously little-known destinations like Sosúa. Online travel
services provide names of "tour guides" and local bars in sex-tourism hot
spots. On the World Sex Guide, a Web site on which sex tourists share information about their trips, one sex tourist wrote that he was impressed by the
availability of "dirt cheap colored girls" in Sosúa, while another gloated,
"When you enter the discos, you feel like you're in heaven! A tremendous
number of cute girls and something for everyone's taste (if you like colored
girls like me)!"

As discussions and pictures of Dominican women proliferate on the Internet
sites—for "travel services" for sex tourists, pen-pal services, and even cyber
classified advertisements in which foreign men "advertise" for Dominican
girlfriends or brides—Dominican women are increasingly often associated with
sexual availability. A number of articles in European magazines and newspapers
portray Dominican women as sexually voracious. The German newspaper
Express even published a seven-day series on the sex trade in Sosúa, called
"Sex, Boozing, and Sunburn," which included this passage: "Just going from
the street to the disco—there isn't any way men can take one step alone. Prostitutes bend over, stroke your back and stomach, and blow you kisses in your
ear. If you are not quick enough, you get a hand right into the fly of your
pants. Every customer is fought for, by using every trick in the book." A photo
accompanying one of the articles in this series shows Dieter, a sex tourist who
has returned to Sosúa nine times, sitting at a German-owned bar wearing a
T-shirt he bought in Thailand; the shirt is emblazoned with the words SEX
TOURIST.

With all the attention in the European press and on the Internet associating
Dominican women with the sex industry, fear of a stigma has prompted many
Dominican women who never have been sex workers to worry that the families
and friends of their European boyfriends or spouses might wonder if they once
were. And since Dominican women's participation in the overseas sex trade has
received so much press coverage in the Dominican Republic, women who have
lived or worked in Europe have become suspect at home. "I know when I tell
people I was really with a folk-dance group in Europe, they don't believe me," a
former dancer admitted. When Sosúans who were not sex workers spoke
casually among themselves of a woman working overseas as a domestic, waitress,
or dancer, they inevitably would raise the possibility of sex work, if only to rule
it out explicitly. One Dominican café owner cynically explained why everyone
assumes that Dominican women working overseas must be sex workers:
"Dominican women have become known throughout the world as prostitutes.
They are one of our biggest exports."...

Marginalized women in a marginalized economy can and do fashion creative strategies to control their economic lives. Globalization and the accompanying transnational phenomena, including sex tourism, do not simply shape everything in their paths. Individuals react and resist. Dominican sex workers use sex, romance, and marriage as means of turning Sosúa's sex trade into a site of opportunity and possibility, not just exploitation and domination. But exits from poverty are rarely as permanent as the sex workers hope; relationships sour, and subsequently, an extended family's only lifeline from poverty disintegrates. For every promise of marriage a tourist keeps, there are many more stories of disappointment. Dominican women's attempts to take advantage of these "walking visas" call attention, however, to the savviness and resourcefulness of the so-called powerless.

PART III

✳

The Structure of Social Institutions

MARGARET L. ANDERSEN AND PATRICIA HILL COLLINS

In the United States and globally, social institutions exert a powerful influence on everyday life. Social institutions are also important channels for societal penalties and privileges. The type of work you do, the structure of your family, whether your religion will be recognized or suppressed, the kind of education you receive, the images you see of yourself in the media, and how you are treated by the state are all shaped by the social institutions of the society where you live. People rely on institutions to meet their needs, although social institutions treat some groups better than others. When a specific institution (such as the economy) fails them, people often appeal to another institution (such as the state) for redress. In this sense, institutions are both sources of support and sources of repression.

The concept of an institution is abstract because there is no thing or object that one can point to as an institution. *Social institutions* are the established societal patterns of behavior organized around particular purposes. The economy is an institution, as are the family, the media, education, and the state. These important societal institutions are examined here. Each is organized around a specific purpose; in the case of the economy, for example, its purpose is the production, distribution, and consumption of goods and services. Within a given institution, there may be various patterns, such as different family structures, but as a whole, institutions have general patterns of behavior that emerge because of the specific societal conditions in which groups live. Institutions do change over time because societal conditions evolve and groups challenge specific institutional structures. Yet institutions are also enduring and persistent, even in the face of active efforts to change them. Institutions confront people from birth and live on after people die.

Across all societies, institutions are patterned by intersections of race, class, gender, sexuality, age, ethnicity, and disability (among others), and, as a result, the effects of these systems of power on social institutions differ from one society to the next. Each society has a distinctive history and institutional configuration of social inequalities. In the United States, race, class, and gender constitute fundamental categories that shape American social institutions that serve as basic conduits for social inequalities. Moreover, social institutions are interconnected. We think the interrelationships among institutions are especially apparent when studied in the context of race, class, and gender. Race, class, and gender oppression rests on a network of interconnected social institutions. Understanding the interconnections among institutions helps us see that we are all part of one historically created system that finds structural form in interrelated social institutions.

Despite their significance, dominant American ideology portrays institutions as neutral in their treatment of different groups; indeed, the liberal framework of the law allegedly makes access to public institutions (such as education and work) gender- and race-blind. Still, institutions differentiate on the basis of race, class, and gender. As the readings included here show, institutions are actually structured on the basis of race, class, and gender relations.

As an example, think of the economy. Economic institutions in American society are founded on capitalism—an economic system based on the pursuit of profit and the principle of private ownership. Such a system creates class inequality because, in simple terms, the profits of some stem from the exploitation of the labor of others. Race and gender further divide the U.S. capitalist economy, resulting in labor and consumer markets that routinely advantage some and disadvantage others. In particular, corporate and government structures create jobs for some while leaving others underemployed or without work.

Those with jobs encounter a dual labor market that includes: (1) a primary labor market characterized by relatively high wages, opportunities for advancement, employee benefits, and rules of due process that protect workers' rights; and, (2) a secondary labor market (where most women and minorities are located) characterized by low wages, little opportunity for advancement, few benefits, and little protection for workers. One result of the dual labor market is the persistent wage gap between men and women (see Figure 1) and between Whites and people of color—even when they have the same level of education. Women tend to be clustered in jobs that employ mostly women workers; such

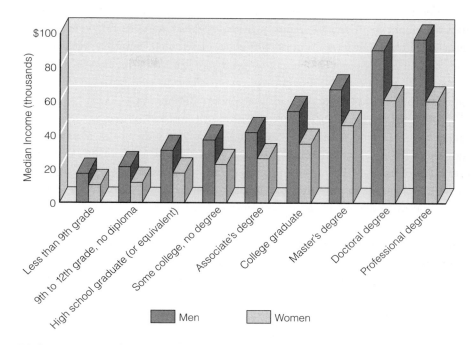

FIGURE 1 Median Income by Educational Attainment
SOURCE: U.S. Census Bureau. 2005. *Historical Income Tables.* www.census.gov.

jobs have been economically devalued as a result. Gender segregation and race segregation intersect in the dual labor market, so women of color are most likely to be working in occupations where most of the other workers are also women of color. At the same time, men of color are also segregated into particular segments of the market. Indeed, there is a direct connection between gender and race segregation and wages because wages are lowest in occupations where women of color predominate (U.S. Department of Labor 2011). This is what it means to say that institutions are structured by intersections of race, class, and gender: institutions are built from and then reflect the historical and contemporary patterns of race, class, and gender relations in society.

As a second example, think of the state. The state refers to the organized system of power and authority in society. This includes the government, the police, the military, and the law (as reflected in both social policy and the civil and criminal justice system). The state is supposed to protect all citizens, regardless of their race, class, and gender (as well as other characteristics, such as disability and age); yet state policies routinely privilege men. This means not only that the majority of powerful people in the state are men (such as elected officials, judges, police, and the military) but, just as important, that the state works to protect men's interests. Policies about reproductive rights provide a good example. They are largely enacted by men but have an important effect on women. Similarly, welfare policies designed to encourage people to work are based on the model of men's experiences because they presume that staying home to care for one's children is not working. In this sense, the state is a gendered institution.

Most people do not examine the institutional structure of society when thinking about intersections of race, class, and gender. The individualist framework of dominant American culture sees race, class, gender, ethnicity, sexuality, and age as attributes of individuals, instead of seeing them as systems of power and inequality that are embedded in institutional structures. People do, of course, have identities of race, class, and gender, and these identities have an enormous impact on individual experience. However, seeing race, class, and gender solely from an individual viewpoint overlooks how profoundly embedded these identities are in the structure of American institutions. Moving historically marginalized groups to the center of analysis clarifies the importance of social institutions as links between individual experience and larger structures of race, class, and gender.

In this section of the book, we examine how social institutions structure systems of power and inequality of race, class, gender, ethnicity, and sexuality that, in turn, structure those very same social institutions. We look at five important social institutions: work and the economic system; families; cultural institutions such as the media that reproduce ideas; education; and the state. Each institution can be shown to have a unique impact on different groups (such as the discriminatory treatment of African American men by the criminal justice system). At the same time, each institution and its interrelationship with other institutions can be seen as specifically structured through intersecting dynamics of race, class, and gender.

WORK AND ECONOMIC TRANSFORMATION

Structural transformations in the global economy have dramatically changed the conditions under which all people work. But, as Teresa Amott and Julie Matthaei ("Race, Class, Gender, and Women's Works") argue, specific social economic forces differentiate work experiences for women and men. Their analysis provides a basic orientation to how jobs in the labor market are arrayed along lines of race, class, and gender. They also identify some of the specific processes that shape the different sectors of work.

We can add to their analysis several major transformations that affect the current character of work. First, the economy is now a global economy; thus, the work experiences in any one nation (the United States included) are deeply linked to the work experiences of those who may be in remote regions of the world. Among other things, this has meant that employers have often exported jobs as a way of cutting costs of labor and to increase profitability. The jobs remaining are, in many cases, either low-wage jobs or jobs requiring a very high degree of education—thus bifurcating the labor market into two classes: those with a high degree of skill and education and those with little opportunity for job mobility.

Second, the consequences of the global economy are exacerbated by the development of new technologies that have made some jobs obsolete and other jobs require a high degree of technological and/or scientific skills. Race, class, and gender frame how individual workers encounter these dual processes of "deskilling" and job export.

Third, the economy in the United States has shifted from being one primarily based on manufacturing jobs to one based on the broadly-defined "service" industry—a term used to include jobs that not only provide direct service (such as cashiers, food preparation workers, janitors, and so forth), but also jobs that are based on information services, not the actual production of goods. Thus, occupations like teaching, financial services, even professional jobs like physicians, lawyers or scientists, are based more on the production and transmission of information than they are on the actual manufacture of goods.

Altogether, these changes have resulted in a very different workforce than one would have seen forty or fifty years ago. The transformation has been underway for some time, but it is having dramatic effects on the character of work in the United States—and other parts of the world. Someone entering this economy as a young person has to have the skills to adapt to these changes or the person may face a lifetime of dead-end work or no work at all—a fact that has plagued many racial-ethnic minorities, such as young, poor, African American men who simply are not positioned in such an economy to get a foot in the door (Wilson 1997).

The growth of the service sector fuels the dual economy described earlier. The dual economy operates by creating high-paying service work for skilled, college-educated workers (accountants, marketing representatives, etc.) and low-paying service work for everyone else (food service workers, nursing home aides, child care workers, and domestic workers, for example). A young man who years ago might have found a decent-paying job in an automobile assembly plant or a steel mill—perhaps even without a high school education—is unlikely in today's economy to do any better than a minimum wage job in a fast food workplace—a job that is not likely to lead to a lifetime of steady employment and economic security.

You can see here that these large-scale, seemingly neutral processes have specific consequences for different race, class, and gender groups. And, as Margaret Andersen points out in "Seeing in 3D: A Race, Class, and Gender Lens on the Economic Downturn," understanding the intersections of race, class, and gender is critical to understanding the serious economic problems facing our nation, such as the recent economic downturn.

These transformations also explain a lot about how work is organized within particular areas of the economy and how structures deeply built into the economy can affect even the simplest things—such as how the unconscious biases associated with people's names may shape opportunities in the labor market, a phenomenon cleverly studied by Marianne Bertrand and Sendhil Mullainathan ("Are Emily and Greg More Employable than Lakisha and Jamal? A Field Experiment on Labor Market Discrimination").

The remaining two articles in this section further detail how the move to a service-based society affects people's experiences at work. Christine L. Williams ("Racism in Toyland") details how jobs within the retail industry are organized in a division of labor that is very much pinned down by the interaction of race, class, and gender. Likewise, Sandra Weissinger's study of women workers in Wal-mart ("Gender Matters, So Do Race and Class: Experiences of Gendered Racism on the Wal-mart Shop Floor") explores how the inequities of race, class, and gender are reflected in the lived experiences of women workers.

FAMILIES

Families are another primary social institution profoundly influenced by intersecting systems of race, class, and gender. Historically, the family has been presumed to be the world of women: the ideology of the family (i.e., dominant belief systems about the family) purports that families are places for nurturing, love, and support—characteristics that have been associated with women. This ideal identifies women with the private world of the family and men with the public sphere of work. Family ideology, of course, only projects an ideal because the large number of working mothers shows that few families actually fit the presumed ideal. Nonetheless, the ideology of the family provides a standard against which all families are judged. Moreover, the ideology of the family is class and race specific; that is, it ignores and distorts the family experiences of African Americans, Latinos, and most Whites.

Bonnie Thornton Dill's essay, "Our Mothers' Grief: Racial-Ethnic Women and the Maintenance of Families," shows that for African American, Chinese American, and Mexican American women, family structure is deeply affected by the relationships of families to the structures of race, class, and gender. Dill's historical analysis of racial-ethnic women and their families examines diverse patterns of family organization directly influenced by a group's placement in the larger political economy. Just as the political economy of the nineteenth century affected women's experience in families, the political economy of the late twentieth century shapes family relations for women and men of all races.

The articles in this section illustrate several points that have emerged from feminist studies of families (Thorne and Yalom 1992). First, the family is not monolithic. The now widely acknowledged diversity among families refutes the idea that there is a normative family, which is White, middle-class, with children, organized around a heterosexual married couple (preferably with the wife not employed), and needing little support from relatives or neighbors. As the experiences of African Americans, Latinos, Asian Americans, lesbians, gays, and others reveal, this so-called "normal" family actually represents a minority experience, as Naomi Gerstel shows in "Rethinking Families and Community: The Color, Class, and Centrality of Extended Kin Ties." Many also presume that families are formed around a heterosexual norm. But as Kath Weston points out in "Straight Is to Gay as Family Is to No Family," gays and lesbians also participate in their families of origin and create families of their own. Claiming sexual identities as gay, lesbian, or bisexual does not mean relinquishing family or, worse yet, being defined as anti-family or as being a threat to family. Gay and lesbian partners often form permanent relationships, often including children. Although such partnerships are not uniformly legally recognized, they are part of the diversity among families. Raising children in diverse family environments so that they appreciate not only their own families, but also diverse ways of forming families, is a challenge in a society that devalues gays and lesbians and people of color.

Family diversity is also apparent in the growing number of multiracial families. But, even with this growth, navigating a world where mono-racial families are the presumed norm can be a challenge for those who marry or date across racial borders. Erica Chito Childs explores this in her study of interracial couples ("Navigating Interracial Borders: Black-White Couples and Their Social Worlds").

Finally, family structures are intricately connected to the race, class, gender structure of work. The boundaries separating family, work, and other institutions are more fluid than was previously believed. No institution is isolated from another. By detailing patterns of domestic work done by new immigrant women of color, Pierrette Hondagneu-Sotelo's article ("Families on the Frontier") illustrates the permeable boundaries between work and family. In essence, the belief in the idealized family form achieved by middle-class families is possible only through the domestic work of women of color. Changes in the economy, for example, affect family structures; moreover, this effect is reciprocal because changes in the family generate changes in the economy. The situation of working mothers challenges this assumption of the separation of work and family. The presumed dichotomy between the public and private spheres is especially seen to be false in light of the experience of women of color. Racial-ethnic families have rarely been provided the protection and privacy that the alleged split between public and private assumes. From welfare policy to reproductive rights, this dichotomy is based on the myth that families are insulated from the society around them.

Recentering one's thinking about the family by understanding the interconnections of race, class, and gender reveals new understandings about families and the myths that have pervaded assumptions about family experience. In particular, ideology that presents families as peaceful, loving, nurturing spaces masks the underlying conflicts that are embedded in family systems. Although we do not examine family violence directly here, we know that violence stems from the race, class, and gender conflicts that are encouraged in a racist, class-based, and sexist society. In addition, we think of family violence not only as that which occurs between family members but also as that which happens to families as they confront the institutional structures of race, class, and gender oppression.

MEDIA AND POPULAR CULTURE

Work, family, and other social institutions are patterned by race, class, and gender, yet many people do not see these patterns and even may claim that they do not exist. The effects of economic restructuring on people of color and women and the differences manifested as Americans try to meet the American family ideal can be difficult to understand. How can race, class, and gender exert such a powerful influence on our everyday lives, yet remain so invisible?

The products of cultural institutions, such as movies, books, television, advertisements in the mass media, the statistics and documents produced by government, and the textbooks, curriculum materials, and teaching methods of schools, explain in part why race, class, and gender relations are obscured. These and other cultural institutions reproduce ideas by identifying which ideas are valuable, which are not, and which should not be heard at all. In this way, the ideas of groups that are privileged within race, class, and gender relations are routinely heard, whereas the ideas of groups who are disadvantaged are silenced. In part, invisibility and silencing result from deeply entrenched ideas about race, class, and gender in the social fabric of the United States. Because beliefs about race, class, and gender are so ordinary, they often go unnoticed. But even in the now ever-present spaces of social media like Facebook and MySpace, images associated with race, class and gender appear. And, as Danah Boyd shows ("White Flight in Networked Publics? How Race and Class Shaped American Teen Engagement with MySpace and Facebook") even patterns of usage in social media can be understood within an intersectional framework. Typically, images in the media are unquestioned, unless examined with a critical eye. Rachel E. Dubrofsky ("*The Bachelor:* Whiteness in the Harem") does this well in her analysis of the television show *The Bachelor.* She equates the structure of this show with the harem: one man surrounded by highly sexualized, exotic women. Her essay shows not only the gender and sexual stereotypes that are embedded in the show, but also how racialized these images are, involving stereotypes of Middle Eastern women, as well as implicitly denigrating African American women and Latinas who, she shows, are routinely eliminated from this sexualized competition. Perhaps having read her analysis, you will not see so-called "reality TV" as so real after all.

Mass media and similar cultural institutions also have the capacity to harm those stigmatized by them. The articles in this section remind us that the ideas produced by these cultural institutions are far from benign because they uphold social policies and affect individuals. Placing his analysis in a global human rights context, Ward Churchill ("Crimes against Humanity") discusses how American national identity has been constructed against the interests of Native Americans. He suggests that while Native American symbolism and imagery is ubiquitous in the United States, the derogatory use of Native American mascots and the like for sports teams stereotypes and stigmatizes contemporary Native Americans. Ironically, while Native Americans continue to experience what Churchill discusses as genocidal policies, distorted ideas about Native American culture continue to be celebrated at college and professional sporting events. Churchill feels so strongly about the damage done to Native Americans by pervasive negative symbolism that he views this practice as criminal.

Certain cultural institutions have gained greater importance in the context of global economic restructuring. With the emergence of a global mass media that circulates via television, music videos, cable, radio, and the Internet to all parts of the globe, racist and sexist ideas can travel quickly. Thus, those who control mass media are in a position to shape the basic ideas of society that come to be seen as truth. Gregory Mantsios's essay "Media Magic: Making Class Invisible" explores how mass media have been central in shaping ideas about social class. His work identifies the growing significance of mass media in framing how people think about race, class, and gender.

Mass media and other cultural institutions can help reproduce and mask social inequalities of race, class, and gender, but these same cultural institutions can be sites for challenging longstanding oppressions. In this sense, cultural institutions are rarely all good or all bad—more often, they are a little bit of both. Julianne Malveaux ("Gladiators, Gazells, and Groupies: Basketball Love and Loathing") also tells us how our passions for various forms of popular culture—in her case, basketball—can catch us in contradictions, confronted, let's say, with the recognition that something we enjoy also involves race, class, and gender inequalities. Confronting and understanding such contradictions in our own engagement with media and popular culture is part of the process of learning about how to challenge the inequities in society.

EDUCATION

American education has recently and frequently been described as an institution in crisis. People worry that children are not learning in school. High school dropout rates, especially among students from poor and working-class families of color, reveal deep problems in the system of education. Schools, in fact, reproduce inequalities of race, class, and gender, at the same time that they reflect the inequalities in the society at-large. Historically, social policies of racial and class segregation have been one important way that schools have fostered inequalities. Yet the elimination of formal segregation does not mean that desegregation as a policy yields educational equity. As Gary Orfield and Chungmei Lee point out in "Historic Reversals, Accelerating Resegregation, and the Need for New Integration Strategies," although actual desegregation policies were generated to attack schooling inequalities more than fifty years ago, racial isolation in schools is actually growing. The promise of school integration, meant not only to provide a good education to all, but also to engage young people in racially integrated relationships, is beginning to seem like a fading dream.

Moreover, segregated schools that are largely schools for poor and minority children are more often than not failing their students. Jeanne Theoharis ("I Hate It When People Treat Me Like a Fxxx-up") vividly describes this, especially from the perspective of students. Moreover, as she argues, students, not schools or society, get the brunt of the blame for the failing education that pervades so many schools.

Schools also serve as sites of military recruitment. In "How a Scholarship Girl Becomes a Soldier: The Militarization of Latina/o Youth in Chicago Public Schools," Gina Perez details how working-class Puerto Rican and Latino youth are targets for military recruitment. The recruitment of racial–ethnic youth is not new, but Perez shows how the practices of military programs such as the Junior Reserve Officer Training Program (JROTC) illustrate intersections of race, class, gender, and sexuality. In the past, the army primarily recruited young Latino men who were considered "at risk" of falling into gangs and drug use. Today, the military also targets young Latinas, arguing that as women, they are "at risk" for unwed teenage pregnancy. Because the military seemingly offers more opportunities to escape poverty than does college, Perez shows how the new push to recruit Latina youth reinforces structures of race, class, and gender inequalities.

THE STATE AND VIOLENCE

Violence in the United States is a pressing social issue. School shootings, snipers, and street criminals dominate the news and project the idea that violence is rampant. Usually violence is depicted as the action of crazed individuals who are socially maladjusted, angry, or desperate. Although this may be the case for some, deeper questions about violence are raised when we think about violence in the context of how state power organizes race, class, and gender relations. Who is perceived as violent shifts, as does who we see as victims of violence. How can we apply our knowledge of race, class, and gender to understanding violence?

Violent acts, the threat of violence, and more generalized policies based on the use of force find organizational homes in state institutions of social control—primarily the police, the military, and the criminal justice system. Intersections of race, class, and gender shape all of these institutions. In "Policing the National Body: Sex, Race, and Criminalization," Jael Silliman demonstrates how intersections of race, class, and gender shape state social policy. Silliman links social policies that are often seen as separate and treated as such. For example, the aggressive policing and the increasing incarceration of African American, Latino, and poor White men, the assault on reproductive rights of women of color, and the search of immigrants of color by the Immigration and Naturalization Service under the pretext of national security all may seem to affect different race, class and gender groups, yet they all are part of an overarching system of state-sanctioned, social control.

Middle and upper class people may escape these policies and, as a result, may perceive the police and the criminal justice system as protecting society. Yet for others, the criminal justice system itself is a source of violence. High rates of incarceration, especially in African American poor communities, as shown by Michelle Alexander in "The Color of Justice," mean that huge numbers of young, African American men experience the state as a form of social control, not an institution of protection.

Moreover, the criminal justice system punishes certain forms of violence while ignoring others. Differential enforcement of state policies, even the presence of policies that define some acts as violent and others, not, mean that state institutions define what counts as violence. For example, race, class, and gender have shaped the very definition of *rape* and the treatment afforded rape survivors and their rapists. In "Rape, Racism, and the Law," Jennifer Wriggins examines how the legal system's treatment of rape has disproportionately targeted Black men for punishment and made Black women especially vulnerable by denying their sexual subordination. Although violence may be experienced individually, it occurs in specific organizational and institutional contexts shaped by race, class, and gender.

Finally, who becomes a target of violence is also shaped by race, class, and gender. Hate crimes are harmful to any group or person experiencing such violence, and those disadvantaged by either race or class or gender are most likely to be victimized by such horrific acts. But, as Doug Meyer shows ("Interpreting and Experiencing Anti-Queer Violence: Race, Class, and Gender Differences among LGBT Hate Crime Victims"), LGBT persons of color find it more difficult than other LGBT victims to have this crime defined as based on their sexuality. Meyer's research illustrates how the intersection of race, class, and gender with sexuality shapes both the definition and handling of this form of violence.

Altogether, the articles in this section show how embracing an intersectional analysis of race, class, gender and sexuality is important to creating an awareness of how society can become more safe, secure, and just.

REFERENCES

Hondagneu-Sotelo, Pierrette. 2001. *Doméstica: Immigrant Workers Cleaning and Caring in the Shadows of Affluence.* Berkeley: University of California Press.

Thorne, Barrie, with Marilyn Yalom. 1992. *Rethinking the Family: Some Feminist Questions*, Rev. ed. Boston: Northeastern University Press.

U.S. Department of Labor. January 2011. *Employment and Earnings.* Washington, D.C.: U.S. Government Printing Office. www.dol.gov

Wilson, William Julius. 1997. *When Work Disappears: The New World of the Urban Poor.* Chicago, IL: University of Chicago Press.

32

Race, Class, Gender, and Women's Works

TERESA AMOTT AND JULIE MATTHAEI

What social and economic factors determine and differentiate women's work lives? Why is it, for instance, that the work experiences of African American women are so different from those of European American women? Why have some women worked outside the home for pay, while others have provided for their families through unpaid work in the home? Why are most of the wealthy women in the United States of European descent, and why are so many women of color poor? …

Throughout U.S. history, economic differences among women (and men) have been constructed and organized along a number of social categories. In our analysis, we focus on the three categories which we see as most central—gender, race-ethnicity, and class with less discussion of others, such as age, sexual preference, and religion. We see these three social categories as interconnected, historical processes of domination and subordination. Thinking about gender, race-ethnicity, and class, then, necessitates thinking historically about power and economic exploitation….

GENDER, RACE-ETHNICITY, AND CLASS PROCESSES: HISTORICAL AND INTERCONNECTED

The concepts of gender, race-ethnicity, and class are neither trans-historical nor independent. Hence, it is artificial to discuss them outside of historical time and place, and separately from one another. At the same time, without such a set of concepts, it is impossible to make sense of women's disparate economic experiences.

Gender, race-ethnicity, and class are not natural or biological categories which are unchanging over time and across cultures. Rather, these categories are socially constructed: they arise and are transformed in history, and themselves transform history. Although societies rationalize them as natural or god-given, ideas of appropriate feminine and masculine behavior vary widely across history and culture. Concepts and practices of race-ethnicity, usually justified by religion

SOURCE: From Teresa Amott and Julie Matthaei, *Race, Gender, and Work: A Multicultural Economic History of Women in the United States*, 2nd ed., pp. 339–348. (New York: South End Press, 1996). Reprinted with permission of South End Press.

or biology, also vary over time, reflecting the politics, economics, and ideology of a particular time and, in turn, reinforcing or transforming politics, economics, and ideology. For example, nineteenth-century European biologists Louis Agassiz and Count Arthur de Gobineau developed a taxonomy of race which divided humanity into separate and unequal racial species; this taxonomy was used to rationalize European colonization of Africa and Asia, and slavery in the United States. Class is perhaps the most historically specific category of all, clearly dependent upon the particular economic and social constellation of a society at a point in time. Still, notions of class as inherited or genetic continue to haunt us, harkening back to earlier eras in which lowly birth was thought to cause low intelligence and a predisposition to criminal activity.

Central to the historical transformation of gender, race-ethnicity, and class processes have been the struggles of subordinated groups to redefine or transcend them. For example, throughout the development of capitalism, workers' consciousness of themselves as workers and their struggles against class oppression have transformed capitalist-worker relationships, expanding workers' rights and powers. In the nineteenth century, educated white women escaped from the prevailing, domestic view of womanhood by arguing that homemaking included caring for the sick and the needy through volunteer work, social homemaking careers, and political organizing. In the 1960s, the transformation of racial-ethnic identity into a source of solidarity and pride was essential to movements of people of color, such as the Black Power and American Indian movements.

Race-ethnicity, gender, and class are interconnected, interdetermining historical processes, rather than separate systems. This is true in two senses, which we will explore in more detail below. First, it is often difficult to determine whether an economic practice constitutes class, race, or gender oppression: for example, slavery in the U.S. South was at the same time a system of class oppression (of slaves by owners) and of racial-ethnic oppression (of Africans by Europeans). Second, a person does not experience these different processes of domination and subordination independently of one another; in philosopher Elizabeth Spelman's metaphor, gender, race-ethnicity, and class are not separate "pop-beads" on a necklace of identity. Hence, there is no generic gender oppression which is experienced by all women regardless of their race-ethnicity or class. As Spelman puts it:

> … in the case of much feminist thought we may get the impression that a woman's identity consists of a sum of parts neatly divisible from one another, parts defined in terms of her race, gender, class, and so on.…
> On this view of personal identity (which might also be called pop-bead metaphysics), my being a woman means the same whether I am white or Black, rich or poor, French or Jamaican, Jewish or Muslim.[1]

The problems of "pop-bead metaphysics" also apply to historical analysis. In our reading of history, there is no common experience of gender across race-ethnicity and class, of race-ethnicity across class and gender lines, or of class across race-ethnicity and gender.

With these caveats in mind, let us examine the processes of gender, class, and race-ethnicity, their importance in the histories of women's works, and some of the ways in which these processes have been intertwined.

Gender

… Gender differences in the social lives of men and women are based on, but are not the same thing as, biological differences between the sexes. Gender is rooted in societies' beliefs that the sexes are naturally distinct and opposed social beings. These beliefs are turned into self-fulfilling prophecies through sex-role socialization: the biological sexes are assigned distinct and often unequal work and political positions, and turned into socially distinct genders.

Economists view the sexual division of labor as central to the gender differentiation of the sexes. By assigning the sexes to different and complementary tasks, the sexual division of labor turns them into different and complementary genders…. The work of males is at least partially, if not wholly, different from that of females, making "men" and "women" different economic and social beings. Sexual divisions of labor, not sexual difference alone, create difference and complementarity between "opposite" sexes. These differences, in turn, have been the basis for marriage in most societies….

The concept of gender certainly helps us understand women's economic histories…. Each racial-ethnic group has had a sexual division of labor which has barred individuals from the activities of the opposite sex. Gender processes do differentiate women's lives in many ways from those of the men in their own racial-ethnic and class group. Further, gender relations in all groups tend to assign women to the intra-familial work of childrearing, as well as to place women in a subordinate position to the men of their class and racial-ethnic group.

But as soon as we have written these generalizations, exceptions pop into mind. Gender roles do not always correspond to sex. Some American Indian tribes allowed individuals to choose among gender roles: a female, for example, could choose a man's role, do men's work, and marry another female who lived out a woman's role. In the nineteenth century, some white females "passed" as men in order to escape the rigid mandates of gender roles. In many of these cases, women lived with and loved other women.

Even though childrearing is women's work in most societies, many women do not have children, and others do not perform their own child care or domestic work. Here, class is an especially important differentiating process. Upper-class women have been able to use their economic power to reassign some of the work of infant care—sometimes even breastfeeding—to lower-class women of their own or different racial-ethnic groups. These women, in turn, may have been forced to leave their infants alone or with relatives or friends. Finally, gender complementarity has not always led to social and economic inequality; for example, many American Indian women had real control over the home and benefited from a more egalitarian sharing of power between men and women.

Since the processes of sex-role socialization are historically distinct in different times and different cultures, they result in different conceptions of appropriate gender behavior. Both African American and Chicana girls, for instance, learn how to be women—but both must learn the specific gender roles which have developed within their racial-ethnic and class group and historical period. For example, for white middle-class homemakers in the 1950s, adherence to the concept of womanhood discouraged paid employment, while for poor Black women it meant employment as domestic servants for white middle-class women. Since racial-ethnic and class domination have differentiated the experiences of women, one cannot assume, as do many feminist theorists and activists, that all women have the same experience of gender oppression—or even that they will be on the same side of a struggle, not even when some women define that struggle as "feminist."

Not only is gender differentiation and oppression not a universal experience which creates a common "women's oppression," the sexual divisions of labor and family systems of people of color have been systematically disrupted by racial-ethnic and class processes. In the process of invasion and conquest, Europeans imposed their notions of male superiority on cultures with more egalitarian forms of gender relations, including many American Indian and African tribes. At the same time, European Americans were quick to abandon their notion of appropriate femininity when it conflicted with profits; for example, slave owners often assigned slave women to backbreaking labor in the fields.

Racial-ethnic and class oppression have also disrupted family life among people of color and the white working class. Europeans interfered with family relations within subordinated racial-ethnic communities through rape and forced cohabitation. Sometimes whites encouraged or forced reproduction, as when slaveowners forced slave women into sexual relations. On the other hand, whites have often used their power to curtail reproduction among peoples of color, and aggressive sterilization programs were practiced against Puerto Ricans and American Indians as late as the 1970s. Beginning in the late nineteenth century, white administrators took American Indian children from their parents to "civilize" them in boarding schools where they were forbidden to speak their own languages or wear their native dress. Slaveowners commonly split up slave families through sale to different and distant new owners. Nevertheless, African Americans were able to maintain strong family ties, and even augmented these with "fictive" or chosen kin. From the mid-nineteenth through the mid-twentieth centuries, many Asians were separated from their spouses or children by hiring policies and restrictions on immigration. Still, they maintained family life within these split households, and eventually succeeded in reuniting, sometimes after generations. Hence, for peoples of color, having children and maintaining families have been an essential part of the struggle against racist oppression. Not surprisingly, many women of color have rejected the white women's movement view of the family as the center of "women's oppression."

These examples reveal the limitations of gender as a single lens through which to view women's economic lives. Indeed, any attempt to understand women's experiences using gender alone cannot only cause misunderstanding, but can also interfere with the construction of broad-based movements against the oppressions experienced by women.

Race-Ethnicity

Like gender, race-ethnicity is based on a perceived physical difference, and ratio-nalized as "natural" or "god-given." But whereas gender creates difference and inequality according to biological sex, race-ethnicity differentiates individuals according to skin color or other physical features.

In all of human history, individuals have lived in societies with distinct languages, cultures, and economic institutions; these ethnic differences have been perpetuated by intermarriage within, but rarely between, societies. However, ethnic differences can exist independently of a conception of race, and without a practice of racial-ethnic domination such as the Europeans practiced over the last three centuries....

Does the concept of race-ethnicity help us understand the economic history of women in the United States? Certainly, ... racial-ethnic domination has been a central force in U.S. history. European colonization of North America entailed the displacement and murder of the continent's indigenous peoples, rationalized by the racist view that American Indians were savage heathens. The economy of the South was based on a racial-ethnic system, in which imported Africans were forced to work for white landowning families as slaves. U.S. military expansion in the nineteenth century brought more lands under U.S. control—the territories of northern Mexico (now the Southwest), the Philippines, and Puerto Rico—incorporating their peoples into the racial-ethnic hierarchy. And from the mid-nineteenth century onward, Asians were brought into Hawaii's plantation system to work for whites as semi-free laborers. In the twentieth century, racial-ethnic difference and inequality have been perpetuated by the segregation of people of color into different and inferior jobs, living conditions, schools, and political positions, and by the prohibition of inter-marriage with whites in some states up until 1967.

Race-ethnicity is a key concept in understanding women's economic histo-ries. But it is not without limitations. First, racial-ethnic processes have never operated independently of class and gender. In the previous section on gender, we saw how racial domination distorted gender and family relations among peo-ple of color. Racial domination has also been intricately linked to economic or class domination. As social scientists Michael Omi ... and Howard Winant ... explain, the early European arguments that people of color were without souls had direct economic meaning:

> At stake were not only the prospects for conversion, but the types of treatment to be accorded them. The expropriation of property, the denial of political rights, the introduction of slavery and other forms of coercive labor, as well as outright extermination, all presupposed a worldview which distinguished Europeans—children of God, human beings, etc.—from "others." Such a worldview was needed to explain why some should be "free" and others enslaved, why some had rights to land and property while others did not.[2]

Indeed, many have argued that racial theories only developed after the economic process of colonization had started, as a justification for white domination of peoples of color.

The essentially economic nature of early racial-ethnic oppression in the United States makes it difficult to isolate whether peoples of color were subordinated in the emerging U.S. economy because of their race-ethnicity or their economic class. Whites displaced American Indians and Mexicans to obtain their land. Whites imported Africans to work as slaves and Asians to work as contract laborers. Puerto Ricans and Filipinas/os were victims of further U.S. expansionism. Race-ethnicity and class intertwined in the patterns of displacement for land, genocide, forced labor, and recruitment from the seventeenth through the twentieth centuries. While it is impossible, in our minds, to determine which came first in these instances—race-ethnicity or class—it is clear that they were intertwined and inseparable.

Privileging racial-ethnic analysis also leads one to deny the existence of class differences, both among whites and among people of color, which complicate and blur the racial-ethnic hierarchy. A racial-ethnic analysis implies that all whites are placed above all peoples of color....

A minority of the dominated race is allowed some upward mobility and ranks economically above whites. At the same time, however, all whites have some people of color below them. For example, there are upper-class Black, Chicana, and Puerto Rican women who are more economically privileged than poor white women; however, there are always people of color who are less economically privileged than the poorest white woman. Finally, class oppression operates among women of the same racial-ethnic group.

A third problem with the analysis of racial domination is that such domination has not been a homogeneous process. Each subordinated racial-ethnic group has been oppressed and exploited differently by whites: for example, American Indians were killed and displaced, Africans were enslaved, and Filipinas/os and Puerto Ricans were colonized. Whites have also dominated whites; some European immigrant groups, particularly Southern and Eastern Europeans, were subjected to segregation and violence. In some cases, people of color have oppressed and exploited those in another group: Some American Indian tribes had African slaves; some Mexicans and Puerto Ricans displaced and murdered Indians and had African slaves. Because of these differences, racial oppression does not automatically bring unity of peoples of color across their own racial-ethnic differences, and feminists of color are not necessarily in solidarity with one another.

To sum up, we see that, as with gender, the concept of race-ethnicity is essential to our analysis of women's works. However, divorcing this concept from gender and class, again, leads to problems in both theory and practice.

Class

... We believe that the concepts of class and exploitation are crucial to understanding the work lives of women in early U.S. history, as well as in the modern, capitalist economy. Up through the nineteenth century, different class relations organized production in different regions of the United States. The South was dominated by slave agriculture; the Northeast by emerging industrial capitalism; the Southwest (then part of Mexico) by the *hacienda* system which carried over many elements of the feudal manor into large-scale production for the market; the rural Midwest by independent family farms that produced on a small scale for the market; and the American Indian West by a variety of tribal forms centered in hunting and gathering or agriculture, many characterized by cooperative, egalitarian economic relations. Living within these different labor systems, and in different class positions within them, women led very different economic lives.

By the late nineteenth century, however, capitalism had become the dominant form of production, displacing artisans and other small producers along with slave plantations and tribal economies. Today, wage labor accounts for over 90 percent of employment; self employment, including family businesses, accounts for the remaining share. With the rise of capitalism, women were brought into the same labor system, and polarized according to the capitalist–wage laborer hierarchy.

At the same time as the wage labor form specific to capitalism became more prevalent, capitalist class relations became more complex and less transparent. Owners of wealth (stocks and bonds) now rarely direct the production process; instead, salaried managers, who may or may not own stock in the company, take on this function. While the capitalist class may be less identifiable, it still remains a small and dominant elite....

Class can be a powerful concept in understanding women's economic lives, but there are limits to class analysis if it is kept separate from race-ethnicity and gender. First, as we saw in the race section above, the class relations which characterized the early U.S. economy were also racial-ethnic and gender formations. Slave owners were white and mostly male, slaves were Black. The displaced tribal economies were the societies of indigenous peoples. Independent family farmers were whites who farmed American Indian lands; they organized production in a patriarchal manner, with women and children's work defined by and subordinated to the male household head and property owner. After establishing their dominance in the pre-capitalist period, white men were able to perpetuate and institutionalize this dominance in the emerging capitalist system, particularly through the monopolization of managerial and other high-level jobs.

Second, the sexual division of labor within the family makes the determination of a woman's class complicated—determined not simply by her relationship to the production process, but also by that of her husband or father. For instance, if a woman is not in the labor force but her husband is a capitalist, then we might wish to categorize her as a member of the capitalist class. But what if that same woman worked as a personnel manager for a large corporation, or as a salesperson in an elegant boutique? Clearly, she derives upper-class status and access to income from her husband, but she is also, in her own right, a worker. Conversely, when women lose their husbands through divorce, widowhood, or desertion, they often change their class position in a downward direction. A second gender-related economic process overlooked by class analysis is the unpaid household labor performed by women for their fathers, husbands, and children—or by other women, for pay.

Third, while all workers are exploited by capitalists, they are not equally exploited, and gender and race-ethnicity play important roles in this differentiation.

Men and women of the same racial-ethnic group have rarely performed the same jobs—this sex-typing and segregation is the labor market form of the sexual division of labor we studied above. Further, women of different racial-ethnic groups have rarely been employed at the same job, at least not within the same workplace or region. This racial-ethnic typing and segregation has both reflected and reinforced the racist economic practices upon which the U.S. economy was built.

Thus, jobs in the labor force hierarchy tend to be simultaneously race-typed and gender-typed. Picture in your mind a registered nurse. Most likely, you thought of a white woman. Picture a doctor. Again, you imagined a person of a particular gender (probably a man), and a race (probably a white person). If you think of a railroad porter, it is likely that a Black man comes to mind. Almost all jobs tend to be typed in such a way that stereotypes make it difficult for persons of the "wrong" race and/or gender to train for or obtain the job. Of course, there are regional and historical variations in the typing of jobs. On the West Coast, for example, Asian men performed much of the paid domestic work during the nineteenth century because women were in such short supply. In states where the African American population is very small, such as South Dakota or Vermont, domestic servants and hotel chambermaids are typically white. Nonetheless, the presence of variations in race-gender typing does not contradict the idea that jobs tend to take on racial-ethnic and gender characteristics with profound effects on the labor market opportunities of job-seekers....

The racial-ethnic and gender processes operating in the labor market have transposed white and male domination from pre-capitalist structures into the labor market's class hierarchy. This hierarchy can be described by grouping jobs into different labor market sectors or segments: "primary," "secondary," and "underground." ... The primary labor market—which has been monopolized by white men—offers high salaries, steady employment, and upward mobility. Its upper tier consists of white-collar salaried or self-employed workers with high status, autonomy, and, often, supervisory capacity. Wealth increases access to this sector, since it purchases elite education and provides helpful job connections....

The lower tier of the primary sector, which still yields high earnings but involves less autonomy, contains many unionized blue-collar jobs. White working-class men have used union practices, mob violence, and intimidation to monopolize these jobs. By World War II, however, new ideologies of worker solidarity, embodied in the mass industrial unions, began to overcome the resistance of white male workers to the employment of people of color and white women in these jobs.

In contrast to both these primary tiers, the secondary sector offers low wages, few or no benefits, little opportunity for advancement, and unstable employment. People of color and most white women have been concentrated in these secondary sector jobs, where work is often part-time, temporary, or seasonal, and pay does not rise with increasing education or experience. Jobs in both tiers of the primary labor market have generally yielded family wages, earnings high enough to support a wife and children who are not in the labor force for pay. Men of color in the secondary sector have not been able to earn enough to support their families, and women of color have therefore participated in wage labor to a much higher degree than white women whose husbands held primary sector jobs.

Outside of the formal labor market is the underground sector, where the most marginalized labor force groups, including many people of color, earn their livings from illegal or quasi-legal work. This sector contains a great variety of jobs, including drug trafficking, crime, prostitution, work done by undocumented workers, and sweatshop work which violates labor standards such as minimum wages and job safety regulations.

NOTES

1. Elizabeth V. Spelman, *Inessential Woman: Problems of Exclusion in Feminist Thought* (Boston: Beacon Press, 1988).
2. Michael Omi and Howard Winant, *Racial Formation in the United States* (New York: Routledge & Kegan Paul, 1986), p. 58.

33

Seeing in 3D

A Race, Class, and Gender Lens on the Economic Downturn

MARGARET L. ANDERSEN

If you have ever gone to a 3D movie, you know you get these funny little glasses that bring out otherwise unseen dimensions to what you see in the film. In ordinary vision, the eye sees two images—one from each eye—and the brain puts the two images together so you see it as one. When an image is projected in three-dimensions, 3D glasses filter the different colors into a three dimensional image, presenting the brain with one image perceived with depth. Thus, in 3D films and with the 3D lenses, you can see objects otherwise invisible to you, perhaps even flying at you! Without 3D lenses, the image will appear blurry, fuzzy, perhaps even difficult to see or confusing, maybe like trying to understand racism or sexism without the knowledge that helps filter what you observe.

The technology of 3D images is akin to understanding a race/class/gender framework in society. With the "lens" of a three-dimensional intersectional framework, you will see new dimensions to any single vision of race or class or gender. Indeed, in a society where race-blind, gender-blind, and class-blind views dominate public understandings of inequality, a three-dimensional view that is anchored in an intersectional analysis of race, class, and gender reveals otherwise hidden aspects of structured inequality.

This essay uses an intersectional analysis to understand the recent economic downturn—a societal event where an intersectional analysis of race, class, and gender has gone largely unnoticed in the public discourse about this historic development—even though these dimensions of inequality are critical to understanding the full consequences of this economic crisis.

As the United States experienced an economic downturn beginning in 2009, daily news headlines proclaimed, "New Jobless Claims Rise Unexpectedly" (*The New York Times*, April 3, 2009); "Jobless Rate Hits 10.2% With More Underemployed" (*The New York Times*, November 11, 2009); "Job Prospects are Poor, Agency Warns" (*The New York Times*, March 20, 2009); and, "Job Market Blues" (*The New York Times*, August 8, 2009). Such headlines suggest that the economy is a neutral force, operating through abstract processes that have little to do with specific group experiences. How different might "the

SOURCE: Copyright © 2012 by Margaret L. Andersen.

economy" seem were headlines to shout, "Women Disadvantaged by Economic Recovery;" or "Recession a National Disaster for African Americans and Latinos/as;" or "Race, Class, and Gender Interact to Produce Inequality"?

The reality is that, as the national unemployment rate rose to 10 percent, it reached a level that has been characteristic of Black unemployment for the past 60 years. Were African American and Latino unemployment used as a measure of the nation's economic well-being, we would have declared ourselves in an ongoing economic crisis for over half a century! Add in the factor of age and you will see that the unemployment rate for African American young people (aged 20-24)—the very point when they are entering the labor market for the first time—is a whopping 26 percent. For Latinos in the same age category, it is 17 percent (U.S. Department of Labor 2011a).

How do we then see the economic crisis that fell upon the entire nation? As *New York Times* columnist Bob Herbert put it, "The unemployment that has wrought such devastation in black communities for decades is now being experienced by a much wider swath of the population. We've been in deep denial about this" (Herbert 2009). The denial that Herbert refers to is characteristic of a society that wants to think of itself as a place where individual effort, not structural advantage and disadvantage, is the basis for anyone's economic well-being. Herbert was writing about race, but race is only one dimension of the economic crisis that has been overlooked in most reports about the economic recession. What about women?

In 2008, then Senator Edward Kennedy worried about women and the economic crisis when he wrote, "Despite their critical role in the workforce and in raising families, women and their vulnerability in economic downturns have received too little focus. [There is a] severe and disproportionate impact of this recession on women and their families. We need to act immediately to restore women's rights to fair pay, provide workers with paid sick days, and shore up programs that help workers and families endure hard times" (Kennedy 2008).

Unfortunately, few have heeded either Herbert's call for action on race or Kennedy's call for action on behalf of women, much less focusing on the intersections of race, gender, and class in analyzing the particular impact of the economic downturn on particular groups. What would we see differently were we to view the economic recession with women and people of color in mind, thus thinking about gender *and* race *and* class simultaneously?

First, you would probably notice the general absence of women in most of the reports being issued about the impact of the recession. There are exceptions and women, of course, are not the only people affected by economic crises. But, from the dominant, public narrative, you would hardly know that women existed. For example: An otherwise excellent report, *State of the Dream 2009: The Silent Depression*, provides compelling detail about the impact of the economic recession on people of color, but with no regard to the specific impact on women and men (United for a Fair Economy 2009). Likewise, in a different report on housing foreclosures, where there is great detail on the impact of housing foreclosures on African Americans, Latinos, and low-income people, there is not one mention of women—even though the report concludes that this economic crisis will result in the largest loss of Black wealth ever witnessed, wiping out a whole generation of home ownership (United for a Fair Economy 2008).

Such reports beg the further question of the recession's impact on women. What do we see when we look at women?

- Women are more worried about money than men (Newport 2009).

- Although men are more likely to be officially unemployed than women, women's unemployment has also risen, particularly when the recession hit the service sector, not just the manufacturing sector (U.S. Department of Labor 2011b).

- Women are more likely to be employed in the public sector than are men; during the recovery, there was more job growth in the private than the public sector; thus, the decline in unemployment for women has been less than that for men (U.S. Department of Labor 2011b).

- Among women heading families, the unemployment rate has grown and is higher than the national unemployment rate and twice as high as that for either married men or married women (Joint Economic Committee 2009).

- Women's wages are also more volatile than men's wages, and women face a much higher risk of seeing large drops in income than do men (Kennedy 2008).

- The number of women among the working poor continues to exceed that of men—working poor being defined as the percentage of those working 27 weeks or more but living below the poverty level. Among such workers, 6.5% of women and 5.6% of men are working poor. Black and White women are more likely to be working poor than are men in their same racial group; for Asian and Hispanic workers, men and slightly more likely to be among the working poor than are women (though the differences are negligible for these latter two groups; U.S. Bureau of Labor Statistics 2010). Thus, while men have been hardest hit by job loss, holding jobs does not necessarily lift women out of poverty.

These are facts about women as an aggregated group, but what about women of color?

- Among Black Americans, unlike other groups, women are a larger share of those employed (U.S. Department of Labor 2011c).

- As the recession unfolded, the growth for unemployment for Hispanic women nearly matched that of White men (Bureau of Labor Statistics, March 2009).

- All women are disproportionately at risk in the current foreclosure crisis, since women are 32% more likely than men to have subprime mortgages (One-third of women, compared to one-fourth of men, have subprime mortgages; and, the disparity between women and men increases in higher income brackets.) Black women earning double an area's median income were nearly five times more likely to receive subprime home purchase mortgages than White men with similar earnings. Latino women earning twice an area's median income were about four times more likely to receive subprime mortgages than White men with similar earnings (Fishbein and Woodall 2006).

- Women of color are the most likely to hold subprime mortgages—five times more likely than men in the same income brackets. This is despite the fact that women on average have slightly higher credit scores than men. African American women are 256 percent more likely to hold a subprime loan than White men (Fishbein and Woodall 2006)!

The intent here is not to ignore men and the impact that the economic crisis is having on them. Instead, it is to be attuned to how gender and race work together to produce economic standing. As one example, when you examine gender with regard to men, while also examining race, you will learn that during the recession the growth in unemployment among Hispanic men actually exceeded that of White men. Such facts are overlooked when looking at either race or gender alone. And, in thinking about men, you might ask yourself would it not improve men's lives if women's economic well-being were improved? The point is that we need to be fully attuned to the realities of gender and race—for both women and men or we ignore the specific impacts that economic change—and thus economic and social policy—have on particular groups.

So why is there not more outrage about the persistent inequalities that women, people of color, and, especially, women of color face? In part, this is because of the persistence of a dominant narrative that remains anchored in a race, gender, and class-blind perspective. Furthermore, the dominant narrative remains focused on a one-dimensional view of individualism, not a three-dimensional view of social structural causes. Thus, the public narrative about economic impacts presumes a White, middle-class, masculinist focus. Public rhetoric about so-called "working families" is a case in point. "Working families" is a phrase that invokes images of heterosexual, married families with two potential earners present. Yet, as we know from volumes of research, such families are no longer the only, nor even, dominant form of family relationship. Moreover, the image of the "working class" is one that connotes White, male, and blue-collar, thus ignoring the vast numbers of women, and especially women of color, who now constitute the actual "working class"—that is, not just those who are employed in blue-collar, industrial work, but those working in the vast low-wage, service sector, such as fast food workers, cleaners, housekeepers, and so forth.

The individualistic perspective of the dominant narrative is also evident in how the causes of the economic recession are generally understood. Several recurrent themes appear to explain the recession:

- that the economic crisis is the result of the greedy behavior of individuals on Wall Street;

- that this crisis resulted from the irresponsible habits of Americans who overspent, did not save enough, and consumed beyond their means.

Ironically, prior to the recession, Americans were said to not be saving enough and to be running up too much debt. Now, to recover people have been encouraged to use credit cards and get the credit market going again. To whose advantage?

▪ that the economic crisis is about job loss. Of course, this is true, but the focus on job creation has also ignored the unpaid labor that people (mostly women and, especially women of color, do)—as if the economy were only about public, paid labor.

It is not that greed and consumer behaviors are irrelevant or that job loss in the paid public sphere is not a problem. The point is that the national discourse blinds us to the gendered, raced, and classed foundations of the economic processes of contemporary capitalism. Instead, public understanding of the economic problems assumes that individual greed and individual deviance have created the recession, thus obscuring the sociological processes that are fundamental to the workings of modern capitalism. Yet, the practices and relations that shape capitalism are fundamentally about race, class, and gender. As sociologist Joan Acker puts it, "gendering and racialization are integral to the creation and recreation of class inequalities and class divisions [and they emerge] ... in complex, multifaceted, boundary-spanning capitalist activities" (Acker 2006: 7). Although the "economy" and "the market" are discussed as if they operate by abstract forces, capitalism is decidedly not neutral in terms of class, race, and gender.

Conclusion: Recovery for Whom?

Most of the public discussion about recovering from the economic crisis has focused (other than on financial institution bailouts) on job creation. But, most of the job creation plans that have stemmed from recovery plans have focused on jobs in skilled labor—the very area of the labor market where women have made the least progress. For example, women remain a mere two percent of the labor force in construction-related jobs as carpenters, electricians, sheet metal workers, highway maintenance workers, and auto mechanics. Alternative job recovery plans could focus on training women in areas such as these where women have historically been underemployed. Instead most innovation still relies on individual initiative. One example is Sarah Lateiner who graduated from college with high honors (Phi Beta Kappa) and a double degree in pre-law and women's studies. Originally planning to go to law school, she changed course and established her own automobile repair shop where she not only repairs cars but also trains women about auto mechanics. She has also started a scholarship for women to enroll in technical skills. In her one way of changing the world, she believes that the entry of women into this largely male-dominated domain is one way of empowering women (Hartman 2010).

Could not financial recovery plans provide incentives for such initiatives? Or, why could a job stimulus package not include enhancing wages in areas of the labor market where women ARE employed? Even in a recession, people still need their children cared for, hotel rooms cleaned, and offices cleaned--all areas of the labor market where women and, especially, women of color work, but at low wages and often with few or no job benefits. The federal stimulus package that was part of the nation's recovery plans did include funds for child care and housing, but to date there have been no plans for increasing wages in fields most populated by women. Although an important change has been the passage of the Lily Ledbetter Act protecting women against discrimination, for most women job segregation, not discrimination, is the depressor of wages.

The race-gender neutral way of understanding economic problems masks the reality of women's lives and blunts the political opportunity for a resurgence of a nationally visible feminist movement—one anchored in what now exists as a wide-ranging base of knowledge about the intersections of race, class, and gender in shaping people's life opportunities. Instead, the current national discourse persists in "othering" people—that is, privileging some, but blaming others for their own plight. I have to believe that, if people knew the facts about gender, race, and class, the national discourse—and national action—would be transformed. Women, for example, have particular needs during times of economic crisis. Some of those needs may complement the needs of men. But other needs are particular to their status as women—and, in the case of women of color—as racial-ethnic women. We need a more three-dimensional agenda for women and for women of color—one that is anchored in the now extensive scholarship on the opportunities allowed and denied because of the connections between race, class, and gender.

REFERENCES

Acker, Joan. 2006. *Class Questions, Feminist Answers.* Lanham, MD: Rowman and Littlefield.

Bureau of Labor Statistics. 2009. www.bls.gov

Fishbein, Allan J., and Patrick Woodall. 2006. "Women are Prime Targets for Subprime Lending: Women are Disproportionately Represented in High-Cost Mortgage Market." Washington, DC: Consumer Federation of America. www.consumerfed.org

Hartman, Steve. 2010. "Female Auto Mechanic Breaks Stereotypes." CBS Evening News, December 6. www.cbsnews.com

Herbert, Bob. 2009. "A Scary Reality." *The New York Times*, August 11, 2009.

Hirshman, Linda. 2008. "Where Are the New Jobs for Women?" *The New York Times*, December 9. www.nytimes.com

Joint Economic Committee. 2009. *Women in the Recession: Working Mothers Face High Rates of Unemployment.* Washington, DC: U.S. Senate, May 28. www.jec.senate.gov

Kennedy, Edward M. 2008. *Taking a Toll: The Effects of Recession on Women.* Washington, DC: Majority Staff of the Committee on Health, Education, Labor and Pensions, United States Senate.

Newport, Frank. 2009. "Worry about Money Peaks with Forty-Somethings." *The Gallup Poll.* New York: Gallup Organization, February 27. www.gallup.com

United for a Fair Economy. 2008. *State of the Dream: Foreclosed*. Boston, MA: United for a Fair Economy. www.faireconomy.org

United for a Fair Economy. 2009. *State of the Dream: The Silent Depression*. Boston, MA: United for a Fair Economy. www.faireconomy.org

U.S. Bureau of Labor Statistics. 2010. *A Profile of the Working Poor, 2008*. Washington, DC: U.S. Department of Labor. www.bls.gov

U.S. Department of Labor. 2011a. *Employment and Earnings*. Washington, DC: U.S. Department of Labor, January. www.dol.gov

U.S. Department of Labor. 2011b. *The Black Labor Force in the Recovery*. Washington, DC: U.S. Department of Labor, June 6. www.dol.gov

U.S. Department of Labor. 2011c. *Women's Employment during the Recovery*. Washington, DC: U.S. Department of Labor, May 3. www.dol.gov

34

Racism in Toyland

CHRISTINE L. WILLIAMS

Not long ago I had to buy a present for a six-year-old. I had at least three choices for where to shop: The Toy Warehouse, a big-box superstore with a vast array of low-cost popular toys; Diamond Toys, a high-end chain store with a more limited range of reputedly high-quality toys; or Tomatoes, a locally owned, neighborhood shop that sells a relatively small, offbeat assortment of traditional and politically correct toys. Can sociology offer any advice to consumers like me?

Unfortunately, many sociologists turn into utilitarian economists when it comes to analyzing shopping, assuming that customer behavior is determined only by price, convenience, and selection. But a number of social factors influence where we choose to shop, including the racial makeup of the store's workers and customers. In my book, *Inside Toyland: Working, Shopping, and Social Inequality* [University of California Press, 2006], I argue that racial inequality (and gender and class inequality as well) influence where we choose to shop, how we shop, and what we buy. The retail industry sustains such inequality through hiring policies that favor certain kinds of workers and advertising aimed at customers from specific racial or ethnic groups.

I noticed the connection between shopping and social inequality while working as a clerk at the Toy Warehouse and Diamond Toys for three months in 2001. These stores belonged to national chains, and both employed about 70 hourly employees. At the warehouse store, I was one of only three white women on the staff; most were African American, Hispanic, or second-generation Asian American. The "guests" (as we were required to call customers) were an amazing mix from every racial and ethnic group and social class. In contrast, only three African Americans worked at the upscale toy store; most clerks were white. Most of the customers were also white and middle to upper class. My experiences taught me to notice racial diversity (or its absence) wherever I shop.

"DON'T SHOP WHERE YOU CAN'T WORK!"

This slogan, popular during the Great Depression, rallied black protesters to demand equal access to jobs in stores, and many chains responded by hiring

SOURCE: From *Contexts*, Vol. 4, Issue 4, pp. 28–32. Copyright © 2005 by the American Sociological Association. Reprinted by permission of the University of California Press.

African Americans in predominately black neighborhoods: "We employ colored salesmen" signs appeared in Sears, Walgreens, and other stores eager for black customers.

Retail work is one of the most integrated occupations in the United States today. The proportions of whites, African Americans, Asian Americans, and Hispanics employed in retail jobs more or less match their representation in the labor force. But these statistics hide segregation at the store level. More than 15 percent of employees in shoe stores and variety stores, for instance, are black, but less than 5 percent of employees are black in stores that sell liquor, gardening equipment, or needlework supplies. Inside stores, there is further segregation by task; Whites usually have the top director and manager positions, and nonwhites have the lowest-paid, often invisible backroom jobs.

In the toy stores where I worked, the two most segregated jobs were the director positions (all white men) and the janitor jobs (all Latinas subcontracted from outside firms). The Toy Warehouse employed mostly African Americans in all the other positions, including cashiers. But over time I noticed that the managers preferred to assign younger and lighter-skinned women to this position. Older African-American women who wanted to work as cashiers had to struggle to get the assignment. Lazelle, for example, who was about 35, had been asking to work as a cashier for the two months she had been working there. She worked as a merchandiser, getting items from the storeroom, pricing items, and checking prices when bar codes were missing. Lazelle finally got her chance at the registers the same day that I started. We set up next to each other, and I noticed with a bit of envy how competent and confident on the register she was. (Later she told me she had worked registers at other stores, including fast-food restaurants.) I told her I had been hoping to get assigned as a merchandiser. I liked the idea of being free to walk around the store, talk with customers, and learn more about the toys. I had mentioned to the manager that I wanted that job, but she made it clear I was destined for cashiering and service desk (and later, to my horror, computer accounting). Lazelle looked at me like I was crazy. Most workers thought merchandising was the worst job in the store because it was so physically taxing. From her point of view, I had gotten the better job, no doubt because of my race, and it seemed to her that I wanted to throw that advantage away. (The manager may also have considered my background and educational credentials in assigning me to particular jobs.)

The preference for lighter-skinned women as cashiers reflects the importance of this job in the store's general operations. In discount stores, customers seldom talk with sales clerks. The cashier is the only person most customers deal with, giving her enormous symbolic—and economic—importance for the corporation. Transactions can break down if clerks do not treat customers as they expect. The preference for white and light-skinned women as cashiers should be understood in this light: In a racist and sexist society, managers generally believe that such women are the most friendly and solicitous, and thus most able to inspire trust and confidence in a commercial transaction.

At the upscale Diamond Toys, virtually all cashiers were white. Unlike the warehouse store, where cash registers were lined up at the front of the store, the upscale store had cash registers scattered throughout the different departments. The preference for white workers in these jobs (and throughout the store) seemed consistent with the marketing of the store's workers as "the ultimate toy experts." In retail work, professional expertise is typically associated with whiteness, much as it is in domestic service.

The purported expertise of salesclerks is one of the great deceptions of the retail industry. Here, where jobs pay little and turnover rates are high (estimated at more than 100 percent per year by the National Retail Federation), many clerks know almost nothing about the products they sell. I knew nothing about toys when I was put behind the cash register, and I received no training on the merchandise at either store. Any advice I gave I literally made up. But at the upscale store I was expected to help customers with their shopping decisions. They frequently asked questions like "What are going to be the hot toys for one-year-olds this Christmas?" or "What one item would you recommend for two sisters of different ages?" One mother asked me to help her pick out a $58 quartz watch for her seven-year-old son. A personal shopper phoned in and asked me to describe the three Britney Spears dolls we carried, help her pick out the "nicest" one, and then arrange to ship it to her employer's niece. Customers asked detailed questions about how the toys were meant to work, and they were especially curious about comparing the merits of the educational toys we offered (I was asked to compare the relative merits of the "Baby Mozart" and the "Baby Bach"). On my first day, I answered a phone call from a customer who asked me to pick out toys for a one-year-old girl and a boy who was two and a half, spend up to $100, and arrange to have the toys gift-wrapped and mailed to their recipients.

At Diamond Toys, most customers didn't mind waiting to talk with me. When the lines were long, they didn't make rude huffing noises or try to make eye contact with their fellow sufferers, as they often did at the Toy Warehouse. I couldn't help but think that the customers—mostly white—were more civil and polite at the upscale store because most of us workers were white. We were presumed to be professional, caring, and knowledgeable, even when we weren't. White customers seemed less respectful of minority service workers than white workers; they were willing to pay more and wait longer for the services of whites because they apparently assumed that whites were more refined and intelligent.

On several occasions at the warehouse store, I saw customers reveal racist attitudes toward my African-American coworkers. One night, after the store had closed, I saw Doris and Selma (fiftyish African Americans) escorting several white customers out. Getting straggling customers to leave the store after closing was always a big chore. Soon after, as I was being audited in the manager's office, Selma came in very upset because one of the women she and Doris escorted out had spit out her chewing gum at her. Doris and Selma were appalled. Doris said to them, "That is really disgusting; how could you do that?" And the woman said to Doris, "What's your name?" like she was going to report her. This got Selma so angry she said, "If you take her name, take mine too," and showed her name-tag. She told the woman that she was never welcome to come back to this store. Selma was very distressed. Talking back to customers was taboo, and she knew she could be fired for what she had said. The manager told her that some people are going to be gross and disgusting and what can you do? Clearly Selma and Doris would not get in trouble over this. But I sensed it was doubly humiliating to have to fear that she might be punished for talking back to a white woman who had spit at her.

Although I suffered from plenty of customer condescension at this store, at least putting up with racism was not part of my job. Once when Tanesha, a 23-year-old African American, was training me at the service desk, two white women elbowed up to the counter to complain to me about the long wait for service (they had waited about five minutes while we were serving another customer). I said something about being in training, and they thought I meant that I was training Tanesha. So they said, "Well, call up someone else to the register!" I said I'd have to ask Tanesha to do this. They demanded that I stop training her for a moment to call another person to the exchange desk. "No you don't understand," I said, "I'm the one who is in training; she knows what she is doing, and she is the only one who can call for backup, and she is in the middle of trying to accommodate this other customer." They seemed embarrassed at having assumed that the white woman was in charge. When they realized their mistake, they looked mortified and stepped back from the counter.

SHOPPING WHILE BLACK

During the civil rights era, equal access to stores was high on the list of demands for racial justice. Before Jim Crow laws were repealed, many stores restricted their facilities to whites only. Black customers often were not allowed to try on clothes, eat at lunch counters, or use public restroom facilities in stores.

Today the worst forms of racism have been eliminated. Gone are the "whites only" signs on restrooms and drinking fountains. But some stores build to exclude. The history of suburban malls is a history of intentional racial segregation. Even today, so-called desirable retail locations are characterized by limited access. In my city of Austin, Texas, local malls have opposed public bus service on the grounds that it would encourage undesirable (nonwhite) patrons.

But open access is not enough to ensure racial diversity. Diamond Toys was located in a racially diverse urban shopping district, next to subway and bus lines, yet nearly all its patrons were white. I didn't fully realize this until one day when Chandrika, an 18-year-old African-American gift wrapper, told me that she thought she saw one of her friends in the store. We weren't allowed to leave our section, so she asked the plainclothes security guard to look around and see if there was a black guy in the store. I asked her about this. Was it so unusual for an African-American teenager to be in the store that one black guy would be so apparent? After all, lots of young men came in to check out the new electronic toys and collectibles. Chandrika assured me that a young black man would definitely stand out.

One way that many stores show hostility to racial/ethnic minorities is through consumer racial profiling. Like racial profiling in police work, this involves detaining, searching, and harassing such people more often than is done for whites, usually because they are suspected of stealing. Some scholars have labeled this potential violation of people's rights "shopping while black." At the stores where I worked, clerks weren't allowed to pursue anyone suspected of stealing; that was the job of the plainclothes security workers. However, relations between customers and clerks sometimes broke down, and I saw double standards in the treatment of whites and minorities.

At the warehouse store, I was told to treat shoppers as if they were my mother (most shoppers at both stores were women). At the service desk, I was told to appease them by honoring all requests for returns, even if the merchandise had been used and worn out. The goal, my manager told me, was to make these shoppers so grateful that they would return to the store and spend $20,000 per child, the amount their marketers claimed was spent on an average child's toys.

In my experience, only middle-class white women could depend on this treatment. Nevertheless, I watched many white women throw fits, loudly complaining of shoddy service and merchandise with comments like "I will never shop in this store again!" Such arrogance no doubt came from being accustomed to having their demands met. On one occasion, when a white woman threw a tantrum because the bike she had ordered was not ready as scheduled, the manager offered her a $25 gift certificate for her troubles. She refused, demanded a refund, and left shouting that she would "never come to this store again." The manager then gathered the entire staff at the front of the store and chewed us out for being disorganized and incompetent.

Members of minority groups who wanted to return used merchandise or needed special consideration were rarely accommodated. The week before the bike incident, I was on a register that broke down in the middle of a credit card transaction. A middle-class black woman in her forties was buying inline skates for her ten-year-old daughter. The receipt came out of the register but not the slip for her to sign, so I had to call a manager, who came over and explained that she needed to go to another register and repeat the transaction. She refused, since it seemed to go through all right and she didn't want to be charged twice. She had to wait more than an hour to get this problem solved, and she wasn't offered any compensation. She didn't yell or make a scene; she waited stoically. I felt sorry for her and went to the service desk to tell a couple of my fellow workers what was happening while the managers tried to resolve it, and I said they should just give her the skates and let her go. My fellow workers thought that was the funniest thing they had ever heard. I said, "What about our policy of letting things go to make sure we keep loyal customers?" But they just laughed at me. Celeste said, "I want Christine to be the manager; she just lets the customers have whatever they want!"

Research has shown that African Americans suffer discrimination in public places, including stores; middle-class whites, on the other hand, are privileged. We do not recognize this precisely because it is so customary. Whites expect first-rate service; when they don't receive it, some feel victimized, even discriminated against, and some throw tantrums when they don't get what they want.

When African-American customers did shout or make a scene, the managers called or threatened to call the police. Each of these instances involved an

African-American man complaining and demanding a refund. Once a young black man, denied a cash refund, threw a toy on the service desk, and accidentally hit the telephone, which flew off the desk and hit me on the side of the head, knocking me to the floor. Within minutes, three police officers arrived and asked me if I wanted to press assault charges. I did not. After all, angry people often threw merchandise on that desk, and what happened had been an accident. But at least I was appeased. At the end of my shift, the manager gave me a "toy buck" for "taking a hit" in the line of duty, which entitled me to a free Coke.

There wasn't much shouting or throwing at the upscale store. It protected itself from conflict by catering to an upper-class clientele, much as a gated community does. This is not to say that diversity always leads to conflict, but it did at the warehouse store because race, class, and gender differences existed under a layer of power differences within the store. Clerks and customers interacted in a context where these differences had been used to shape marketing agendas, hiring practices, and labor policies—all of which benefited specific groups (especially middle-class white men and women).

WHAT ABOUT TOMATOES?

There are alternatives to shopping at large chain stores, although their numbers are dwindling. The store I'm calling Tomatoes is a small, family-owned business in an upper-middle-class neighborhood on a busy street with lots of pedestrian traffic. It's been in the neighborhood for 25 years, owned by the same family. It sells an offbeat assortment of toys, including many traditional items like kites and wooden blocks, and a variety of toys I would call "politically correct." It didn't carry Barbie, for example, but it did have "Get Real Girls," female action figures that look like G.I. Joe's sisters. I laughed when I saw a pack of plastic "multi-cultural" food, including spaghetti, sushi, a taco, and a bagel (all marked "made in China").

Working conditions at the store seemed very relaxed compared to what I had experienced. The owner wore shorts and a Hawaiian shirt, and the workers dressed like punk college students, including weirdly dyed hair, piercings, and tattoos. They didn't wear uniforms (as we did at the other two stores). One clerk wore her T-shirt hiked up in a knot in the front and stuffed under her bra in the back. Clerks seemed to be on a friendly, first-name basis with several of the customers, who were mostly middle-class white women.

Although I didn't get hired at Tomatoes, after several visits I noticed social patterns in the store's organization. The owner and the manager were both white men, and all the clerks were young white women. The owner was the only one who was near my age (mid-40s). You can never be sure why you aren't hired, but my impression is that I wasn't young enough or hip enough to work there. Although Tomatoes allowed more autonomy and self-expression than the stores where I worked, race, class, and gender inequality were as much a part of the social organization there as in other retail stores.

CONCLUSION

I ended up buying my gift at Tomatoes—a children's book written and auto-graphed by Marge Piercy. My decision reflects my identity and my social relationships. But what are the implications of my choice for social inequality? My purchase supported a store that was organized around racial exclusion, gender segregation, and class distinctions.

Everyone has to shop in our consumer society, yet the way shopping is organized bolsters social divisions. The racism of shopping is reflected in labor practices, store organization, and the guidelines, explicit or unspoken, for relations between clerks and customers. When deciding where and how to shop, consumers should be aware of what their choices imply with regard to racial justice and equality. Although an individual shopper can do little to change the overall social organization of shopping, raising awareness of the inequalities that our choices support must be a first step in imagining and then creating a better alternative.

35

Are Emily and Greg More Employable Than Lakisha and Jamal?

A Field Experiment on Labor Market Discrimination

BY MARIANNE BERTRAND AND SENDHIL MULLAINATHAN

Every measure of economic success reveals significant racial inequality in the U.S. labor market. Compared to Whites, African-Americans are twice as likely to be unemployed and earn nearly 25 percent less when they are employed (Council of Economic Advisers, 1998). This inequality has sparked a debate as to whether employers treat members of different races differentially. When faced with observably similar African-American and White applicants, do they favor the White one? Some argue yes, citing either employer prejudice or employer perception that race signals lower productivity. Others argue that differential treatment by race is a relic of the past, eliminated by some combination of employer enlightenment, affirmative action programs and the profit-maximization motive. In fact, many in this latter camp even feel that stringent enforcement of affirmative action programs has produced an environment of reverse discrimination. They would argue that faced with identical candidates, employers might favor the African-American one. Data limitations make it difficult to empirically test these views. Since researchers possess far less data than employers do, White and African-American workers that appear similar to researchers may look very different to employers. So any racial difference in labor market outcomes could just as easily be attributed to differences that are observable to employers but unobservable to researchers.

To circumvent this difficulty, we conduct a field experiment that builds on the correspondence testing methodology that has been primarily used in the past to study minority outcomes in the United Kingdom. We send resumes in response to help-wanted ads in Chicago and Boston newspapers and measure callback for interview for each sent resume. We experimentally manipulate perception of race via the name of the fictitious job applicant. We randomly assign very White-sounding names (such as Emily Walsh or Greg Baker) to half the resumes and very African-American-sounding names (such as Lakisha Washington or Jamal Jones) to the other half. Because we are also interested in how credentials affect the racial gap in callback, we experimentally vary the quality of the resumes used in response

SOURCE: Bertrand, Marianne, and Sendhil Mullainathan. "Are Emily and Greg More Employable than Lakisa and Jamal: A Field Experiment on Labor Market Discrimination." *American Economic Review* 94 (September): 991–1013. Copyright © 2004 American Economic Review. Reprinted by permission.

to a given ad. Higher-quality applicants have on average a little more labor market experience and fewer holes in their employment history; they are also more likely to have an e-mail address, have completed some certification degree, possess foreign language skills, or have been awarded some honors. In practice, we typically send four resumes in response to each ad: two higher-quality and two lower-quality ones. We randomly assign to one of the higher- and one of the lower-quality resumes an African-American-sounding name. In total, we respond to over 1,300 employment ads in the sales, administrative support, clerical, and customer services job categories and send nearly 5,000 resumes. The ads we respond to cover a large spectrum of job quality, from cashier work at retail establishments and clerical work in a mail room, to office and sales management positions.

We find large racial differences in callback rates. Applicants with White names need to send about 10 resumes to get one callback whereas applicants with African-American names need to send about 15 resumes. This 50-percent gap in callback is statistically significant. A White name yields as many more callbacks as an additional eight years of experience on a resume. Since applicants' names are randomly assigned, this gap can only be attributed to the name manipulation.

Race also affects the reward to having a better resume. Whites with higher-quality resumes receive nearly 30-percent more callbacks than Whites with lower-quality resumes. On the other hand, having a higher-quality resume has a smaller effect for African-Americans. In other words, the gap between Whites and African-Americans widens with resume quality. While one may have expected improved credentials to alleviate employers' fear that African-American applicants are deficient in some unobservable skills, this is not the case in our data.

The experiment also reveals several other aspects of the differential treatment by race. First, since we randomly assign applicants' postal addresses to the resumes, we can study the effect of neighborhood of residence on the likelihood of callback. We find that living in a wealthier (or more educated or Whiter) neighborhood increases callback rates. But, interestingly, African-Americans are not helped more than Whites by living in a "better" neighborhood. Second, the racial gap we measure in different industries does not appear correlated to Census-based measures of the racial gap in wages. The same is true for the racial gap we measure in different occupations. In fact, we find that the racial gaps in callback are statistically indistinguishable across all the occupation and industry categories covered in the experiment. Federal contractors, who are thought to be more severely constrained by affirmative action laws, do not treat the African-American resumes more preferentially; neither do larger employers or employers who explicitly state that they are "Equal Opportunity Employers." In Chicago, we find a slightly smaller racial gap when employers are located in more African-American neighborhoods.

I. PREVIOUS RESEARCH

With conventional labor force and household surveys, it is difficult to study whether differential treatment occurs in the labor market.[1] Armed only with survey data, researchers usually measure differential treatment by comparing the labor market performance of Whites and African-Americans (or men and women) for which they observe similar sets of skills. But such comparisons can be quite misleading. Standard labor force surveys do not contain all the characteristics that employers observe when hiring, promoting, or setting wages. So one can never be sure that the minority and nonminority workers being compared are truly similar from the employers' perspective. As a consequence, any measured differences in outcomes could be attributed to these unobserved (to the researcher) factors.

This difficulty with conventional data has led some authors to instead rely on pseudo-experiments.[2] Claudia Goldin and Cecilia Rouse (2000), for example, examine the effect of blind auditioning on the hiring process of orchestras. By observing the treatment of female candidates before and after the introduction of blind auditions, they try to measure the amount of sex discrimination. When such pseudo-experiments can be found, the resulting study can be very informative; but finding such experiments has proven to be extremely challenging.

A different set of studies, known as audit studies, attempts to place comparable minority and White actors into actual social and economic settings and measure how each group fares in these settings.[3] Labor market audit studies send comparable minority (African-American or Hispanic) and White auditors in for interviews and measure whether one is more likely to get the job than the other.[4] While the results vary somewhat across studies, minority auditors tend to perform worse on average: they are less likely to get called back for a second interview and, conditional on getting called back, less likely to get hired.

These audit studies provide some of the cleanest nonlaboratory evidence of differential treatment by race. But they also have weaknesses.... First, these studies require that both members of the auditor pair are identical in all dimensions that might affect productivity in employers' eyes, except for race. To accomplish this, researchers typically match auditors on several characteristics (height, weight, age, dialect, dressing style, hairdo) and train them for several days to coordinate interviewing styles. Yet, critics note that this is unlikely to erase the numerous differences that exist between the auditors in a pair.

Another weakness of the audit studies is that they are not double-blind. Auditors know the purpose of the study.... This may generate conscious or subconscious motives among auditors to generate data consistent or inconsistent with their beliefs about race issues in America. As psychologists know very well, these demand effects can be quite strong. It is very difficult to insure that auditors will not want to do "a good job." Since they know the goal of the experiment, they can alter their behavior in front of employers to express (indirectly) their own views. Even a small belief by auditors that employers treat minorities differently can result in measured differences in treatment. This effect is further magnified by the fact that auditors are not in fact seeking jobs and are therefore more free to let their beliefs affect the interview process.

Finally, audit studies are extremely expensive, making it difficult to generate large enough samples to understand nuances and possible mitigating factors. Also, these budgetary constraints worsen the problem of mismatched auditor pairs. Cost considerations force the use of a limited number of pairs of auditors, meaning that any one mismatched pair can easily drive the results. In fact, these studies generally tend to find significant differences in outcomes across pairs.

Our study circumvents these problems. First, because we only rely on resumes and not people, we can be sure to generate comparability across race. In fact, since race is randomly assigned to each resume, the same resume will sometimes be associated with an African-American name and sometimes with a White name. This guarantees that any differences we find are caused solely by the race manipulation. Second, the use of paper resumes insulates us from demand effects. While the research assistants know the purpose of the study, our protocol allows little room for conscious or subconscious deviations from the set procedures. Moreover, we can objectively measure whether the randomization occurred as expected. This kind of objective measurement is impossible in the case of the previous audit studies. Finally, because of relatively low marginal cost, we can send out a large number of resumes. Besides giving us more precise estimates, this larger sample size also allows us to examine the nature of the differential treatment from many more angles.

★★★

Our results indicate that for two identical individuals engaging in an identical job search, the one with an African-American name would receive fewer interviews. Does differential treatment within our experiment imply that employers are discriminating against African-Americans. (whether it is rational, prejudice-based, or other form of discrimination)? In other words, could the lower callback rate we record for African-American resumes *within our experiment* be consistent with a racially neutral review of the *entire pool* of resumes the surveyed employers receive?

In a racially neutral review process, employers would rank order resumes based on their quality and call back all applicants that are above a certain threshold. Because names are randomized, the White and African-American resumes we send should rank similarly on average. So, irrespective of the skill and racial composition of the applicant pool, a race-blind selection rule would generate equal treatment of Whites and African-Americans. So our results must imply that employers use race as a factor when reviewing resumes, which matches the legal definition of discrimination....

II. CONCLUSION

This paper suggests that African-Americans face differential treatment when searching for jobs and this may still be a factor in why they do poorly in the labor market. Job applicants with African-American names get far fewer callbacks for each resume they send out. Equally importantly, applicants with African-American names find it hard to overcome this hurdle in callbacks by improving their observable skills or credentials.

Taken at face value, our results on differential returns to skill have possibly important policy implications. They suggest that training programs alone may not be enough to alleviate the racial gap in labor market outcomes. For training to work, some general-equilibrium force outside the context of our experiment would have to be at play. In fact, if African-Americans recognize how employers reward their skills, they may rationally be less willing than Whites to even participate in these programs.

NOTES

1. See Joseph G. Altonji and Rebecca M. Blank (1999) for a detailed review of the existing literature on racial discrimination in the labor market.
2. William A. Darity, Jr., and Patrick L. Mason (1998) describe an interesting non-experimental study. Prior to the Civil Rights Act of 1964, employment ads would explicitly state racial biases, providing a direct measure of differential treatment. Of course, as Arrow (1998) mentions, discrimination was at that time "a fact too evident for detection."
3. Michael Fix and Marjery A. Turner (1998) provide a survey of many such audit studies.
4. Earlier hiring audit studies include Jerry M. Newman (1978) and Shelby J. McIntyre et al. (1980). Three more recent studies are Harry Cross et al. (1990), Franklin James and Steve W. DelCastillo (1991), and Turner et al. (1991). Heckman and Peter Siegelman (1992), Heckman (1998), and Altonji and Blank (1999) summarize these studies. See also David Neumark (1996) for a labor market audit study on gender discrimination.

REFERENCES

Council of Economic Advisers. *Changing America: Indicators of social and economic well-being by race and Hispanic origin.* September 1998, http://www.gpoaccess.gov/eop/ca/pdfs/ca.pdf.

Goldin, Claudia and Rouse, Cecilia. "Orchestrating Impartiality: The Impact of Blind Auditions on Female Musicians." *American Economic Review,* September 2000, *90*(4), pp. 715–41.

36

Gender Matters. So Do Race and Class

Experiences of Gendered Racism on the Wal-Mart Shop Floor[*]

SANDRA E. WEISSINGER

In this case study, inequitable access to power found within one particular institution, Wal-Mart stores, is examined. As the largest corporation in the world—
with revenues larger than those accumulated by Switzerland and outselling Target,
Home Depot, Sears, Kmart, Safeway, and Kroger combined—the company is
quite powerful both in the United States and abroad (Litchenstein 2006). In the
U.S., 1.3 million are employed (Rathke 2006) within the four thousand stores
across the nation (Litchenstein 2006). Because the lives of so many converge
there, Wal-Mart has the potential to mirror many of the oppressive practices
observed in social relationships outside of the stores. For this reason, the statements
given by Wal-Mart employees are telling of the robust and adaptive nature of
discriminatory practices, regardless of geographic location. Within Wal-Mart, the
multidimensional nature of inequality can be illuminated as gender, race, and class
intersect and shape the experiences of the women that work at these stores.

CASE BACKGROUND

In 2004, 1.6 million plaintiffs were granted class action status, making theirs the
largest sex discrimination case seen in U.S. courts.[**] The suit, Dukes v. Wal-
Mart Stores Inc., began when current and former female Wal-Mart employees
from across the United States claimed that the retailer discriminated against them
in terms of access to promotions, wages similar to their male colleagues, and
their job tasks. Individual employees examined their personal issues and developed a sociological imagination (Mills 2000 [1959])—the ability to see that the

---

[*]This essay is dedicated to Ms. Betty Dukes and the nearly two million women who were
courageous enough to stand in the gap for other employees experiencing similar treatment.
[**]Editors' note: In June 2011, the U.S. Supreme Court ruled against the women who
filed a class action suit against Wal-Mart.
SOURCE: Wessinger, Sandra, "Gender Matters, So Do Race And Class: Experiences of
Gendered Racism on the Wal-Mart Shop Floor" *Humanity and Society* 33 (November)
pp. 341–362. Copyright © 2009 Association for Humanist Sociology. Reprinted
by permission.

treatment they were subject to was shared, at least in part, by those with similar biographical characteristics. For example, when her supervisor referred to her as a "Mexican princess" in front of her peers, Gina Espinoza-Price (a plaintiff) was humiliated. She also realized that others were being treated in a similar unprofessional fashion.

Wal-Mart's defense lawyers posed several challenges to the plaintiffs; the most important was whether their case qualified as a class action suit. If the case had not qualified, plaintiffs would have had to file individual suits against the corporation. At best, this could produce rulings that benefited an individual plaintiff but failed to bring about large-scale changes to Wal-Mart's employment and promotions procedures that would benefit all employees. This suit is the impetus behind changes in Wal-Mart's human resources departments nationwide. All employees can now view and apply for all positions. They are also given the opportunity to officially declare their career goals in their computerized personnel file (Featherstone 2004).

The plaintiff for whom the case is named, Betty Dukes, is still employed at Wal-Mart. Dukes has stated that her motivation to continue speaking about employment practices at Wal-Mart comes from her religious background as a Christian minister. According to Dukes, "I am participating in this case in order to insure that young women such as my nieces and other women are treated fairly at every Wal-Mart store. The time has surely come for equality for women" (Rosen 2004). Dukes's statement is unique in that she sees her actions today as connected to the lives of others who will come after her. Although the other plaintiffs did not specifically list similar motivations, all of the plaintiffs made claims concerning inequality. These claims have been bolstered by the statistical findings of Drogin (2003), who found that across geographic locations, women working for Wal-Mart earn less than male employees who occupy the same positions. In addition, Drogin found that women were promoted into management positions at lower and slower rates than male employees.

The Dukes suit rightly identifies the effects of sex-based discrimination that are generalizable across women's experiences within Wal-Mart stores. To provide another vantage point from which to understand the effects of the discrimination experienced by these employees (and perhaps those at corporations that seek to model their businesses after Wal-Mart), I contend that discrimination due to biological sex differences alone does not explain the range of the plaintiffs' experiences or properly gauge individual and social damages caused when one must navigate treacherous environments on a daily basis. Rather, I argue, those individuals who are targeted for mistreatment experience such treatment as equally raced, classed, and gendered people whose lives exist within a web of intersecting and relational inequalities. Therefore, a reexamination of what plaintiffs of the class action suit said about their employment experiences (when guided by a gender primacy framework) is necessary so that the multiple and differing daily work experiences of the plaintiffs can be illuminated, adding to sociological knowledge concerning the work lives of women across standpoints. Reexamination is needed not simply to see difference in lives, but to reveal the persistence of inequality and the multiple ways discriminatory work atmospheres are maintained.

According to the files Wal-Mart made available to the court, women worked for the company longer and received better reviews from supervisors, yet earned less money and received fewer promotions when compared to male employees.

Women's Experiences Across Race and Class

… Scholarship about the social locations of women has addressed not only the lived experiences of women of color, but also those of white women across class and geographic location. Enlarging our understanding of the differences between women, sociologists have documented the ways in which white women act as oppressors as well as the oppressed—benefiting from certain race and/or class positions. As explained by Blee (2002) and Frankenberg (1993), one experience of marginalization does not automatically inspire empathy in white women toward other oppressed communities. Whiteness, as all categories of difference, is socially constructed through carrying real consequences that bear upon individuals' everyday lives. "In a white-dominated society, whiteness is invisible. Race is something that adheres to others as a mark of otherness, a stigma of difference. It is not perceived by those sheltered under the cloak of privilege that masks its own existence" (Blee 2002:56). In short, white women can be marginalized due to gender and class, but experience certain benefits because of their race.

Methodology

… In this work I identify six biographical characteristics most plaintiffs mentioned in their declarations: gender, race, class, family makeup, geographic location, and age. To be clear, I do not argue that one area of oppression or facet of one's biography is more important than another. My argument is not one that posits that there is a monolithic or authentic experience either. Certainly, lived experiences can be similar and should be juxtaposed against one another for analysis. Rather, I argue that each employee had rich and unique work experiences resulting from the intersecting characteristics that make up their biographies.

… Past studies have failed to examine women's work experiences outside of the gender lens. Questioning how the multiple features of each plaintiff's biography influence and shape the stories they tell about work life at Wal-Mart is at the heart of this work.

Findings

Insider Outsider Statuses

Very few white plaintiffs mentioned their race but when they did, it was in relation to the race of others. These statements were made when the plaintiff reflected on "backstage" (Picca and Feagin 2007), private conversations they engaged in with white male supervisors. The failure of white women plaintiffs to mention race, except when noting the difference between themselves and people of color, illustrates that "whiteness is unmarked because of the pervasive nature of white domination" (Blee 2002:56). To illuminate this argument, the following excerpts show how two white women who held positions of authority at Wal-Mart stores addressed race.

The first quote is from Ms. Lorie Williams, the youngest declarant in this sample (twenty-six years old) and a single mother who supports her child [working] as a "Front End Manager," the lowest ranking management position in the occupational hierarchy at the stores.

> In 1996, I became the front-end manager. With no training, I was single-handedly responsible for hiring door greeters, cart pushers, over sixty new cashiers, and preparing the entire front end in order to transition the Collierville Wal-Mart into a Supercenter. Almost immediately, Co-Manager Doug Ayerst and new Store Manager Robert Hayes (he replaced Wes Grab in 1996) began criticizing the way in which I was managing the front end. They repeatedly complained to me about problems in the front end but did not give me any practical management advice and often gave inconsistent instructions. Store Manager Hayes once told me that the problem was that there were "too many damn women in the front end." On another occasion, Store Manager Hayes and Co-Manager Jim Belcoff pulled me aside to tell me that I needed to "whiten up [the staff of] the front end." Both of these statements seemed to indicate that they wanted me to stop hiring women and African-American cashiers. When I asked why, they indicated that the staff was "intimidating" to the clientele. I tried to explain that I did not believe this was true and that I was hiring the individuals whom I believed were the most qualified applicants. They were unresponsive.

… In the second example, the words of Melissa Howard, a 35-year-old white woman who had risen to the rank of Store Manager while raising her biracial child in the Midwest, are examined.

> … In July 1997, I was promoted to the position of Store Manager in Marysville, Kansas. I drove with my daughter's father, who is African-American, to find a place to live there. When we arrived in town, we were treated hostilely by a clerk in the local Wal-Mart store, the town realtor, and by a number of prospective landlords because we were a mixed-race family. At the last rental we visited, the landlord told us that my daughter was not safe and that we needed to be out of town by dark. I knew then that I could not move my family to this place. I called the Regional Personnel Manager Gary Coward, explained what had happened and asked to be placed as a store manager anywhere else. He refused and told me that I would have to go to Marysville as planned or accept a demotion and return as an assistant manager to the 86th Store in Indianapolis, one of the worst stores in the area. I took the demotion … This was particularly humiliating because my employees in Plainfield had just thrown me a big going-away party to celebrate my promotion.

Melissa Howard was chastised by a male supervisor not just because of her gender, but because she broke a racialized stereotype of white women's chasteness (Collins 2000:132–134). By breaking this stereotype, her colleagues and supervisors saw her as a lower class white. As a result of this construction, the unequal ways individuals have access to power are produced and upheld (Brown 1995; DeVault 1999:27). As a white woman in an interracial relationship, Melissa Howard was punished and seen as an outsider by the white people with whom she interacted. It was as though she created the problems she and her family experienced in Kansas.

In both examples, these white women were treated as incompetent and at fault for the problems they experienced. In turn, this affected their work. In these examples it is clear that while these women faced oppression due to their gender, rules and relationships to power are established for individuals based on the multiple, and differently valued, statuses they hold. Melissa Howard seemingly rose above the occupational glass ceiling noted by many plaintiffs. However, her failure to adhere to dating rules held by those with power, including landlords, realtor, and the Wal-Mart Regional Manager, created experiences of discrimination. As an insider due to her race, but a subordinate due to gender and class, Lorie Williams was able to attain a lower-level management position, but was chastised and belittled because she hired men of color and women.

These stories illustrate the lived experiences of white women who work at Wal-Mart. To succeed, they were socialized to accept and replicate discrimination against people of color. Similarly, workers are trained to follow rules of "respectability" concerning their actions. Insiders like Lorie Williams who hold beliefs that make them outsiders, need support so they will continue to believe that the long-term benefits of their actions will outweigh the short-term punishments they endure for failing to replicate discriminatory practices.

The Continuing Significance of Race

... In the following statement, it is clear subordinates also draw on racial and gendered privileges to gain power, even in instances in which they occupy lower class positions. In her statement, Ms. Jennifer Johnson, a Black woman with a community college degree, describes her interpersonal relationships with store staff.

> In 1991, Mr. Pshek was promoted to District Manager in a different district. He told me that if I was willing to relocate to a store in his district, that he would promote me to Assistant Manager, a position that I had sought for some time. I agreed, although I knew that Scott Schwalback, Mark Melatesta, and a male named Rick had been promoted to Assistant Manager without having to switch districts. I was transferred to the Eustis, Florida store as an hourly employee. I worked in that store as a Department Manager for approximately five months without receiving the promotion to Assistant Manager that Mr. Pshek had promised. Finally, I spoke with Bob Hart, Regional Vice President, about the situation, and he told me I had to wait another three to fourth months to be promoted to Assistant Manager. I was finally promoted in February 1992. Both Scott Schwalback and Rick (a Department Manager in Furniture) became Department Managers after I had, since

> I had worked for Wal-Mart for longer, but they were both promoted to
> Assistant Manager positions before I was. Thus, I worked much longer
> as a Department Manager before being promoted to Assistant Manager
> than similarly qualified men had. Shortly after I was promoted to Co-
> Manager, one of the male Assistant Managers who reported to me was
> disrespectful and avoided doing work I assigned him. I spoke to my
> Store Manager Kevin Robinson about it. He told me that the man was
> upset that I had been promoted instead of him, and said "you have two
> strikes against you: 1) you're a woman; 2) you're black."

Jennifer Johnson described the years of obstacles she navigated to gain
promotions. Even with her experience and dedication, others reduced her
accomplishments and used them to treat her with disdain. For her and other
women of color, working in these stores means that they must not only do
their jobs well, but they must also navigate a workplace where others use them
as scapegoats for their own frustrations, block them from training opportunities,
or sabotage them by not doing work these women of color assign them. In
addition, those with the power to chastise offenders and influence workplace
discussions of inequality did nothing. Therefore, women of color had to find
their own ways to overcome inequality.

The above example illustrates blatant, hostile racism. But racism is not always
aggressively expressed. Feagin (1991) argues after a lifetime of experiencing discrimination, people of color develop a special lens through which to recognize
even subtle discrimination. For example, joking can be a subtle medium through
which people of color experience belittlement and struggles for dominance.
Ms. Gina Espinoza-Price, an energetic, innovative woman, worked her way up
to become the District Manager of Mexico. She provided an example of how a
white man used joking to force others to acknowledge the labels he created for
them and to show dominance. Because he knew he would not face a penalty for
his actions, he was able to set a behavioral example for other men under his
supervision.

> Male Photo Division Management behaved in ways that demeaned and
> belittled women and minorities. In fall 1996, there was a Photo District
> Manager meeting in Valencia, California. Wal-Mart had just hired a
> second female Photo District Manager for the western region, Linda
> Palmer. During dinner, Jeff Gwartney introduced all of the District
> Managers to Ms. Palmer using nicknames for the minorities and
> women. I was introduced as Gina, "the little Mexican princess." I was
> very offended by Mr. Gwartney's comment and left the dinner early.
> Throughout the meeting, men made sexual statements and jokes that
> I thought were very offensive. For example, a flyer with an offensive
> joke about women being stupid was left on my belongings. In February
> 1997, during an evaluation, I complained to One-hour Photo Divi-
> sional Manager Joe Lisuzzo about harassment based on gender at the
> previous Photo District Manager meeting. He replied that he would
> take care of it. I knew from trainings on Wal-Mart's sex harassment

policy given by Wal-Mart Legal Department employee Canetta Ivy that company policy mandates that when someone complains of sexual harassment, an investigation must begin within twenty-four hours. Therefore, I expected to be interviewed as a part of an investigation. I was never called. A couple of weeks later, in March 1997, I saw Mr. Lisuzzo at a meeting. I asked him if he had been conducting an investigation of my sexual harassment complaint. He replied that it was being taken care of. I was never aware of any action taken in response to my complaint. Six weeks after complaining about sexual harassment, I was terminated.

Gina Espinoza-Price's statement highlights blatant examples of racism and sexism as well as subtle, institutional examples of discrimination as carried out by individuals who occupied similar class positions, but different status levels. Mr. Lisuzzo's handling of the sexual harassment claim filed by the plaintiff is an example of how the concerns of people are ignored, requiring victims of harassment and discrimination to waste emotional energy rationalizing their own reactions and creating coping tactics to help them interact with hostile colleagues. In addition, it shows how racist joking is often coupled with class and gender.

Class Matters

… Just as some women enjoy privilege and opportunities based on their race, class also shapes their experiences and actions. The following excerpt demon-strates the difficulties and frustrations of women of color who are single mothers trying to support their families with Wal-Mart wages.

… Ms. Uma Jean Minor, a single mother living in Alabama, described the need to remain gainfully employed.

> As a single mother, I could not support my family on this wage [paid by Wal-Mart] and was forced to take a second, full-time day job at Food Fair. At the time, I planned to work this second job only until I was able to move to a higher paying position at Wal-Mart. I had no idea that it would take me another seven years to obtain a management position at Wal-Mart and that I would be working two full-time jobs for this entire period.

In this example, Uma Jean Minor notes that her Wal-Mart wages were not suf-ficient to lift her out of poverty, an argument echoed by living wage supporters throughout the United States (see McCarthy and Ciokajlo 2006; Talbott and Dolby 2003; Warren 2005). Though she shows a tremendous amount of will power, motivation, and agency, her actions would not be sustainable for most women in her position as a single mother of five. Missing from her declaration are the stories of fear, if not hardship, she lived through raising five children and working two full time jobs. In addition, we do not know if she had a network of kin or fictive kin on which she could rely. Therefore, we can only guess at the psychological and material effects discrimination had on her and her children.

Clearly, the effects of sex-based discrimination at Wal-Mart shape the lives of women and their families in different and important ways.

... For female heads of household whose families depend on their earnings, the wages, difficult decisions about how to navigate work, and workplace stress factor into whether it is worthwhile to work for poverty wages. Studies like that of Edin and Lein (1997) have demonstrated that the benefits lost by taking a low-wage job are detrimental to these women's children, who lose state funded health care, housing assistance, and food support. For women like the defendants, work provides another burden; because they need these jobs, they must sometimes sacrifice their dignity and morals in the face of oppressive supervisors (like Mary Crawford) and glass ceilings (like Uma Jean Minor) for the sake of their children.

DISCUSSION AND CONCLUSION

... These plaintiffs did not experience the waste of their talent, energy, and potential in the same way. Rather, it can be observed that the Dukes case is a collection of varying vignettes through which social scientists can observe how some women have more access to power and resources, leaving others to struggle because of their marginalized positions in the power hierarchy. To be clear, although some of the women at Wal-Mart have suffered because of sexism, gender abuse takes multiple forms allowing certain women privilege (even in their oppression) because power is not distributed equally.

REFERENCES

Blee, Kathleen M. 2002. *Inside Organized Racism: Women in the Hate Movement.* Berkeley: University of California Press.

Brown, Elsa B. 1995. " 'What has Happened Here': The Politics of Difference in Women's History and Feminist Politics." In D.C. Hine, W. King and L. Reed (eds). *We Specialize in the Wholly Impossible: A Reader in Black Women's History,* pp. 39–54. Brooklyn, NY: Carlson Publishing.

Collins, Patricia H. 2000. *Black Feminist Thought: Knowledge, Consciousness and the Politics of Empowerment.* New York: Routledge.

DeVault, Marjorie L. 1999. *Liberating Method: Feminism and Social Research.* Philadelphia: Temple University Press.

Drogin, Richard. 2003. *Statistical Analysis of Gender Patterns in Wal-Mart Workforce.* Retrieved from www.walmartclass.com on 22 Sept 2008.

Edin, Kathryn and Lein, Laura. 1997. *Making Ends Meet: How Single Mothers Survive Welfare and Low-Wage Work.* New York: Russell Sage Foundation.

Feagin, Joe R. 1991. "The Continuing Significance of Race: Antiblack Discrimination in Public Places." *American Sociological Review* 56(1): 101–116.

Featherstone, Liza. 2004. *Selling Women Short*. New York: Basic Books.

Ferber, Abby L. 1999. "What White Supremacists Taught a Jewish Scholar about Identity." *The Chronicle of Higher Education* May 7: B6–B7.

Frankenberg, Ruth. 1993. *White Women, Race Matters: The Social Construction of Whiteness*. Minneapolis: University of Minnesota Press.

Litchenstein, Nelson. 2006. "Wal-Mart: A Template for Twenty-First Century Capitalism." Chapter 1 in Nelson Litchenstein (ed). *Wal-Mart: The Face of Twenty-First Century Capitalism*. New York: The New Press.

McCarthy, Brendan, and Mickey Ciokajlo. 2006. "Clerics Slam Big-Box Wage Law: Ordinance Would Chase Jobs From City, They Contend." *Chicago Tribune*, July 18, 2006. Accessed from Lexis Nexis August 5, 2007.

Mills, C. Wright. 2000[1959]. *The Sociological Imagination*. Oxford: Oxford University Press.

Picca, Leslie Houts, and Feagin, Joe R. 2007 *Two-Faced Racism: Whites in the Backstage and Frontstage*. New York: Routledge.

Rathke, Wade. 2006. "A Wal-Mart Workers Association? An Organizing Plan" in Litchenstein, Nelson (ed). *Wal-Mart: The Face of Twenty-First Century Capitalism*. New York: The New Press.

Rosen, Ruth. 2004. *Big-Box Battle: A Review of Selling Women Short: The Landmark Battle for Workers' Rights at Wal-Mart, by Liza Featherstone*. Retrieved from http://www.longviewinstitute.org/research/rosen/walmart/sellingwomenshort on October 14, 2008.

Talbott, Madeline, and Doby, Michael. 2003. "Using the Big Box Living Wage Ordinance to Keep Wal-Mart Out of the Cities." *Social Policy* 34(2/3): 23–28.

Warren, Dorian T. 2005. "Wal-Mart Surrounded: Community Alliances and Labor Politics in Chicago." *New Labor Forum* 14(3): 17–23.

37

Our Mothers' Grief
Racial-Ethnic Women and the Maintenance of Families

BONNIE THORNTON DILL

REPRODUCTIVE LABOR[1] FOR WHITE WOMEN IN EARLY AMERICA

In eighteenth- and nineteenth-century America, the lives of white[2] women in the United States were circumscribed within a legal and social system based on patriarchal authority. This authority took two forms: public and private. The social, legal, and economic position of women in this society was controlled through the private aspects of patriarchy and defined in terms of their relationship to families headed by men. The society was structured to confine white wives to reproductive labor within the domestic sphere. At the same time, the formation, preservation, and protection of families among white settlers was seen as crucial to the growth and development of American society. Building, maintaining, and supporting families was a concern of the State and of those organizations that prefigured the State. Thus, while white women had few legal rights as women, they were protected through public forms of patriarchy that acknowledged and supported their family roles of wives, mothers, and daughters because they were vital instruments for building American society....

In colonial America, white women were seen as vital contributors to the stabilization and growth of society. They were therefore accorded some legal and economic recognition through a patriarchal family structure....

Throughout the colonial period, women's reproductive labor in the family was an integral part of the daily operation of small-scale family farms or artisan's shops. According to Kessler-Harris (1981), a gender-based division of labor was common, but not rigid. The participation of women in work that was essential to family survival reinforced the importance of their contributions to both the protection of the family and the growth of society.

Between the end of the eighteenth and mid-nineteenth century, what is labeled the "modern American family" developed. The growth of industrialization and an urban middle class, along with the accumulation of agrarian wealth among Southern planters, had two results that are particularly pertinent to this

SOURCE: From *Journal of Family History* 13 (1988): 415–431. Reprinted by permission.

discussion. First, class differentiation increased and sharpened, and with it, distinctions in the content and nature of women's family lives. Second, the organization of industrial labor resulted in the separation of home and family and the assignment to women of a separate sphere of activity focused on childcare and home maintenance. Whereas men's activities became increasingly focused upon the industrial competitive sphere of work, "women's activities were increasingly confined to the care of children, the nurturing of the husband, and the physical maintenance of the home" (Degler 1980, p. 26).

This separate sphere of domesticity and piety became both an ideal for all white women as well as a source of important distinctions between them. As Matthaei (1982) points out, tied to the notion of wife as homemaker is a definition of masculinity in which the husband's successful role performance was measured by his ability to keep his wife in the homemaker role. The entry of white women into the labor force came to be linked with the husband's assumed inability to fulfill his provider role.

For wealthy and middle-class women, the growth of the domestic sphere offered a potential for creative development as homemakers and mothers. Given ample financial support from their husband's earnings, some of these women were able to concentrate their energies on the development and elaboration of the more intangible elements of this separate sphere. They were also able to hire other women to perform the daily tasks such as cleaning, laundry, cooking, and ironing. Kessler-Harris cautions, however, that the separation of productive labor from the home did not seriously diminish the amount of physical drudgery associated with housework, even for middle-class women.... In effect, household labor was transformed from economic productivity done by members of the family group to home maintenance; childcare and moral uplift done by an isolated woman who perhaps supervised some servants.

Working-class white women experienced this same transformation but their families' acceptance of the domestic code meant that their labor in the home intensified. Given the meager earnings of working-class men, working-class families had to develop alternative strategies to both survive and keep the wives at home. The result was that working-class women's reproductive labor increased to fill the gap between family need and family income. Women increased their own production of household goods through things such as canning and sewing; and by developing other sources of income, including boarders and homework. A final and very important source of other income was wages earned by the participation of sons and daughters in the labor force. In fact, Matthaei argues that "the domestic homemaking of married women was supported by the labors of their daughters" (1982, p. 130)....

Another way in which white women's family roles were socially acknowledged and protected was through the existence of a separate sphere for women. The code of domesticity, attainable for affluent women, became an ideal toward which nonaffluent women aspired. Notwithstanding the personal constraints placed on women's development, the notion of separate spheres promoted the growth and stability of family life among the white middle class and became the basis for working-class men's efforts to achieve a family wage, so that they could keep their wives at home. Also, women gained a distinct sphere of authority and expertise that yielded them special recognition.

During the eighteenth and nineteenth centuries, American society accorded considerable importance to the development and sustenance of European immigrant families. As primary laborers in the reproduction and maintenance of family life, women were acknowledged and accorded the privileges and protections deemed socially appropriate to their family roles. This argument acknowledges the fact that the family structure denied these women many rights and privileges and seriously constrained their individual growth and development. Because women gained social recognition primarily through their membership in families, their personal rights were few and privileges were subject to the will of the male head of the household. Nevertheless, the recognition of women's reproductive labor as an essential building block of the family, combined with a view of the family as the cornerstone of the nation, distinguished the experiences of the white, dominant culture from those of racial ethnics.

Thus, in its founding, American society initiated legal, economic, and social practices designed to promote the growth of family life among European colonists. The reception colonial families found in the United States contrasts sharply with the lack of attention given to the families of racial-ethnics. Although the presence of racial-ethnics was equally as important for the growth of the nation, their political, economic, legal, and social status was quite different.

REPRODUCTIVE LABOR AMONG RACIAL-ETHNICS IN EARLY AMERICA

Unlike white women, racial-ethnic women experienced the oppressions of a patriarchal society but were denied the protections and buffering of a patriarchal family. Their families suffered as a direct result of the organization of the labor systems in which they participated.

Racial-ethnics were brought to this country to meet the need for a cheap and exploitable labor force. Little attention was given to their family and community life except as it related to their economic productivity. Labor, and not the existence or maintenance of families, was the critical aspect of their role in building the nation. Thus they were denied the social structural supports necessary to make *their* families a vital element in the social order. Family membership was not a key means of access to participation in the wider society. The lack of social, legal, and economic support for racial-ethnic families intensified and extended women's reproductive labor, created tensions and strains in family relationships, and set the stage for a variety of creative and adaptive forms of resistance.

African-American Slaves

Among students of slavery, there has been considerable debate over the relative "harshness" of American slavery, and the degree to which slaves were permitted or encouraged to form families. It is generally acknowledged that many slave-owners found it economically advantageous to encourage family formation as a way of reproducing and perpetuating the slave labor force. This became increasingly true after 1807 when the importation of African slaves was explicitly prohibited. The existence of these families and many aspects of their functioning, however, were directly controlled by the master. In other words, slaves married and formed families but these groupings were completely subject to the master's decision to let them remain intact. One study has estimated that about 32 percent of all recorded slave marriages were disrupted by sale, about 45 percent by death of a spouse, about 10 percent by choice, with the remaining 13 percent not disrupted at all (Blassingame 1972, pp. 90–92). African slaves thus quickly learned that they had a limited degree of control over the formation and maintenance of their marriages and could not be assured of keeping their children with them. The threat of disruption was perhaps the most direct and pervasive cultural assault[3] on families that slaves encountered. Yet there were a number of other aspects of the slave system which reinforced the precariousness of slave family life.

In contrast to some African traditions and the Euro-American patterns of the period, slave men were not the main provider or authority figure in the family. The mother–child tie was basic and of greatest interest to the slaveowner because it was critical in the reproduction of the labor force.

In addition to the lack of authority and economic autonomy experienced by the husband-father in the slave family, use of the rape of women slaves as a weapon of terror and control further undermined the integrity of the slave family…. The slave family, therefore, was at the heart of a peculiar tension in the master-slave relationship. On the one hand, slaveowners sought to encourage familial ties among slaves because, as Matthaei (1982) states: "… these provided the basis of the development of the slave into a self-conscious socialized human being" (p. 81). They also hoped and believed that this socialization process would help children learn to accept their place in society as slaves. Yet the master's need to control and intervene in the familial life of the slaves is indicative of the other side of this tension. Family ties had the potential for becoming a competing and more potent source of allegiance than the slavemaster himself. Also, kin were as likely to socialize children in forms of resistance as in acts of compliance.

It was within this context of surveillance, assault, and ambivalence that slave women's reproductive labor took place. She and her menfolk had the task of preserving the human and family ties that could ultimately give them a reason for living. They had to socialize their children to believe in the possibility of a life in which they were not enslaved. The slave woman's labor on behalf of the family was, as Davis (1971) has pointed out, the only labor the slave engaged in that could not be directly appropriated by the slaveowner for his own profit. Yet, its indirect appropriation, as labor crucial to the reproduction of the slaveowner's labor force, was the source of strong ambivalence for many slave women. Whereas some mothers murdered their babies to keep them from being slaves, many sought within the family sphere a degree of autonomy and creativity denied them in other realms of the society. The maintenance of a distinct African-American culture is testimony to the ways in which slaves maintained a degree of cultural autonomy and resisted the creation of a slave family that only served the needs of the master.

Gutman (1976) provides evidence of the ways in which slaves expressed a unique Afro-American culture through their family practices. He provides data on naming patterns and kinship ties among slaves that flies in the face of the dominant ideology of the period. That ideology argued that slaves were immoral and had little concern for or appreciation of family life.

Yet Gutman demonstrated that within a system which denied the father authority over his family, slave boys were frequently named after their fathers, and many children were named after blood relatives as a way of maintaining family ties. Gutman also suggested that after emancipation a number of slaves took the names of former owners in order to reestablish family ties that had been disrupted earlier. On plantation after plantation, Gutman found considerable evidence of the building and maintenance of extensive kinship ties among slaves. In instances where slave families had been disrupted, slaves in new communities reconstituted the kinds of family and kin ties that came to characterize Black family life throughout the South. These patterns included, but were not limited to, a belief in the importance of marriage as a long-term commitment, rules of exogamy that included marriage between first cousins, and acceptance of women who had children outside of marriage. Kinship networks were an important source of resistance to the organization of labor that treated the individual slave, and not the family, as the unit of labor (Caulfield 1974).

Another interesting indicator of the slaves' maintenance of some degree of cultural autonomy has been pointed out by Wright (1981) in her discussion of slave housing. Until the early 1800s, slaves were often permitted to build their housing according to their own design and taste. During that period, housing built in an African style was quite common in the slave quarters. By 1830, however, slaveowners had begun to control the design and arrangement of slave housing and had introduced a degree of conformity and regularity to it that left little room for the slave's personalization of the home. Nevertheless, slaves did use some of their own techniques in construction and often hid it from their masters....

Housing is important in discussions of family because its design reflects sociocultural attitudes about family life. The housing that slaveowners provided for their slaves reflected a view of Black family life consistent with the stereotypes of the period. While the existence of slave families was acknowledged, it certainly was not nurtured. Thus, cabins were crowded, often containing more than one family, and there were no provisions for privacy. Slaves had to create their own....

Perhaps most critical in developing an understanding of slave women's reproductive labor is the gender-based division of labor in the domestic sphere. The organization of slave labor enforced considerable equality among men and women. The ways in which equality in the labor force was translated into the family sphere is somewhat speculative....

We know, for example, that slave women experienced what has recently been called the "double day" before most other women in this society. Slave narratives (Jones 1985; White 1985; Blassingame 1977) reveal that women had primary responsibility for their family's domestic chores. They cooked (although on some plantations meals were prepared for all of the slaves), sewed, cared for their children, and cleaned house, all after completing a full day of labor for the master. Blassingame (1972) and others have pointed out that slave men engaged in hunting, trapping, perhaps some gardening, and furniture making as ways of contributing to the maintenance of their families. Clearly, a gender-based division of labor did exist within the family and it appears that women bore the larger share of the burden for housekeeping and child care....

Black men were denied the male resources of a patriarchal society and therefore were unable to turn gender distinctions into female subordination, even if that had been their desire. Black women, on the other hand, were denied support and protection for their roles as mothers and wives and thus had to modify and structure those roles around the demands of their labor. Thus, reproductive labor for slave women was intensified in several ways: by the demands of slave labor that forced them into the double-day of work; by the desire and need to maintain family ties in the face of a system that gave them only limited recognition; by the stresses of building a family with men who were denied the standard social privileges of manhood; and by the struggle to raise children who could survive in a hostile environment.

This intensification of reproductive labor made networks of kin and quasi-kin important instruments in carrying out the reproductive tasks of the slave community. Given an African cultural heritage where kinship ties formed the basis of social relations, it is not at all surprising that African American slaves developed an extensive system of kinship ties and obligations (Gutman 1976; Sudarkasa 1981). Research on Black families in slavery provides considerable documentation of participation of extended kin in child rearing, childbirth, and other domestic, social, and economic activities (Gutman 1976; Blassingame 1972; Genovese and Miller 1974)....

With individual households, the gender-based division of labor experienced some important shifts during emancipation. In their first real opportunity to establish family life beyond the controls and constraints imposed by a slavemaster, family life among Black sharecroppers changed radically. Most women, at least those who were wives and daughters of able-bodied men, withdrew from field labor and concentrated on their domestic duties in the home. Husbands took primary responsibility for the fieldwork and for relations with the owners, such as signing contracts on behalf of the family. Black women were severely criticized by whites for removing themselves from field labor because they were seen to be aspiring to a model of womanhood that was considered inappropriate for them. This reorganization of female labor, however, represented an attempt on the part of Blacks to protect women from some of the abuses of the slave system and to thus secure their family life. It was more likely a response to the particular set of circumstances that the newly freed slaves faced than a reaction to the lives of their former masters. Jones (1985) argues that these patterns were "particularly significant" because at a time when industrial development was introducing a labor system that divided male and female labor, the freed Black family was establishing a pattern of joint work and complementary tasks between males and females that was reminiscent of the preindustrial American families. Unfortunately, these former slaves had to do this without the institutional supports that white farm families had in the midst of a sharecropping system that deprived them of economic independence.

Chinese Sojourners

An increase in the African slave population was a desired goal. Therefore, Africans were permitted and even encouraged at times to form families subject to the authority and whim of the master. By sharp contrast, Chinese people were explicitly denied the right to form families in the United States through both law and social practice. Although male laborers began coming to the United States in sizable numbers in the middle of the nineteenth century, it was more than a century before an appreciable number of children of Chinese parents were born in America. Tom, a respondent in Nee and Nee's (1973) book, *Longtime Californ'*, says: "One thing about Chinese men in America was you had to be either a merchant or a big gambler, have lot of side money to have a family here. A working man, an ordinary man, just can't!" (p. 80).

Working in the United States was a means of gaining support for one's family with an end of obtaining sufficient capital to return to China and purchase land. The practice of sojourning was reinforced by laws preventing Chinese laborers from becoming citizens, and by restrictions on their entry into this country. Chinese laborers who arrived before 1882 could not bring their wives and were prevented by law from marrying whites. Thus, it is likely that the number of Chinese-American families might have been negligible had it not been for two things: the San Francisco earthquake and fire in 1906, which destroyed all municipal records; and the ingenuity and persistence of the Chinese people who used the opportunity created by the earthquake to increase their numbers in the United States. Since relatives of citizens were permitted entry, American-born Chinese (real and claimed) would visit China, report the birth of a son, and thus create an entry slot. Years later the slot could be used by a relative or purchased. The purchasers were called "paper sons." Paper sons became a major mechanism for increasing the Chinese population, but it was a slow process and the sojourner community remained predominantly male for decades.

The high concentration of males in the Chinese community before 1920 resulted in a split-household form of family....

The women who were in the United States during this period consisted of a small number who were wives and daughters of merchants and a larger percentage who were prostitutes. Hirata (1979) has suggested that Chinese prostitution was an important element in helping to maintain the split-household family. In conjunction with laws prohibiting intermarriage, Chinese prostitution helped men avoid long-term relationships with women in the United States and ensured that the bulk of their meager earnings would continue to support the family at home.

The reproductive labor of Chinese women, therefore, took on two dimensions primarily because of the split-household family form. Wives who remained in China were forced to raise children and care for in-laws on the meager remittances of their sojourning husband. Although we know few details about their lives, it is clear that the everyday work of bearing and maintaining children and a household fell entirely on their shoulders. Those women who immigrated and worked as prostitutes performed the more nurturant aspects of reproductive labor, that is, providing emotional and sexual companionship for men who were far from home. Yet their role as prostitute was more likely a means of supporting their families at home in China than a chosen vocation.

The Chinese family system during the nineteenth century was a patriarchal one wherein girls had little value. In fact, they were considered only temporary members of their father's family because when they married, they became members of their husband's families. They also had little social value: girls were sold by some poor parents to work as prostitutes, concubines, or servants. This saved the family the expense of raising them, and their earnings also became a source of family income. For most girls, however, marriages were arranged and families sought useful connections through this process.

With the development of a sojourning pattern in the United States, some Chinese women in those regions of China where this pattern was more prevalent would be sold to become prostitutes in the United States. Most, however, were married off to men whom they saw only once or twice in the 20- or 30-year period during which he was sojourning in the United States. Her status as wife ensured that a portion of the meager wages he earned would be returned to his family in China. This arrangement required considerable sacrifice and adjustment on the part of wives who remained in China and those who joined their husbands after a long separation....

Despite these handicaps, Chinese people collaborated to establish the opportunity to form families and settle in the United States. In some cases it took as long as three generations for a child to be born on United States soil....

Chicanos

Africans were uprooted from their native lands and encouraged to have families in order to increase the slave labor force. Chinese people were immigrant laborers whose "permanent" presence in the country was denied. By contrast, Mexican-Americans were colonized and their traditional family life was disrupted by war and the imposition of a new set of laws and conditions of labor. The hardships faced by Chicano families, therefore, were the result of the United States colonization of the indigenous Mexican population, accompanied by the beginnings of industrial development in the region. The treaty of Guadalupe Hidalgo, signed in 1848, granted American citizenship to Mexicans living in what is now called the Southwest. The American takeover, however, resulted in the gradual displacement of Mexicans from the land and their incorporation into a colonial labor force (Barrera 1979). In addition, Mexicans who immigrated into the United States after 1848 were also absorbed into the labor force.

Whether natives of Northern Mexico (which became the United States after 1848) or immigrants from Southern Mexico, Chicanos were a largely peasant population whose lives were defined by a feudal economy and a daily struggle on the land for economic survival. Patriarchal families were important instruments of community life and nuclear family units were linked together through an elaborate system of kinship and godparenting. Traditional life was characterized by hard work and a fairly distinct pattern of sex-role segregation....

As the primary caretakers of hearth and home in a rural environment, *Las Chicanas* labor made a vital and important contribution to family survival....

Although some scholars have argued that family rituals and community life showed little change before World War I (Saragoza 1983), the American conquest of Mexican lands, the introduction of a new system of labor, the loss of Mexican-owned land through the inability to document ownership, plus the transient nature of most of the jobs in which Chicanos were employed, resulted in the gradual erosion of this pastoral way of life. Families were uprooted as the economic basis for family life changed. Some immigrated from Mexico in search of a better standard of living and worked in the mines and railroads. Others who were native to the Southwest faced a job market that no longer required their skills and moved into mining, railroad, and agricultural labor in search of a means of earning a living. According to Camarillo (1979), the influx of Anglo[4] capital into the pastoral economy of Santa Barbara rendered obsolete the skills of many Chicano males who had worked as ranchhands and farmers prior to the urbanization of that economy. While some women and children accompanied their husbands to the railroad and mine camps, they often did so despite prohibitions against it. Initially many of these camps discouraged or prohibited family settlement.

The American period (post-1848) was characterized by considerable transiency for the Chicano population. Its impact on families is seen in the growth of female-headed households, which was reflected in the data as early as 1860. Griswold del Castillo (1979) found a sharp increase in female-headed households in Los Angeles, from a low of 13 percent in 1844 to 31 percent in 1880. Camarillo (1979, p. 120) documents a similar increase in Santa Barbara from 15 percent in 1844 to 30 percent by 1880. These increases appear to be due not so much to divorce, which was infrequent in this Catholic population, but to widowhood and temporary abandonment in search of work. Given the hazardous nature of work in the mines and railroad camps, the death of a husband, father or son who was laboring in these sites was not uncommon. Griswold del Castillo (1979) reports a higher death rate among men than women in Los Angeles. The rise in female-headed households, therefore, reflects the instabilities and insecurities introduced into women's lives as a result of the changing social organization of work.

One outcome, the increasing participation of women and children in the labor force was primarily a response to economic factors that required the modification of traditional values....

Slowly, entire families were encouraged to go to railroad workcamps and were eventually incorporated into the agricultural labor market. This was a response both to the extremely low wages paid to Chicano laborers and to the preferences of employers who saw family labor as a way of stabilizing the work-force. For Chicanos, engaging all family members in agricultural work was a means of increasing their earnings to a level close to subsistence for the entire group and of keeping the family unit together....

While the extended family has remained an important element of Chicano life, it was eroded in the American period in several ways. Griswold del Castillo (1979), for example, points out that in 1845 about 71 percent of Angelenos lived in extended families and that by 1880, fewer than half did. This decrease in extended families appears to be a response to the changed economic conditions and to the instabilities generated by the new sociopolitical structure. Additionally, the imposition of American law and custom ignored and ultimately undermined some aspects of the extended family. The extended family in traditional Mexican life consisted of an important set of familial, religious, and community obligations. Women, while valued primarily for their domesticity, had certain legal and prop-erty rights that acknowledged the importance of their work, their families of origin and their children....

In the face of the legal, social, and economic changes that occurred during the American period, Chicanas were forced to cope with a series of dislocations in traditional life. They were caught between conflicting pressures to maintain traditional women's roles and family customs and the need to participate in the economic support of their families by working outside the home. During this period the preservation of some traditional customs became an important force for resisting complete disarray....

Of vital importance to the integrity of traditional culture was the perpetua-tion of the Spanish language. Factors that aided in the maintenance of other aspects of Mexican culture also helped in sustaining the language. However, entry into English-language public schools introduced the children and their families to systematic efforts to erase their native tongue....

Another key factor in conserving Chicano culture was the extended family network, particularly the system of *compadrazgo* or godparenting. Although the full extent of the impact of the American period on the Chicano extended family is not known, it is generally acknowledged that this family system, though lacking many legal and social sanctions, played an important role in the preser-vation of the Mexican community (Camarillo 1979, p. 13). In Mexican society, godparents were an important way of linking family and community through respected friends or authorities. Named at the important rites of passage in a child's life, such as birth, confirmation, first communion, and marriage, *compa-drazgo* created a moral obligation for godparents to act as guardians, to provide financial assistance in times of need, and to substitute in case of the death of a parent. Camarillo (1979) points out that in traditional society these bonds cut across class and racial lines....

The extended family network—which included godparents—expanded the support groups for women who were widowed or temporarily abandoned and for those who were in seasonal, part-, or full-time work. It suggests, therefore, the potential for an exchange of services among poor people whose income did not provide the basis for family subsistence.... This family form is important to the continued cultural autonomy of the Chicano community.

CONCLUSION: OUR MOTHERS' GRIEF

Reproductive labor for Afro-American, Chinese-American, and Mexican-American women in the nineteenth century centered on the struggle to maintain family units in the face of a variety of cultural assaults. Treated primarily as individual units of labor rather than as members of family groups, these women labored to maintain, sustain, stabilize, and reproduce their families while working in both the public (productive) and private (reproductive) spheres. Thus, the concept of reproductive labor, when applied to women of color, must be modified to account for the fact that labor in the productive sphere was required to achieve even minimal levels of family subsistence. Long after industrialization had begun to reshape family roles among middle-class white families, driving white women into a cult of domesticity, women of color were coping with an extended day. This day included subsistence labor outside the family and domestic labor within the family. For slaves, domestics, migrant farm laborers, seasonal factory-workers, and prostitutes, the distinctions between labor that reproduced family life and which economically sustained it were minimized. The expanded workday was one of the primary ways in which reproductive labor increased.

Racial-ethnic families were sustained and maintained in the face of various forms of disruption. Yet racial-ethnic women and their families paid a high price in the process. High rates of infant mortality, a shortened life span, the early onset of crippling and debilitating disease provided some insight into the costs of survival.

The poor quality of housing and the neglect of communities further increased reproductive labor. Not only did racial-ethnic women work hard outside the home for a mere subsistence, they worked very hard inside the home to achieve even minimal standards of privacy and cleanliness. They were continually faced with disease and illness that directly resulted from the absence of basic sanitation. The fact that some African women murdered their children to prevent them from becoming slaves is an indication of the emotional strain associated with bearing and raising children while participating in the colonial labor system.

We have uncovered little information about the use of birth control, the prevalence of infanticide, or the motivations that may have generated these or other behaviors. We can surmise, however, that no matter how much children were accepted, loved, or valued among any of these groups of people, their futures in a colonial labor system were a source of grief for their mothers. For those children who were born, the task of keeping them alive, of helping them to understand and participate in a system that exploited them, and the challenge of maintaining a measure—no matter how small—of cultural integrity, intensified reproductive labor.

Being a racial-ethnic woman in nineteenth-century American society meant having extra work both inside and outside the home. It meant having a contradictory relationship to the norms and values about women that were being generated in the dominant white culture. As pointed out earlier, the notion of separate spheres of male and female labor had contradictory outcomes for the nineteenth-century whites. It was the basis for the confinement of women to the household and for much of the protective legislation that subsequently developed. At the same time, it sustained white families by providing social acknowledgment and support to women in the performance of their family roles. For racial-ethnic women, however, the notion of separate spheres served to reinforce their subordinate status and became, in effect, another assault. As they increased their work outside the home, they were forced into a productive labor sphere that was organized for men and "desperate" women who were so unfortunate or immoral that they could not confine their work to the domestic sphere. In the productive sphere, racial-ethnic women faced exploitative jobs and depressed wages. In the reproductive sphere, however, they were denied the opportunity to embrace the dominant ideological definition of "good" wife or mother. In essence, they were faced with a double-bind situation, one that required their participation in the labor force to sustain family life but damned them as women, wives, and mothers because they did not confine their labor to the home. Thus, the conflict between ideology and reality in the lives of racial-ethnic women during the nineteenth century sets the stage for stereotypes, issues of self-esteem, and conflicts around gender-role prescriptions that surface more fully in the twentieth century. Further, the tensions and conflicts that characterized their lives during this period provided the impulse for community activism to jointly address the inequities, which they and their children and families faced.

NOTES

1. The term *reproductive labor* is used to refer to all of the work of women in the home. This includes but is not limited to: the buying and preparation of food and clothing, provision of emotional support and nurturance for all family members, bearing children, and planning, organizing, and carrying out a wide variety of tasks associated with their socialization. All of these activities are necessary for the growth of patriarchal capitalism because they maintain, sustain, stabilize, and *reproduce* (both biologically and socially) the labor force.

2. The term *white* is a global construct used to characterize peoples of European descent who migrated to and helped colonize America. In the seventeenth century, most of these immigrants were from the British Isles. However, during the time period covered by this article, European immigrants became increasingly diverse. It is a limitation of this article that time and space does not permit a fuller discussion of the variations in the white European immigrant experience. For the purposes of the argument made herein and of the contrast it seeks to draw between the experiences of mainstream (European) cultural groups and that of racial/ethnic minorities, the differences among European settlers are joined and the broad similarities emphasized.

3. Cultural assaults, according to Caulfield (1974), are benign and systematic attacks on the institutions and forms of social organization that are fundamental to the maintenance and flourishing of a group's culture.

4. This term is used to refer to white Americans of European ancestry.

REFERENCES

Barrera, Mario. 1979. *Race and Class in the Southwest*. South Bend, IN: Notre Dame University Press.

Blassingame, John. 1972. *The Slave Community: Plantation Life in the Antebellum South*. New York: Oxford University Press.

Blassingame, John. 1977. *Slave Testimony: Two Centuries of Letters, Speeches, Interviews, and Autobiographies*. Baton Rouge, LA: Louisiana State University Press.

Camarillo, Albert. 1979. *Chicanos in a Changing Society*. Cambridge, MA: Harvard University Press.

Caulfield, Mina Davis. 1974. "Imperialism, the Family, and Cultures of Resistance." *Socialist Review* 4(2)(October): 67–85.

Davis, Angela. 1971. "The Black Woman's Role in the Community of Slaves." *Black Scholar* 3(4)(December): 2–15.

Degler, Carl. 1980. *At Odds*. New York: Oxford University Press.

Genovese, Eugene D., and Elinor Miller, eds. 1974. *Plantation, Town, and County: Essays on the Local History of American Slave Society*. Urbana: University of Illinois Press.

Griswold del Castillo, Richard. 1979. *The Los Angeles Barrio: 1850–1890*. Los Angeles: The University of California Press.

Gutman, Herbert. 1976. *The Black Family in Slavery and Freedom: 1750–1925*. New York: Pantheon.

Hirata, Lucie Cheng. 1979. "Free, Indentured, Enslaved: Chinese Prostitutes in Nineteenth-Century America." *Signs* 5 (Autumn): 3–29.

Jones, Jacqueline. 1985. *Labor of Love, Labor of Sorrow*. New York: Basic Books.

Kessler-Harris, Alice. 1981. *Women Have Always Worked*. Old Westbury: The Feminist Press.

Matthaei, Julie. 1982. *An Economic History of Women in America*. New York: Schocken Books.

Nee, Victor G., and Brett de Bary Nee. 1973. *Longtime Californ'*. New York: Pantheon Books.

Saragoza, Alex M. 1983. "The Conceptualization of the History of the Chicano Family: Work, Family, and Migration in Chicanos." Research Proceedings of the Symposium on Chicano Research and Public Policy. Stanford, CA: Stanford University, Center for Chicano Research.

Sudarkasa, Niara. 1981. "Interpreting the African Heritage in Afro-American Family Organization." Pp. 37–53 in *Black Families*, edited by Harriette Pipes McAdoo. Beverly Hills, CA: Sage Publications.

Wright, Deborah Gray. 1985. *Ar'n't I a Woman?: Female Slaves in the Plantation South*. New York: W. W. Norton.

Wright, Gwendolyn. 1981. *Building the Dream: A Social History of Housing in America*. New York: Pantheon Books.

38

Rethinking Families and Community

The Color, Class, and Centrality of Extended Kin Ties

NAOMI GERSTEL

* * *

THE FOCUS ON NUCLEAR FAMILIES

When talking about family obligations and solidarities, social critics of both the left and the right typically focus on the ties between spouses and between parents and their children, especially when those children are young. Take no less a figure than Barack Obama. When he talks about the value of families (which he does fairly often), he—like most other politicians—tends to focus on the nuclear family. Recently he said: "Of all the rocks upon which we build our lives, we are reminded today that family is the most important." Obama elaborated by saying: "too many fathers are missing" and pointed out that half of all black children live in single-parent households. He concluded by saying that "the foundations of our community are weaker because of it."

Instead of asking whether his claims about the effects of having a father in the house—whether moral or empirical—are true, note the irony of Obama's emphasis on the intact nuclear family. This emphasis flies in the face of his own much celebrated experience as a man raised by a single mother and grandparents, with a famously absent father, who nonetheless attended an Ivy League college and a prestigious law school before becoming senator and president. Little has been made of the fact that he now lives with his mother-in-law. Few note that the White House now contains an extended family. Obama's much cited defense of the nuclear family but his rare mention of the extended family contradicts his own experience. Many others share his claims and stated assumptions.

… Another example comes from a very different sort of organization: Focus on the Family, the group that ran the antiabortion ad in the recent Super Bowl. The group is led by evangelical James Dobson and emphasizes that we need to return to what he calls the "traditional family." What is that family? Husband and wife, parents and young children. It is not just that Dobson leaves out single parents and gay parents; he also leaves out adult children, siblings, aunts, uncles, cousins, and grandparents.

SOURCE: Gertsel, Naomi. 2011. "Rethinking Families and Community: The Color, Class, and Centrality of Extended Kin Ties." *Sociological Forum* 26 (March): 1–20. Copyright © 2011 John Wiley and Sons, Inc. Reprinted by permission.

We can even look at the focus of the U.S. government. Although the U.S. Census defines a family as those individuals related by blood, marriage, or adoption who reside together, a recent 2010 report from the U.S. Census addresses the effects of unemployment and the recession on families and households even more narrowly. The report suggested: "These statistics show us that families are having a difficult time during this recession" (U.S. Census Bureau News, 2010:1). But the report was entirely about husbands and wives (or ex-husbands and ex-wives) and parents and young children. There were no parents of adult children, no sisters or brothers, no aunts, no uncles, no grandparents (although as this article will show, these are the very people on whom those who lose their jobs often rely).

These examples should make it clear that there is a lot of talk about family these days, but that talk is focused narrowly—on marriage and nuclear families, that is, husbands and wives, or parents and minor children.

THE REALITY OF FAMILIES TODAY: FAMILY EXPERIENCE AND PRACTICES

This focus on marriage and the nuclear family contains strong racial and ethnic— as well as class—biases. On the one hand, beginning with race, researchers find that blacks and Latinos/as have lower rates of marriage and higher rates of single motherhood than non-Hispanic whites, which is why many social critics say their families are more disorganized than white families. But this is the case *only* if what we mean by family is limited to marriage and the nuclear family. This very narrow view of families, when translated into policy, often penalizes the very people it purports to help.

The image of the independent nuclear family is particularly misleading for understanding the lives of people of color and poor people. Using the National Survey of Families and Households, Natalia Sarkisian, Mariana Gerena, and I find that minority individuals—in particular, blacks and Latino/as—rely on extended kin more than do whites. For example, looking at co-residence, approximately 40% of adult blacks and about a third of Latino/as—compared to under a fifth of whites—share households with relatives other than partners or young children. Similar patterns exist for living near relatives: Over half of blacks and Latino/as compared to only about a third of whites live within two miles of kin. Similar patterns exist for visiting: these same data show blacks and Latino/as visit kin more frequently than do whites. As for giving and receiving care, blacks and Latino/as are also more likely to rely on their relatives. Although they are less likely to give money and emotional support, blacks and Latino/as are more likely than whites to give relatives hands-on practical assistance, such as help with household work and childcare, with rides, and running errands. The same racial and ethnic disparities hold for receiving care: blacks and Latinos/as are much more likely than whites to receive such practical care from a wide range of kin.

These racial and ethnic differences are especially pronounced among women. However, black and Latino men are also involved with kin, in contrast to what is implied by pervasive images of minority men. In fact, black and Latino men are more likely than white men to live near relatives and to stay in touch with them, although white men are more likely to give and receive large-scale financial support (Sarkisian, 2007).

As these data suggest, if we only consider the care provided by husbands and wives to each other or by parents to their young children, we are missing much of the experience of families. We are especially likely to miss the experience of families of African Americans and Latino/as. Worse, this focus reinforces what Dorothy Smith (1993) called "SNAF—the Standard North American Family"—an ideological code that derides, stigmatizes, or abandons those who do not or cannot follow it.

WHY DO EXTENDED KIN MATTER?
STRATEGIES OF SURVIVAL

These comparisons suggest that an emphasis on the nuclear family is inadequate to describe the realities of family life. But why are extended kin more important among people of color? Analysis of the National Survey of Families and Households suggests these differences exist because of economic differences among racial groups: blacks and Latino/as less often have the economic resources that allow the kind of privatization that the nuclear family entails. Extended kinship, then, is a survival strategy in the face of economic difficulties.

Individuals' lack of economic resources increases their need for help from kin *and* boosts their willingness to give help in return. White, black, and Latino/a individuals with the same amount of income and education have similar patterns of involvement with their extended families. For example, just like poor minorities, impoverished whites more often engage in mutual, practical help with extended kin than do their wealthier counterparts. Just like middle-class whites, middle-class blacks and Latino/as are more likely to exchange advice or share money with relatives than are their poorer counterparts. It is the relative economic deprivation of racial/ethnic minorities that leads to higher levels of extended family involvement. Primarily because blacks and Latinos/as have less money and education than whites, they are more likely to give and receive help from kin.

As this analysis suggests, kinship is a strategy for survival. That is, helping others with practical matters and receiving help from them is a *class*-based strategy for survival. Class trumps race in this regard.

★★★

THE SOCIAL IMPLICATIONS
OF A FOCUS ON MARRIAGE

Thus far, this article has argued that an emphasis on the nuclear family both misses a great deal of what goes on in many families and may denigrate the family life of the poor and near poor. There are still other implications of this argument: marriage and the nuclear family cut off both women and men from extended and fictive kin. Marriage and the nuclear family do so for the wealthy and the poor, the middle class and the working class.

This argument is a significant departure from classical accounts. Theorists beginning with Durkheim have viewed marriage as fostering attachment to wider communities and heightening investment in social ties ([1897] 1966).[1] For many observers, the nuclear family is not only the basic unit of society, but marriage is also the foundation of community. This article suggests otherwise by analyzing how marriage and the nuclear family actually detract from rather than bolster ties to other kin.[2]

Married people—women as well as men—are less involved with their extended kin than those who are never married or previously married.[3] This includes many different types of involvement, including co-residence, frequency of conversations and visiting, and giving as well as receiving different types of care. Not only are the married less likely to live with relatives than the unmarried, but they are also are less likely to visit relatives with whom they do not share a home. Married adult children take care of elderly parents less often than their unmarried siblings; they are less likely to take them places or help them with household chores. Compared to those without a spouse, married adult children have less contact, and exchange less help, with their sisters and brothers. Married mothers, whether black, Latina, or white, are often unable to obtain help from kin in the way that their single counterparts can. They are also less involved with their neighbors and friends.

One might argue this difference between those married and those not results from the fact that the married are older or because married children are more likely to have children of their own. But that is not the case. These differences between the married and unmarried exist both among parents of young children and among the childless. Moreover, the disparities exist across race and class.

… Marriage clearly has troublesome implications for the community that are often overlooked. As the population ages, the greediness of marriage deprives more elderly parents—who, ironically, have often pressed their children to marry—of the help and support that they want and need. Marriage can also generate excessive burdens on those who are single, as they are expected to provide the care that their married siblings do not. Although marriage is greedy across race and class, because those with fewer economic resources are more likely to rely on extended kin, this is for them a particularly costly outcome. Thus, not

---

1. See, for example, Slater (1963).
2. These analyses are based on the analyses of a number of different data sets—some national, some local, some qualitative, some quantitative—over the last couple of decades. See Gallagher and Gerstel (1993), Gerstel (1988), Gerstel and Sarkisian (2006a, 2007, 2010a), and Sarkisian and Gerstel (2008).
3. See Gerstel and Sarkisian (2010b) for an analysis of ties of the married, never married, and previously married. We find that both unmarried groups have more social ties than the married but the previously married have fewer ties than the never married—suggesting that marriage continues to exert some effect after marriages dissolve.

only is the focus on marriage a narrow vision, but it may actually detract from the very resources—rooted outside the nuclear family and marriage—on which Americans depend. In addition, reduced ties to others may put a strain on marriage itself. As marriage becomes the only place where individuals look for support and comfort, marriages may become fragile.

SOCIAL POLICIES AND THE CONSTITUTION OF FAMILIES

Much U.S. policy today is based on a set of assumptions about the priority of marriage and the nuclear family. In contrast, the extended family is all but invisible. This invisibility makes public policy less effective than it would be otherwise. Policy based on the nuclear family works to the benefit of those already privileged. At the same time, it turns family into a privilege rather than a right.

... Social policies that overlook extended family obligations may introduce, reproduce, or even increase class and racial/ethnic inequalities. To create unbiased, inclusive policy, we must attend to the realities and practices of families and the care they actually give and receive rather than confine our policies to narrow biological or conjugal conceptions of who should count as family.

REFERENCES

Armenia, Amy, and Naomi Gerstel. 2006. "Family Leaves, the FMLA, and Gender Neutrality: The Intersection of Race and Gender," *Social Science Research* 35: 871–891.

Badgett, M. V. Lee. 2009. *When Gay People Get Married.* New York: New York University Press.

Carlson, Marcia, Sara McLanahan, and Paula England. 2004. "Union Formation in Fragile Families," *Demography* 41: 237–261.

Clawson, Dan, Naomi Gerstel, and Jillian Crocker. 2008. "Employers Meet Families: Gender, Class, and Paid Work Hour Differences Among Four Occupations," *Social Indicators Research* December.

Coser, Lewis, with Rose Coser. 1974. *Greedy Institutions: Patterns of Undivided Commitment.* New York: Free Press.

Crocker, Jillian, Ming Li, Naomi Gerstel, and Dan Clawson. 2010. "Unofficial Flexibility at a Nursing Home: Nurses, Nursing Assistants, and Variations in Work Schedules and Hours," paper presented at the Eastern Sociological Society Meetings, Boston, MA, March.

Duggan, Lisa. 2003. *The Twilight of Equality?: Neoliberalism, Cultural Politics and the Attack on Democracy.* New York: Beacon Press Books.

Durkheim, Emile. 1966. *Suicide: A Study in Sociology*, John A. Spaulding and George Simpson (trans.). New York: Free Press (Orig. pub. 1897).

Edin, Kathyrn, and Maria Kefalsas. 2007. *Promises I Can Keep: Why Poor Women Put Motherhood Before Marriage.* Berkeley: University of California Press.

Edin, Kathryn, Laura Tach, and Ronald Mincy. 2009. "Claiming Fatherhood: Race and the Dynamics of Paternal Involvement Among Unmarried Men," *Annals of the American Academy of Political and Social Science* 621: 149–177.

Elving, Ronald. 1995. *Conflict and Compromise: How Congress Makes the Law*. New York: Touchstone.

Elwert, Felix, and Nicholas A. Christakis. 2006. "Widowhood and Race," *American Sociological Review* 71: 16–41.

Essig, Laurie, and Lynn Owens. 2009. "What if Marriage Is Bad for Us?" *Chronicle of Higher Education* October 5.

Feldblum, Chai, and Robin Appleberry. 2006. "Legislatures, Agencies, Courts and Advocates: How Laws Are Made, Interpreted and Modified," in Marcie Pitt-Catsouphes, Ellen Kossek, and Stephen Sweet (eds.), *The Work and Family Handbook: Multi-Disciplinary Perspectives, Methods and Approaches*: pp. 627–650. New York: Lawrence Erlbaum.

Fornby, Paula, and Andrew Cherlin. 2007. "Family Instability and Child Well-Being," *American Sociological Review* 72: 181–204.

Furstenberg, Frank. 2009. "If Moynihan Had Only Known: Race, Class, and Family Change in the Late 20th Century," *Annals of the American Academy of Political Science and Social Science* 621: 94–110.

Gallagher, Sally, and Naomi Gerstel. 1993. "The Work of Kinkeeping and Friend Keeping: The Effects of Marriage," *Gerontologist* 33: 675–681.

Gerson, Kathleen. 2009. "Changing Lives, Resistant Institutions: A New Generation Negotiates Gender, Work, and Family Change," *Sociological Forum* 24(4): 745–753.

Gerstel, Naomi. 1988. "Divorce, Gender and Social Integration," *Gender & Society* 2: 343–367.

Gerstel, Naomi. 2000. "The Third Shift: Gender, Employment, and Care Work Outside the Home," *Qualitative Sociology* 23: 467–483.

Gerstel, Naomi, and Amy Armenia. 2009. "Giving and Taking Family Leaves: Right or Privilege," *Yale Journal of Law and Feminism* 21: 161–184.

Gerstel, Naomi, Amy Armenia, and Coady Wing. 2010. "Family Leave: Right or Privilege," *New York Times* January 25.

Gerstel, Naomi, Dan Clawson, and Dana Huyser. 2007. "Explaining Job Hours of Physicians, Nurses, EMTs and Nursing Assistants: Gender, Class, Jobs and Family," in B. Rubin (ed.), *Workplace Temporalities*. Amsterdam: JAI, Elsevier Press.

Gerstel, Naomi, and Harriet Gross. 1984. *Commuter Marriage*. New York: Guilford Press.

Gerstel, Naomi, and Kate McGonagle. 1999. "Job Leaves and the Limits of the Family and Medical Leave Act: The Effects of Gender, Race and Family," *Work and Occupations* 26: 510–534.

Gerstel, Naomi, and Natalia Sarkisian. 2006a. "Marriage: The Good, the Bad, and the Greedy," *Contexts* 5: 16–22.

Gerstel, Naomi, and Natalia Sarkisian. 2006b. "Sociological Perspective on Families and Work: The Import of Gender, Race, Ethnicity, and Class," in M. P. Catsouphes, E. Kossek, and S. Sweet (eds.), *The Work and Family Handbook: Multi-Disciplinary Perspectives, Methods, and Approaches*: pp. 237–266. New York: Routledge.

Gerstel, Naomi, and Natalia Sarkisian. 2007. "Adult Children's Relationship to Their Parents: The Greediness of Marriage," in Timothy J. Owens and J. Jill Suitor (eds.),

Advances in the Life Course Volume 7: Interpersonal Relations Across the Life Course: pp. 153–188. New York: Elsevier.

Gerstel, Naomi, and Natalia Sarkisian. 2008. "The Color of Family Ties: Race, Class, Gender, and Extended Family Involvement," in Stephanie Coontz, Maya Parson, and Gabrielle Rayley (eds.), *American Families: A Multicultural Reader*: pp. 447–453. New York: Routledge.

Gerstel, Naomi, and Natalia Sarkisian. 2010a. "Marriage Reduces Social Ties," in Barbara Risman (ed.), *Families as They Really Are*. New York: Norton.

Gerstel, Naomi, and Natalia Sarkisian. 2010b. "Marriage and Social Integration: Building Block or Greedy Institution?" paper under review.

Goodwin, Jeff. 1997. "The Libidinal Constitution of a High Risk Social Movement: Affectual Ties and Solidarity in the Huk Rebellion, 1946–1954," *American Sociological Review* 62: 53–69.

Hansen, Karen. 2005. *The Not-So-Nuclear Family*. New Brunswick, NJ: Rutgers University Press.

Hochschild, Arlie. 1989. *The Second Shift: Working Parents and the Revolution at Home*. New York: Viking Penguin.

Landale, Nancy, R. S. Oropesa, and C. Bradatan. 2006. "Hispanic Families in the United States: Family Structure and Process in an Era of Family Change," in Marta Tienda (ed.), *Hispanics and the Future of America*: pp. 138–178 Washington, DC: National Academy Press.

Lareau, Annette. 2003. *Unequal Childhoods: Class, Race and Family Life*. Berkeley: University of California Press.

McAdam, Doug. 1986. "Recruitment to High-Risk Activism: The Case of Freedom Summer," *American Journal of Sociology* 92: 64–90.

Nelson, Margaret. 2005. *The Social Economy of Single Motherhood: Raising Children in Rural America*. New York: Routledge.

Raley, R. Kelly. 2000. "Recent Trends and Differentials in Marriage and Cohabitation: The United States," in Linda Waite (ed.), *The Ties that Bind: Perspectives on Marriage and Cohabitation*: pp. 19–39. New York: Aldine de Gruyter.

Riley, Matilda, and J. W. Rile. 1993. "Connections: Kin and Cohort," in V. Bengston and A. Aschenbaum (eds.), *The Changing Contract Across Generations*. New York: Aldine De Gruyter.

Sampson, Robert J., John Laub, and Christopher Wimer. 2006. "Does Marriage Reduce Crime? A Counterfactual Approach to Within Individual Causal Effects?" *Criminology* 44: 465–508.

Sarkisian, Natalia. 2007. "Street Men, Family Men: Race and Men's Extended Family Involvement," *Social Forces* 86: 763–794.

Sarkisian, Natalia, Mariana Gerena, and Naomi Gerstel. 2006. "Extended Family Ties Among Mexicans, Puerto Ricans, and Whites: Superintegration or Disintegration?" *Family Relations* 55: 331–344.

Sarkisian, Natalia, Mariana Gerena, and Naomi Gerstel. 2007. "Extended Family Integration Among Euro and Mexican Americans: Ethnicity, Gender, and Class," *Journal of Marriage and Family* 69: 40–54.

Sarkisian, Natalia, and Naomi Gerstel. 2005. "Kin Support Among Blacks and Whites: Race and Family Organization," *American Sociological Review* 69: 812–837.

Sarkisian, Natalia, and Naomi Gerstel. 2008. "Till Marriage Do Us Part: Adult Children's Relationship with Their Parents," *Journal of Marriage and Family* 70: 360–376.

Shows, Carla, and Naomi Gerstel. 2009. "Fathering, Class, and Gender: A Comparison of Physicians and EMTs.," *Gender & Society* 23: 161–187.

Slater, Phillip. 1963. "On Social Regression," *American Sociological Review* 28: 339–364.

Smith, Dorothy. 1993. "The Standard North American Family," *Journal of Family Issues* 14: 50–65.

Stack, Carol. 1974. *All My Kin: Strategies for Survival in a Black Community*. New York: Harper & Row.

Stone, Pamela. 2007. *Opting Out? Why Women Really Quit Careers and Head Home*. Berkeley: University of California Press.

Swarz, Teresa. 2009. "Intergenerational Family Relations in Adulthood: Patterns, Variations, and Implications in the Contemporary United States," *Annual Review of Sociology* 35: 191–212.

U.S. Census Bureau News. 2010. *Census Bureau Reports Families with Children Increasingly Face Unemployment*. Washington, DC: U.S. Department of Commerce.

Wiltfang, Gregory, and Doug McAdam. 1991. "Distinguishing Cost and Risk in Sanctuary Activism," *Social Forces* 69: 987–1010.

39

Straight Is to Gay as Family Is to No Family

KATH WESTON

IS "STRAIGHT" TO "GAY" AS "FAMILY" IS TO "NO FAMILY"?

For years, and in an amazing variety of contexts, claiming a lesbian or gay identity has been portrayed as a rejection of "the family" and a departure from kinship. In media portrayals of AIDS, Simon Watney ... observes that "we are invited to imagine some absolute divide between the two domains of 'gay life' and 'the family,' as if gay men grew up, were educated, worked and lived our lives in total isolation from the rest of society." Two presuppositions lend a dubious credence to such imagery: the belief that gay men and lesbians do not have children or establish lasting relationships, and the belief that they invariably alienate adoptive and blood kin once their sexual identities become known. By presenting "the family" as a unitary object, these depictions also imply that everyone participates in identical sorts of kinship relations and subscribes to one universally agreed-upon definition of family.

Representations that exclude lesbians and gay men from "the family" invoke what Blanche Wiesen Cook ... has called "the assumption that gay people do not love and do not work," the reduction of lesbians and gay men to sexual identity, and sexual identity to sex alone. In the United States, sex apart from heterosexual marriage tends to introduce a wild card into social relations, signifying unbridled lust and the limits of individualism. If heterosexual intercourse can bring people into enduring association via the creation of kinship ties, lesbian and gay sexuality in these depictions isolates individuals from one another rather than weaving them into a social fabric. To assert that straight people "naturally" have access to family, while gay people are destined to move toward a future of solitude and loneliness, is not only to tie kinship closely to procreation, but also to treat gay men and lesbians as members of a nonprocreative species set apart from the rest of humanity....

It is but a short step from positioning lesbians and gay men somewhere beyond "the family"—unencumbered by relations of kinship, responsibility, or

SOURCE: From Kath Weston, *Families We Choose: Lesbians, Gays, Kinship* (New York: Columbia University Press, 1991), pp. 22–29. Reprinted by permission of Columbia University Press.

affection—to portraying them as a menace to family and society. A person or group must first be outside and other in order to invade, endanger, and threaten. My own impression from fieldwork corroborates Frances Fitzgerald's ... observation that many heterosexuals believe not only that gay people have gained considerable political power, but also that the absolute number of lesbians and gay men (rather than their visibility) has increased in recent years. Inflammatory rhetoric that plays on fears about the "spread" of gay identity and of AIDS finds a disturbing parallel in the imagery used by fascists to describe syphilis at mid-century, when "the healthy" confronted "the degenerate" while the fate of civilization hung in the balance....

A long sociological tradition in the United States of studying "the family" under siege or in various states of dissolution lent credibility to charges that this institution required protection from "the homosexual threat."...

... By shifting without signal between reproduction's meaning of physical procreation and its sense as the perpetuation of society as a whole, the character-ization of lesbians and gay men as nonreproductive beings links their supposed attacks on "the family" to attacks on society in the broadest sense. Speaking of parents who had refused to accept her lesbian identity, a Jewish woman explained, "They feel like I'm finishing off Hitler's job." The plausibility of the contention that gay people pose a threat to "the family" (and, through the family, to ethnicity) depends upon a view of family grounded in heterosexual relations, combined with the conviction that gay men and lesbians are incapable of procreation, parenting, and establishing kinship ties.

Some lesbians and gay men ... had embraced the popular equation of their sexual identities with the renunciation of access to kinship, particularly when first coming out. "My image of gay life was very lonely, very weird, no family," Rafael Ortiz recollected. "I assumed that my family was gone now—that's it." After Bob Korkowski began to call himself gay, he wrote a series of poems in which an orphan was the central character. Bob said the poetry expressed his fear of "having to give up my family because I was queer." When I spoke with Rona Bren after she had been home with the flu, she told me that whenever she was sick, she relived old fears. That day she had remembered her mother's grim prediction: "You'll be a lesbian and you'll be alone the rest of your life. Even a dog shouldn't be alone."

Looking backward and forward across the life cycle, people who equated their adoption of a lesbian or gay identity with a renunciation of family did so in the double-sided sense of fearing rejection by the families in which they had grown up, and not expecting to marry or have children as adults. Although few in num-bers, there were still those who had considered "going straight" or getting married specifically in order to "have a family." Vic Kochifos thought he understood why:

> It's a whole lot easier being straight in the world than it is being gay....
> You have built-in loved ones: wife, husband, kids, extended family. It
> just works easier. And when you want to do something that requires
> children, and you want to have a feeling of knowing that there's gonna
> be someone around who cares about you when you're 85 years old, there
> are thoughts that go through your head, sure. There must be. There's a
> way of doing it gay, but it's a whole lot harder, and it's less secure.

Bernie Margolis had been sexually involved with men since he was in his teens, but for years had been married to a woman with whom he had several children. At age 67 he regretted having grown to adulthood before the current discussion of gay families, with its focus on redefining kinship and constructing new sorts of parenting arrangements.

> I didn't want to give up the possibility of becoming a family person. Of having kids of my own to carry on whatever I built up.... My mother was always talking about [how] she's looking forward to the day when she would bring her children under the canopy to get married. It never occurred to her that I wouldn't be married. It probably never occurred to me either.

The very categories "good family person" and "good family man" had seemed to Bernie intrinsically opposed to a gay identity. In his fifties at the time I interviewed him, Stephen Richter attributed never having become a father to "not having the relationship with the woman." Because he had envisioned parenting and procreation only in the context of a heterosexual relationship, regarding the two as completely bound up with one another, Stephen had never considered children an option.

Older gay men and lesbians were not the only ones whose adult lives had been shaped by ideologies that banish gay people from the domain of kinship. Explaining why he felt uncomfortable participating in "family occasions," a young man who had no particular interest in raising a child commented, "When families get together, what do they talk about? Who's getting married, who's having children. And who's not, okay? Well, look who's not." Very few of the lesbians and gay men I met believed that claiming a gay identity automatically requires leaving kinship behind. In some cases people described this equation as an outmoded view that contrasted sharply with revised notions of what constitutes a family.

Well-meaning defenders of lesbian and gay identity sometimes assert that gays are not inherently "anti-family," in ways that perpetuate the association of heterosexual identity with exclusive access to kinship. Charles Silverstein ..., for instance, contends that lesbians and gay men may place more importance on maintaining family ties than heterosexuals do because gay people do not marry and raise children. Here the affirmation that gays and lesbians are capable of fostering enduring kinship ties ends up reinforcing the implication that they cannot establish "families of their own," presumably because the author regards kinship as unshakably rooted in heterosexual alliance and procreation. In contrast, discourse on gay families cuts across the politically loaded couplet of "pro-family" and "anti-family" that places gay men and lesbians in an inherently antagonistic relation to kinship solely on the basis of their nonprocreative sexualities. "Homosexuality is not what is breaking up the Black family," declared Barbara Smith ..., a black lesbian writer, activist, and speaker at the 1987 Gay and Lesbian March on Washington. "Homophobia is. My Black gay brothers and my Black lesbian sisters are members of Black families, both the ones we were born into and the ones we create."

At the height of gay liberation, activists had attempted to develop alternatives to "the family," whereas by the 1980s many lesbians and gay men were struggling to legitimate gay families as a form of kinship.... Gay or chosen families might incorporate friends, lovers, or children, in any combination. Organized through ideologies of love, choice, and creation, gay families have been defined through a contrast with what many gay men and lesbians ... called "straight," "biological," or "blood" family. If families we choose were the families lesbians and gay men created for themselves, straight family represented the families in which most had grown to adulthood.

What does it mean to say that these two categories of family have been defined through contrast? One thing it emphatically does *not* mean is that heterosexuals share a single coherent form of family (although some of the lesbians and gay men doing the defining believed this to be the case). I am not arguing here for the existence of some central, unified kinship system vis-à-vis which gay people have distinguished their own practice and understanding of family. In the United States, race, class, gender, ethnicity, regional origin, and context all inform differences in household organization, as well as differences in notions of family and what it means to call someone kin.

In any relational definition, the juxtaposition of two terms gives meaning to both. Just as light would not be meaningful without some notion of darkness, so gay or chosen families cannot be understood apart from the families lesbians and gay men call "biological," "blood," or "straight." Like others in their society, most gay people ... considered biology a matter of "natural fact." When they applied the terms "blood" and "biology" to kinship, however, they tended to depict families more consistently organized by procreation, more rigidly grounded in genealogy, and more uniform in their conceptualization than anthropologists know most families to be. For many lesbians and gay men, blood family represented not some naturally given unit that provided a base for all forms of kinship, but rather a procreative principle that organized only one possible *type* of kinship. In their descriptions they situated gay families at the opposite end of a spectrum of determination, subject to no constraints beyond a logic of "free" choice that ordered membership. To the extent that gay men and lesbians mapped "biology" and "choice" onto identities already opposed to one another (straight and gay, respectively), they polarized these two types of family along an axis of sexual identity.

The following chart recapitulates the ideological transformation generated as lesbians and gay men began to inscribe themselves within the domain of kinship.

What this chart presents is not some static substitution set, but a historically motivated succession. To move across or down the chart is to move through

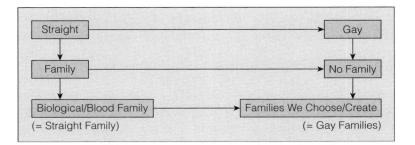

time. Following along from left to right, time appears as process, periodized with reference to the experience of coming out. In the first opposition, coming out defines the transition from a straight to a gay identity. For the person who maintains an exclusively biogenetic notion of kinship, coming out can mark the renunciation of kinship, the shift from "family" to "no family" portrayed in the second opposition. In the third line, individuals who accepted the possibility of gay families after coming out could experience themselves making a transition from the biological or blood families in which they had grown up to the establishment of their own chosen families.

Moving from top to bottom, the chart depicts the historical time that inaugurated contemporary discourse on gay kinship. "Straight" changes from a category with an exclusive claim on kinship to an identity allied with a specific kind of family symbolized by biology or blood. Lesbians and gay men, originally relegated to the status of people without family, later lay claim to a distinctive type of family characterized as families we choose or create. While dominant cultural representations have asserted that straight is to gay as family is to no family (lines 1 and 2), at a certain point in history gay people began to contend that straight is to gay as blood family is to chosen families (lines 1 and 3).

What provided the impetus for this ideological shift? Transformations in the relation of lesbians and gay men to kinship are inseparable from sociohistorical developments: changes in the context for disclosing a lesbian or gay identity to others, attempts to build urban gay "community," cultural inferences about relationships between "same-gender" partners, and the lesbian baby boom associated with alternative (artificial) insemination…. If … kinship is something people use to act as well as to think, then its transformations should have unfolded not only on the "big screen" of history, but also on the more modest stage of day-to-day life, where individuals have actively engaged novel ideological distinctions and contested representations that would exclude them from kinship.

40

Navigating Interracial Borders

Black-White Couples and Their Social Worlds

ERICA CHITO CHILDS

The 1967 Academy Award-winning movie *Guess Who's Coming to Dinner* concluded with a warning from a white father to his daughter and her "Negro" fiancé. That same year, the Supreme Court overturned any laws against interracial marriage as unconstitutional. Yet how does the contemporary U.S. racial landscape compare? In this ever-changing world of race and color, where do black-white couples fit, and has this unimaginable opposition disappeared?

While significant changes have occurred in the realm of race relations largely from the civil rights struggle of the 1960s, U.S. society still has racial borders. Most citizens live, work, and socialize with others of the same race—as if living within borders, so to speak—even though there are no longer legal barriers such as separate facilities or laws against intermarriage. Yet if these largely separate racial worlds exist, what social world(s) do black-white couples live in and how do they navigate these racial borders? Even more important, how do white communities and black communities view and respond to black-white couples? In other words, do they navigate the racial borders by enforcing, ignoring, or actively trying to dismantle them? My goal is to explore these issues to better understand the contemporary beliefs and practices surrounding black-white couples.... My data come from varied sources, including Web sites, black-white couples, Hollywood films, white communities, and black communities....

My own story also brought me to this research. The social world of black-white couples is the world I navigate. From my own experiences, I have seen the ways most whites respond to an interracial relationship. Growing up white, second-generation Portuguese in a predominantly white Rhode Island suburb, race never was an issue, or at least not one I heard about. After moving to Los Angeles during high school and beginning college, I entered into a relationship with an African American man (who I eventually married), never imagining what it would bring. My family did not disown me or hurl racial slurs. Still, in many ways I learned what it meant to be an "interracial couple" and how this was not what my family, community, or countless unknown individuals had scripted for me. Not many whites ever said outright that they were opposed to the relationship, yet their words and actions signaled otherwise.

SOURCE: From *Navigating Interracial Borders: Black-White Couples and Their Social World* by Erica Chito Childs. Copyright © 2005. Reprinted by permission of Rutgers University Press.

One of the most telling examples occurred a few years into our relationship. An issue arose when my oldest sister's daughter wanted to attend her prom with an African American schoolmate she was dating. My sister and her husband refused to let him in the house the night of the prom or any other time because, they said, he was "not right" for her. It was clear to everyone, however, that skin color was the problem. To this day, my niece will tell you that her parents would never have accepted her with a black man. Yet my sister and her family never expressed any opposition to my relationship and even seemed supportive, in terms of inviting us over to their house, giving wedding and holiday gifts, and so forth. Although my sister never openly objected to my relationship, she drew the line with her daughter—quite literally enforcing a racial boundary to protect her daughter and family from blackness. For me, this personal story and the countless stories of other interracial couples point to the necessity of examining societal attitudes, beliefs, images, and practices regarding race and, more specifically, black-white relations. Interracial couples—because of their location on the line between white and black—often witness or bring forth racialized responses from both whites and blacks. As with my sister, opposition may exist yet is not visible until a close family member or friend becomes involved or wants to become involved interracially....

INTERRACIAL RELATIONSHIPS AS A MINER'S CANARY

It is these community and societal responses, as well as the images and beliefs produced and reproduced about these unions that provide the framework within which to understand the issue of interracial couplings. Underlying these responses and images is a racial ideology, or, in other words, a dominant discourse, that posits interracial couples and relationships as deviant. Still, the significance of these discourses and what exactly they reveal about race in society can be hard to see. For some of us the effects of race are all too clear, while for others race—and the accompanying advantages and disadvantages—remain invisible. As a white woman, it was only through my relationship, and raising my two children, that I came to see how race permeates everything in society. Being white yet now part of a multiracial family, I experienced, heard, and even thought things much differently than before, primarily because whites and blacks responded to me differently than before. I think of the metaphor of the "miner's canary"—the canaries miners use to alert them to a poisonous atmosphere. In *The Miner's Canary: Enlisting Race, Resisting Power, Transforming Power,* Lani Guinier and Gerald Torres argue that the experiences of racial minorities, like the miner's canary, can expose the underlying problems in society that ultimately affect everyone, not just minorities. In many ways, the experiences of black-white couples are a miner's canary, revealing problems of race that otherwise can remain hidden, especially to whites. The issues surrounding interracial couples—racialized/sexualized stereotypes, perceptions of difference, familial opposition, lack of community acceptance—should not be looked at as individual problems, but rather as a reflection of the larger racial issues that divide the races. Since interracial couples exist on the color-line in society—a "borderland" between white and black—their experiences and the ways communities respond to these relationships can be used as a lens through which we can understand contemporary race relations....

LIFE ON THE BORDER: NARRATIVES OF BLACK-WHITE COUPLES

From 1999 to 2001, I interviewed fifteen black-white heterosexual couples who were referred to me through personal and professional contacts, and some of whom I encountered randomly in public. They ranged in age from twenty to sixty-nine and all were in committed relationships of two to twenty-five years. Nine were married. The couples' education levels varied. All respondents had finished high school or its equivalent; twenty-one respondents had attended some college and/or had received a bachelor's degree; and four respondents had advanced degrees. The socioeconomic status of the couples ranged from working class to upper middle class. The respondents included a college student, waitress, manager, factory worker, university professor, social worker, salesperson, and postal worker. All couples lived in the northeastern United States, from Maine to Pennsylvania, yet many of the couples had traveled extensively and had lived in other parts of the country, including California, Florida, and the South.

I interviewed the couples together, since I was interested in their experiences, accounts, narratives, and the ways they construct their lives and create their "selves" and their identities as "interracial couples." The interviews lasted for two to three hours, and I ended up with more than forty hours of interview data.... These accounts are seen not only as "descriptions, opinions, images, or attitudes about race relations but also as 'systems of knowledge' and 'systems of values' in their own right, used for the discovery and organization of reality."

THE SEPARATE WORLDS OF WHITES AND BLACKS

To explore the larger cultural and sociopolitical meanings that black-white couplings have for both the white and black communities in which they occur, a significant portion of this work is based on original qualitative research in white communities and black communities about their ideas, beliefs, and views on interracial sexuality and marriage. Community research was conducted to further explore the responses to interracial couples that are found in social groups and communities—family, friends, neighbors, religious groups, schools, etc. The ways that these couples provide the occasion for groups to express and play out their ideas and prejudices about race and sex are integral to understanding the social construction of interracial couples....

A black person and a white person coming together has been given many names—miscegenation, amalgamation, race mixing, and jungle fever—conjuring up multiple images of sex, race, and taboo. Black–white relationships and marriages have long been viewed as a sign of improving race relations and assimilation, yet these unions have also been met with opposition from both white and black communities. Overall, there is an inherent assumption that interracial couples are somehow different from same-race couples. Within the United States, the responses to black–white couplings have ranged from disgust to curiosity to endorsement, with the couples being portrayed as many things—among them, deviant, unnatural, pathological, exotic, but always sexual. Even the way that couples are labeled or defined as "interracial" tells us something about societal expectations. We name what is different. For example, a male couple is more likely to be called a "gay couple" than a gender-mixed couple is to be called a "heterosexual couple."

Encompassed by the history of race relations and existing interracial images, how do black–white couples view themselves, their relationships, and the responses of their families and communities? And how do they interpret these familial and community responses? Black–white couples, like all of us, make meaning out of their experiences in the available interpretive frameworks and often inescapable rules of race relations in this country....

Black–white couples come together across the boundaries of race and perceived racial difference seemingly against the opposition of their communities. This is not to say, however, that the couples are free from racialized thinking, whether it be in their use of color-blind discourse or their own racial preferences, such as to date only interracially or to live in all-white neighborhoods. Nonetheless, these couples create multiracial families, not only creating multiracial families of their own but also changing the racial dynamic of the families from which they come. What significance does this have for the institution of family, and how does this play out for the white and black families to whom it occurs?

It might be expected that the family is the source of the greatest hostility toward interracial relationships. It is in families that the meanings and attachments to racial categories are constructed and learned; one's family is often "the most critical site for the generation and reproduction of racial formations." This includes who is and is not an acceptable marriage partner. In white and black families, certain discourses are used when discussing black–white relationships that reproduce the image of these unions as different, deviant, even dangerous. Interracial relationships and marriage often bring forth certain racialized attitudes and beliefs about family and identity that otherwise may have remained hidden. The ways that white and black families understand and respond to black–white interracial couples and the racialized discourses they use are inextricably tied to ideas of family, community, and identity. White and black families' (and communities') interpretations and responses to interracial couples are part of these available discourses on race and race relations in our society. Many times, black–white couples provide the occasion for families to express and play out their ideas and prejudices about race and sex, which is integral to understanding the social construction of "interracial couples" within America today.

ALL IN THE FAMILY: WHITE FAMILIES

Among whites, the issue of interracial marriage is often a controversial topic, and even more so when they are asked to discuss their own views of their family's views. The white community respondents in this study were hesitant to discuss their personal views on family members becoming involved interracially. One strategy used by the respondents was to discuss other families they knew rather than their own views. For example, during the white focus group interviews, the first responses came from two individuals who had some experience or knowledge of interracial couples or families. Sara discussed a friend who adopted two "very dark" black children and how white people would stare at her and the children in public. Anne mentioned her niece who married a black man and said "the family is definitely against it," which causes them problems.

In group interviews with white college students, a number of the students also used stories about other interracial families they knew to explain why their family would prefer they marry someone of the same race. One college student said her family would have a problem with her marrying interracially, explaining that their opinion is based on their experiences with an interracial family in their neighborhood. She had babysat for this family and, according to her, they had "social issues because the dad was *real dark* and the mom was white, and the kids just had major issues." Her choice of words reveals the importance of color and the use of a discursive strategy such as referring to the children as a problem and not the relationship itself....

White "Concern" and Preference

For the white college students interviewed, the role of family is key and certainly influences their decisions not to date interracially. The majority of white students expressed verbally or by raising their hand that their parents, white parents in general, would have a difficult time with an interracial partner for a number of reasons. Yet their parents' opposition was described in nonracial terms, much like their own views. For example, one white male student said, "All parents find something wrong [with the person their child chooses] if they're not exactly as they imagined. My parents would be *surprised* because [an interracial relationship] is not what they are expecting, so it would be difficult in that sense." Many students simply stated something to the effect of "my parents aren't prejudiced; they just wouldn't want me to marry a black guy."

Other students described their families' views against interracial dating as based on the meanings of family and marriage in general. For example, one male student stated that his parents would have a problem with an interracial relationship, because "they brought me up to date within [my] race. They're not like racist, but [they say] just keep with your own culture, be proud of who you are and carry that on to your kids." Some students cited the difference between dating and marriage. As one female college student said, "It becomes more of an issue when you get to be juniors or seniors, because you start thinking long-term, about what your parents will say, and dating a black person just isn't an option." Another student described dating interracially in similar terms: "Dating's not an issue, they always encouraged [me] to interact with all people. For the future and who you're gonna spend the rest of your life with, there's a difference between being with someone of the same race and someone of a different race, [interracial marriage] is not like it's wrong but … just that it would be too difficult."

In one of the discussions of family with the white college students, an interesting incident occurred. One of the male students, who was vocal throughout the focus group, sat listening to the other students and looking around the room. Finally he interrupted the discussion: "Are you kidding me? … parents would shit, they'd have a freaking heart attack. [*He dramatically grabs the front of his shirt and imitates a growling father's voice*]. "Uhh, son, how could you do this to us, the family?" Everyone laughed at his performance, but this student's use of humor to depict his father (or a white father in general) having "a heart attack" is an interesting discursive strategy. Although he presented it in a comical way, his statement does not seem unrealistic to the group. Often, jocular speech is used to convey a serious message in order to avoid being labeled racist or prejudiced.

When asked why their families would respond in these ways to an interracial relationship, the group largely cited the "opposition of the larger society" as the reason why they and/or their family personally would prefer that their family not become involved interracially. Among the white college students, parents were described as "concerned about how difficult it would be."…

IT'S A FAMILY AFFAIR: BLACK FAMILIES

Black families, like white families, can operate as a deterrent to interracial relationships. A family member becoming involved with a white individual is seen as problematic on a number of levels, yet the black families often raised different issues than the white families, such as the importance of "marrying black" and the negative meanings attached to becoming involved interracially. These issues figured prominently among the black partners, the black community respondents, and even black popular culture.

Among the black college students and community respondents, a main issue was the emphasis on marrying within their race, explicitly identifying race as the issue, unlike the white communities. Most students discussed how their parents would have a problem. One black female student stated, "My family would outright disown me" (which received a number of affirmations from the group). Another college student commented on the beliefs her family instilled in her: "My family raised me to be very proud of who I am, a black woman, and they instilled in me the belief that I would never want to be with anyone but a strong black man." Other students described incidents or comments they had heard that let them know they were expected to marry black. Similarly, Leslie, a black community respondent, stated she was raised by her parents to date anyone, but they were "adamant about me marrying black." She added that they told her not to "even think about marrying interracially." Allen also stated that his parents told him that "high school dating is fine but not marriage, ohhh nooo!" Couples like Gwen and Bill, Chris and Victoria, and others also recounted how the black partner's family had more difficulty accepting the relationship when they realized that the couple was getting married as opposed to just dating.

A significant piece of black familial opposition involves the perceived racism of whites in the larger society. Black families were described as having a hard time accepting a family member getting involved with a white person because of lingering racism and a distrust in whites in general. For example, one black college student stated that her family would have a problem if she brought home a white man, "because they would always be wondering what his family was saying, you know, do they talk about me behind my back?" Other students' responses echoed these views, such as one student who remarked, "My mom would have a problem with it. She just doesn't trust white people."

Also, black community members such as Alice and Jean argued that their opposition to a family member getting involved with a white person was based on the belief that the white individual (or their white family/neighborhood) would mistreat their family member. All but two of the black community members expressed the concern that since white society is racist there is no reason to become involved with a white person....

The black families related to this study objected to having a white person in their family and intimate social circles because they viewed whites as the "enemy" and their presence as a sign that the black partner is not committed to his or her community or family. Not surprisingly, black individuals opposed interracial marriage much more than interracial dating, primarily because marriage represents a legitimation of the union and formally brings the white partner into the family.

Despite such opposition, the black college students argued that black families are still more accepting than white families....

While white families discourage their family members from engaging in interracial relationships to maintain white privilege, black families discourage these unions to maintain the strength and solidarity of black communities. Black families view interracial relationships as a loss in many ways—the loss of individuals to white society, the weakening of families and communities, and the devaluing of blackness....

FAMILY MATTERS

... The familial responses discussed throughout the research clearly demonstrate how images of oneself and one's family [are ...] linked to the concept of race and otherness, especially within the construction of families, both white and black.

White families often objected to the idea of a member dating interracially, not because they met the black individual and were confronted with overwhelming "racial" differences, but because they were merely responding based on their ideas and beliefs about interracial relationships and blacks in general....

Being in an interracial relationship often brings forth problems within families, since white and black families overwhelmingly want to remain monoracial....

This ... highlights how difficult it is to negotiate issues of race and family. Discussing race and, more important, views on interracial relationships in nonracial terms often makes it even more difficult for individuals to challenge and confront their family's views. By using phrases such as "It's not my personal preference" or "I just worry about the problems you will face in society," families and individuals are able to oppose interracial relationships without appearing prejudiced or racist. A family member's opposition, however, is also tied to the issue of identity—the identity of the family and the individuals involved. Families express concern over the identities of the biracial children who will be produced, and in many ways there is a tendency for white families to worry that the children will be "too black," and the black families worry about the children being "more white." The white individuals who enter into a relationship with a black person are seen as "less white" and as tainting the white family, while the black individuals who get involved are seen as "not black enough" and as leaving their blackness behind. All of these fears and beliefs demonstrate the centrality of race to the constructions of families and identities and, more important, the socially constructed nature of race, if one's relationship can change one's "race" in this society still divided by racial boundaries. Ultimately, black-white couples and biracial children are forced to exist somewhere in between, with or without their families.

41

Families on the Frontier

From Braceros in the Fields to Braceras in the Home

PIERRETTE HONDAGNEU-SOTELO

Throughout the United States, a plethora of occupations today increasingly ... rely on the work performed by Latina and Asian immigrant women. Among these are jobs in downgraded manufacturing, jobs in retail, and a broad spectrum of service jobs in hotels, restaurants, hospitals, convalescent homes, office buildings, and private residences. In some cases, such as in the janitorial industry and in light manufacturing, jobs have been re-gendered and re-racialized so that jobs previously held by U.S.-born white or black men are now increasingly held by Latina immigrant women. Jobs in nursing and paid domestic work have long been regarded as "women's jobs," seen as natural outgrowths of essential notions of women as care providers. In the late twentieth-century United States, however, these jobs entered the global marketplace, and immigrant women from developing nations around the globe were increasingly represented in them. In major metropolitan centers around the country, Filipina and Indian immigrant women make up a sizable proportion of HMO nursing staffs—a result due in no small part to deliberate recruitment efforts. Caribbean, Mexican, and Central American women increasingly predominate in low-wage service jobs, including paid domestic work.

This diverse gendered labor demand is quite a departure from patterns that prevailed in the western United States only a few decades ago. The relatively dramatic transition from the explicit demand for *male* Mexican and Asian immigrant workers to demand that today includes women has its roots in a changing political economy. From the late nineteenth century until 1964, the period during which various contract-labor programs were in place, the economies of the Southwest and the West relied on primary, extractive industries. As is well-known, Mexican, Chinese, Japanese, and Filipino immigrant workers, primarily men, were recruited for jobs in agriculture, mining, and railroads. These migrant workers were recruited and incorporated in ways that mandated their long-term separation from their families of origin.

As the twentieth century turned into the twenty-first, the United States was once again a nation of immigration. This time, however, immigrant labor is not

SOURCE: From Hondagenu-Sotelo, Pierrette, pp. 259–273 in *Latinos: Remaking America* edited by Marcelo M. Suarez and Mariela M. Paez. Copyright © 2002 University of California Press. Reprinted with permission.

involved in primary, extractive industry. Agribusiness continues to be a financial leader in the state of California, relying primarily on Mexican immigrant labor and increasingly on indigenous workers from Mexico, but only a fraction of Mexican immigrant workers are employed in agriculture. Labor demand is now extremely heterogeneous and is structurally embedded in the economy of California (Cornelius 1998). In the current period, which some commentators have termed postindustrial, business and financial services, computer and other high-technology firms, and trade and retail prevail alongside manufacturing, construction, hotels, restaurants, and agriculture as the principal sources of demand for immigrant labor in the western United States.

As the demand for immigrant women's labor has increased, more and more Mexican and (especially) Central American women have left their families and young children behind to seek employment in the United States. Women who work in the United States in order to maintain their families in their countries of origin constitute members of new transnational families. And because these arrangements are choices that the women make in the context of very limited options, they resemble apartheid-like exclusions. These women work in one nation-state but raise their children in another. Strikingly, no formalized, temporary contract-labor program mandates these separations. Rather, this pattern is related to the contemporary arrangements of social reproduction in the United States.

WHY THE EXPANSION IN PAID DOMESTIC WORK?

... The exponential growth in paid domestic work is due in large part to the increased employment of women, especially married women with children, to the underdeveloped nature of U.S. childcare centers, and to patterns of U.S. income inequality and global inequalities. National and global trends have fueled this growing demand for paid domestic services. Increasing global competition and new communications technologies have led to work speed-ups in all sorts of jobs, and the much-bemoaned "time bind" has hit professionals and managers particularly hard (Hochschild 1997). Meanwhile, normative middle-class ideals of child rearing have been elaborated (consider the proliferation of soccer, music lessons, and tutors). At the other end of the age spectrum, greater longevity among the elderly has prompted new demands for care work.

... Social reproduction consists of those activities that are necessary to maintain human life, daily and intergenerationally. This includes how we take care of ourselves, our children and elderly, and our homes. Social reproduction encompasses the purchasing and preparation of food, shelter, and clothing; the routine daily upkeep of these, such as cooking, cleaning, and laundering; the emotional care and support of children and adults; and the maintenance of family and community ties. The way a society organizes social reproduction has far-reaching consequences not only for individuals and families, but also for macro-historical processes (Laslett and Brenner 1989).

Many components of social reproduction have become commodified and outsourced in all kinds of new ways. Today, for example, not only can you purchase fast-food meals, but you can also purchase, through the Internet, the home delivery of customized lists of grocery items. Whereas mothers were once available to buy and wrap Christmas presents, pick up dry cleaning, shop for groceries, and wait around for the plumber, today new businesses have sprung up to meet these demands—for a fee....

GLOBAL TRENDS IN PAID DOMESTIC WORK

Just as paid domestic work has expanded in the United States, so too it appears to have grown in many other postindustrial societies, in the "newly industrialized countries" of Asia, in the oil-rich nations of the Middle East, in Canada, and in parts of Europe. In paid domestic work around the globe, Caribbean, Mexican, Central American, Peruvian, Sri Lankan, Indonesian, Eastern European, and Filipina women—the latter in disproportionately large numbers—predominate. Worldwide, paid domestic work continues its long legacy as a racialized and gendered occupation, but today, divisions of nation and citizenship are increasingly salient.

The inequality of nations is a key factor in the globalization of contemporary paid domestic work. This has led to three outcomes. (1) Around the globe, paid domestic work is increasingly performed by women who leave their own nations, their communities, and often their families of origin to do the work. (2) The occupation draws not only women from the poor socioeconomic classes, but also women who hail from nations that colonialism has made much poorer than those countries where they go to do domestic work. This explains why it is not unusual to find college-educated women from the middle class working in other countries as private domestic workers. (3) Largely because of the long, uninterrupted schedules of service required, domestic workers are not allowed to migrate as members of families.

Nations that "import" domestic workers from other countries do so using vastly different methods. Some countries have developed highly regulated, government-operated, contract-labor programs that have institutionalized both the recruitment and the bonded servitude of migrant domestic workers. Canada and Hong Kong provide paradigmatic examples of this approach. Since 1981 the Canadian federal government has formally recruited thousands of women to work as live-in nannies/housekeepers for Canadian families. Most of these women came from Third World countries in the 1990s (the majority came from the Philippines, in the 1980s from the Caribbean), and once in Canada, they must remain in live-in domestic service for two years, until they obtain their landed immigrant status, the equivalent of the U.S. "green card."...

Similarly, since 1973 Hong Kong has relied on the formal recruitment of domestic workers, mostly Filipinas, to work on a full-time, live-in basis for Chinese families. Of the 150,000 foreign domestic workers in Hong Kong in 1995,

130,000 hailed from the Philippines, and smaller numbers were drawn from Thailand, Indonesia, India, Sri Lanka, and Nepal (Constable 1997, 3).... Filipina domestic workers in Hong Kong are controlled and disciplined by official employment agencies, employers, and strict government policies. Filipinas and other foreign-born domestic workers recruited to Hong Kong find themselves working primarily in live-in jobs and bound by two-year contracts that stipulate lists of job rules, regulations for bodily display and discipline (no lipstick, nail polish, or long hair; submission to pregnancy tests; etc.), task timetables, and the policing of personal privacy. Taiwan has adopted a similarly formal and restrictive government policy to regulate the incorporation of Filipina domestic workers (Lan 2000).

In this global context, the United States remains distinctive, because it takes more of a laissez-faire approach to the incorporation of immigrant women into paid domestic work. No formal government system or policy exists to legally contract foreign domestic workers in the United States. Although in the past, private employers in the United States were able to "sponsor" individual immigrant women who were working as domestics for their green cards using labor certification (sometimes these employers personally recruited them while vacationing or working in foreign countries), this route is unusual in Los Angeles today. Obtaining legal status through labor certification requires documentation that there is a shortage of labor to perform a particular, specialized occupation. In Los Angeles and in many parts of the country today, a shortage of domestic workers is increasingly difficult to prove. And it is apparently unnecessary, because the significant demand for domestic workers in the United States is largely filled not through formal channels of foreign recruitment but through informal recruitment from the growing number of Caribbean and Latina immigrant women who are *already* legally or illegally living in the United States. The Immigration and Naturalization Service, the federal agency charged with enforcement of migration laws, has historically served the interests of domestic employers and winked at the employment of undocumented immigrant women in private homes.

As we compare the hyper-regulated employment systems in Hong Kong and Canada with the more laissez-faire system for domestic work in the United States, we find that although the methods of recruitment and hiring and the roles of the state in these processes are quite different, the consequences are similar. Both systems require the incorporation as workers of migrant women who can be separated from their families.

The requirements of live-in domestic jobs, in particular, virtually mandate this. Many immigrant women who work in live-in jobs find that they must be on call during all waking hours and often throughout the night, so there is no clear line between working and nonworking hours. The line between job space and private space is similarly blurred, and rules and regulations may extend around the clock. Some employers restrict the ability of their live-in employees to receive phone calls, entertain friends, attend evening ESL classes, or see boyfriends during the workweek. Other employers do not impose these sorts of restrictions, but because their homes are located in remote hillsides, suburban enclaves, or gated communities, live-in nannies/housekeepers are effectively restricted from participating in anything resembling social life, family life of their own, or public culture.

These domestic workers—the Filipinas working in Hong Kong or Taiwan, the Caribbean women working on the East Coast, and the Central American and Mexican immigrant women working in California—constitute the new braceras. They are literally "pairs of arms," disembodied and dislocated from their families and communities of origin, and yet they are not temporary sojourners. In the section that follows, I suggest some of the dimensions of this phenomenon and present several illustrative career trajectories.

THE NEW BRACERAS AND TRANSNATIONAL MOTHERHOOD

What are the dimensions of this phenomenon that I refer to as the new braceras and the new forms of transnational motherhood? Precise figures are not available. No one has counted the universe of everyone working in private paid domestic work, which continues to be an informal sector—an "under the table" occupation. Accurately estimating the numbers in immigrant groups that include those who are poor and lack legal work authorization is always difficult. Several indicators, however, suggest that the dimensions are quite significant. My nonrandom survey of 153 Latina immigrant domestic workers, which I conducted at bus stops, at ESL evening classes, and in public parks where Latina nannies take their young charges, revealed the following. Approximately 75 percent of the Latina domestic workers had children of their own, and a startling 40 percent of the women with children had at least one of their children "back home" in their country of origin.

... [A]n estimated 40 to 50 percent of Central American and Mexican women leave their children in their countries of origin when they migrate to the United States. They believe the separation from their children will be temporary, but physical separation may endure for long, and sometimes undetermined, periods of time. Job constraints, legal-status barriers, and perceptions of the United States as a dangerous place to raise children explain these long-term separations....

Private domestic workers who hail primarily from Mexico, Central America, and the Caribbean hold various legal statuses. Some, for example, are legal, permanent residents or naturalized U.S. citizens, many of them beneficiaries of the Immigration Reform and Control Act's amnesty-legalization program, which was enacted in 1986. Most Central American women, who prevail in the occupation, entered the United States after the 1982 cutoff date for amnesty-legalization, so they did not qualify for legalization. Throughout the 1990s, a substantial proportion of Central Americans either remained undocumented or held a series of temporary work permits, granted to delay their return to war-ravaged countries.

Their precarious legal status as an "illegal" or "undocumented immigrant" discouraged many immigrant mothers from migrating with their children or bringing them to the United States. The ability of undocumented immigrant parents to bring children north was also complicated in the late 1990s by the militarization of the U.S.-Mexico border through the implementation of various border-control programs (such as Operation Gatekeeper). As Jacqueline Hagan and Nestor Rodriguez compellingly show, the U.S.-Mexico border has become a zone of danger, violence, and death. This not only made it difficult to migrate with children, it also made it difficult to travel back and forth to visit family members "back home." Bound by precarious legal status in the United States, which might expose them to the risk of deportation and denial of all kinds of benefits, and by the greater danger and expense in bringing children to the United States, many immigrant mothers opted not to travel with their children or send for them. Instead, they endure long separations.

Many immigrant parents of various nationalities also view the United States as a highly undesirable place to raise children. Immigrant parents fear the dangers of gangs, violence, drugs, and second-rate schools to which their children are likely to be exposed in poor, inner-city neighborhoods. They are also appalled by the way immigration often weakens generational authority. As one Salvadoran youth put it, "Here I do what I please, and no one can control me" (Menjivar 2000, 213).

... Bringing children north after long periods of separation may entail unforeseen "costs of transnationalism," as Susan Gonzalez-Baker suggests. Just as migration often rearranges gender relations between spouses, so too, does migration prompt challenges to familiar generational relations between parents and offspring.

CONCLUSION

What do these developments mean for a twenty-first century research agenda on Latino families? Clearly, they force us to rethink monolithic understandings of Latino families, familism, and sentimentalized notions of motherhood. There are also many remaining empirical questions about the impact of these transnational processes on children and youth. How does the length of separation affect processes of adaptation? How does the age at which the children are reunited with their "bracera" mothers affect family social relations? How do these migration processes affect adolescent identity and school performance? Answering questions such as these will require incorporating children and youths into research as active agents in migration.

The trajectories described here pose an enormous challenge to those who would celebrate any and all instances of transnationalism. These migration patterns alert us to the fact that the continued privatization of social reproduction among the American professional and managerial class has broad repercussions for the social relations among new Latina immigrants and their families. Similarly, strong emotional ties between mothers and children, which we might dismiss as personal subjectivities, are fueling massive remittances to countries such as

El Salvador and the Philippines. Our most fundamental question remains unanswered: Who will continue to pick up the cost of raising the next generation? Surely, we can hope for a society wherein Latina immigrant women and children are not the first to bear those costs.

BIBLIOGRAPHY

Constable, Nicole. 1997. *Maid to Order in Hong Kong: Stories of Filipina Workers.* Ithaca and London: Cornell University Press.

Cornelius. Wayne. 1998. "The Structural Embeddedness of Demand for Mexican Immigrant Labor: New Evidence from California." In Márcelo M. Suarez-Orozco (ed.), *Crossings: Mexican Immigration in Interdisciplinary Perspectives.* Cambridge, MA: Harvard University, David Rockefeller Center for Latin American Studies, pp. 113–44.

Hochschild, Arlie. 1997. *The Time Bind: When Work Becomes Home and Home Becomes Work.* New York: Metropolitan Books. Henry Holt.

Lan, Pei-chia. 2000. "Global Divisions, Local Identities: Filipina Migrant Domestic Workers and Taiwanese Employers." Dissertation, Northwestern University.

Laslett, Barbara, and Johanna Brenner. 1989. "Gender and Social Reproduction: Historical Perspectives." *Annual Review of Sociology* 15: 381–404.

Menjivar, Cecilia. 2000. *Fragmented Ties: Salvadoran Immigrant Networks in America.* Berkeley: University of California Press.

42

White Flight in Networked Publics?
How Race and Class Shaped American Teen Engagement with MySpace and Facebook

DANAH BOYD

In a historic small town outside Boston, I interviewed a group of teens at a small charter school that included middle-class students seeking an alternative to the public school and poorer students who were struggling in traditional schools. There, I met Kat, a white 14-year-old from a comfortable background. We were talking about the social media practices of her classmates when I asked her why most of her friends were moving from MySpace to Facebook. Kat grew noticeably uncomfortable. She began simply, noting that *"MySpace is just old now and it's boring."* But then she paused, looked down at the table, and continued.

> *"It's not really racist, but I guess you could say that. I'm not really into racism, but I think that MySpace now is more like ghetto or whatever."*—Kat

On that spring day in 2007, Kat helped me finally understand a pattern that I had been noticing throughout that school year. Teen preference for MySpace or Facebook went beyond simple consumer choice; it reflected a reproduction of social categories that exist in schools throughout the United States. Because race, ethnicity, and socio-economic status shape social categories, the choice between MySpace and Facebook became racialized. This got reinforced as teens chose to self-segregate across the two sites, just as they do in schools.

After Kat told me that MySpace was *"like ghetto,"* I asked her if people at her school were still using MySpace and she hesitantly said yes. Her discomfort in discussing the topic was palpable and it became clear that she didn't know how to talk about race or the social divisions she recognized in her school.

> *"The people who use MySpace—again, not in a racist way—but are usually more like ghetto and hip hop rap lovers group."*—Kat

In trying to distinguish those who use MySpace from those who use Facebook, Kat uses a combination of spatial referents and taste markers that she knows have racial overtones. While Kat does not identify as a racist, her life and social world are shaped by race. Her small school is divided, with poorer students and teens of

SOURCE: From danah boyd in *Race After the Internet* edited by Lisa Nakamura and Peter Chow-White (Routledge). Copyright © 2011 by danah boyd. Reprinted by permission of the author.

color in different classes than the white students from wealthier backgrounds. Kat's friends are primarily white and her classes and activities are primarily filled with white students.

Kat's use of the term "ghetto" simultaneously references spatial and taste-based connotations. On the one hand, *the* ghetto is a part of a city historically defined by race and class (Wacquant, 1997). On the other hand, *being* ghetto refers to a set of tastes that emerged as poor people of color developed fashion and cultural artifacts that proudly expressed their identity. Just as physical spaces and tastes are organized around and shaped by race and class, so too are digital environments.

This article explores a division that emerged between MySpace and Facebook among American teens during the 2006–2007 school year. At the beginning of the year, a common question in American schools was: "Are you on MySpace?" By the end of the year, the question had shifted to "MySpace or Facebook?" As Facebook started gaining momentum, some teenagers switched from MySpace to Facebook. Others joined Facebook without having ever been on MySpace. Still others chose to adopt both. During this period, MySpace did not lose traction. Teens continued to flock to the site, opting for MySpace in lieu of or in addition to Facebook. To complicate matters more, some teens who had initially adopted MySpace began to switch to Facebook.

Slowly, a distinction emerged. Those who adopted MySpace were from different backgrounds and had different norms and values than those who adopted Facebook. White and more affluent individuals were more likely to choose and move to Facebook. Even before statistical data was available, the language teens used to describe each site and its users revealed how the division had a race and class dimension to it. Given these dynamics and Kat's notion of MySpace as ghetto, one way to conceptualize the division that unfolded is through the lens of white flight. In essence, many of the same underlying factors that shaped white city dwellers' exodus to the suburbs—institutional incentives and restrictions, fear and anxiety, social networks, and racism—also contribute to why some teens were more likely to depart than others.

What distinguishes adoption of MySpace and Facebook among American teens is not cleanly about race or class, although both are implicated in the story at every level. The division can be seen through the lens of taste and aesthetics, two value-laden elements that are deeply entwined with race and class. It can also be seen through the network structures of teen friendship, which are also directly connected to race and class. And it can be seen through the language that teens—and adults—use to describe these sites, language like Kat's that rely on racial tropes to distinguish the sites and their users. The notion that MySpace may be understood as a digital ghetto introduces an analytic opportunity to explore the divisions between MySpace and Facebook—and namely, the movement of some teens from MySpace to Facebook—in light of the historic urban tragedy produced by white flight. Drawing parallels between these two events sheds light on how people's engagement with technology reveals social divisions and the persistence of racism.

The data and analysis used in this article stems from four years of ethnographic fieldwork examining the role that social media play in the everyday lives of American teens. From 2004–2009, I interviewed and observed teens in diverse communities across 17 different states, spent over 2,000 hours observing online practices, and analyzed 10,000 randomly selected MySpace profiles. The quotes in this article stem from both online data and a subset of the 103 formal semi-structured interviews I conducted. I also use online commentary about this division, including blog comments and news articles. I illustrate how distinctions in social network site adoption and the perceptions teens—and adults—have about these sites and their users reflect broader narratives of race and class in American society.

TEEN ADOPTION OF MYSPACE AND FACEBOOK

The first social network site was neither MySpace nor Facebook, but these sites emerged as the two most popular social network sites in the United States. MySpace launched in 2003 on the heels of Friendster, an earlier social network site that was notably popular among 20–30-something urban dwellers in major urban U.S. cities. Although individual teenagers joined MySpace early on, teens became a visible demographic on the site in 2004. Most early adopter teens learned about MySpace through one of two paths: bands or older family members.

Teens who learned of MySpace through bands primarily followed indie rock music or hip-hop, the two genres most popular on MySpace early on. MySpace allowed teens to connect with and follow their favorite bands. Early adopter teens who were not into music primarily learned about the site from a revered older sibling or cousin who was active in late-night culture. For these teens, MySpace was cool because cool elders thought so.

Teenagers who joined MySpace began proselytizing the site to their friends. Given its popularity among musicians and late-night socialites, joining MySpace became a form of (sub) cultural capital. Teens, especially those in urban settings, tend to look to the 20 to 30-something crowd for practices that they can emulate. MySpace's early popularity among teens was tightly entwined with its symbolic reference to maturity, status, and freedom in the manner espoused by urban late-night culture. While teens often revere the risky practices of a slightly older cohort, many adults work to actively dissuade them from doing so. By propagating and glorifying 20-something urban cultural practices and values, MySpace managed to alienate parents early on.

With little mass media coverage of MySpace before News Corporation acquired the company in mid-2005, many teens learned of the site through word-of-mouth networks, namely friends at school, church, activities, and summer camp, as well as from older family members. Given its inception in the Los Angeles region, West Coast teens found MySpace before East Coast teens, and urban teens joined before suburban or rural teens. The media coverage that followed the acquisition further escalated growth among teens.

Immediately after News Corporation bought MySpace, much of the media coverage focused on the bands. After adults began realizing how popular MySpace was with teens, news media became obsessed with teen participation and the potential dangers they faced. This media coverage was both a blessing and a curse for MySpace. On one hand, some teens joined the site because media sold it as both fashionable among teens and despised by parents. On the other hand, some teens avoided joining because of the perceived risks and parents began publicly demonizing the site.

As MySpace simultaneously appealed to and scared off U.S. teens, other social network sites started gaining traction with different demographics. Most did not appeal to teenagers en masse, although niche groups of teens did join many different sites. In 2004, Facebook launched, targeting college students. Originally, access to Facebook was intentionally limited. Facebook started as a Harvard-only social network site before expanding to support all Ivy League schools and then top-tier colleges and then a wider array of colleges. Because of its background, some saw Facebook as an "elite" social network site. The "highbrow" aura of Facebook appealed to some potential participants while repelling others.

The college-centered nature of Facebook quickly appealed to those teenagers who saw college, and thus Facebook access, as a rite of passage. They were aware of the site through older family members and friends who had already graduated high school and gone off to college. Before access became readily available, college-bound teens began coveting entrance. For many, access to the social world of college became a marker of status and maturity. Even those who had MySpace accounts relished the opportunity to gain access to the college-only Facebook as a rite of passage.

In September 2005, Facebook began slowly supporting high schools. While this gave some teens access, the processes in place for teens to join and be validated were challenging, creating a barrier to entry for many potential participants. Those who managed to join were often from wealthier schools where the validation process was more solidified or quite motivated—typically because they wanted to communicate with close friends in college.

Facebook finally opened access to all in September 2006. Sparking a wave of teen adoption, this is the origin point of teens self-sorting into MySpace and Facebook. The segment of teens that initially flocked to Facebook was quite different from those who were early adopters of MySpace. Yet, in both cases, the older early adopters shaped early teen engagement, both in terms of influencing adoption and defining the norms. As teens engaged, they developed their own norms stemming from those set forth by the people they already knew on the site.

While plenty of teens chose to participate on both sites, I began noticing that those teens who chose one or the other appeared to come from different backgrounds. Subculturally identified teens appeared more frequently drawn to MySpace while more mainstream teens tended towards Facebook. Teens from less-privileged backgrounds seemed likely to be drawn to MySpace while those headed towards elite universities appeared to be head towards Facebook. Racial and ethnic divisions looked messier, tied strongly to socio-economic factors, but I observed that black and Latino teens appeared to preference MySpace while white and Asian teens seemed to privilege Facebook.

In observing these patterns in multiple communities in the US, I found myself uncertain as to whether or not they could be generalized.

Marketing research firm Nielsen Claritas reported that wealthy individuals are 25% more likely to use Facebook while less affluent individuals are 37% more likely to be on MySpace (Hare, 2009). In the same year, S. Craig Watkins (2009) published his qualitative and survey data with college students, revealing a racial and ethnic division in preference as well as anti-MySpace attitudes by collegiate Facebook users that parallel those of high school students. While there is no definitive longitudinal statistical data tracking the division amongst teens, these studies provide a valuable backdrop to the perceptions teens have about the sites and their users.

THE ORGANIZATION OF TEEN FRIENDSHIP

There's an old saying that "birds of a feather flock together." Personal networks tend to be rather homogeneous, as people are more likely to befriend those like them.

Youth often self-segregate by race, even in diverse schools (Moody, 2001). While it is easy to lament racial segregation in friendships, there are also social and psychological benefits to racial and ethnic clustering. Tatum (1997) argues that self-segregation is a logical response to the systematized costs of racism; connecting along lines of race and ethnicity can help youth feel a sense of belonging, enhance identity development, and help youth navigate systematic racism. Still, as Bonilla-Silva (2003) has highlighted, people's willingness to accept and, thus expect, self-segregation may have problematic roots and contribute to ongoing racial inequality.

When I asked teens why race defines their friendships, they typically shrugged and told me that it's just the way it is. As Traviesa, a Hispanic 15-year-old from Los Angeles, explained,

> *"If it comes down to it, we have to supposedly stick with our own races....*
> *That's just the unwritten code of high school nowadays."*—Traviesa

Race was not an issue only in major metropolitan communities. Heather, a white 16-year-old in Iowa, told me that her school was not segregated, but then she went on to mark people by race, noting that, *"the black kids are such troublemakers."* This conflicting message—refusing to talk about race explicitly while employing racial language in conversation—was common in my interviews. While there was no formal segregation in Heather's school, like the de facto residential segregation that continues to operate in many American cities, the black teens in her predominantly white school stuck together socially and were stereotyped by the white teens.

Another way of looking at teen friendships is through the lens of social categories and group labels. Many of the teens I interviewed had language to demarcate outcasts (e.g., *"gothics," "nerds," "Dirty Kids,"* etc.) and identify groups of peers by shared activity (e.g., *"band kids," "art kids," "cheerleaders,"* etc.). Often, labels come with a set of stereotypes. For example, Heather explained:

> *"You've got the pretties, which are the girls that tan all the time. They put on excessive makeup. They wear the short skirts, the revealing shirts, that kind of things. Then you've got the guys who are kind of like that, dumb as rocks by the way."*—Heather

Youth subcultures can be seen as an extension of social categories; what differentiates them typically concerns identification. While teens often identify with particular subcultures, social categories are more frequently marked by others.

Social categories serve to mark groups and individuals based on shared identities. Social categories develop in ways that reproduce social distinctions. Unlike class, race and ethnicity are often made visible—albeit, blurred—in the labels youth use. In my fieldwork, I found that clearly dominant racial groups went unmarked, but labels like *"the blacks," "the Chinese people," "the Hispanics," "the Mexicans," "the white people,"* and so forth were regularly employed to define social groupings. In other cases, and in part because they are aware that using such categories could be perceived as racist, teens used substitutes that more implicitly mark race and class-based difference. For example, the word *"urban"* signals *"black"* when referring to a set of tastes or practices. Similarly, some of the labels teens use have racial implications, such as *"Dirty Kids," "gangstas,"* and *"terrorists."* While not all Dirty Kids are white, not all gangstas are black, and not all terrorists are of Middle Eastern descent, they are overwhelmingly linked in teens' minds. Race and class are also often blurred, especially in situations where the logic of stratification may not be understood by teens but appears visible through skin color.

Social categories and race-based labels are also used to mark physical turf in the lunchroom and beyond. Often, this becomes a way in which youth self-segregate. Keke, a 16-year-old black girl in Los Angeles, described in detail where students in her racially diverse school physically gathered during lunch and between classes:

> *"The hallways is full of the Indians, and the people of Middle Eastern decent. They in the hallways and by the classrooms. The Latinos, they all lined up on this side. The blacks is by the cafeteria and the quad.... Then the outcasts, like the uncool Latinos or uncool Indians, the uncool whites, they scattered."*—Keke

Each ethnic and racial group had its gathering spot, but only one had a name: *"Disneyland"* is an area in the public yard where *"the white people"* gathered. While Keke is probably unaware that Disneyland is, as Avila (2004: 3) puts it, "the archetypical example of a postwar suburban order," the notion that students in an urban Los Angeles school label the turf where white people gather by referencing the Orange County suburban theme park known for its racial and ethnic caricatures is nonetheless poignant.

Like the school yard, online environments are often organized by identity and social categories. In some cases, this is explicit. Social network sites like Black Planet, Asian Avenue, and MiGente explicitly target audiences based on race and ethnicity. While neither MySpace nor Facebook are explicitly defined in terms of race, they too are organized by race. Most participants self-segregate when connecting with their pre-existing networks without been fully aware of the social divisions that unfold. Yet, when teens are asked explicitly about who participates where, racial terms emerge.

THE NETWORK EFFECTS OF MYSPACE
AND FACEBOOK ADOPTION

Like school lunchrooms and malls, social network sites are another space where youth gather to socialize with peers. Teens joined social network sites to be with their friends. Given social divisions in both friendship patterns and social spaces, it is unsurprising that online communities reflect everyday social divisions.

Teens provide a variety of different explanations for why they chose MySpace or Facebook. Some argued that it was a matter of personal preference having to do with the features or functionality. For example, Jordan, a biracial Mexican-white 15-year-old from Austin prefers Facebook because it allows unlimited photos. Conversely, Anindita, an Indian-American 17-year-old from Los Angeles values MySpace's creative features:

> "Facebook's easier than MySpace but MySpace is more complex … You can add music, make backgrounds and layouts, but Facebook is just plain white and that's it."—Anindita

Teens also talked about their perception of the two sites in relation to their values and goals. Cachi, an 18-year-old Puerto Rican girl from Iowa uses both MySpace and Facebook, but she sees them differently:

> "Facebook is less competitive than MySpace. It doesn't have the Top 8 thing or anything like that, or the background thing."—Cachi

Safety—or rather the perception of safety—also emerged as a central factor in teen preference. While teens believed Facebook was safer, they struggled to explain why. Tara, a Vietnamese-American 16-year-old from Michigan said,

> "[Facebook] kind of seemed safer, but I don't know like what would make it safer, like what main thing. But like, I don't know, it just seems like everything that people say, it seems safer."—Tara

Teens' fear of MySpace as 'unsafe' undoubtedly stems from the image portrayed by the media, but it also suggests a fear of the 'other.'

By far, the most prominent explanation teens gave for choosing one or the other is the presence of their friends. Teens choose to use the social network site that their friends use. Kevin, a white 15-year-old in Seattle, explains:

> "I'm not big on Facebook; I'm a MySpace guy. I have a Facebook and I have some friends on it, but most of my friends don't check it that often so I don't check it that often."—Kevin

When teens choose to adopt both, what distinguishes one from the other often reflects distinct segments of their social network. For example, Red, a white 17-year-old from Iowa has a profile on both sites, but

> *"the only reason I still have my MySpace is because my brother's on there."*—Red

Even teens who prefer the features and functionality of one site use the other when that's where their friends are. Connor, a white 17-year-old from Atlanta, says that he personally prefers MySpace because there's *"too much going on"* on Facebook.

> *"It's like hug me and poke me … what does that even mean?"*—Connor

Yet, Connor signs into Facebook much more frequently than MySpace *"because everybody's on Facebook."*

Social network site adoption took the form of a social contagion spreading through pre-existing peer networks. For some teens, the presence of just one friend was enough of an incentive to participate; others only joined once many of their friends were present. Once inside, teens encouraged their friends to participate. MySpace and Facebook have network effects: they are more valuable when more friends participate. Some teens went so far as to create accounts for resistant friends in order to move the process along. As word of each site spread, adoption hopped from social group to social group through pre-existing networks for teens. In choosing to go where their friends were, teens began to self-segregate along the same lines that shape their social relations more broadly: race and ethnicity, socio-economic status, education goals, lifestyle, subcultural affiliation, social categories, etc.

TASTES, AESTHETICS, AND SOCIAL STATUS

For many teens, embracing MySpace or Facebook is seen as a social necessity. Which site is "cool" depends on one's cohort. Milo, an Egyptian 15-year-old from Los Angeles, joined MySpace because it was *"the thing"* in his peer group but another girl from the same school, Korean 17-year-old Seong, told me that Facebook was the preferred site among her friends.

What is socially acceptable and desirable differs across social groups. One's values and norms are strongly linked with one's identity membership. When working class individuals eschew middle class norms in preference for the norms and expectations of their community, they reproduce social class. The idea that working class individuals should adopt middle class norms is fundamentally a middle class notion; for many working class individuals, the community and its support trump potential upwards mobility. Norms also differ across racial and ethnic groups and are reinforced as people of color seek to identify with their racial and ethnic background.

While what is seen as cool can be differentiated by group, there is also a faddish nature to the process. Seong preferred Facebook because it was *"exclusive."* She moved from Xanga to MySpace to Facebook as each new site emerged, preferring to adopt what was new rather than stay on a site as it became widely embraced. Conversely, white 15-year-old Summer from Michigan rejected the idea of switching to Facebook simply because it was new. She preferred to be where her peers were, but she noted that the *"designer class of people"* in her school joined Facebook because they felt the need to have *"the latest thing."* In this way, subcultural capital influenced the early adoption of Facebook; it was fashionable to some simply because of its newness.

The construction of "cool" is fundamentally about social status among youth. Teenagers both distinguish themselves through practices of consumption, fashion, and attitudes and assess others through these markers. Yet, neither tastes nor attitudes nor cultural consumption practices are adopted randomly. Race and class shape practices and the social agendas around race and class also drive them.

Online, status markers take on new form but in ways that are reminiscent of offline practices.

In an environment where profiles serve as "digital bodies," profile personalization can be seen a form of digital fashion. Teens' Facebook and MySpace profiles reflect their taste, identity, and values. Through the use of imagery and textual self-expressions, teens make race, class, and other identity markers visible. As Nakamura (2008) has argued, even in the most constrained online environments, participants will use what's available to them to reveal identity information in ways that make race and other identity elements visible.

In describing what was desirable about the specific sites, teens often turned to talk about aesthetics and profile personalization. Teens' aesthetics shaped their attitudes towards each site. In essence, the *"glitter"* produced by those who *"pimp out"* their MySpaces is seen by some in a positive light while others see it as *"gaudy,"* *"tacky,"* and *"cluttered."* While Facebook fans loved the site's aesthetic minimalism, others viewed this tone as *"boring,"* *"lame,"* and *"elitist."* Catalina, a white 15-year-old from Austin, told me that Facebook is better because

"Facebook just seems more clean to me."—Catalina

What Catalina sees as cleanliness, Indian-American 17-year-old Anindita from Los Angeles, labels simplicity; she recognizes the value of simplicity, but she prefers the *"bling"* of MySpace because it allows her to express herself.

The extensive options for self-expression are precisely what annoy some teens. Craig Pelletier, a 17-year-old from California, complained that,

"these tools gave MySpacers the freedom to annoy as much as they pleased. Facebook was nice because it stymied such annoyance, by limiting individuality. Everyone's page looked pretty much the same, but you could still look at pictures of each other. The MySpace crowd felt caged and held back because they weren't able to make their page unique."—Craig

Craig believes the desire to personalize contributed to his peers' division between MySpace and Facebook.

Teens who preferred MySpace lamented the limited opportunities for creative self-expression on Facebook, but those who preferred Facebook were much more derogatory about the style of profiles in MySpace. Not only did

Facebook users not find MySpace profiles attractive, they argued that the styles produced by MySpace users were universally ugly. While Facebook's minimalism is not inherently better, conscientious restraint has been one marker of bourgeois fashion. On the contrary, the flashy style that is popular on MySpace is often marked in relation to "bling-bling," a style of conspicuous consumption that is associated with urban black culture and hip-hop. To some, bling and flashy MySpace profiles are beautiful and creative; to others, these styles are garish. While style preference is not inherently about race and class, the specific styles referenced have racial overtones and socio-economic implications. In essence, although teens are talking about style, they are functionally navigating race and class.

Taste is also performed directly through profiles; an analysis of "taste statements" in MySpace combined with the friend network reveals that distinctions are visible there. The importance of music to MySpace made it a visible vector of taste culture. Youth listed their musical tastes on their profiles and attached songs to their pages. While many genres of music were present on MySpace, hip-hop stood out, both because of its salience amongst youth and because of its racial connotations. Although youth of all races and ethnicities listen to hip-hop, it is most commonly seen as a genre that stems from black culture inside urban settings. Narratives of the ghetto and black life dominate the lyrics of hip-hop and the genre also serves as a source of pride and authenticity in communities that are struggling for agency in American society. For some, participating in this taste culture is a point of pride; for others, this genre and the perceived attitudes that go with it are viewed as offensive. Although MySpace was never about hip-hop, its mere presence became one way in which detractors marked the site.

Taste and aesthetics are not universal, but deeply linked to identity and values. The choice of certain cultural signals or aesthetics appeals to some while repelling others. Often, these taste distinctions are shaped by class and race and, thus, the choice to mark Facebook and MySpace through the language of taste and aesthetics reflect race and class.

A NETWORKED EXODUS

After posting my controversial blog essay about the distinction between MySpace and Facebook, teens began to contact me with their own stories. Anastasia, a 17-year-old from New York, emailed me to explain that it wasn't simply a matter of choice between the two sites; many of her peers simply moved from MySpace to Facebook. Until now, I have focused on the choice that teens make to adopt MySpace or Facebook. But Anastasia's right: there is also movement as teens choose to leave one social network site and go to the other. By and large, teens did not leave Facebook and go to MySpace. Rather, a subset of teens left MySpace to go to Facebook. This can be partially explained as an issue of fads, with teens leaving MySpace to go to the "new" thing. But even if this alone could explain the transition, it does not explain why some teens were more likely to switch than others. Anastasia argues that, at least in her school, who participated can be understood in terms of social categories:

> *"My school is divided into the 'honors kids,' (I think that is self-explanatory), the 'good not-so-honors kids,' 'wangstas,' (they pretend to be tough and black but when you live in a suburb in Westchester you can't claim much hood), the 'latinos/hispanics,' (they tend to band together even though they could fit into any other groups) and the 'emo kids' (whose lives are allllllways filled with woe).*
>
> *We were all in MySpace with our own little social networks but when Facebook opened its doors to high schoolers, guess who moved and guess who stayed behind … The first two groups were the first to go and then the 'wangstas' split with half of them on Facebook and the rest on MySpace… I shifted with the rest of my school to Facebook and it became the place where the 'honors kids' got together and discussed how they were procrastinating over their next AP English essay."*—Anastasia

The social categories Anastasia uses reflect racial, ethnic, and class divisions in her school. Anastasia's description highlights how structural divisions in her school define what plays out on MySpace and Facebook. Movement from MySpace to Facebook further magnifies already existing distinctions. In California, 17-year-old Craig blogged about the movement in his school, using the language of taste, class, and hierarchy.

> *"The higher castes of high school moved to Facebook. It was more cultured, and less cheesy. The lower class usually were content to stick to MySpace. Any high school student who has a Facebook will tell you that MySpace users are more likely to be barely educated and obnoxious. Like Peet's is more cultured than Starbucks, and Jazz is more cultured than bubblegum pop, and like Macs are more cultured than PC's, Facebook is of a cooler caliber than MySpace."*—Craig

In his description, Craig distinguishes between what he sees as highbrow and lowbrow cultural tastes, using consumption patterns to differentiate classes of people and describe them in terms of a hierarchy. By employing the term "caste," Craig uses a multicultural metaphor with ethnic and racial connotations that runs counter to the supposed class mobility available in U.S. society. In doing so, he's locating his peers in immutable categories and tying tastes to them. While Craig may not have intended to imply this, his choice of the term "caste" is nonetheless interesting.

MYSPACE: A DIGITAL GHETTO?

One provocative way of reflecting on the networked movement from MySpace to Facebook is through the lens of "white flight." The term "white flight" refers to the exodus of white people from urban American centers to the suburbs during the 20th century. This simplistic definition obscures the racial motivations of those who left, the institutionalized discrimination that restricted others from leaving, and the ramifications for cities and race relations. Many who left did so to avoid racial integration in communities and schools. Not everyone could leave. Although the suburbs were touted as part of the "American Dream," families of color were often barred directly explicitly by ethnically exclusive restrictions on housing developments or indirectly by discriminatory lending practices. Suburbs were zoned to limit low-income housing and rentals, thereby limiting who could afford to move there. What followed was urban decay. Governmental agencies reduced investments in urban communities, depopulation lowered property values and shrunk tax bases, and unemployment rose as jobs moved to the suburbs. The resultant cities were left in disrepair and the power of street gangs increased. Through "white flight," racial identities were reworked as spaces were reconfigured.

Given the formalized racism and institutionalized restrictions involved in urban white flight, labeling teen movement from MySpace to Facebook as "digital white flight" may appear to be a problematic overstatement. My goal is not to dismiss or devalue the historic tragedy that white racism brought to many cities, but to offer a stark framework for seeing the reproduction of social divisions in a society still shaped by racism.

Needless to say, the frame of "white flight" only partially works, but the metaphor provides a fertile backdrop to address the kinds of language I heard by youth. It also provides a fruitful framework for thinking of the fear and moral panic surrounding MySpace. Fear of risk and perception of safety are salient in discussions of ghettos. Many whites fled the city, believing it crime-ridden, immoral, and generally unsafe. While outsiders are rarely targets of violence in the inner-city, the perception of danger is widespread and the suburbs are commonly narrated as the safe alternative. The same holds for MySpace. Fears concerning risks on MySpace are overstated at best and more often outright misunderstood. Yet, they are undoubtedly widespread. In contrast, Facebook's origin as a gated community and parents' belief that the site is private and highly monitored reflect the same values signaled by the suburbs.

The network segmentation implied by a "digital white flight" also helps explain why, two years later, news media behaved as though MySpace was dead. Quite simply, white middle-class journalists didn't know anyone who still used MySpace. On May 4, 2009, the New York Times ran a story showing that MySpace and Facebook usage in the U.S. had nearly converged (with Facebook lagging slightly behind MySpace); the title for this article was "Do You Know Anyone Still on MySpace?" Although the article clearly stated that the unique visitors were roughly equal, the headline signaled the cultural divide. The New York Times staff was on Facebook and assumed their readers were too. This article generated 154 comments from presumably adult readers. Some defended MySpace, primarily by pointing to its features, the opportunity for connecting, and the cultural relevance of musicians and bands. Many more condemned MySpace, bemoaning its user interface, spam, and outdated-ness. Yet, while only two MySpace fans used condescending language to describe Facebook (*"Facebook is very childish"* and *"Facebook is for those who live in the past"*), dozens of MySpace critics demeaned MySpace and its users. Some focused on the perception that MySpace was filled with risky behavior:

> *"MySpace become synonymous with hyper-sexual, out of control teens, wild partying 20-somethings, 30–40 somethings craving attention, sexual predators on the hunt, and generally un-cool personal behavior from a relatively small, but highly visible number of users."*

Others used labels, stereotypes, and dismissive language to other those who preferred MySpace, often suggesting a class based distinction:

> *"My impression is that MySpace is for the riffraff and Facebook is for the landed gentry."*

> *"Compared to Facebook, MySpace just seems like the other side of the tracks— I'll go there for fun, but I wouldn't want to live there."*

> *"my impression is [MySpace is] for tweens, high school kids that write emo poetry, and the proletariat. once the younger demo goes to college, they shift to facebook. the proletariat? everyone knows they never go to college!"*

Just as those who moved to the suburbs looked down upon those who remained in the cities, so too did Facebook users demean those on MySpace. This can be seen in the attitudes of teens I interviewed, the words of these commenters, and the adjectives used by the college students Watkins (2009) interviewed. The language used in these remarks resembles the same language used throughout the 1980s to describe city dwellers: dysfunctional families, perverts and deviants, freaks and outcasts, thieves, and the working class. Implied in this is that no decent person could possibly have a reason to dwell in the city or on MySpace. While some who didn't use MySpace were harshly critical of the site, others simply forgot that it existed. They thought it to be irrelevant, believing that no one lived there anymore simply because no one they knew did.

To the degree that some viewed MySpace as a digital ghetto or as being home to the cultural practices that are labeled as ghetto, the same fear and racism that underpinned much of white flight in urban settings is also present in the perception of MySpace. The fact that many teens who left MySpace for Facebook explained their departure as being about features, aesthetics, or friendship networks does not disconnect their departure from issues of race and class. Rather, their attitude towards specific aesthetic markers and features is shaped by their experiences with race and class. Likewise, friendship networks certainly drove the self-segmentation, but these too are shaped by race such that departure logically played out along race lines. The explanations teens gave for their decisions may not be explicitly about race, ethnicity, or class, but they cannot be untangled from them, just as fear-based narratives about the "ghetto" cannot be considered without also accounting for race, ethnicity, and class.

In some senses, the division in the perception and use of MySpace and Facebook seems obvious given that we know that online environments are a reflection of everyday life. Yet, the fact that such statements are controversial highlights a widespread techno-utopian belief that the internet will once and for all eradicate inequality and social divisions. What unfolded as teens adopted MySpace and Facebook suggests that this is not the case. Neither social media nor its users are colorblind simply because technology is present. The internet mirrors and magnifies everyday life, making visible many of the issues we hoped would disappear, including race and class-based social divisions in American society.

BIBLIOGRAPHY

Avila, Eric. 2004. "Popular Culture in the Age of White Flight: Film Noir, Disneyland, and the Cold War (Sub)Urban Imaginary." *Journal of Urban History*, 31(1): 3–22.

Bonilla-Silva, Eduardo. 2003. *Racism without Racists: Color-blind Racism and the Persistence of Racial Inequality in the United States.* Lanham, MD: Rowman and Littlefield.

Hare, Breanna. 2009. "Does Class Decide Online Social Networks?" *CNN*, October 24. http://www.cnn.com/2009/TECH/10/24/tech.networking.class/

Moody, James. 2001. "Race, School Integration, and Friendship Segregation in America." *American Journal of Sociology* 107(3), 679–716.

Nakamura, Lisa. 2008. *Digitizing Race: Visual Cultures of the Internet.* Minneapolis: University of Minnesota Press.

Perry, Pamela. 2002. *Shades of White: White Kids and Racial Identities in High School.* Durham, NC: Duke University Press.

Shrock, Andrew and danah boyd. 2009. "Online Threats to Youth: Solicitation, Harassment, and Problematic Content." *Enhancing Child Safety & Online Technologies* (eds. John Palfrey, Dena Sacco, danah boyd). Carolina Academic Press.

Tatum, Beverly. 1997. *'Why Are All the Black Kids Sitting Together in the Cafeteria' and other Conversations About Race.* New York: Basic Books.

Wacquant, Loic J.D. 1997. "Three Pernicious Premises in the Study of the American Ghetto." *International Journal of Urban and Regional Research*, 21(2): 341–353.

Watkins, S. Craig. 2009. *The Young & the Digital: What the Migration to Social-Network Sites, Games, and Anytime, Anywhere Media Means for Our Future.* Boston, MA: Beacon Press.

43

The Bachelor

Whiteness in the Harem

RACHEL E. DUBROFSKY

Over the course of the first season of *American Idol* in the summer of 2002, more people voted by phone to help select a winner than voted in the 2000 U.S. presidential election (Albiniak, 2002, p. 22). Contemporary reality-based programming has captured the attention of TV viewers in the United States, and television scholars have responded accordingly. However, little work on reality-based shows featuring the activities of everyday people focuses on women, although women figure centrally in the reality-based romance genre, or on the intersection of women and race, though people of color figure more prominently in reality-based shows than in scripted shows. This article outlines how *The Bachelor* is "raced": The series is a context in which only white people find romantic partners, while women of color work to facilitate the coupling of white people. I examine racial stereotypes, racial references, and the ways that narrative structures oppressively regulate the television text.

I base my noting of race on visible racial markers and comments by participants about their racial background. Some women are marked by their dark skin or physical features as women of color and are treated as such on the series. Other women who are not explicitly marked physically as women of color but are described with a specific ethnic heritage—for instance, Latin American descent—are marked by the series sometimes as women of color, and sometimes not. I do not put the word "white" in quotation marks here, although quotation marks would properly emphasize the term's instability.

I argue that the structure, mise-en-scène, and setting of *The Bachelor* evoke the Westernized trope of the Eastern harem (Ahmed, 1982; Shohat & Stam, 1994), enforcing timeworn racist gender dynamics and duplicating the trope's imperialist, Orientalist, and oppressive racist structures (Said, 1978). The very form of *The Bachelor* naturalizes the desire of white men for women of color as a means of preparing for union with their ultimate partners, white women. In *The Bachelor,* whiteness is an implicit prerequisite for finding a mate. While many of the white women do not find love with the bachelor, they may be the center of the storyline for one or more episodes. In fact, the more spectacularly the white women fail to become the bachelor's partner, the more screen time they get. This is not the case for women

SOURCE: From *Critical Studies in Media Communication*, Vol. 23, No. 1, March 2006, pp. 39–56. Reprinted by permission of the Taylor & Francis Group.

of color, who work only to frame the narrative about white people forming a romantic union.

THE BACHELOR'S SUCCESS

Originally aired in March 2002, *The Bachelor* was one of the first reality-based shows to focus on romance. It has proved the most enduring, with eight eight-week seasons aired to date. In *The Bachelor,* a man (a different white man each season) selects one woman from among 25 eligible women to be his potential bride. Based on information given in the first episode of seasons one, two, and six, the bachelors for those seasons were selected over other applicants because they were well-rounded, wanted marriage, had an established career, and boasted a great personality. Bachelor Bob Guiney was selected for season three because he was the most popular (among audiences) of the men rejected on the first season of *The Bachelorette.* Other bachelors were approached by producers to star on the series: Andrew Firestone from season four is heir to the Firestone fortune; Jesse Palmer from season five is a football player for the New York Giants; Charlie O'Connell from season seven is a minor Hollywood actor; and Travis Stork from season eight is an emergency doctor. In season two's special *The Bachelor Revealed* (no specials of this kind aired during other seasons), Chris Harrison, the host (who is white), revealed that when selecting from female applicants who applied for the series, producers "scrutinize" every aspect: "She must be single and between the ages of 21 and 35. She must be adventurous, ready for marriage. She should be intelligent. She should be ambitious."

In each season, the bachelor goes on a series of dates with the women (some one-on-one dates with "lucky" contestants and some group dates, during which individual women vie for his attention). At the end of each episode, at least one participant is eliminated during a "rose ceremony," during which the bachelor presents a rose to each of the women he wants to keep for the next week. In the first few rose ceremonies, the bachelor eliminates several women at once; when the group has been narrowed to four, he eliminates the women one at a time, until he selects his final choice. Thus far none of the bachelors have married the woman [he] selected; only two of the seven couples remain together at the time of this writing (at the time of this writing season eight had not ended yet).

Each season also includes two specials. In the *The Women Tell All* special, the second-to-last episode aired in the season, before the finale, the eliminated women discuss their experiences on the show. When this special airs, the audience does not yet know whom the bachelor has chosen. The *After the Final Rose* special is the last episode in the season, airing after the finale (where the bachelor makes his final selection). In this special, the couple is reunited....

Because the series uses the rhetoric of realism (through its form and content, the activities of "real" people doing "real" things), it naturalizes the constructions of race and romance it promotes to its audience. This article breaks down some of this "common-sense" rhetoric to ask how TV naturalizes the process of a white man and woman as the final pair. How is it that none of the ways women of color behave provides access to the central narrative? How does the harem structure situate race and experiences with racial "others" as both a necessary part of a white person's journey to finding a mate and a challenge to be overcome?

REALITY-BASED PROGRAMMING

The term "reality-based television" refers broadly to shows that are unscripted, though most have a very specific structure (with set tasks and events for each episode). The term purposely implies that the shows are based on reality without suggesting that they *are* reality—emphasizing the constructedness not only of reality-based programming but also of TV representations more generally. My analysis assumes that what occurs on reality-based shows is a constructed fiction, like the action on scripted shows, with the twist that real people create the fiction of the series. In other words, reality-based shows use footage of "real" people in "real" situations to create a fictional text, while scripted shows use a script to create the action. What happens on reality-based shows is not, of course, a representation of what "really" happened. The narrative is constructed by TV workers, sometimes using a tiny percentage of the footage actually shot. Like Stam (2000), I approach characters not as "real" people but rather as discursive constructions, and I look at "accents" and "intonations" … discernible in the televisual voice, examining which ambient ethnic voices are "heard" and which are elided or distorted. I assume, therefore, that the story about race in *The Bachelor* is assembled through the editing process. The production process shapes the final product, the "voices," accents," and intonations."

WOMEN OF COLOR IN "THE BACHELOR"

While women of color appear on the show, they do not thrive. In the first season, all four women of color in the initial pool were eliminated by the third week. In the second season, the last two of the three women of color voluntarily left during the second week. The only woman of color on the third season was eliminated in the first week. All three woman of color on the fourth season were eliminated by the sixth week. On the fifth season, all four women of color were eliminated by the second week. On the sixth season, the last of the three women of color in the original pool was eliminated by the fourth week. However, a Cuban American woman, Mary, joined the series on the third episode of the sixth season, and she was the woman bachelor Byron Velvick finally chose as his mate. This was the first and only time the series added women in the middle of a season (a white woman was added along with Mary). Mary had appeared in season four, when the series marked her explicitly as Cuban American. In season six her ethnicity was not mentioned until the second-to-last episode. I argue that while Mary was marked as a woman of color in season four, she was effectively "whitened" in season six. Because she was not marked physically as a woman of color, the series could represent her ethnicity in a mutable fashion. The only woman of color on the seventh season was eliminated at the first rose ceremony. The three women of color on the eighth season were eliminated in the first episode.

MAKING TOO MUCH OF TOO LITTLE?

Focusing on women of color in *The Bachelor* is tricky: Because they figure so little, I risk seeming to be making "too much" of "too little." Indeed, nearly all women are eventually eliminated each season. Perhaps it is statistically insignificant that only one woman who was marked as a woman of color (and not marked as such in the season when the bachelor proposed to her) ended up with a bachelor. Possibly most of the women of color never "clicked" romantically with the bachelors.

The series invites us to consider race within the logic of relational choice, rather than within the logic of representation or of production. What is crucial is how consistently the series marks the presence of the women of color. Appadurai suggests that universities often encourage diversity on "the principle that more difference is better" but frequently fail to create "a habitus where diversity is at the heart of the apparatus itself" (1996, p. 26). This is also true of television. *The Bachelor* habitus is such that women of color exist but are mostly irrelevant to the dominant narrative, except to the extent that their actions work to frame the white women's access to this narrative, or to frame the white bachelor's journey to finding his ideal mate. In other words, the show's racism is not overt. At times it is difficult to pin down. This is what Hall calls "inferential racism": when racist representations are unspoken and naturalized, making the racist premises upon which the representations rely difficult to bring to the surface (2003, p. 91).

FRAMING WHITE WOMEN'S ILLEGITIMATE BEHAVIOR

Women of color on *The Bachelor* verify the behavior of white women and thus re-center whiteness. Women of color generally frame the white women's illegitimate behavior—their excessive emotional behavior—thus highlighting that these particular women are unsuitable matches for the bachelor. For example, Anindita, a South Asian American woman who voluntarily left at the second rose ceremony on the second season, caused quite a stir during a group date with bachelor Aaron Buerge and four white women by confronting Christi about her negative feelings about Suzanne. The central focus of this scene, however, is on how hurt Christi was by the day's events. In fact, during the confrontation, Anindita's voice was heard but she appeared only in the corner of the frame. What was apparent was Christi's emotion, given the close-ups of her teary face. While Anindita exposed the tension between the two women— even inciting it—she was never central to this narrative, much less to the romance narrative with the bachelor. Anindita's role was to highlight Christi's excessive emotionality.

Karin, an African American woman, acted to frame and re-center the presence of a white woman, Lee-Ann, in the fourth season. Although Karin made it to the fourth rose ceremony, she was never seen interacting with the bachelor. That she remained on the show for so long perhaps indicates some interaction with Bob, yet the focus was on Bob's interactions with the other women, never with her. In fact, the series referred to her in only three ways: as beautiful (by Bob and the other women); as high-maintenance (by the other women); and as a great friend (by Lee-Ann).

Karin had little or no screen time. Although the other women referred to Karin as high-maintenance and commented on her lack of enthusiasm for the day's group date at a water-park, this received little attention. Nor did her friendship with Lee-Ann, a much disliked fellow-participant, emerge as an issue. The absence of any fuss over Karin's marked lack of enthusiasm or her closeness to a detested participant is curious, considering that whenever any of the (white) women set themselves apart from the main action of the series (by being disdainful of activities, aloof, self-absorbed, and so forth), this became a focus of at least one episode and thereafter marked the woman negatively. In essence, although the series presented Karin behaving like some of the white women, she did not signify in the same ways.

Karin's most notable moment was when she tried to comfort Lee-Ann, who was furious with the other women and with the bachelor. At this point, we heard Karin speak for longer (and not very long at that) than she had at any other time. Lee-Ann had been the dramatic center of much of the last four episodes (because she set herself apart from the other women and incurred their disdain). Karin's interactions with her foregrounded this drama. To some extent, … Karin was very visible but had no voice. Comments about Karin repeatedly referenced her beauty, yet she herself rarely spoke and little attention was paid to her behavior.

Frances, an Asian American woman on the second season (who left voluntarily), figured only nominally on the show, but she framed the over-emotional behavior of a white woman, Heather. The series did not show Frances interacting with the bachelor on her only date with him (a group date), although she did narrate a lot of the action on the date. However, in *The Women Tell All* special, Frances became the catalyst for an emotional outburst from Heather. Frances joked about how Heather cooked lots of good food for the women in an attempt to fatten them up. While all the women laughed, the next shot showed Heather bursting into tears. When the host asked what was wrong, Heather claimed Frances's comment hurt her. Frances looked confused but apologized. Again, a woman of color's fleeting presence served to frame the characterization of a white woman as excessively emotional.

If the explicit aim of the series is to promote romance (ideally leading to marriage, or at least to a long-term commitment) and winners are those who succeed in becoming the star of the romance narrative, then the overriding message is that women of color do not count. They are positioned as neither legitimate nor illegitimate romantic partners for the bachelor. Yet the women of color are a vital part of the story of two white people finding a partner: They verify the bachelor's choices by highlighting the unsuitability of certain white women; they serve as window dressing for the white women, giving them a "special flavor, an added spice" (Hooks, 1992, p. 157).

CHOOSING TO LEAVE THE SHOW

Despite the show's avowed attempt to offer *all* the women the opportunity for love with the bachelor and the power to make choices, these are seriously circumscribed when it comes to women of color. Gray (1995) explains that what he calls assimilationist programs "celebrate racial invisibility and color blindness ... [by integrating] individual black characters into hegemonic white worlds void of any hint of African American traditions, social struggle, racial conflicts, and cultural difference" (p. 85). In these texts, in contrast to the hegemonic status of whiteness, Gray adds: "Blackness simply works to reaffirm, shore up, and police the cultural and moral boundaries of the existing racial order. From the privileged angle of their normative race and class positions, whites are portrayed as sympathetic advocates for the elimination of prejudice" (1995, p. 87). The overriding assimilationist paradigm of *The Bachelor* is apparent in the very set-up of the series: By pairing white men with women of color as open to the possibility of romance, the series constructs them as implicitly willing to engage in interracial relationships. On the surface, the series operates as if color does not matter, as if people in the series (and implicitly the makers of the show) are neutral when it comes to racial differences, or cultural differences read as racial ones, and will treat everyone as if these differences do not exist. Everyone, ostensibly, can compete to win the rewards of the show— finding a romantic partner. The suggestion is that white women and women of color have access to the same choices, will benefit from the same rewards, and suffer the same consequences for the choices they make. However, the choices and opportunities afforded women of color do not allow access to the central romance narrative, as they do for white women. The actions of women of color do not bring them closer to winning the bachelor's heart, though it is often through the actions of women of color that white women lose access to the bachelor's heart....

HAREM STRUCTURE

Shohat suggests that one way to look at texts is to explore their "*inferential ethnic presences,* that is, the various ways in which ethnic cultures penetrate the screen without always literally being represented by ethnic and racial themes or even characters" (1991, p. 223). While the premise of the series, to find the bachelor a long–term romantic partner, may not immediately bring to mind the concept of the harem, the very set-up of *The Bachelor* implicitly references the harem: one man with 25 beautiful women who live in the same quarters and are always at the bachelor's disposal. In fact, the women have little to do but lounge around and wait to share time with the bachelor....

The women on *The Bachelor* are often so similar in appearance (e.g., body size and skin color) that, at least visually, they may at times appear to be interchangeable. The structure of the show is such that the supply of willing women—willing to make themselves accessible to this one man, and no other man, for the duration of the show—is endless. The series adds to the harem structure the notion of choice: The women's willingness is bolstered by the paradigm of choice since the women are told that they can choose either to leave or to stay and accept the will of the bachelor (who will decide if they stay or leave)....

The décor of the bachelor pad mimics this construction of the harem with its sumptuous, boudoir-like furniture; the array of sitting rooms with stuffed couches, throw rugs, and oversized pillows; and wall hangings in rich dark colors. Of course, the requisite hot-tub and swimming pool are nestled in a lush garden, with many verdant private settings for midnight trysts. The harem décor persists throughout the show. Almost every date is located near a pool and a hot tub (or this is the final destination of the date), and many private rooms (with candles, carpets, pillows, and a fireplace) are conducive to intimacy.

The bachelor sometimes has the opportunity to share intimacies with several women in a single night. For example, on a group date he may spend some time with all the women together and then divide his time between kissing one woman in a secluded garden area, "making-out" with another in an intimate boudoir-like room, and embracing yet a third in a hot tub. The sexually charged atmosphere is enhanced by the "far-Eastern," "Orientalist" themes of the activities as well as the setting, many requiring participants to stay low to the ground, ever ready for a "tumble in the hay." In season one, for example, bachelor Alex Michael and Amanda enjoyed sushi in a private room with vaguely "Oriental" red tapestries on the wall; they sat at a low table with a carved-out floor for their feet (it looked like they were sitting on the floor). After dinner, they rolled over onto the floor and "made out." Then they moved into another private room, donned kimonos, and gave each other full-body massages. In season four, bachelor Bob and three of the women went to a karaoke pajama party. Wearing negligée-like attire, the women sang to Bob and lounged together on deep plush red cushions. Bob "made out" with one of the women in the more private, curtained-off rooms in the back, adorned with a bed and pillows in deep shades of red and an "Oriental"-looking red tapestry on one wall. On another date, Bob and several women went to a private cabaret club where the women went on stage and danced for him in erotic poses. In the third season, bachelor Andrew took a few of the women to share Ethiopian food, which they ate with their hands off a low table, sitting on the floor in a room decorated with "Middle-Eastern"-looking tapestries. At the end of the meal, they lounged on cushions on the floor and were entertained by belly dancers who encouraged the women to dance for Andrew. In the fifth season, bachelor Jesse took three women on a date to an opulent tent set outdoors where they were greeted by a live elephant. The set designers decked out the tent in "Middle-Eastern"-looking décor, with gilded red pillows strewn across the floor, gilded red tapestries hanging from the walls, a low table at the center of the room, and "Oriental" rugs layered on the ground. "Middle-Eastern"-sounding music played in the background. Jesse spent part of the date doing what one might expect in this setting: rolling around and "making out" with one of the women. Over the course of the date, he kissed all three women, either in the tent or in secluded areas outside.

The harem structure was even more explicit in the sixth season, when bachelor Byron moved into a guest house in the garden behind the bachelorette pad where the women lived, giving him unlimited access to the women. In episode four, for example, he organized a pajama party at the women's house, providing the lingerie for the women to wear. While the women had access to the bachelor, Byron decided just how much: in the fifth episode, for example, Byron insisted that Jayne, who came to visit him, leave his quarters because, as he told us in voice-over, he did not want to "complicate" things by having any of the women spend the night. When the bachelor is not present, the activities of the women are staged for the pleasure of "the scopic privilege of the master in an exclusionary space inaccessible to other men" (Shohat & Stam, 1994, p. 164): We witness them navigating the difficulties of living together while they frolic about in bikinis, prance from hot tub to pool, play lawn games, drink lots of alcohol, and, of course, gossip and fight with one another.

The harem décor accentuates the sexual possibilities at the bachelor's disposal and emphasizes that the women are to be ever-ready and willing to make use of the cushy privacy (cameras notwithstanding) afforded by this setting. The space of *The Bachelor* is one where sexual dilettantism is to be celebrated and explored, part of the Western myth about the Eastern harem, a place of sexual excess, of limitless pleasure for Western men (Ahmed, 1982; Alloula, 1986). The bachelor is the Sheik of this realm, the white playboy who has arrived in a dark land to frolic before returning to more serious ventures and to his white leading lady. On the road to finding a mate, the bachelor has numerous opportunities for lusty forays with many women who await the pleasure of being conquered by him.

EXOTIC STOPS EN ROUTE TO FINDING A MATE

The permeation of the "Eastern" via the implicit (never explicitly referenced) trope of the harem is emblematic of the way *The Bachelor* works through both implicit and explicit issues having to do with race: Racial diversity frames the central narrative of two white people forming a romantic union. Diversity ultimately works to maintain pedagogies of whiteness, while situating diversity as essential in defining and highlighting whiteness. ... The white couple can fully enjoy and partake in the sensual rituals of the East (when in Rome do as the Romans do), while in the end distancing themselves from the very same eroticized culture that enabled them to find each other. People of color in *The Bachelor* are part of the backdrop, the setting, of the otherwise white narrative. While the bachelor may frolic with women of color during his sojourn in the harem, in the end, he leaves the harem with his chosen white partner....

THE WHITE HERO AND HIS MISTRESS

The sexual gallivanting of the bachelor is "otherized" within the larger romance narrative of the series; surely the bachelor is to use this time to sow his wild oats with the express purpose of settling down with one woman at the end, presumably in a home with a white picket fence and very few Persian carpets or red pillows and certainly no elephant grazing on the front lawn. Male protagonists in *The Bachelor,* like the protagonists in classic Hollywood harem films, explore the evils of the Westernized version of the Eastern harem before overcoming these excesses in the interest of sustaining a monogamous union. The bachelors transgress Western monogamous norms via the trope of the Western version of the Eastern harem in order to reaffirm Western norms. A romantic union for a white man can only be found with one white woman; being in love means not lusting after others. The harem experience and the women who are part of this process are, to some extent, the grotesque but necessary other (Stallybrass and White, 1986).

Just as the harem narrative is embedded in the Western narrative of monogamous love, portrayals of people of color are embedded in the defining of whiteness. Through the trope of the Eastern harem, *The Bachelor*'s racist, imperialist structure tells a story about the romantic heterosexual union of two white people; it relies on the myth of white man's ability to conquer the dark Other to find his way towards the ideal white mate. Women of color and white women serve different purposes in this Orientalist set-up. White women can rise above the harem structure to become the bachelor's chosen woman, for whom he will forsake the harem's pleasures. Unless they can be whitened, however, women of color cannot.

By locating the harem structure in the United States (in Hollywood, the harem remains in the East but at the end the white couple returns to the West), the series reinscribes not only imperialist desires, but also the historical power dynamic of slavery in the United States. Given the history of slavery in the United States, and the relationship of women of color to their white masters— as sexual slaves, as mothers to their illegitimate children, but never as legitimate romantic partners (wives) or as equal citizens—the positioning of women of color in *The Bachelor* is particularly disturbing. Here again women of color fulfill the time worn roles of satisfying the sexual desires of white men and taking care of white men (by highlighting the inappropriate and potentially dangerous behaviors of white women who want to marry white men). *The Bachelor* tells a very specific story about whiteness, where whiteness is essential to finding a romantic partner. Once women of color help the white heroes find one another, they must disappear into the background—just as the harem structure disappears to let the white master and his chosen mistress take center stage.

REFERENCES

Ahmed, L. (1982). Western ethnocentrism and perceptions of the harem. *Feminist Studies,* 8, 521–534.

Albiniak, P. (2002, September 9). Ideal, not idle summer for Fox. *Broadcasting and Cable,* p. 22.

Alloula, M. (1986). *The colonial harem.* Minneapolis, MN: University of Minnesota Press.

Appadurai, A. (1996). Diversity and disciplinarity as cultural artifacts. In C. Nelson & D. Parameshwar Gaonkar (Eds.), *Disciplinarity and dissent in cultural studies* (pp. 23–36). New York: Routledge.

Gray, H. (1995). *Watching race: Television and the struggle for "blackness."* Minneapolis, MN: University of Minnesota Press.

Hall, S. (2003). The whites of their eyes. In G. Dines & J. M. Humez (Eds.), *Gender, race, and class in media: A text-reader* (pp. 89–93) (2nd ed.). London: Sage Publications.

Hooks, B. (1992). Madonna: Plantation mistress or soul sister. *Black looks: Race and representation* (pp. 157–164). Toronto: Between the Lines.

Said, E. W. (1978). *Orientalism.* New York: Vintage Books.

Shaheen, J. G. (2001). *Reel bad Arabs: How Hollywood vilifies a people.* New York: Olive Branch Press.

Shohat, E. (1991). Ethnicities-in-relation: Toward a multicultural reading of American cinema. In L. D. Friedman (Ed.), *Unspeakable images: Ethnicity and the American cinema* (pp. 215–250). Chicago: University of Illinois Press.

Shohat, E., & Stam, R. (1994). *Unthinking Eurocentrism: Multiculturalism and the media.* New York: Routledge.

Stallybrass, P., & White, A. (1986). *Transgression: The politics and poetics of transgression.* Ithaca, NY: Cornell University Press.

Stam, R. (2000). Bakhtin, polyphony, and ethnic/racial representation. In L. D. Friedman (Ed.), *Unspeakable images: Ethnicity and the American cinema* (pp. 251–276). Chicago: University of Illinois Press.

44

Crimes Against Humanity

WARD CHURCHILL

If nifty little "pep" gestures like the "Indian Chant" and the "Tomahawk Chop" are just good clean fun, then let's spread the fun around, shall we?

During the past couple of seasons, there has been an increasing wave of controversy regarding the names of professional sports teams like the Atlanta "Braves," Cleveland "Indians," Washington "Redskins," and Kansas City "Chiefs." The issue extends to the names of college teams like Florida State University "Seminoles," University of Illinois "Fighting Illini," and so on, right on down to high school outfits like the Lamar (Colorado) "Savages." Also involved have been team adoption of "mascots," replete with feathers, buckskins, beads, spears, and "warpaint" (some fans have opted to adorn themselves in the same fashion), and nifty little "pep" gestures like the "Indian Chant" and "Tomahawk Chop."

A substantial number of American Indians have protested that use of native names, images, and symbols as sports team mascots and the like is, by definition, a virulently racist practice. Given the historical relationship between Indians and non-Indians during what has been called the "Conquest of America," American Indian Movement leader (and American Indian Anti-Defamation Council founder) Russell Means has compared the practice to contemporary Germans naming their soccer teams the "Jews," "Hebrews," and "Yids," while adorning their uniforms with grotesque caricatures of Jewish faces taken from the Nazis' anti-Semitic propaganda of the 1930s. Numerous demonstrations have occurred in conjunction with games—most notably during the November 15, 1992 match-up between the Chiefs and Redskins in Kansas City—by angry Indians and their supporters.

In response, a number of players—especially African Americans and other minority athletes—have been trotted out by professional team owners like Ted Turner, as well as university and public school officials, to announce that they mean not to insult but to honor native people. They have been joined by the television networks and most major newspapers, all of which have editorialized that Indian discomfort with the situation is "no big deal," insisting that the whole thing is just "good, clean fun." The country needs more such fun, they've argued, and "a few disgruntled Native Americans" have no right to undermine the nation's enjoyment of its leisure time by complaining. This is especially the case, some have argued, "in hard times like these." It has even been contended that

SOURCE: From Ward Churchill, "Crimes Against Humanity," *Z Magazine* 6 (March 1993): 43–47. Reprinted by permission of the author.

Indian outrage at being systematically degraded—rather than the degradation itself—creates "a serious barrier to the sort of intergroup communication so necessary in a multicultural society such as ours."

Okay, let's communicate. We are frankly dubious that those advancing such positions really believe their own rhetoric, but, just for the sake of argument, let's accept the premise that they are sincere. If what they say is true, then isn't it time we spread such "inoffensiveness" and "good cheer" around among *all* groups so that *everybody* can participate *equally* in fostering the round of national laughs they call for? Sure it is—the country can't have too much fun or "inter-group involvement"—so the more, the merrier. Simple consistency demands that anyone who thinks the Tomahawk Chop is a swell pastime must be just as hearty in their endorsement of the following ideas—by the logic used to defend the defamation of American Indians—should help us all really start yukking it up.

First, as a counterpart to the Redskins, we need an NFL team called "Niggers" to honor Afro-Americans. Half-time festivities for fans might include a simulated stewing of the opposing coach in a large pot while players and cheerleaders dance around it, garbed in leopard skins and wearing fake bones in their noses. This concept obviously goes along with the kind of gaiety attending the Chop, but also with the actions of the Kansas City Chiefs, whose team members—prominently including black team members—lately appeared on a poster looking "fierce" and "savage" by way of wearing Indian regalia. Just a bit of harmless "morale boosting," says the Chiefs' front office. You bet.

So that the newly formed Niggers sports club won't end up too out of sync while expressing the "spirit" and "identity" of Afro-Americans in the above fashion, a baseball franchise—let's call this one the "Sambos"—should be formed. How about a basketball team called the "Spearchuckers"? A hockey team called the "Jungle Bunnies"? Maybe the "essence" of these teams could be depicted by images of tiny black faces adorned with huge pairs of lips. The players could appear on TV every week or so gnawing on chicken legs and spitting watermelon seeds at one another. Catchy, eh? Well, there's "nothing to be upset about," according to those who love wearing "war bonnets" to the Super Bowl or having "Chief Illiniwik" dance around the sports arenas of Urbana, Illinois.

And why stop there? There are plenty of other groups to include. Hispanics? They can be "represented" by the Galveston "Greasers" and San Diego "Spics," at least until the Wisconsin "Wetbacks" and Baltimore "Beaners" get off the ground. Asian Americans? How about the "Slopes," "Dinks," "Gooks," and "Zipperheads?" Owners of the latter teams might get their logo ideas from editorial page cartoons printed in the nation's newspapers during World War II: slant-eyes, buck teeth, big glasses, but nothing racially insulting or derogatory, according to the editors and artists involved at the time. Indeed, this Second World War—vintage stuff can be seen as just another barrel of laughs, at least by what current editors say are their "local standards" concerning American Indians.

Let's see. Who's been left out? Teams like the Kansas City "Kikes," Hanover "Honkies," San Leandro "Shylocks," Daytona "Dagos," and Pittsburgh "Polacks" will fill a certain social void among white folk. Have a religious belief?

Let's all go for the gusto and gear up the Milwaukee "Mackerel Snappers" and Hollywood "Holy Rollers." The Fighting Irish of Notre Dame can be rechristened the "Drunken Irish" or "Papist Pigs." Issues of gender and sexual preferences can be addressed through creation of teams like the St. Louis "Sluts," Boston "Bimbos," Detroit "Dykes," and the Fresno "Fags." How about the Gainesville "Gimps" and Richmond "Retards," so the physically and mentally impaired won't be excluded from our fun and games?

Now, don't go getting "overly sensitive" out there. None of this is demeaning or insulting, at least not when it's being done to Indians. Just ask the folks who are doing it, or their apologists like Andy Rooney in the national media. They'll tell you in fact they *have* been telling you—that there's been no harm done, regardless of what their victims think, feel, or say. The situation is exactly the same as when those with precisely the same mentality used to insist that Step 'n' Fetchit was okay, or Rochester on the *Jack Benny Show,* or Amos and Andy, Charlie Chan, the Frito Bandito, or any of the other cutesy symbols making up the lexicon of American racism. Have we communicated yet?

Let's get just a little bit real here. The notion of "fun" embodied in rituals like the Tomahawk Chop must be understood for what it is. There's not a single non-Indian example used above which can be considered socially acceptable in even the most marginal sense. The reasons are obvious enough. So why is it different where American Indians are concerned? One can only conclude that, in contrast to the other groups at issue, Indians are (falsely) perceived as being too few, and therefore too weak, to defend themselves effectively against racist and otherwise offensive behavior.

Fortunately, there are some glimmers of hope. A few teams and their fans have gotten the message and have responded appropriately. Stanford University, which opted to drop the name "Indians" from Stanford, has experienced no resulting drop-off in attendance. Meanwhile, the local newspaper in Portland, Oregon, recently decided its long-standing editorial policy prohibiting use of racial epithets should include derogatory team names. The Redskins, for instance, are now referred to as "the Washington team," and will continue to be described in this way until the franchise adopts an inoffensive moniker (newspaper sales in Portland have suffered no decline as a result).

Such examples are to be applauded and encouraged. They stand as figurative beacons in the night, proving beyond all doubt that it is quite possible to indulge in the pleasure of athletics without accepting blatant racism into the bargain.

NUREMBERG PRECEDENTS

On October 16, 1946, a man named Julius Streicher mounted the steps of a gallows. Moments later he was dead, the sentence of an international tribunal composed of representatives of the United States, France, Great Britain, and the Soviet Union having been imposed. Streicher's body was then cremated, and—so horrendous were his crimes thought to have been—his ashes dumped into an unspecified German river so that "no one should ever know a particular place to go for reasons of mourning his memory."

Julius Streicher had been convicted at Nuremberg, Germany, of what were termed "Crimes Against Humanity." The lead prosecutor in his case—Justice Robert Jackson of the United States Supreme Court—had not argued that the defendant had killed anyone, nor that he had personally committed any especially violent act. Nor was it contended that Streicher had held any particularly important position in the German government during the period in which the so-called Third Reich had exterminated some 6,000,000 Jews, as well as several million Gypsies, Poles, Slavs, homosexuals, and other untermenschen (subhumans).

The sole offense for which the accused was ordered put to death was in having served as publisher/editor of a Bavarian tabloid entitled *Der Sturmer* during the early-to-mid 1930s, years before the Nazi genocide actually began. In this capacity, he had penned a long series of virulently anti-Semitic editorials and "news" stories, usually accompanied by cartoons and other images graphically depicting Jews in extraordinarily derogatory fashion. This, the prosecution asserted, had done much to "dehumanize" the targets of his distortion in the mind of the German public. In turn, such dehumanization had made it possible—or at least easier—for average Germans to later indulge in the outright liquidation of Jewish "vermin." The tribunal agreed, holding that Streicher was therefore complicit in genocide and deserving of death by hanging.

During his remarks to the Nuremberg tribunal, Justice Jackson observed that, in implementing its sentences, the participating powers were morally and legally binding themselves to adhere forever after to the same standards of conduct that were being applied to Streicher and the other Nazi leaders. In the alternative, he said, the victorious allies would have committed "pure murder" at Nuremberg—no different in substance from that carried out by those they presumed to judge—rather than establishing the "permanent benchmark for justice" which was intended.

Yet in the United States of Robert Jackson, the indigenous American Indian population had already been reduced, in a process which is ongoing to this day, from perhaps 12.5 million in the year 1500 to fewer than 250,000 by the beginning of the 20th century. This was accomplished, according to official sources, "largely through the cruelty of [Euro-American] settlers," and an informal but clear governmental policy which had made it an articulated goal to "exterminate these red vermin," or at least whole segments of them.

Bounties had been placed on the scalps of Indians—any Indians—in places as diverse as Georgia, Kentucky, Texas, the Dakotas, Oregon, and California, and had been maintained until resident Indian populations were decimated or disappeared altogether. Entire peoples such as the Cherokee had been reduced to half their size through a policy of forced removal from their homelands east of the Mississippi River to what were then considered less preferable areas in the West.

Others, such as the Navajo, suffered the same fate while under military guard for years on end. The United States Army had also perpetrated a long series of wholesale massacres of Indians at places like Horseshoe Bend, Bear

River, Sand Creek, the Washita River, the Marias River, Camp Robinson, and Wounded Knee.

Through it all, hundreds of popular novels—each competing with the next to make Indians appear more grotesque, menacing, and inhuman—were sold in the tens of millions of copies in the U.S. Plainly, the Euro-American public was being conditioned to see Indians in such a way as to allow their eradication to continue. And continue it did until the Manifest Destiny of the U.S.—a direct precursor to what Hitler would subsequently call Le bens raum politik (the politics of living space)—was consummated.

By 1900, the national project of "clearing" Native Americans from their land and replacing them with "superior" Anglo-American settlers was complete: the indigenous population had been reduced by as much as 98 percent while approximately 97.5 percent of their original territory had "passed" to the invaders. The survivors had been concentrated, out of sight and mind of the public, on scattered "reservations," all of them under the self-assigned "plenary" (full) power of the federal government. There was, of course, no Nuremberg-style tribunal passing judgment on those who had fostered such circumstances in North America. No U.S. official or private citizen was ever imprisoned—never mind hanged—for implementing or propagandizing what had been done. Nor had the process of genocide afflicting Indians been completed. Instead, it merely changed form.

Between the 1880s and the 1980s, nearly half of all Native American children were coercively transferred from their own families, communities, and cultures to those of the conquering society. This was done through compulsory attendance at remote boarding schools, often hundreds of miles from their homes, where native children were kept for years on end while being systematically "deculturated" (indoctrinated to think and act in the manner of Euro-Americans rather than Indians). It was also accomplished through a pervasive foster home and adoption program—including "blind" adoptions, where children would be permanently denied information as to who they were/are and where they'd come from—placing native youths in non-Indian homes.

The express purpose of all this was to facilitate a U.S. governmental policy to bring about the "assimilation" (dissolution) of indigenous societies. In other words, Indian cultures as such were to be caused to disappear. Such policy objectives are directly contrary to the United Nations 1948 Convention on Punishment and Prevention of the Crime of Genocide, an element of international law arising from the Nuremberg proceedings. The forced "transfer of the children" of a targeted "racial, ethnical, or religious group" is explicitly prohibited as a genocidal activity under the Convention's second article.

Article II of the Genocide Convention also expressly prohibits involuntary sterilization as a means of "preventing births among" a targeted population. Yet, in 1975, it was conceded by the U.S. government that its Indian Health Service (IHS), then a subpart of the Bureau of Indian Affairs (BIA), was even then conducting a secret program of involuntary sterilization that had affected approximately 40 percent of all Indian women. The program was allegedly discontinued, and the IHS was transferred to the Public Health Service, but no one was punished. In 1990, it came out that the IHS was inoculating Inuit children in Alaska with Hepatitis-B vaccine. The vaccine had already been banned by the World Health Organization as having a demonstrated correlation with the HIV-Syndrome which is itself correlated to AIDS. As this is written, a "field test" of Hepatitis-A vaccine, also HIV-correlated, is being conducted on Indian reservations in the northern plains region.

The Genocide Convention makes it a "crime against humanity" to create conditions leading to the destruction of an identifiable human group, as such. Yet the BIA has utilized the government's plenary prerogatives to negotiate mineral leases "on behalf of" Indian peoples paying a fraction of standard royalty rates. The result has been "super profits" for a number of preferred U.S. corporations. Meanwhile, Indians, whose reservations ironically turned out to be in some of the most mineral-rich areas of North America, which makes us, the nominally wealthiest segment of the continent's population, live in dire poverty.

By the government's own data in the mid-1980s, Indians received the lowest annual and lifetime per capita incomes of any aggregate population group in the United States. Concomitantly, we suffer the highest rate of infant mortality, death by exposure and malnutrition, disease, and the like. Under such circumstances, alcoholism and other escapist forms of substance abuse are endemic in the Indian community, a situation which leads both to a general physical debilitation of the population and a catastrophic accident rate. Teen suicide among Indians is several times the national average.

The average life expectancy of a reservation-based Native American man is barely 45 years; women can expect to live less than three years longer.

Such itemizations could be continued at great length, including matters like the radioactive contamination of large portions of contemporary Indian Country, the forced relocation of traditional Navajos, and so on. But the point should be made: genocide, as defined in international law, is a continuing fact of day-to-day life (and death) for North America's native peoples. Yet there has been—and is— only the barest flicker of public concern about, or even consciousness of, this reality. Absent any serious expression of public outrage, no one is punished and the process continues.

A salient reason for public acquiescence before the ongoing holocaust in Native North America has been a continuation of the popular legacy, often through more effective media. Since 1925, Hollywood has released more than 2,000 films, many of them rerun frequently on television, portraying Indians as strange, perverted, ridiculous, and often dangerous things of the past. Moreover, we are habitually presented to mass audiences one-dimensionally, devoid of recognizable human motivations and emotions; Indians thus serve as props, little more. We have thus been thoroughly and systematically dehumanized.

Nor is this the extent of it. Everywhere, we are used as logos, as mascots, as jokes: "Big-Chief" writing tablets, "Red Man" chewing tobacco, "Winnebago" campers, "Navajo" and "Cherokee" and "Pontiac" and "Cadillac" pickups and automobiles. There are the Cleveland "Indians," the Kansas City "Chiefs," the Atlanta "Braves" and the Washington "Redskins" professional sports teams—not to mention those in thousands of colleges, high schools, and elementary schools across the country—each with their own degrading caricatures and parodies of Indians and/or things Indian. Pop fiction continues in the same vein, including an unending stream of New Age manuals purporting to expose the inner works of indigenous spirituality in everything from pseudo-philosophical to do-it-yourself styles. Blond yuppies from Beverly Hills amble about the country claiming to be reincarnated 17th century Cheyenne Ushamans ready to perform previously secret ceremonies.

In effect, a concerted, sustained, and in some ways accelerating effort has gone into making Indians unreal. It is thus of obvious importance that the American public begin to think about the implications of such things the next time they witness a gaggle of face-painted and war-bonneted buffoons doing the "Tomahawk Chop" at a baseball or football game. It is necessary that they think about the implications of the grade-school teacher adorning their child in turkey feathers to commemorate Thanksgiving. Think about the significance of John Wayne or Charlton Heston killing a dozen "savages" with a single bullet the next time a western comes on TV. Think about why Land-o-Lakes finds it appropriate to market its butter with the stereotyped image of an "Indian princess" on the wrapper. Think about what it means when non-Indian academics profess—as they often do—to "know more about Indians than Indians do themselves." Think about the significance of charlatans like Carlos Castaneda and Jamake Highwater and Mary Summer Rain and Lynn Andrews churning out "Indian" bestsellers, one after the other, while Indians typically can't get into print.

Think about the real situation of American Indians. Think about Julius Streicher. Remember Justice Jackson's admonition. Understand that the treatment of Indians in American popular culture is not "cute" or "amusing" or just "good, clean fun."

Know that it causes real pain and real suffering to real people. Know that it threatens our very survival. And know that this is just as much a crime against humanity as anything the Nazis ever did. It is likely that the indigenous people of the United States will never demand that those guilty of such criminal activity be punished for their deeds. But the least we have the right to expect—indeed, to demand—is that such practices finally be brought to a halt.

45

Media Magic
Making Class Invisible

GREGORY MANTSIOS

O f the various social and cultural forces in our society, the mass media is arguably the most influential in molding public consciousness. Americans spend an average twenty-eight hours per week watching television. They also spend an undetermined number of hours reading periodicals, listening to the radio, and going to the movies. Unlike other cultural and socializing institutions, ownership and control of the mass media is highly concentrated. Twenty-three corporations own more than one-half of all the daily newspapers, magazines, movie studios, and radio and television outlets in the United States.[1] The number of media companies is shrinking and their control of the industry is expanding. And a relatively small number of media outlets is producing and packaging the majority of news and entertainment programs. For the most part, our media is national in nature and single-minded (profit-oriented) in purpose. This media plays a key role in defining our cultural tastes, helping us locate ourselves in history, establishing our national identity, and ascertaining the range of national and social possibilities. In this essay, we will examine the way the mass media shapes how people think about each other and about the nature of our society.

The United States is the most highly stratified society in the industrialized world. Class distinctions operate in virtually every aspect of our lives, determining the nature of our work, the quality of our schooling, and the health and safety of our loved ones. Yet remarkably, we, as a nation, retain illusions about living in an egalitarian society. We maintain these illusions, in large part, because the media hides gross inequities from public view. In those instances when inequities are revealed, we are provided with messages that obscure the nature of class realities and blame the victims of class-dominated society for their own plight. Let's briefly examine what the news media, in particular, tells us about class.

ABOUT THE POOR

The news media provides meager coverage of poor people and poverty. The coverage it does provide is often distorted and misleading.

SOURCE: From *Race, Class, and Gender in the United States: An Integrated Study*, 4th ed. Paula Rothenberg, ed. (New York: St. Martin's Press, 1998). Reprinted with permission of the author.

The Poor Do Not Exist

For the most part, the news media ignores the poor. Unnoticed are forty million poor people in the nation—a number that equals the entire population of Maine, Vermont, New Hampshire, Connecticut, Rhode Island, New Jersey, and New York combined. Perhaps even more alarming is that the rate of poverty is increasing twice as fast as the population growth in the United States. Ordinarily, even a calamity of much smaller proportion (e.g., flooding in the Midwest) would garner a great deal of coverage and hype from a media usually eager to declare a crisis, yet less than one in five hundred articles in the *New York Times* and one in one thousand articles listed in the *Readers Guide to Periodic Literature* are on poverty. With remarkably little attention to them, the poor and their problems are hidden from most Americans.

When the media does turn its attention to the poor, it offers a series of con-tradictory messages and portrayals.

The Poor Are Faceless

Each year the Census Bureau releases a new report on poverty in our society and its results are duly reported in the media. At best, however, this coverage empha-sizes annual fluctuations (showing how the numbers differ from previous years) and ongoing debates over the validity of the numbers (some argue the number should be lower, most that the number should be higher). Coverage like this desensitizes us to the poor by reducing poverty to a number. It ignores the human tragedy of poverty—the suffering, indignities, and misery endured by millions of children and adults. Instead, the poor become statistics rather than people.

The Poor Are Undeserving

When the media does put a face on the poor, it is not likely to be a pretty one. The media will provide us with sensational stories about welfare cheats, drug addicts, and greedy panhandlers (almost always urban and Black). Compare these images and the emotions evoked by them with the media's treatment of middle-class (usually white) "tax evaders," celebrities who have a "chemical dependency," or wealthy businesspeople who use unscrupulous means to "make a profit." While the behavior of the more affluent offenders is considered an "impropriety" and a deviation from the norm, the behavior of the poor is considered repugnant, indic-ative of the poor in general, and worthy of our indignation and resentment.

The Poor Are an Eyesore

When the media does cover the poor, they are often presented through the eyes of the middle class. For example, sometimes the media includes a story about community resistance to a homeless shelter or storekeeper annoyance with pan-handlers. Rather than focusing on the plight of the poor, these stories are about middle-class opposition to the poor. Such stories tell us that the poor are an inconvenience and an irritation.

The Poor Have Only Themselves to Blame

In another example of media coverage, we are told that the poor live in a personal and cultural cycle of poverty that hopelessly imprisons them. They routinely center on the Black urban population and focus on perceived personality or cultural traits that doom the poor. While the women in these stories typically exhibit an "attitude" that leads to trouble or a promiscuity that leads to single motherhood, the men possess a need for immediate gratification that leads to drug abuse or an unquenchable greed that leads to the pursuit of fast money. The images that are seared into our mind are sexist, racist, and classist. Census figures reveal that most of the poor are white not Black or Hispanic, that they live in rural or suburban areas not urban centers, and hold jobs at least part of the year.[2] Yet, in a fashion that is often framed in an understanding and sympathetic tone, we are told that the poor have inflicted poverty on themselves.

The Poor Are Down on Their Luck

During the Christmas season, the news media sometimes provides us with accounts of poor individuals or families (usually white) who are down on their luck. These stories are often linked to stories about soup kitchens or other charitable activities and sometimes call for charitable contributions. These "Yule time" stories are as much about the affluent as they are about the poor: they tell us that the affluent in our society are a kind, understanding, giving people—which we are not. The series of unfortunate circumstances that have led to impoverishment are presumed to be a temporary condition that will improve with time and a change in luck.

Despite appearances, the messages provided by the media are not entirely disparate. With each variation, the media informs us what poverty is not (i.e., systemic and indicative of American society) by informing us what it is. The media tells us that poverty is either an aberration of the American way of life (it doesn't exist, it's just another number, it's unfortunate but temporary) or an end product of the poor themselves (they are a nuisance, do not deserve better, and have brought their predicament upon themselves).

By suggesting that the poor have brought poverty upon themselves, the media is engaging in what William Ryan has called "blaming the victim."[3] The media identifies in what ways the poor are different as a consequence of deprivation, then defines those differences as the cause of poverty itself. Whether blatantly hostile or cloaked in sympathy, the message is that there is something fundamentally wrong with the victims—their hormones, psychological make up, family environment, community, race, or some combination of these—that accounts for their plight and their failure to lift themselves out of poverty.

But poverty in the United States is systemic. It is a direct result of economic and political policies that deprive people of jobs, adequate wages, or legitimate support. It is neither natural nor inevitable: there is enough wealth in our nation to eliminate poverty if we chose to redistribute existing wealth or income. The plight of the poor is reason enough to make the elimination of poverty the nation's first priority. But poverty also impacts dramatically on the non-poor. It has a dampening effect on wages in general (by maintaining a reserve army of unemployed and underemployed anxious for any job at any wage) and breeds crime and violence (by maintaining conditions that invite private gain by illegal means and rebellion-like behavior, not entirely unlike the urban riots of the 1960s). Given the extent of poverty in the nation and the impact it has on us all, the media must spin considerable magic to keep the poor and the issue of poverty and its root causes out of the public consciousness.

ABOUT EVERYONE ELSE

Both the broadcast and the print news media strive to develop a strong sense of "we-ness" in their audience. They seek to speak to and for an audience that is both affluent and like-minded. The media's solidarity with affluence, that is, with the middle and upper class, varies little from one medium to another. Benjamin DeMott points out, for example, that the *New York Times* understands affluence to be intelligence, taste, public spirit, responsibility, and a readiness to rule and "conceives itself as spokesperson for a readership awash in these qualities."[4] Of course, the flip side to creating a sense of "we," or "us," is establishing a perception of the "other." The other relates back to the faceless, amoral, undeserving, and inferior "underclass." Thus, the world according to the news media is divided between the "underclass" and everyone else. Again the messages are often contradictory.

The Wealthy Are Us

Much of the information provided to us by the news media focuses attention on the concerns of a very wealthy and privileged class of people. Although the concerns of a small fraction of the populace, they are presented as though they were the concerns of everyone. For example, while relatively few people actually own stock, the news media devotes an inordinate amount of broadcast time and print space to business news and stock market quotations. Not only do business reports cater to a particular narrow clientele, so do the fashion pages (with $2,000 dresses), wedding announcements, and the obituaries. Even weather and sports news often have a class bias. An all news radio station in New York City, for example, provides regular national ski reports. International news, trade agreements, and domestic policies issues are also reported in terms of their impact on business climate and the business community. Besides being of practical value to the wealthy, such coverage has considerable ideological value. Its message: the concerns of the wealthy are the concerns of us all.

The Wealthy (as a Class) Do Not Exist

While preoccupied with the concerns of the wealthy, the media fails to notice the way in which the rich as a class of people create and shape domestic and foreign policy. Presented as an aggregate of individuals, the wealthy appear without special interests, interconnections, or unity in purpose. Out of public view are the class interests of the wealthy, the interlocking business links, the concerted actions to preserve their class privileges and business interests (by running for public office, supporting political candidates, lobbying, etc.). Corporate lobbying is ignored, taken for granted, or assumed to be in the public interest. (Compare this with the media's portrayal of the "strong arm of labor" in attempting to defeat trade legislation that is harmful to the interests of working people.) It is estimated that two-thirds of the U.S. Senate is composed of millionaires.[5] Having such a preponderance of millionaires in the Senate, however, is perceived to be neither unusual nor antidemocratic; these millionaire senators are assumed to be serving "our" collective interests in governing.

The Wealthy Are Fascinating and Benevolent

The broadcast and print media regularly provide hype for individuals who have achieved "super" success. These stories are usually about celebrities and superstars from the sports and entertainment world. Society pages and gossip columns serve to keep the social elite informed of each others' doings, allow the rest of us to gawk at their excesses, and help to keep the American dream alive. The print media is also fond of feature stories on corporate empire builders. These stories provide an occasional "insider's" view of the private and corporate life of industrialists by suggesting a rags to riches account of corporate success. These stories tell us that corporate success is a series of smart moves, shrewd acquisitions, timely mergers, and well thought out executive suite shuffles. By painting the upper class in a positive light, innocent of any wrongdoing (labor leaders and union organizations usually get the opposite treatment), the media assures us that wealth and power are benevolent. One person's capital accumulation is presumed to be good for all. The elite, then, are portrayed as investment wizards, people of special talent and skill, who even their victims (workers and consumers) can admire.

The Wealthy Include a Few Bad Apples

On rare occasions, the media will mock selected individuals for their personality flaws. Real estate investor Donald Trump and New York Yankees owner George Steinbrenner, for example, are admonished by the media for deliberately seeking publicity (a very un-upper class thing to do); hotel owner Leona Helmsley was caricatured for her personal cruelties; and junk bond broker Michael Milkin was condemned because he had the audacity to rob the rich. Michael Parenti points out that by treating business wrongdoings as isolated deviations from the socially beneficial system of "responsible capitalism," the media overlooks the features of the system that produce such abuses and the regularity with which they occur. Rather than portraying them as predictable and frequent outcomes of corporate power and the business system, the media treats abuses as if they were isolated and atypical. Presented as an occasional aberration, these incidents serve not to challenge, but to legitimate, the system.[6]

The Middle Class Is Us

By ignoring the poor and blurring the lines between the working people and the upper class, the news media creates a universal middle class. From this perspective, the size of one's income becomes largely irrelevant: what matters is that most of "us" share an intellectual and moral superiority over the disadvantaged. As *Time* magazine once concluded, "Middle America is a state of mind."[7] "We are all middle class," we are told, "and we all share the same concerns": job security, inflation, tax burdens, world peace, the cost of food and housing, health care, clean air and water, and the safety of our streets. While the concerns of the wealthy are quite distinct from those of the middle class (e.g., the wealthy worry about investments, not jobs), the media convinces us that "we [the affluent] are all in this together."

The Middle Class Is a Victim

For the media, "we" the affluent not only stand apart from the "other"—the poor, the working class, the minorities, and their problems—"we" are also victimized by the poor (who drive up the costs of maintaining the welfare roles), minorities (who commit crimes against us), and by workers (who are greedy and drive companies out and prices up). Ignored are the subsidies to the rich, the crimes of corporate America, and the policies that wreak havoc on the economic well-being of middle America. Media magic convinces us to fear, more than anything else, being victimized by those less affluent than ourselves.

The Middle Class Is Not a Working Class

The news media clearly distinguishes the middle class (employees) from the working class (i.e., blue collar workers) who are portrayed, at best, as irrelevant, outmoded, and a dying breed. Furthermore, the media will tell us that the hardships faced by blue collar workers are inevitable (due to progress), a result of bad luck (chance circumstances in a particular industry), or a product of their own doing (they priced themselves out of a job). Given the media's presentation of reality, it is hard to believe that manual, supervised, unskilled, and semiskilled workers actually represent more than 50 percent of the adult working population.[8] The working class, instead, is relegated by the media to "the other."

In short, the news media either lionizes the wealthy or treats their interests and those of the middle class as one in the same. But the upper class and the middle class do not share the same interests or worries. Members of the upper class worry about stock dividends (not employment), they profit from inflation and global militarism, their children attend exclusive private schools, they eat and live in a royal fashion, they call on (or are called upon by) personal physicians, they have few consumer problems, they can escape whenever they want from environmental pollution, and they live on street and travel to other areas under the protection of private police forces.[*][9]

The wealthy are not only a class with distinct life-styles and interests, they are a ruling class. They receive a disproportionate share of the country's yearly income, own a disproportionate amount of the country's wealth, and contribute a disproportionate number of their members to governmental bodies and decision-making groups—all traits that William Domhoff, in his classic work *Who Rules America,* defined as characteristic of a governing class.[10]

This governing class maintains and manages our political and economic structures in such a way that these structures continue to yield an amazing proportion of our wealth to a minuscule upper class. While the media is not above referring to ruling classes in other countries (we hear, for example, references to Japan's ruling elite),[11] its treatment of the news proceeds as though there were no such ruling class in the United States.

Furthermore, the news media inverts reality so that those who are working class and middle class learn to fear, resent, and blame those below, rather than those above them in the class structure. We learn to resent welfare, which accounts for only two cents out of every dollar in the federal budget (approximately $10 billion) and provides financial relief for the needy,[**] but learn little about the $11 billion the federal government spends on individuals with incomes in excess of $100,000 (not needy),[12] or the $17 billion in farm subsidies, or the $214 billion (twenty times the cost of welfare) in interest payments to financial institutions.

Middle-class whites learn to fear African Americans and Latinos, but most violent crime occurs within poor and minority communities and is neither inter-racial[†] nor interclass. As horrid as such crime is, it should not mask the destruction and violence perpetrated by corporate America. In spite of the fact that 14,000 innocent people are killed on the job each year, 100,000 die prematurely, 400,000 become seriously ill, and 6 million are injured from work-related accidents and diseases, most Americans fear government regulation more than they do unsafe working conditions.

Through the media, middle-class—and even working-class—Americans learn to blame blue collar workers and their unions for declining purchasing power and economic security. But while workers who managed to keep their jobs and their unions struggled to keep up with inflation, the top 1 percent of American families saw their average incomes soar 80 percent in the last decade.[13]

---

[*]The number of private security guards in the United States now exceeds the number of public police officers. (Robert Reich, "Secession of the Successful," *New York Times Magazine,* February 17, 1991, p. 42.)

[**]A total of $20 billion is spent on welfare when you include all state funding. But the average state funding also comes to only two cents per state dollar.

[†]In 92 percent of the murders nationwide the assailant and the victim are of the same race (46 percent are white/white, 46 percent are black/black), 5.6 percent are black on white, and 2.4 percent are white on black. (FBI and Bureau of Justice Statistics, 1985–1986, quoted in Raymond S. Franklin, *Shadows of Race and Class,* University of Minnesota Press, Minneapolis, 1991, p. 108.)

Much of the wealth at the top was accumulated as stockholders and corporate executives moved their companies abroad to employ cheaper labor (56 cents per hour in El Salvador) and avoid paying taxes in the United States. Corporate America is a world made up of ruthless bosses, massive layoffs, favoritism and nepotism, health and safety violations, pension plan losses, union busting, tax evasions, unfair competition, and price gouging, as well as fast buck deals, financial speculation, and corporate wheeling and dealing that serve the interests of the corporate elite, but are generally wasteful and destructive to workers and the economy in general.

It is no wonder Americans cannot think straight about class. The mass media is neither objective, balanced, independent, nor neutral. Those who own and direct the mass media are themselves part of the upper class, and neither they nor the ruling class in general have to conspire to manipulate public opinion. Their interest is in preserving the status quo, and their view of society as fair and equitable comes naturally to them. But their ideology dominates our society and justifies what is in reality a perverse social order—one that perpetuates unprecedented elite privilege and power on the one hand and widespread deprivation on the other. A mass media that did not have its own class interests in preserving the status quo would acknowledge that inordinate wealth and power undermines democracy and that a "free market" economy can ravage a people and their communities.

NOTES

1. Martin Lee and Norman Solomon, *Unreliable Sources*, Lyle Stuart (New York, 1990), p. 71. See also Ben Bagdikian, *The Media Monopoly*, Beacon Press (Boston, 1990).

2. Department of Commerce, Bureau of the Census, "Poverty in the United States: 92," *Current Population Reports, Consumer Income*, Series P60–185, pp. xi, xv, 1.

3. William Ryan, *Blaming the Victim*, Vintage (New York, 1971).

4. Benjamin Demott, *The Imperial Middle*, William Morrow (New York, 1990), p. 123.

5. Fred Barnes, "The Zillionaires Club," *The New Republic*, January 29, 1990, p. 24.

6. Michael Parenti, *Inventing Reality*, St. Martin's Press (New York, 1986), p. 109.

7. *Time*, January 5, 1979, p. 10.

8. Vincent Navarro, "The Middle Class—A Useful Myth," *The Nation*, March 23, 1992, p. 1.

9. Charles Anderson, *The Political Economy of Social Class*, Prentice Hall (Englewood Cliffs, N.J., 1974), p. 137.

10. William Domhoff, *Who Rules America*, Prentice Hall (Englewood Cliffs, N.J., 1967), p. 5.

11. Lee and Solomon, *Unreliable Sources*, p. 179.

12. *Newsweek*, August 10, 1992, p. 57.

13. *Business Week*, June 8, 1992, p. 86.

46

Gladiators, Gazelles, and Groupies
Basketball Love and Loathing

JULIANNE MALVEAUX

The woman in me loves the sheer physical impact of a basketball game. I don't dwell on the fine points, the three-point shot or the overtime, and can't even tell you which teams make me sweat (though I could have told you twenty years ago). I can tell you, though, how much I enjoy watching mostly black men engage in an elegantly (and sometimes inelegantly) skilled game, and how engrossing the pace and physicality can be. In my younger days, I spent hours absorbed by street basketball games, taking the subway to watch the Rucker Pros in upper Manhattan, hanging on a bench near the basketball courts on West Fourth Street in the Village. I didn't consider myself a basketball groupie— I was a nerd who liked the game. Still, there is something about muscled, scantily clad men (the shorts have gotten longer in the 1990s) that spoke to me. Once I even joked that watching basketball games was like safe sex in the age of AIDS. Yes, the woman in me, even in my maturity, loves the physical impact of a basketball game, a game where muscles strain, sweat flies, and intensity demands attention.

The feminist in me abhors professional basketball and the way it reinforces gender stereotypes. Men play, women watch. Men at the center, women at the periphery, and eagerly so. Men making millions, women scheming to get to some of the millions through their sex and sexuality. The existence of the Women's National Basketball Association hardly ameliorates my loathing for the patriarchal underbelly of the basketball sport, since women players are paid a scant fraction of men's pay and attract a fraction of the live and television audience. To be sure, women's basketball will grow and develop and perhaps even provide men with role models of what sportsmanship should be. I am also emphatically clear that basketball isn't the only patriarchy in town. Despite women's participation in politics it, too, is a patriarchy, with too many women plying sex and sexuality as their stock in trade, as the impeachment imbroglio of 1998–99 reminds us.

The race person in me has mixed feelings about basketball. On one hand, I enjoy seeing black men out there making big bank, the kind of bank they can't make in business, science, or more traditional forms of work. On the other hand,

SOURCE: From Todd Boyd and Kenneth L. Shropshire, eds., *Basketball Jones: America Above the Rim*, pp. 51–58. Copyright © 2000 New York University Press. Reprinted with permission.

I'm aware of the minuscule odds that any high school hoopst'
Michael Jordan. If some young brothers spent less time on th'
in the labs, perhaps the investment of time in chemistry would y.~
mega–millions that professional basketball does. This is hardly the sole tau..
the youngsters with NBA aspirations. It is also the responsibility of coaches to
introduce a reality check to those young people whose future focus is exclusively
on basketball. And it is society's responsibility to make sure there are opportunities
for young men, especially young African American men, outside basketball. The
athletic scholarship should not be the sole passport to college for low–income
young men. Broader access to quality education must be a societal mandate;
encouraging young African American men to focus on higher education is critical
to our nation's fullest development.

Professional basketball bears an unfortunate similarity to an antebellum plantation, with some coaches accustomed to barking orders and uttering racial
expletives to get maximum performance from their players. While few condone
the fact that Latrell Sprewell, then of the Golden State Warriors, choked his coach
P. J. Carlesimo in 1997, many understand that rude, perhaps racist, but certainly
dismissive treatment motivated his violent action. The race of coaches, writers,
and commentators, in contrast to that of the players, often has broader racial
implications. White players whose grammar needs a boost often get it from sports-
writers, who make them sound far more erudite than they are. The same writers
quote black players with exaggerated, cringe–producing broken English—"dis,"
"dat," "dese," and "dose"—almost a parody of language. White commentators
frequently remark on the "natural talent" of black players, compared to the skill
of white ones. Retired white players are more likely to get invitations to coach
and manage than are their black counterparts.

Despite this plantation tension, there are those who tout the integration in
basketball as something lofty and desirable and the sport itself as one that teaches
discipline, teamwork, and structure. In *Values of the Game,* former New Jersey
senator and 2000 presidential candidate Bill Bradley describes the game as one
of passion, discipline, selflessness, respect, perspective, courage, and other virtues.
Though Bradley took a singular position as a senator in lifting up the issue of
police brutality around the Rodney King beating, his record on race matters is
otherwise relatively unremarkable. But Bradley gets credit for being far more
racially progressive than he actually is because he, a white man, played for the
New York Knicks for a decade, living at close quarters with African American
men.

As onerous as I find the basketball plantation, the race person in me wonders
why I am so concerned about the gladiators. Few if any of them exhibit anything
that vaguely resembles social consciousness. Michael Jordan, for example, passed
on the opportunity to make a real difference in the 1990 North Carolina Senate
race between the former Charlotte mayor and Democratic candidate, the African
American architect Harvey Gantt, and the ultra conservative and racially manip-
ulative Republican senator Jesse Helms, preferring to save his endorsements for
Nike. Charles Barkley once arrogantly crowed that he was not a role model,
ignoring the biblical adage that much is expected of those to whom much is

given. Few of the players, despite their millions, invest in the black community and in black economic development. Magic Johnson is a notable exception, with part ownership in a Los Angeles-based black bank and development of a theater chain that has the potential for revitalizing otherwise abandoned inner-city communities. Johnson's example notwithstanding, basketball players are more likely to make headlines for their shenanigans than for giving back to the community.

With my mixed feelings, love and loathing, why pay attention to basketball at all? Why not tune it out as forcefully as I tune out anything else I'm disinterested in? The fact is that it is nearly impossible to tune out, turn off, or ignore basketball. It is a cultural delimiter, a national export, a medium through which messages about race, gender, and power are transmitted not only nationally but also internationally. As Walter LaFeber observes in his book, *Michael Jordan and the New Global Capitalism,* Mr. Jordan's imagery has been used not only to sell the basketball game and Nike shoes but to make hundreds of millions of dollars for both Mr. Jordan and the companies he represents. Jordan is not to be faulted for brokering his skill into millions of dollars. Still, the mode and methods of his enrichment make the game of basketball a matter of intellectual curiosity as one explores the way in which capitalism and popular culture intersect. Jordan's millions, interesting as they may be, are less interesting than the way in which media have created his iconic status in both domestic and world markets and offered him up as a role model of the reasonable, affable African American man, the antithesis of "bad boy" trash talkers who so frequently garner headlines.

From a gender perspective, too, there is another set of questions. What would a woman have to do to achieve the same influence, iconic status, and bankability as Jordan? Is it even conceivable, in a patriarchal world, that a woman could earn, gain, or be invented in as iconic a status as a Michael Jordan? A woman might sing—but songbirds come and go and aren't often connected with international marketing of consumer products. She might dance—but that, too, would not turn into international marketability. She might possess the dynamic athleticism of the tennis-playing Williams sisters, Venus and Serena, or she might have the ethereal beauty of a Halle Berry. In either case, her product identification would solely be connected with "women's" products—hair, beauty, and women's sports items. The commercial heavy lifting has been left to the big boys, or to one big boy in particular, Michael Jordan.

To be sure, the public acceptance of women's sports has changed. Thanks to Title IX, women's sports get better funding and more attention at the college level than they did only decades ago. After several fits and starts, the professional women's basketball league seems to be doing well, although not as well as the NBA. In time this may change, but my informal survey of male basketball fans suggests that some find women's basketball simply less entertaining than the NBA. "When women start dunking and trash talking, more people will start watching," said a man active in establishing Midnight Basketball teams in urban centers. If this is the case, then it raises questions both about women's basketball and about the tension between male players and officials about on-court buffoonery. Do officials kill the goose that laid the golden egg when they ask a Dennis Rodman to "behave himself"?

My ire is not about gender envy; it is about imagery and the replication of gender-oppressive patterns by using basketball, especially, to connect an image of sportsmanship, masculinity, with product identification. In a commercial context that promotes this connection, there is both a subtle and a not-so-subtle message about the status of women.

There is also a set of subtle messages about race and race relations that emerges when African American men are used in the same way that African American women were used in the "trade cards" of the early twentieth century. In the patriarchal context, hoopsters become hucksters while the invisible (white) male role as power broker is reinforced. In too many ways the rules and profit of the game reinforce the rules, profit, and history of American life, with iconic gladiators serving as symbols and servants of multinational interests. Black men who entertain serve as stalkers for white men who measure profits. Neutered black men can join their white colleagues in chachinging cash registers but can never unlock the golden handcuffs and the platinum muzzles that limit their ability to generate independent opinions. Women? Always seen, never heard. Ornamental or invisible assets. Pawns in an unspoken game.

A subtext of the basketball culture relegates women, especially African American women, to a peripheral, dependent, and soap-operatic role. These women are sometimes seen as sources of trouble (as in the imbroglio with the Washington then Bullets, now Wizards, where apparently false rape charges spotlighted the lives of two players and perhaps changed the long-term composition of the team). Or they are depicted (for example, in a 1998 *Sports Illustrated* feature) as predatory sperm collectors, whose pregnancies are part of a plot" to collect child support from high-earning hoopsters.

There are aspects to some women's behavior in relationship with basketball stars that are hardly laudable. (Some of this is detailed in the fictional *Homecourt Advantage,* written by two basketball significant others.) At the same time, it is clear that this behavior is more a symptom than a cause of predatory patriarchy. After all, no one forces a player into bed or into a relationship with an unscrupulous woman. But conditions often make a player irresistible to women whose search for legitimacy comes through attachment to a man.

The disproportionate attention and influence that the basketball lifestyle gives some men makes them thoroughly irresistible to some women. This begs the question: In a patriarchy, just what can women do to attain the same influence and irresistibility? In a patriarchy, male games are far more intriguing than female games. Women, spectators, can watch and attach themselves to high-achieving gladiators. There is little that they can do, in the realm of sport, capitalism, and imagery, to gain the same access that men have. Again, this is not restricted to basketball but is a reflection of gender roles in our society. One might describe this as the "Hillary Rodham Clinton dilemma." Does a woman gain more power and influence by developing her own career or by attaching herself to a powerful man?

This dilemma may be a short-term one, given changes in the status of women in our society. Still, a range of scholars have noted that while women have come a long way, with higher incomes, higher levels of labor-force participation, and increased representation in executive ranks, at the current pace, income equality (which may not be the equivalent of the breakdown of patriarchy) will occur in the middle of the twenty-first century.

✱ Those who depict the basketball culture have been myopic in their focus on the men who shoot hoops, because their coverage has ignored the humanity of the women who hover around hoopsters like moths drawn to a flame. Those writers who are eager to write about the "paternity ward" or about "Darwin's athletes" might also ask how the gender relations in basketball replicate or differ from gender relations in our society. Unfortunately, gender relations in basketball are too often the norm in society. Men play, women watch. To the extent that the basketball culture is elevated, women's roles are denigrated. Why are we willing to endow male hoopsters with a status that no woman can attain in our society? Sedentary men sing the praises of male hoopsters, sit enthralled and engrossed by their games, thus elevating a certain form of achievement in our society. This behavior is seen as normal, even laudable. Yet it takes women, often the daughters or sisters of these enthralled men, out of the iconic, high-achievement mix and pushes them to the periphery of a culture that reveres the athletic achievement derived from the combination of basketball prowess and patriarchy.

The combination of prowess and patriarchy denigrates the groupie who is attracted by the bright lights, but it also dehumanizes the swift gazelle, the gladiator, and the hoopster, whose humanity is negated by his basketball identity. Media coverage depicts young men out of control, trash-talking, wild-walking, attention-grabbing icons. Men play, women watch. Were there other center stages, this would be of limited interest. But the fact is that this stage is one from which other stages reverberate. Men play, women watch, in politics, economics, technology, and sports. And it is "blown up" into international cultural supremacy in basketball.

This woman watches, with love and loathing, the way basketball norms reinforce those that exist in our society. It is a triumph of patriarchy, this exuberance of masculine physicality. It is a reminder to women that feminism notwithstanding, we have yet to gain economic and cultural equivalency with Michael Jordan's capitalist dominance. While we should not, perhaps, ask men to walk away from arenas in which they can dominate, we must ask ourselves why there is no equivalent space for us; why the basketball tenet that men play, women watch, reverberates in so many other sectors of our society.

REFERENCES

Bradley, Bill. *Values of the Game*. New York: Artisan Press, 1998.

Ewing, Rita, and Crystal McCrary. *Homecourt Advantage*. New York: Avon Books, 1998.

Frey, Darcy. *The Last Shot: City Street, Basketball Dreams*. Boston: Houghton Mifflin, 1994.

Hoberman, John. *Darwin's Athletes: How Sport Has Damaged Black America and Preserved the Myth of Race*. New York: Mariner Books, 1997.

LaFeber, Walter. *Michael Jordan and the New Global Capitalism*. New York: W. W. Norton & Co., 1999.

Wahl, Grant, and L. Jon Wertheim. "Paternity Ward." *Sports Illustrated*, May 4, 1998, at 62.

47

Historic Reversals, Accelerating Resegregation, and the Need for New Integration Strategies

GARY ORFIELD AND CHUNGMEI LEE

Although the principle of "separate but equal" in schools and other public facilities is now unconstitutional, schools are now rapidly re-segregating in the face of court decisions in recent years. White students are thus highly segregated from other groups, and African Americans and Latinos are highly likely to be attending poorly supported "majority-minority" schools.

American schools, resegregating gradually for almost two decades, are now experiencing accelerating isolation and this will doubtless be intensified by the recent decision of the U.S. Supreme Court. In 2007, the Supreme Court handed down its first major decision on school desegregation in 12 years in the Louisville and Seattle cases.[1] A majority of a divided Court told the nation both that the goal of integrated schools remained of compelling importance but that most of the means now used voluntarily by school districts are unconstitutional. As a result, most voluntary desegregation actions by school districts must now be changed or abandoned. As educational leaders and citizens across the country try to learn what they can do, and decide what they will do, we need to know how the nation's schools are changing, what the underlying trends are in the segregation of American students, and what the options are they might consider.

The Supreme Court struck down two voluntary desegregation plans with a majority of the Justices holding that individual students may not be assigned or denied a school assignment on the basis of race in voluntary plans even if the intent is to achieve integrated schools—and despite the fact that the locally designed plans actually fostered integration. A majority of the Justices, on a Court that divided 4-4-1 on the major issues, also held that there are compelling reasons for school districts to seek integrated schools and that some other limited techniques such as choosing where to build schools are permissible. In the process, the Court reversed nearly four decades of decisions and regulations which had permitted and even required that race be taken into account because of the

SOURCE: Orfield, Gary and Chungmei Lee. 2007. "Historic Reversals, Accelerating Resegregation, and the Need for New Integration Strategies." Civil rights Project/ Proyecto Derechos Civiles. UCLA. Reprinted with permission. www.civilrightsproject. ucla.edu

earlier failure of desegregation plans that did not do that. The decision also called into question magnet and transfer plans affecting thousands of American schools and many districts. In reaching its conclusion the Court's majority left school districts with the responsibility to develop other plans or abandon their efforts to maintain integrated schools. The Court's decision rejected the conclusions of several major social science briefs submitted by researchers and professional associations which reported that such policies would foster increased segregation in schools that were systematically unequal and undermine educational opportunities for both minority and white students. The Court's basic conclusion, that it was unconstitutional to take race into account in order to end segregation, represented a dramatic reversal of the rulings of the civil rights era which held that race must be taken into account to the extent necessary to end racial separation.

The trends ... are those of increasing isolation and profound inequality. The consequences become larger each year because of the growing number and percentage of nonwhite and impoverished students and the dramatic relationships between educational attainment and economic success in a globalized economy. Almost nine-tenths of American students were counted as white in the early 1960s, but the number of white students fell 20 percent from 1968 to 2005, as the baby boom gave way to the baby bust for white families, while the number of blacks increased 33 percent and the number of Latinos soared 380 percent amid surging immigration of a young population with high birth rates. The country's rapidly growing population of Latino and black students is more segregated than they have been since the 1960s and we are going backward faster in the areas where integration was most far-reaching. Under the new decision, local and state educators have far less freedom to foster integration than they have had for the last four decades. The Supreme Court's 2007 decision has sharply limited local control in this arena, which makes it likely that segregation will further increase.

Compared to the civil rights era we have a far larger population of "minority" children and a major decline in the number of white students. Latino students, who are the least successful in higher education attainment, have become the largest minority population. We are in the last decade of a white majority in American public schools and there are already minorities of white students in our two largest regions, the South and the West. When today's children become adults, we will be a multiracial society with no majority group, where all groups will have to learn to live and work successfully together. School desegregation has been the only major policy directly addressing this need and that effort has now been radically constrained.

The schools are not only becoming less white but also have a rising proportion of poor children. The percentage of school children poor enough to receive subsidized lunches has grown dramatically. This is not because white middle class students have produced a surge in private school enrollment; private schools serve a smaller share of students than a half century ago and are less white. The reality is that the next generation is much less white because of the aging and small family sizes of white families and the trend is deeply affected by immigration from Latin American and Asia. Huge numbers of children are growing up in families with very limited resources, and face an economy with deepening inequality of income distribution, where only those with higher education are securely in the middle class. It is a simple statement of fact to say that the country's future depends on finding ways to prepare groups of students who have traditionally fared badly in American schools to perform at much higher levels and to prepare all young Americans to live and work in a society vastly more diverse than ever in our past. Some of our largest states will face a decline in average educational levels in the near future as the racial transformation proceeds if the educational success of nonwhite students does not improve substantially.

From the "excellence" reforms of the Reagan era and the Goals 2000 project of the Clinton Administration to the No Child Left Behind Act of 2001, we have been trying to focus pressure and resources on making the achievement of minority children in segregated schools equal. The record to date justifies deep skepticism. On average, segregated minority schools are inferior in terms of the quality of their teachers, the character of the curriculum, the level of competition, average test scores, and graduation rates. This does not mean that desegregation solves all problems or that it always works, or that segregated schools do not perform well in rare circumstances, but it does mean that desegregation normally connects minority students with schools which have many potential advantages over segregated ghetto and barrio schools especially if the children are not segregated at the classroom level.

Desegregation is often treated as if it were something that occurred after the *Brown* decision in the 1950s. In fact, serious desegregation of the black South only came after Congress and the Johnson Administration acted powerfully under the 1964 Civil Rights Act; serious desegregation of the cities only occurred in the 1970s and was limited outside the South. Though the Supreme Court recognized the rights of Latinos to desegregation remedies in 1973, there was little enforcement as the Latino numbers multiplied rapidly and their segregation intensified.

Resegregation, which took hold in the early 1990s after three Supreme Court decisions from 1991 to 1995 limiting desegregation orders, is continuing to grow in all parts of the country for both African Americans and Latinos and is accelerating the most rapidly in the only region that had been highly desegregated—the South. The children in United States schools are much poorer than they were decades ago and more separated in highly unequal schools. Black and Latino segregation is usually double segregation, both from whites and from middle class students. For blacks, more than a third of a century of progress in racial integration has been lost—though the seventeen states which had segregation laws are still far less segregated than in the 1950s when state laws enforced apartheid in the schools and the massive resistance of Southern political leaders delayed the impact of *Brown* for a decade. For Latinos, whose segregation in many areas is now far more severe than when it was first measured nearly four decades ago, there never was progress outside of a few areas and things have been getting steadily worse since the 1960s on a national scale. Too often Latino students face triple segregation by race, class, and language. Many of these segregated black and Latino schools have now been sanctioned for not meeting the requirements of No Child Left Behind and segregated high poverty schools account for most of the "dropout factories" at the center of the nation's dropout crisis....

One would assume that a nation which now has more than 43 percent non-white students, but where judicial decisions are dissolving desegregation orders and fostering increasing racial and economic isolation must have discovered some way to make segregated schools equal since the future of the country will depend on the education of its surging nonwhite enrollment which already accounts for more than two students of every five. You would suppose that it must have identified some way to prepare students in segregated schools to live and work effectively in multiracial neighborhoods and workplaces since experience in many racially and ethnically divided societies show that deep social cleavages, especially subordination of the new majority, could threaten society and its basic institutions. Those assumptions would be wrong. The basic judicial policies are to terminate existing court orders, to forbid most race-conscious desegregation efforts without court orders, and to reject the claim that there is a right to equal resources for the segregated schools. Not only do the federal courts not require either integration or equalization of segregated schools but this means that they forbid state and local officials to implement most policies that have proven effective in desegregating schools. State and local politics will determine what, if anything, happens in terms of equalizing resources between segregated schools and privileged schools....

While the courts are terminating desegregation plans, statistics show steadily increasing separation. After three decades of preparing reports on trends in segregation in American schools, the most disturbing element of this year's report is the finding that the great success of the desegregation battle—turning Southern education, which was still 98 percent segregated in 1964, into the most desegregated part of the nation—is being rapidly lost. These new data show that the South has lost the leadership it held as the most desegregated region for a third of a century, even as the region becomes majority nonwhite and faces a dramatic Latino immigration. It took decades of struggle to achieve desegregated schools in the South, our most populous region, and no one would have predicted during the civil rights era that leaders in some Southern communities once forced to desegregate with great difficulty would, in the early 21st century, wish to remain desegregated but be forbidden to maintain their plans by federal courts. Yet that is exactly what happened in a number of districts.

The basic educational policy model in the post-civil rights generation assumes that we can equalize schools without dealing with segregation through testing and accountability. It is nearly a quarter century since the country responded to the Reagan Administration's 1983 report, "A Nation at Risk," warning of dangerous shortcomings in American schools and demanding that "excellence" policies replace the "equity" policies of the 1960s. Since then almost every state has adopted the recommendations for the more demanding tests and accountability and more required science and math classes the report recommended. Congress and the last three Presidents have established national goals for upgrading and equalizing education. The best evidence indicates that these efforts have failed, both the Goals 2000 promise of equalizing education for nonwhite students by 2000 and the NCLB promise of closing the achievement gap with mandated minimum yearly gains so that everyone would be proficient by 2013. In fact, the previous progress in narrowing racial achievement gaps from the 1960s well into the 1980s has ended and most studies find that there has been no impact from NCLB on the racial achievement gap. These reforms have been dramatically less effective in that respect than the reforms of the 1960s and '70s, including desegregation and anti-poverty programs. On some measures the racial achievement gaps reached their low point around the same time as the peak of black–white desegregation in the late 1980s.

Although the U.S. has some of the best public schools in the world, it also has too many far weaker than those found in other advanced countries. Most of these are segregated schools which cannot get and hold highly qualified teachers and administrators, do not offer good preparation for college, and often fail to graduate even half of their students. Although we have tried many reforms, often in confusing succession, public debate has largely ignored the fact that racial and ethnic separation continues to be strikingly related to these inequalities. As the U.S. enters its last years in which it will have a majority of white students, it is betting its future on segregation. The data coming out of the No Child Left Behind tests and the state accountability systems show clear relationships between segregation and educational outcomes but this fact is rarely mentioned by policy makers.

The fact of resegregation does not mean that desegregation failed and was rejected by Americans who experienced it. Of course the demographic changes made full desegregation with whites more difficult, but the major factor, particularly in the South, was that we stopped trying. Five of the last seven Presidents actively opposed urban desegregation and the last significant federal aid for desegregation was repealed 26 years ago in 1981. The last Supreme Court decision expanding desegregation rights was handed down in 1973, more than a third of a century ago, one year before a decision rejecting city–suburban desegregation. This second decision in 1974 meant that desegregation was impossible in much of the North since the large majority of white students in many areas were already in the suburbs and stable desegregation was impossible within city boundaries, as Justice Thurgood Marshall accurately predicted in his dissent in the 1974 *Milliken v. Bradley* decision.

The *Milliken* decision could be seen as the return of the doctrine of "separate but equal" for urban school children in a society where four of five Americans live in metropolitan areas. The problem is that the Supreme Court held in the 1973 *Rodriguez* case that there is no federal right to an equal education, so "separate but equal" could not be enforced either. With the 2007 rejection of most of the techniques that have preserved a modicum of desegregation by voluntary local action, the doctrine is basically one of separation and local political control, except if local governments want to pursue voluntary integration strategies, which are now largely prohibited....

One of the deepest ironies of this period is that never before has there been more evidence about the inequalities inherent in segregated education, the potential benefits for both nonwhite and white students, and the ways in which those benefits could be maximized. The evidence submitted to the Supreme Court regarding the Louisville and Seattle cases was many times more compelling than that the Court relied on in striking down the segregation system of the South in 1954. This evidence does not claim that desegregation will eliminate inequalities, since those are based on social and economic issues that reach far beyond the schools but it does show that the policies provide important benefits in both educational attainment and life chances—and that there are no harms and some large benefits for white as well nonwhite students, and for society and its institutions. Yet we are dismantling plans that actually work in favor of an alternative, double and triple segregation, that has never worked on any substantial scale....

Nearly 40 years after the assassination of Dr. Martin Luther King, Jr., we have now lost almost all the progress made in the decades after his death in desegregating our schools. It was very hard won progress that produced many successes and enabled millions of children, particularly in the South to grow up in more integrated schools. Though it was often imperfectly implemented and sometimes poorly designed, school integration was, on average, a successful policy, linked to a period of social mobility and declining gaps in achievement and school completion and improved attitudes and understanding among the races. The experience under No Child Left Behind and similar high stakes testing and accountability policies that ignore segregation has been deeply disappointing and the evidence from those tests shows the continuing inequality of segregated schools even after many years of fierce pressure and sanctions on those schools and students.

It is time to think very seriously about the central proposition of the *Brown* decision, that segregated education is "inherently unequal" and think about how we can begin to regain the ground that has been lost. The pioneers whose decades of investigations and communication about the conditions of racial inequality helped make the civil rights revolution possible a half century ago should not be honored merely by naming schools and streets or even holidays after them but should be remembered as a model of the work that must be done, as many times as necessary, for as long as it takes, to return to the promise of truly equal justice under law in our schools, to insist that we have the kind of schools that can build and sustain a successful profoundly multiracial society....

NATIONAL SEGREGATION TRENDS

Across the country, segregation is high for all racial groups except Asians. While white students are attending schools with slightly more minority students than in the past, they remain the most isolated of all racial groups: the average white student attends schools where 77 percent of the student enrollment is white. Black and Latino students attend schools where more than half of their peers are black and Latino (52% and 55% respectively), a much higher representation than one would expect given the racial composition of the nation's public schools and substantially less than a third of their classmates are white. Whites had been even more segregated back in 1990, when they constituted a significantly larger share of the total enrollment.

WHITES: STILL THE MOST SEGREGATED

Though white students in 2005–6 were in schools with more minority students than in the past, they were still the most segregated population, being in schools that were 77 percent white, on average, in a country with 57 percent white students. Almost no attention has been given in the discussion of desegregation strategies and neighborhood schools about the consequences of ending city- and county-wide desegregation plans for white students living in city and inner suburban areas. In the absence of desegregation plans, much of the racial contact that exists is accounted for either by the small but significant number of whites in heavily minority schools or reflects the temporary diversity produced by residential racial transition as blacks and Latinos move very rapidly into some sectors of suburbia. A transitional neighborhood is a highly unstable process of a sort all too familiar during the decades when residential resegregation converted thousands of white city neighborhoods to minority communities. Under neighborhood schools or magnet school plans without desegregation guidelines more of these urban white students are going to end up isolated in high poverty, very high minority schools, a process that could well undermine some stably integrated residential areas and further limit the options of poor whites if choice plans are not operating. Unrestricted choice plans in the past have often accelerated residential resegregation when white students from integrated neighborhoods transferred out to whiter schools, helping tip the neighborhood school toward resegregation and making the neighborhood less attractive for white home seekers. When the courts and federal civil rights officials prohibited choice plans without desegregation standards in the 1960s they were very conscious of these problems and often found unrestricted choice strategies to be contributors to segregation. Now, as a result of the recent Court decision, we will have more such plans.

ASIANS: THE MOST INTEGRATED STUDENTS

When considering issues of immigration, the success of Asian students is often compared to the academic challenges facing Latino students. One of the significant differences is the level of segregation. Asian students are in schools where, on average, less than a fourth of fellow students are Asian and, since Asians speak many languages, they are far less likely to be in a school where their language is a major factor. Asians typically attend schools that are 48 percent white, compared to 32 percent for Latinos. However, despite the fact that Asians represent only five percent of the total student enrollment, the average Asian attends a school that is 24 percent Asian.

Likely due to their high residential integration and relatively small numbers outside the West, Asian students, on average, are the most integrated group and the group which attends school where their own ethnicity is least represented. Asians are also the most integrated racial group in residential patterns. U.S. immigration policies have tended to produce a very highly educated immigration from Asia. When educated middle class migrations have taken place from Latin

America, such as the first wave of Cuban migration, their experience has been similar to the average Asian experience, but most Latino immigration is of people with far fewer resources and lower levels of education.

The Asian experience, however, is a complex one. Although on average Asians are more educated and have higher family incomes than whites, some Asian groups, particularly refugee Indochinese populations who entered after the Vietnam War experience very different patterns of education and mobility, much more similar to those of typical disadvantaged Latino immigrants. Particularly in the West where Asians already outnumber African Americans and are a very visible presence in the schools it will be increasingly important to understand these differences.

DESEGREGATION TRENDS FOR BLACK STUDENTS

As previously mentioned, national statistics for black students show very slow progress the first decade after *Brown*, then a substantial decline in black segregation from white from the mid-60s through the early 1970s. There was gradual improvement through most of the 1980s, but then a reversal and a steady gradual rise in segregation since the early 1990s, a rise which is accelerating in the South. In terms of enrollment in majority white schools, most of the progress from urban desegregation has now been lost. The level of extreme segregation of black students in schools with 0–10 percent whites, however, remains far lower than it was before the civil rights era, though it also is rising. Table 1 shows a sharp rise in the percentage of black students in majority nonwhite schools since the 1980s and by far the largest increase takes place in the South.

T A B L E 1   **Percentage of Black Students in Predominantly (>50%) Minority Schools by Region, 1968–2007**

Region	1968	1980	1988	1991	2005
South	81	57	57	60	72
Border	72	59	60	59	70
Northeast	67	80	77	75	78
Midwest	77	70	70	70	72
West	72	67	67	69	77
US Total	77	63	63	66	73

SOURCE: U.S. Department of Education office for Civil Rights data in Orfield, Public School Desegregation in the United States, 1980–1; 1988–9, 1991-2; 2005–6 NCES Common Core of Data.

LATINO SEGREGATION

On a national level, the segregation of Latino students has grown the most since the civil rights era. Since the early 1970s, the period in which the Supreme Court recognized Latinos' right to desegregation there has been an uninterrupted national trend toward increased isolation. Latino students have become, by some measures, the most segregated group by both race and poverty and there are increasing patterns of triple segregation—ethnicity, poverty and linguistic isolation. No national administration has made a serious effort to desegregate Latinos and there have been few court orders addressing this problem, the most important of which have now been terminated—those in Denver and Las Vegas. In comparative terms, by 2005 Latinos were most likely to be in schools with less than half whites (78%) and in intensely segregated schools (39%).

NOTE

1. *Parents Involved In Community Schools V. Seattle School District No. 1 Et Al.* June 28, 2007.

48

"I Hate It When People Treat Me Like a Fxxx-up"

Phony Theories, Segregated Schools, and the Culture of Aspiration among African American and Latino Teenagers

JEANNE THEOHARIS

Public discourse on urban education has been overtaken by a discussion of values. It has become common sense to bemoan the declining value of education within urban Black and Latino communities, to assume a priori that students who value education succeed in school and that those with poor values drop out. This misinformed discussion about values not only takes the responsibility for schools away from the society that creates them and places it solely on students and their parents but distorts the regard for education held in the African American and Latino communities. Moreover, it caricatures criticism of the unequal structures of schooling as devaluing education itself.

Many commentators, while purporting to speak authoritatively on urban teenagers, do not spend much time listening to how young people actually think about and frame issues of schooling. This chapter foregrounds the perspectives of a group of African American and Latino high school students who attend a deeply segregated public high school in Los Angeles on the value and nature of their own educations. Using a set of journal writings students did for me over the course of the 2004 spring semester, I look at how young people write about their goals and aspirations, about good teaching and the use of testing, and about the structures of schooling today. Their writings demonstrate how profoundly students value education, how deeply they wish to be successful academically, and how much they hope to make their families proud.

Interweaving their beliefs and values with an analysis of the schooling they receive, I show how these young people are expected to be responsible (about attending school, doing their homework, making plans for college) while the

SOURCE: Theoharis, Jeanne. "I Hate it When People Treat me Like a Fxxx-up." Pp. 71–93 in *Our Schools Suck: Students Talk Back to a Segregated Nation on the Failures of Urban Education*, edited by Gaston Alonso, Noel S. Anderson, Celina Su, and Jeanne Theoharis. Copyright ©2009 New York University Press. Reprinted with permission.

school district does not have to demonstrate an equal level of responsibility in providing adequate classes and ample and excellent materials, clean bathrooms, sufficient college counselors, and a productive learning environment. Indeed, by showing the ways students hold up the promise of education, this chapter seeks to counter the incessant public lamentations that these students are unreachable and unteachable by demonstrating their hunger for substantive learning. While these students did not always choose to do their homework and attend class, this stemmed less from their devaluing of education than from a loss of confidence in themselves and the school to do right by them. Finally, this chapter demonstrates the ways this overwhelming focus on goals, hard work, and individualism is taken up by students themselves as over and over they are quick to blame themselves and narrate their talents through a frame that holds themselves—and often only themselves—responsible for the quality and success of their schooling.

THE CASE STUDY: FREMONT HIGH SCHOOL

Considered one of the most troubled schools in Los Angeles and placed on the state list of failing schools, Fremont High School is not a flagship of Los Angeles Unified School District (LAUSD). In the bottom 10 percent of the state, Fremont was one of thirteen schools in LAUSD (a district with over six hundred schools) audited by the state in 2003. With approximately five thousand students, Fremont operates year round on three tracks. A response to severe overcrowding, year-round schooling maximizes the use—and thus capacity—of the building because the rooms are used every weekday of the year (and often at night and on the weekends). On this three-track system, students go to school for four months and then are off for two months in scattered rotation. The level of disruption caused by students and teachers coming on and off track, the shuffling of rooms and textbooks, and periodic interruptions to the daily schedule is treated as routine, though it actually regularly impedes instruction. B Track is perhaps the worst pedagogically: their May–September, November–March schedule means that midway through each semester (following the traditional calendar) students change courses. Attending school all summer, students are not eligible for any of the city's summer jobs/internships programs, and the yearly testing occurs just a couple of weeks after B Track has come back to school.

Because year-round schooling deprives students of seventeen class days per year, this instructional time is made up by lengthening the school day by seventy-one minutes. Thus school begins at 7:35 in the morning and does not end until 3:25 in the afternoon. With six sixty-one-minute periods a day, students and teachers are exhausted by the afternoon. Everyone looks forward to the one day each month when the day is shortened to a blissful six and a half hours to give time for a staff meeting. Perhaps what is most galling is that multitrack schedules are not customary throughout the city, according to the research of UCLA's Institute for Democracy, Education and Access (IDEA), but are used predominantly in public schools that educate Latino and African American students. This disparity, never publicly accounted for, seemingly reflects an assumption that certain students' schooling is fungible; excellence is not essential in these schools, and other logistical considerations can take precedence.

Not only the school but also the classes are overcrowded. Class size averages thirty-five students, with many class rosters climbing well into the mid-forties (or even fifties) in 2003–4. While no teacher or administrator would ever wish it, what makes this situation even moderately manageable is dropouts: if some students do not show up, the class is smaller and better for the rest of the students as well as for the teacher. Thus, for classes to function, the system banks on— indeed needs—truancy. Trying to get past the numbers games that districts like LAUSD often play to mask their dropout rates, IDEA has calculated what they term the "disappearance rate" in L.A.'s schools. At Fremont, they calculate, for every hundred students who start freshman year, only thirty-two finish four years later and a mere sixteen have had full access to and completed the A–G academic curriculum that makes a student eligible for admission to a four-year college.

At Fremont there are very limited spaces in the Advanced Placement classes (on average one section is typically offered per subject per track). Even though many teachers supplement their classes with college preparatory material and increasingly the school emphasizes preparing every student for college, not enough courses are offered to allow every student to become eligible for a four-year college. Despite being a huge school, the variety of elective course offerings at Freemont is limited and tends toward vocational classes like sewing, cosmetology, and auto mechanics.

Fremont employs only one college counselor (separate from the other guidance counselors) for a student body of five thousand. She receives a teacher's salary, not a year-long salary, even though she works year round.

School overcrowding also corresponds to limited classroom resources. In 2003–4 there were not enough textbooks for every student, only a classroom set, so the textbooks could never go home for homework. Indeed, the textbooks were collected at the end of each track semester so they could be juggled among the tracks. This often resulted in students going without books for the first week or two of each term, and teachers lost the first week or two of instructional time because they had not received their allocation of books. (This has recently been partially addressed through the out-of-court settling of the *Williams* case discussed below.)

While the original brick and courtyard structure built in the 1920s is quite lovely, the tremendous growth at Fremont in the 1980s and 1990s required the construction of a great deal more cheap classroom space now called "the bungalows." With its prefab gray architecture and few windows, its similarity to a juvenile detention facility is striking to visitors and does not create an inviting learning environment. Because the school is used all the time, it is rundown; roaches, ants, and rats make periodic appearances in class. Since the school is always in session, repairs have to be done during classes, producing a tremendous amount of noise, dirt, and disruption. The bathrooms are often dirty, and students are not allowed to go to the bathroom during lunch or during class.

Police, both LAUSD and LAPD, are a constant and visible presence at Fremont, but no metal detectors greet students entering the school. Instead, a fence and locked gates surround the school, and a complex bureaucracy and surveillance system regulates anyone wishing to enter or leave the school. The school is entirely locked in during the school day; even the parking lots are gated shut. With students periodically paraded through the halls in handcuffs, the police presence at the school is palpable.

The connection between cultural values and educational success lacks empirical foundation. Relying heavily on the transcendent power of rugged individualism, this culture talk is remarkably detached from the kind of schooling many young people are receiving. Yet the current fixation with values makes it important to recognize these students' commitment to education. Their journals provide a new and revealing window on their cultural values with which the public is so concerned. Most students wrote about school as a way to get ahead in the world, and many saw it as a source of personal pleasure. When the second semester began in March, a number of students wrote about how "glad" and "excited" they were to be back in school after the two-month break. Over the course of the semester they wrote extensively about class and their education more generally, about teachers, college, and their plans for the future. Despite images of urban students being antischool, educational concerns abounded in their journals. Again and again, they wrote about hard work being the road to success, about the importance of a good attitude and sustained effort, about the centrality of education to achieve their goals. Berating themselves for not applying themselves enough, a number of students attributed their problems to their own motivation and vowed to adopt new and better attitudes. Overwhelmingly, these students believed that hard work was rewarded and that they could achieve whatever they put their minds to. They did not always make the choice to come to school or complete their work—but that reflected less a decline in their valuing of education than a declining confidence in themselves, in their teachers, or in the school to do right by them.

What becomes clear from reading hundreds of pages of these Fremont students' writings is that scholars and journalists convinced of the indifference of urban students to school have not really listened to students. Popular talk of declining values does not reflect how real students explain their ambitions and fears, their thoughts on good teachers, their favorite classes, and their goals for the future. From student surveys in the National Educational Longitudinal Survey, Duke University scholars Phillip Cook and Jens Ludwig have found that Black students approach the educational system in much the same way as white students. Black students are as likely to believe they will get a college degree as white students, drop out at the same rates as whites after controlling for family circumstances, cut classes and miss school at the same rates as whites, and have parents who attend school meetings and inquire about their progress with the same frequency as whites. More recently, sociologist Karolyn Tyson and economist William Darity, in an ethnographic study conducted in North Carolina, concluded that Black and white students are fundamentally similar in their desire to succeed academically and in their positive feelings about themselves when they do. When anti-intellectual attitudes appear among white students, they found, "it is seen as inevitable, but when the same dynamic is observed among black students it is pathologized as racial neurosis." In sum, white teenagers are allowed to adopt poses, complain about school, and go through different emotional stages without a parade of sociologists and public figures lamenting a culture of failure—while these very same behaviors from Black and Latino teenagers are treated as a sign of cultural dysfunction and seen as intrinsic to their nature.

... Rather than asserting that academic success was "white," students complained about how some teachers, counselors, and other adults did not believe they were capable of success. A reserved and talented Chicano student, Rodrigo wrote, "I hate it when people treat me like a fxxx-up. I'm not stupid and I'm not a little kid. If people give me time and space, I can do things at my own pace.... I do plan on going to college but with all this pressure how can I succeed." Indeed, Rodrigo challenged the disparagement at the heart of this discussion of declining values, highlighting the impact such negativity had on his ability to achieve his goal of going to college. Interestingly, "fxxx-up" is Rodrigo's own choice of spelling; it suggests that he wanted to make his point forcefully but did not want to be impolite and swear outright in his journal.

If some journalists and academics have heard students make statements about educational success being "white" (as I never did), they need to reflect on why students might be saying this. Such an ideology seems to be coming more from adults and a larger society that often does not hold out much possibility for these young people. Most students were not resigned to such thinking and, at times, actively resisted it, believing they would succeed through education despite what others said about them. Moreover, students respond in different ways to being excluded from the rewards of the educational system—and these comments may be a way to shield themselves from the pain of that exclusion. To say that education is "white" reflects a critique of the ways that education actually operates in U.S. society. Our students were well aware of the much finer schools to be found in L.A.'s more exclusive (and whiter) neighborhoods across town, the kinds of college opportunities such schools opened for students, and the social meanings of the differences between Fremont and those more excellent schools. In the face of such differences, some students picked up on readily available explanations for their own self-protection—they "didn't really care about school anyway," since it was hard to be convinced that the school always cared about them. In addition, when students say, "School sucks," analysts too often read it as an expression of these young people's "attitude" about school rather than a factual, if vernacular, description of the conditions of their schools.

Certainly, peer pressure exists. Peer pressure against doing well flares up in all high schools across this country (academic success never being a mechanism for teenage social success). Being a "nerd" carries a great social cost in the suburbs and the city. Teenagers across the country spend a great deal of energy trying to be part of the crowd, and succeeding academically is often stigmatized because it marks a student as different. Still, the assertion that problems of urban schools stem from negative peer pressure fails to recognize that most students in these schools believe intensely in the worth of education and hope to succeed academically.

... In fact, many students looked down on those who did not value the pursuit of knowledge. As Johanna, a similarly argumentative Latina who periodically cut class herself, explained, "People in school seem to come for the fuck of it. Like just to flash what they have, not to learn."

Moreover, getting good grades or associating with the "smart kids" did not seem to automatically compromise students' popularity. As Trevor, an articulate African American young man, wrote about his old school: "I was semi popular ... because I knew different people like the smart kids, a little [of the] rockers." Not a single student wrote about being laughed at for doing his or her work or excelling in class—though a number recalled incidents of being laughed at for how they dressed or for not being able to speak English well. Others recalled fondly the public recognition of doing well in school, often calling up memories from elementary school.

Many students worried that their classmates would think they were stupid or laugh at them for having difficulty reading or writing and thus often resisted reading aloud or turning in their essays because of these fears. Indeed, the fear of looking stupid was much more palpable than the fear of looking smart.

The young people in our classes possessed a great deal of faith in the fairness of the system. Their ideas about education correspond to popular American ideologies about success through education. Hopeful and determined, they were informed by the messages the media and society continually espouse: hard work and good attitude are what it takes. They stressed the barriers to their success but also their determination to overcome them. They were simultaneously excited and nervous about the prospects their future holds. In figuring out what the future would bring, they consciously constructed themselves as students who would go to college, who had the right to big dreams and to imagine the possibilities for themselves.

... Indeed, the number of journals written explicitly on goals, hard work, and motivation—and the constant exhortations of wanting to be someone when they grow up—may differentiate these Fremont students from their more affluent counterparts. It is hard to imagine students in Beverly Hills or Santa Monica feeling that they have to demonstrate over and over their determination to succeed in life and go on for pages about how hard they work. The number of journals on the subjects of hard work and motivation attests to the ways these students knew how they were viewed and attempted to write against these images...

One could argue, then, that these Fremont students had an abundance of educational values—that they had been forced, by their circumstances, to be more articulate about their values and more focused in their goals and motivation than middle-class high school students would have to be. But such articulateness offers little protection in a post-civil rights society where students ... are expected to want to go to college but are not provided enough courses to do so and are expected to wait in line day after day to see a college counselor.

One of the greatest difficulties of teaching at a school like Fremont is this paradox. American society supports schools like Fremont that warehouse students but somehow expects young people not to act as if they are being warehoused. As teachers, we constantly exhorted students to work hard, aim high, and take responsibility for themselves while the school system treated them as undeserving of excellence and evaded responsibility for the parade of permanent subs, insufficient books, crowded conditions, and unbearable temperatures in many classrooms. We told them to set goals for themselves, not to let others define their worth, but in doing so, we risked being complicit in a system that works because students then blame themselves for their own failure.

... TWENTY-FIRST CENTURY SEGREGATION

> The bathrooms at my school are really dirty. Out of the four bathrooms for girls, one bathroom, the one of the first floor of the main building is open, and the rest are locked.... Sometimes the bathrooms don't have toilet paper, and they always smell really bad. Sometimes I don't use the restrooms because they are so dirty.

—Cindy Barragan, student at Fremont, declaration for the *Williams* case

To be told by adults that you are not sacred or special; to be treated like a fxxx-up or an alien; to be given an education that, as often as not, reinforces your lack of social importance; to be schooled in overcrowded classrooms without adequate books and materials or clean bathrooms while the nation proclaims that segregation has ended; to wait in line day after day to meet with a college counselor, all the while being scolded for not holding the proper value of education; to have your beliefs and values publicly and loudly lied about and to have those lies be the justification for your inferior education—that is what segregation looks like in twenty-first-century America.

Part of the appeal of believing that urban students do not care about education is that the responsibility for change lies with them and not the rest of the nation. This culture talk works effectively to invalidate a call for social change, to cast socioeconomic factors as having "limited explanatory power," as Orlando Patterson suggests, and to individualize the task of school failure and success. The conditions in schools like Fremont parallel the unequal conditions of Black schools pre-*Brown*. Yet if students are seen to be less committed to their education they can be cast as deserving of (or at least not particularly harmed by) the limited resources and personnel in urban schools like Fremont. If students are believed to be the problem with urban education, then the American public is not at fault in maintaining a segregated system of schooling that produces over-crowded, under-resourced high schools like Fremont where one counselor serves the entire student body, class sizes climb into the forties, the school day is made too long in order to accommodate a perverse three-track system, and there are not enough books and materials to go around. This ideology creates a Catch-22 for young people themselves. Criticized for not valuing education when they actually do, they too hold themselves, and often *only* themselves, accountable for the quality of their own education and are quick to engage in self-blame for the successes and failures of their own schooling.

In a political climate that increasingly scapegoats African American and Latino youth for the problems facing schools and attributes to them the decline of American values and decency, these young people's writings strikingly challenge this portrayal. They demonstrate ... the racial malice that is at the heart of our present system of unequal schooling and in the claims that the problems with schools stem from the values of students. These young people insist on the right to represent their own humanity—on their entitlement to the rights and privileges of American society. By believing that hard work leads to opportunity and that society rewards those who try hard and play by its rules, they reveal the lie of this ideology and the cynicism behind its widespread trumpeting. And they celebrate—and long for—rigorous classes, committed teachers, decent, well-appointed classrooms, and the opportunity to go to college, to fulfill their dreams for the future and their aspirations of educational success.

49

Across the Great Divide
Crossing Classes and Clashing Cultures

BARBARA JENSEN

The blonde curls of Shelly's home permanent stuck to the tears on her face as she dashed from the classroom. "Oh God, I'm so sorry," she cried out. Just twenty minutes earlier she had been in the midst of an animated class discussion in a college course she liked, the psychology of women. Shelly had never thought about being a woman much before; she found it exciting and comforting to do so.

The class was having a discussion about relationships between women and men. The subject was intimacy, and the students were discussing some of the different ways men and women understand and express it. Shelly felt she was starting to understand some of the problems in her marriage. Maybe she could make things better. She was eager and animated in the discussion.

But something went wrong. Shelly was talking about the declining intimacy in her marriage and how college "made things really weird" between her and her husband. It wasn't just his complaining about the time she was gone; he was starting to make fun of her studying, saying she was turning into a "geek" and an "egghead." She told the class, "He even picked up a textbook and threw it against the wall, smashing the spine of a $65 book!" Then he hollered, "This shit means more to you than me and the kids!" and stomped out of the house. She said that later, when they "talked it out," he said *she* wasn't any fun anymore, *she* wasn't interested in anything. The class laughed out loud, because in class she was interested in everything. Encouraged, she exclaimed, "I couldn't believe it! That's just what I think about him! He's the one ... I'm interested in things now that I never even *thought* of before, you know what I mean? I asked him, 'What am I not interested in?' and he said, 'Bowling with Georgie and Bill and watching TV'! Like I have time for that now! Like *he* has shown any interest at all in all the things I've been studying."

Shelly's eyes blurred with tears and she fell silent; her pale skin was flushed. A couple of older women in the class started to talk, gently and with warmth, about how they had had to leave their husbands because they needed to "find themselves" and "get a new start." A forty-something woman offered that her spouse really wanted her to go to school, and that Shelly deserved to have that

SOURCE: "Across the Great Divide" by Barbara Jensen. Copyright © 2004 by Barbara Jensen. From Michael Zweig, ed., *What's Class Got to Do With It? American Society in the Twenty-First Century*, pp. 168–183. Copyright © 2004 by Cornell University. Reprinted by permission of the publisher, Cornell University Press. All rights reserved.

support. A man, who was on the board of a battered women's shelter, emphasized that she had a *right* to expect that support, that men have to learn to give women the things they have always had. He went on to warn her about "offender psychology" and how "they can't stand for their women to be independent, that's how they keep control." The other women from blue-collar backgrounds were uncharacteristically quiet.

"But that's not it!" Shelly insisted in frustration. "You don't understand ...," she trailed off, struggling for words and understanding. "He, he's a good husband, you know…. He was my *only* support at first … when my family was lecturing me about my duty to him and the kids. He was great, he—"

A woman who had identified herself as a former battered woman and the man who worked for the shelter exchanged glances with each other. Shelly saw this and scrambled to undo the impression she had given: "That's not it! He really doesn't mind me going to school. I know how it must sound … he doesn't normally yell and he's never hit me or even thrown anything like that before, you know? My girlfriends always envy me because he's so sweet and he's great with the kids and he's *so* handsome, I mean … he always knows what to say to people, I mean, not *college* people, but … you know, regular people. And it never really got stale, I mean, I was still crazy about him until … until … I don't know…"

Shelly stumbled to a halt and fell silent. Just when someone else was about to speak she blurted out, "I love him! When I think of losing him …" Her eyes teared up and she started shaking her head. "It's like the whole world is turned upside down!" Tears streamed [*sic*] down her red-hot face and she ran from the classroom to the lavatory down the hall.

Shelly is a college student at a small, urban university that mostly serves "returning" older and first-generation college students. She is close to her extended Swedish American and German American farm family and is the first one to go to college. Her husband and friends are all working class. She never really thought about going to college before her boss said she might lose her job if she didn't and that the company would pay for it. To her surprise, she loves it. She eagerly reads the class materials; she finds it surprisingly easy to talk in class, and other people often seem to appreciate what she has to say. She suspects she talks way too much, but she "just get[s] so excited." She wonders how could she not know before that she "loves ideas," as another woman in class put it. She was thirty-two years old and had had two children. "Where was I all those years?" she asked once in class.

After the class had left, Shelly came back in to apologize to me. She assured me that in more than two years of college, she had never behaved so "unprofessional before." She apologized a few more times. Her shoulders sank, deflated. She bowed her head and stared at her shoes. "Maybe he's right, maybe I don't belong here." She was embarrassed, and afraid.

Shelly is experiencing a confusing, exciting, and debilitating situation both in her outer life and within her. She is by turns excited, lost, elated, angry, bewildered, shameful, grateful, and "numb." All of a sudden, her past won't cohere with her present, her future has become uncertain; nothing quite "fits right" anymore. Shelly knows that no one she knew before seems to understand what she is going through, and some even resent it. That night she realized that her new friends don't understand either. She is in the midst of a working class "crossover" experience, something she never expected when she went back to school to "get my piece of paper" so she could keep her job as a legal secretary. She had no idea what she was getting into, she had no idea she would fall in love with a new world. She certainly didn't know that she might actually begin to *become someone else*. Though she is delighted with all the new things she has learned, nothing she has learned in this new world helps explain her situation to her. With no language or concepts to bridge or even explain this experience, she is falling prey to the contradictions within it.

There is suffering in this private passage, unvoiced and unseen, a particularly confused suffering in the midst of outward success. This struggle to figure out "who I am anymore," as Shelly once put it, the crossover's collection of contradictory experiences, emotions, and values are the subject of this [article]. I have come to believe Shelly's struggle constitutes a particular inner and outer (psychological and sociological) constellation that many working class people who enter the middle class experience. The psychological similarities among "upwardly mobile" working class people are striking to me. So is the invisible and "privatized" nature of this potentially painful experience. I am a (counseling and community) psychologist, a teacher of first-generation college students, and a person from the working class who has spent my adult life jostling back and forth between different worlds.

Like me, [many] people … have bumped uneasily between professional middle and working class cultures. We engage (or struggle to avoid engaging) with these often opposing worldviews. This often creates a state of *cognitive dissonance,* or an inner clashing of values and experiences that create emotional and mental confusion. Common emotional reactions to this are anger, shame, sorrow (loss), "impostor syndrome," and substance abuse. These are often so muffled as to be invisible to crossovers themselves. Common behavioral reactions I have seen are distancing, resisting, and creating/bridging.

I believe that central to the "crossover experience" is an existential dilemma. By "existential" I mean a problem of existence: of living one's life, of how best to live, and of the human need to make meaning in and of our lives. And central to this dilemma, though not its only feature, is the presence of cultural differences between the professional middle class and working class people. There are stories, sacrifices, and secret shame that have no ear and precious little voice. The hearing and seeing of these cultural differences—the ability to see *outside* the cultural biases of the professional middle class—is crucial to any meaningful understanding of working class life. Without this, all the well-meaning "solidarity" one may feel for the working class is ineffective. "I feel like they're always talking down to me," said one of my working class students, who is active in the political Left, "but maybe I *am* stupid, because, honestly, half the time I don't know what the hell they're talking about."

People in or from the professional/managerial class will likely be the vehicle of change for the "upwardly mobile" working class person in higher education, job promotions, marriage, psychotherapy, and other crossover experiences. They can show Shelly how to write and speak in Standard English, how to put her napkin in her lap instead of on the table, and how to negotiate with difficult clients. But they can't tell her where she's been and how it has made her who she is, or where it is she might be going.

In this article, I address the less obvious ways that class hurts working class people in higher education (and other avenues of upward mobility). I point toward unfair, unjust advantage and disadvantage that cuts across lines of gender, ethnicity, and "race." In higher education (as elsewhere), this unacknowledged crossover challenge serves to exclude working class people from certain opportunities and privileges, even from their own inner lives. Their counterparts, from the professional middle class, find in higher education the cultural rules, values, language, and community mores that are familiar to them. Working class people must do psychosocial back flips through a maze of new rules, new values, and new language. My concern is twofold: I am worried about the Shellys, and I am worried about the society we all live in that creates, mystifies, and personalizes unequal opportunity and the cultural (as well as economic) domination of one class of people over another. The painful distance between the ideal and the real is felt by those who fall between the cracks; working class crossovers bear it as personal "stupidity," lack of "ambition," "failure," and even psychopathology (depression, substance abuse, and more). These constitute significant invisible costs that working class crossovers are forced to pay. Visibility and voice are the first practical antidotes to this invisible identity crisis.

"SURVIVOR GUILT" AND CULTURAL COLLISION

Working class crossovers are likely to be completely invisible to people from the professional middle class, because middle class people have learned to assume their inner and outer lives are "normal." If you have learned to walk and talk middle class well enough to "fake it," middle class people will assume you have always been one of them, at least if you have white skin. Successful crossovers can't necessarily help you either. As likely as not, they have already been "made safe," … via the cultural processes and decisions they have gone through to get where they are.

The *invisibility* and the *unconsciousness* of the crossover experience, in my view, can make it painful, debilitating, even devastating. The dilemma manifests in a multitude of so-called (and genuine) personal problems. If crossovers are not conscious of the cross-class experience, the problems it creates can hide behind many personal perspectives. For Shelly, it is a marriage problem. For someone else it is a problem with her "unenlightened" parents. For yet another it is a "chemical imbalance." For many it is a compulsion to ditch class or get "loaded," or to suddenly "blow off" an important exam. Maybe it is simply having "the blues" all the time. For marriage problems, depression, chemical abuse, fear of success, and family-of-origin problems there is, at least, a certain amount of collective wisdom about coping, changing, treating, managing. In my experience, the process of moving from the working class to the professional middle class is a highly personalized and tangled mess of psychological, sociological, and cultural confusion. As philosopher Ludwig Wittgenstein said, that which one has no language for is often not even perceived....

For Shelly, who was previously happy with her husband and her working class life, there is a blazing new star on her horizon, a life of the mind. It complicates things because she wants both to keep her working class roots and to develop her intellectual abilities. She loves her husband, and she can barely stand the strain of not "doing it together." The ambivalence in her relationship is a mirror of her own gathering ambivalence, her own feeling of being "torn"— *torn not just between success and failure in college but between two different notions of what it means to succeed in life.*

I, and others, have tried to articulate what seem to be some of the valuable and central features of working class cultures and how these contrast with middle class culture.... Here's my snapshot: working class people are raised with a more here-and-now sensibility, in activities and worldview; individuality (but not necessarily self) is downplayed in favor of a powerful sense of community and loyalty, and an internal sense of "belonging." Working class cultures also tend to be more embedded in ethnic (non–Anglo) traditions, and so are more diverse by nature. Conversely, middle class culture is more homogeneous (and Anglo) and tends to put a premium on individual accomplishments, on the achievement of planned (and publicly recognized) goals in general, and on earning self-definition by way of these achievements—what I called "becoming." These are fundamental differences in outlooks and approaches to life....

As rules start to change for the crossover, family problems abound. Some parents push their children toward "good, clean jobs" as a way to show love and to make their own difficult work lives feel more meaningful. But their reward for this sacrifice is sometimes poignant. If their children succeed, more likely than not, they have adopted the culture, style, and *classism* of the professional/managerial class. Many parents shrink back in shame and confusion while the children they worked so hard to send to school become cultural strangers to them. They fear what is too often true; their children have become embarrassed by their "low class," "backward" family....

DOMINATION AND CULTURAL "CAPITAL"

My mother resisted my evangelical efforts to improve her perfectly good and colorful English. She was a fighter, and the struggle I'm describing is a matter not simply of *different* cultures but of one dominating the other. Is it any wonder that working class families do not easily surrender their children to the people who they know help make their own lives difficult? In the world of work the professional/managerial class is employed by the very wealthy to inflict appalling abuse and neglect on "lower class" workers. Working class people do not have equality of economic opportunity....

This domination also happens, in more genteel settings, by way of what Bourdieu calls "cultural capital." Professional middle class social style, language, and knowledge constitute a kind of social currency. People who have learned these things can use it for entrance into, and access to some amount of power in, the academy (as in business and government). Cultural barriers may be as effective in shutting out working class people as are the (significant) economic ones, perhaps more so. I have said elsewhere that most working class people's native tongue is more metaphoric than literal, more personal and particular than abstract and universal. It is more implicit than explicit, more for members of a defined social group, also more pithy, colorful, and narrative. It reflects cultural differences from the middle class. It is the opposite of how students are expected to write and speak to get good grades in school. This makes trying to "make it" in school considerably more difficult. Indeed, successful working class students are not necessarily "making it" in the sense that their parents, partners, and former peers understand that term. Cultural difference and prejudice against working class culture combine to frustrate the "upwardly" mobile student.

To succeed in higher education (and, often, in a middle class marriage) working class people must learn to adopt and represent middle class culture as their own. This culture does not grant dual citizenship. You must "leave behind" your "low class" ways, your "bad" English, your values of humility and inclusion (don't show off and be a "big shot," because it says you think you're better than others), and much more—not least the people you love! In early adulthood there are developmental tasks of differentiation at play that I suspect help fuel the leap the young crossover student is trying to make. But it is a cruel and unsuspected consequence to have that process set up a chasm that may never be bridged again.

CONCLUDING WITH CREATION

... Somewhere at the center of all these arguments and abstractions sits Shelly, hiding and crying in a bathroom stall at a midwestern state university. It would be wrong-headed to try to tell her what her own decision and solutions might be. What we *can* do is clarify what those decision points are by seeing her dilemma more clearly. We can start by illuminating and validating both her past and her present. If she can see her and her husband's dilemma as a clash of cultures rather than a battle of good and bad, better and worse, normal and abnormal, she may even be able to avoid choosing decisively *between* the cultures. If she has someone to talk with about how she might reconcile them in her own life, she and her family might move forward in a way right for her and them.

Working class cultures have many humane, healthy, and life-giving qualities for which people from the middle class pine and search, at no small consequence to their bank accounts. Like most counseling psychologists, I have spent a part of my career helping both "failures" and "successes" from the professional middle class improve their mental health. We help them, often, to embrace a kind of humanity that values warmth over brilliance, "connectedness" over competition, and that helps them to find a self that exists in spite of personal achievements or failures. I do not intend to romanticize what can be, in many ways, a difficult working class life with limited options, but it is also easy for me to remember, and enjoy to this day, many positive aspects of working class life....

What college *should* give Shelly (and what it gives her middle class peers) is additions to herself, as Phillips points out—not subtractions from herself, or a "transformation." What she needs is an *integration* of new abilities and awarenesses, not a compartmentalized half-self plagued with doubt and addicted to success. This work is "creating" or "bridging," a third response to the clashing of cultures and the resulting confusion. If this is difficult, if it seems contradictory to people (like me) who long for a classless society, we still need to apply ourselves to the task.

50

How a Scholarship Girl Becomes a Soldier

The Militarization of Latina/o Youth in Chicago Public Schools

GINA M. PÉREZ

Like many working-class communities, Chicago Puerto Ricans have had a complicated relationship with the United States military, an institution that represents, in a particularly visible way, Puerto Rico's unresolved political status, one that is contested by community activists. Recent broad-based political mobilization in both Chicago and New York City demanding the end of naval operations on the island of Vieques, as well as a smaller, less visible presence of island residents supporting the navy's role on the island, is just one example of the complex responses Puerto Ricans have to the military. The military also is an institution that is understood to be (and is often successfully used as) an important avenue of social mobility for families and households with limited resources and opportunities available to them, particularly for young women who often are sources of productive and reproductive labor critical to a household's survival....

In what follows, I draw on ethnographic and interview data to discuss how Puerto Rican and Latina/o youth in Chicago are implicated in an increasingly militarized world. Military programs like the Junior Reserve Officer Training Program (JROTC) are not new (it was founded in 1916) and have long targeted Puerto Rican youth. In recent years, however, these programs have operated with new intensity, appearing more frequently in urban schools with an explicit goal of targeting populations deemed "at risk" in "less affluent large urban schools." It is not surprising that this "at risk" group includes young men who are allegedly in danger of falling into gangs or drug use. What is alarming, however, is the extent to which these programs are now targeting young women who are deemed "at risk" because of the possibility that they might become unwed teenage mothers. The gendered dynamics through which militarization is occurring on the homefront are in need of critical attention if we are to understand how poor and working-class youth and their families struggle and strategize to make ends meet....

SOURCE: From *Identities: Global Studies in Culture and Power*, 13 pp. 53–72. Copyright © 2006. Reprinted by permission of the Taylor & Francis Group.

CHICAGO: A GLOBAL LATINO CITY

With more than three-quarters of a million Latina/o residents, Chicago is home to the third largest and one of the most diverse Latina/o populations in the country. It also is the only place where large numbers of Mexicans and Puerto Ricans of several generations live, marry, and work side by side....

Chicago Latinas/os, however, are also some of the city's most impoverished residents, with nearly twenty-four percent of Latinas/os living in poverty. Chicago Puerto Ricans are the poorest of all Chicago Latinos. With limited employment opportunities, primarily in the low-wage service sector, Puerto Ricans in Chicago index the highest poverty rates in the city at 33.8 percent, compared to thirty-three percent for the city's African-American families.[1] In contrast with Cubans and South Americans, Puerto Ricans and Mexicans have lower educational and average income levels and are concentrated in the low-wage service sector and as operatives.[2] Moreover, studies by the Latino Institute, The Chicago Urban League, and Northern Illinois University demonstrate that none of the jobs requiring only a high school diploma—and even many of those demanding some post-secondary education—pay a living wage for a family with dependent children.[3] According to the Latino Institute (1994), more than seventy-five percent of Puerto Ricans are employed in those sectors of the economy. In short, while transnational investment and the loss of manufacturing jobs in the Chicago area have rendered Latinos in general much poorer than a decade earlier, Puerto Ricans specifically remain the most economically disadvantaged group in all of Chicago. It is no wonder that many Latina/o youth consider military service as one of the few opportunities that will guarantee employment, provide them with marketable skills, and serve as a vehicle for achieving economic stability.

JROTC, MILITARY SERVICE, AND
MAKING ENDS MEET

Beginning at a very early age, military service becomes one of the most appealing and seemingly sure options for many poor and working-class Latina/o families. Bombarded by television advertisements, visits by recruiters in junior high and high schools, and the prospect of receiving JROTC programs' financial incentives (both immediate and promises of money for the future), poor children and particularly children of color quickly come to consider the military—rather than college—as an employment venue after high school. Money, promises of free education, training, discipline, honor, and respect seduce young men and women into the military ranks.

Chicago schools provide particularly fertile ground for cultivating military careers after high school. Nearly eighty-five percent of Chicago public school students come from low-income families and are either African American or Latina/o.[4] They also are increasingly enmeshed in what Pauline Lipman has identified as "stratified academic programs," whereby African-American and

Latina/o high school students attend schools with "limited offerings of advanced courses and new vocational academies, basic skills transitional high schools" or public military academies rather than the "[n]ew academically selective magnet schools and programs, mainly located in largely white upper-income and/or gentrifying neighborhoods" (Lipman 2003: 81). The Chicago public schools, for example, lead the nation with more than 10,000 students participating in a wide range of expanding public school military programs. Chicago is home to two school-wide military academies—Chicago Military Academy and Carver Military Academy—both affiliated with the United States Army and located on Chicago's South Side; eight military academies within regular high schools (four affiliated with the Army, two with the Marines, and one each with the Navy and the Air Force); a middle school military academy in the predominately Mexican neighborhood of Little Village; and part-time JROTC in forty-three Chicago high schools and after-school programs at seventeen middle schools.[5] ...

In Chicago, JROTC is part of the Education to Careers (ETC) program, whose mission is to "equip students to successfully transition into postsecondary education, advanced training, and the workplace." ... High school students may participate in JROTC to satisfy the Chicago public high school career education requirement or, in some schools, to satisfy the physical education requirement for graduation. JROTC also offers many extracurricular activities, including honor guard, competitions, service, field trips, and opportunities to visit military installations. Many students explain that it is precisely these curricular and extra-curricular offerings that make JROTC appealing. Almost all of the young Latina/o and African-American JROTC participants I interviewed in one Chicago high school also cited "being treated with respect" (especially while wearing the cadet uniform) both as a reason for joining, as well as one of the greatest advantages of participating in, JROTC.

Students' concern with respect is no small matter, since most reside in poor and working-class (and, often, slowly gentrifying) neighborhoods regarded in local media as dangerous and are enmeshed in racialized policing practices aimed at containing suspect youth.[6] Latina/o and African-American youth are painfully aware of how their bodies are read. Thus, wearing a military uniform is, perhaps, one way of negotiating the racialized systems of surveillance that not only operate within their neighborhoods, but also within their own schools....

Even when young people are not in formal high schools, the military continues to shape many of their educational and employment opportunities in other ways. For example, when nineteen-year-old Marvin Polanco[7] entered a G.E.D. class I taught at a Puerto Rican cultural center in Chicago, he regaled me with stories about his experience in Lincoln's Challenge, a military boarding school affiliated with the Illinois National Guard.[8] Shortly after dropping out of Clemente High School, Marvin's mother signed him up for the five-month program, which promised to instill discipline, prepare him for the G.E.D., and pay him $2,200 after successfully graduating from the school. "But they tricked my mom," Marvin told me solemnly one day after class. He never passed the G.E.D., a failure he attributes to the fact that they required him to exercise, run, and engage in other physical training immediately before the exam, leaving him exhausted and unable to finish the test. Marvin received less than $500 when he finished the program. "They try to brainwash you there," he explained another day, referring to the deeply militarized atmosphere and training each cadet receives while in residence. Between the physical discipline imposed by the officers and the danger of getting jumped by gang members also participating in the program, Marvin said he often was afraid he wouldn't get out alive. After a few months in the program, he ran away from the school and returned home, begging his mother not to send him back to the school and promising to get his G.E.D. somewhere in Chicago.

Many of my male G.E.D. students wanted to go into the military once they completed their G.E.D. When they would approach me about this, I encouraged them to consider enrolling in one of Chicago's city colleges and eventually transferring to either University of Illinois–Chicago (UIC) or Northeastern University, which both had programs specifically targeting and supporting incoming students with their G.E.D. With the help of the Centro's counselors and staff, we put many students in touch with representatives from these universities, invited them on a tour of one of the city colleges, and helped organize a campus visit to a small liberal arts college to whet their educational appetites. Michael, one of the Centro's counselors, was extremely invested in pushing our students to consider college. As a G.E.D. graduate himself and current student at Northeastern, he knew firsthand the strong push toward military service and vocational training, and he worked tirelessly with me to provide the students with alternatives.

Considering the military as an option, however, infected even the brightest of students. Nineteen-year-old Eddie Vélez, for example, a smart student who delighted in being a good student and a math wiz after years of being told he was not, arrived late to G.E.D. class one day, announcing that he met with an army recruiter who explained to him how the G.I. Bill would pay for his college education. Eddie was seriously considering this option and his enthusiasm spread to Frankie, a sixteen-year-old student who was also interested in joining the military. I was furious and gathered the class together and asked them to think about why the military aggressively recruited them and other poor people of color and not the mostly white affluent Northwestern University students I also taught on the city's northern suburbs. Initially stunned, my students responded passionately about the "honor of serving your country," telling me about their fathers, grandfathers, uncles, and brothers who had all served time in the armed forces. They were angry with me for being so upset and critical of what for many of them had been small steps toward social and economic mobility. "My father and *grandfather* were in the army," Eddie insisted. "It's a good job. And that's what I need. I need to get out of my house and *away* from the drugs and get into a good place. I can't concentrate when I'm at home, you know. They're smokin', drinkin', and I can't concentrate. Even if I go to college, I still have to go *home*. Why do you think I work so many hours and *like* to come to school? I need to get out of that environment." The class was silent, and a few students nodded their heads.

The idea of using the military as a haven or as a way to avoid troubled households and communities is not lost on those who run JROTC, as well as some of the programs' staunchest advocates.... The idea of the benefits JROTC programs provide for "inner-city" and "at-risk" youth is so entrenched that the program's Pentagon funding [was] expected to increase more than fifty percent of its current budget of \$215 million in 2001 to \$326 million by 2004.[9] Much of this money goes to local school districts eager to implement JROTC programs, since doing so guarantees the infusion of federal funds that many proponents argue can be a strategy for broadening school curriculum and securing more money for already financially strapped schools....

GENDER, SEXUALITY, AND THE QUEST FOR AUTONOMY

Young men, of course, are not alone in considering the military as a viable educational and employment option. Increasingly, young girls swell the JROTC ranks, a phenomenon that allegedly demonstrates the military's egalitarian gender roles.... In Chicago public schools, young women outnumber young men in Army JROTC programs by a slim margin; they also comprise nearly half of all JROTC program participants, with Latinas and African-American women constituting the overwhelming majority.

Perhaps the appeal of more egalitarian gender relations partially explains young women's interest in JROTC. Ethnographic evidence, however, suggests that young girls' decisions to enter JROTC programs are slightly more complicated. Latina adolescents, for example, emphasize how household economic imperatives, as well as parental pressure and encouragement from friends and kin, influence their ideas about JRTOC and military service. Participating in JROTC's extracurricular programs may also offer freedom and autonomy otherwise unavailable to some young girls who are expected to abide by culturally prescribed norms of behavior requiring them to be in the home. Some parents support their daughter's participation in JROTC because of the program's promise to cultivate in its participants the values of structure, discipline, and honor. As parents grow increasingly concerned with their adolescent daughter's sexuality, JROTC participation can be regarded as another way of preserving their daughter' s chastity and honor, while simultaneously reinforcing their critical role in their household's economic survival.

Thus, while military involvement allegedly keeps young men out of gangs, for young women it allegedly functions as a way to mitigate unwed teen pregnancy, providing necessary discipline for "unruly," "unpredictable" women's sexuality. Chicana and Puerto Rican scholars have demonstrated how culturally prescribed notions of female chastity and virginity among Latinas frequently result in family policing of young unmarried women's bodies by observing them, being vigilant of their movements and activities, and designating the household as the appropriate place for young women (Souza 2002; Trujillo 1991; Zavella 1997). These concerns with female chastity and purity also characterize poor and working-class *puertorriqueñas* whose families (usually mothers) not only monitor young women's sexual behavior, but who also depend on them for the economic survival of their households.

As Caridad Souza (2002) has shown in her research among working-class Puerto Rican women in Queens, young women's reproductive work is critical to households whose survival depends on family and community solidarity through the exchange of goods, services, and the pooling of resources across multiple kin and non-kin households. In this context, young women are expected to be *en la casa* (inside the home or family) for important material and cultural reasons: Without their reproductive work, adult women—mothers, grandmothers, and aunts—are easily overwhelmed by (and perhaps unable to meet) the reproductive demands within their own households and kin networks. Being *una muchacha de la casa* also means abiding by the cultural norms of respectability, chastity, and family honor valued by the community. To be *de la calle* (outside the home or literally "on the streets") is to be transgressive, sexually promiscuous, and dangerous (Souza 2002: 35).

For young Latinas, therefore, participating in JROTC programs offers both a way for them to contribute to their household economies as well as the possibility of creating a space for them to exercise some autonomy, giving them "legitimate" and "respectable" reasons to be outside of the home in ways that conform to cultural expectations of young women. In this way, young Puerto Rican women creatively construct ways of fulfilling their culturally prescribed gendered responsibilities, although they do so by participating in a hyper-masculinized social context that regards female sexuality as a problem in need of discipline and control (Enloe 2000). Seventeen-year-old Jasmín Rodríguez's story provides a glimpse into these competing demands and dreams that shape young women's thinking of JROTC and their sense of obligation to their families.

Throughout middle and high school, Jasmín has been a model student, consistently making honor roll, being ranked at the top of her high school class, and actively participating in school-sponsored community projects, writing contests, and youth-leadership programs. Jasmín has also been the recipient of summer scholarship programs targeting young women of color to attend summer-long programs at prestigious universities like Princeton and Cornell. She is an avid reader with a keen sense of social justice and inequality, characteristics that led her to announce at the age of ten that she was going to go to Harvard Law School one day so she could help make the world a better place. As a junior at one of Chicago's public high schools, she scored well on her PSATs and began receiving letters from four-year colleges and universities, whetting her desire to go away to college, to be the first woman in her family to do so. Certainly, how she is going to pay for college weighs heavily on her mind, but with the assurance of her high school counselors and mentors, Jasmín was confident that she would be able to get enough money from scholarships and financial aid to pay for college without having to rely on her mother for help.

Jasmín's enthusiasm for college has spread to her younger sister, Myrna, who also talks excitedly about college. Together with their college degrees, they talk confidently that they will soon be financially secure enough to help support their mother, Aida Rodríguez, and the rest of their family [who] barely survive month to month on the pooled resources of Aida's salary as a high school clerk and their older sister's and brother's part-time jobs as a cashier and a security guard at O'Hare Airport. Both Jasmín and Myrna believe college is a sure way to economic security and to helping their mother achieve her dream of finally owning a modest home for their family one day. Jasmín's determination to go to college has been such a defining feature of her personality that I was amazed when, in the Spring of 2002, she told me she had signed up for the military. When she made the announcement at her nephew's first birthday party, she glowed with excitement and even seemed to take pleasure in the shocked response her decision provoked. When I asked her why she had changed her mind and why she was now considering the military, she repeated several times, "I just had to do it. It was the right decision. I'm really going to do this *and* go to college. And JROTC will help pay." When I assured her that based on her grades and family's financial need she would be able to pay for college without relying on money from military service, she replied, "Yeah, I know, but this is just another way to be sure I can pay for college. And they also give you money now. And we really need it."

Aida was equally confused by her daughter's decision. Aida takes enormous pride in Jasmín's accomplishments, and the walls and shelves in their small second story apartment, covered with mostly Jasmín's diplomas, awards, and trophies recognizing her academic achievements, are a clear sign of Aida's pride in and admiration for her daughter. Jasmín's decision, therefore, was tremendously distressing to Aida, who could not explain why Jasmín wanted to join JROTC and serve in the military after graduation. Aida had encouraged her oldest daughter, Milly, to participate in JROTC in high school and Aida even hoped Milly would go into the military after graduation, assuring her that doing so would benefit her financially and educationally. Aida feared that Milly wouldn't have the same kinds of options Jasmín had and military service, she assured me on many occasions, was also a way to keep Milly "out of trouble" and to prevent her from making the same mistakes Aida made as a young girl, namely becoming a teenaged single mother. Like many Puerto Rican mothers, Aida worried that Milly and her other daughters would *meterse las patas* (to mess up, and in this case, to get pregnant) as teenagers and she continuously strategized ways of protecting her daughters, including contemplating sending Milly to Puerto Rico when she was fourteen years old. Jasmín, however, was different and carried the hope and responsibility of being the first woman in her family to go away and graduate from college.

Jasmín understood her mother's hopes for her. But she was also painfully aware of her family's precarious economic situation that became even more tenuous with the birth of Milly's son, who brought an incredible amount of joy to the family. When I interviewed Jasmín about her decision later that summer, she provided a narrative filled with concerns with money, the need for security, and a desire to protect and serve both her nation and her family. She explained how one morning she was approached by the JROTC sergeant at her high school to take the ASVAB. His praise of her high scores, his relationship with her older sister, and his promise of $2,000 if she participated in a summer boot camp that would prepare her for her service in the army reserves all highlight a critical feature of successful JROTC programs: Their strength derives from a trusting and respectful relationship between the school commanding officer and the cadets. In June 2004 I observed, for example, the ease with which young men and women cadets interacted with their commanding officer, and his sensitive understanding of each cadet's struggles and hopes for the future. Many of the cadets I interviewed had siblings or cousins who also participated in JROTC, and one young man, Johnny, explained how he convinced his younger sister to join. When I asked him why, Johnny jokingly offered that since he is older and has "rank" in his unit, his younger sister would have to "show me respect" and defer to him in ways that she refused to do when at home. While Johnny admitted this was only one reason for persuading his sister to participate, his comments speak to the role of kinship and social networks in expanding the program's ranks. Over time, JROTC commanders meet these family members and friends and are able to draw on those networks not only to support his cadets, but also to approach young people who otherwise may not consider JROTC or military service after high school. It was precisely these networks that facilitated Jasmín's conversations with one of the JROTC officers in the school. She explains, "He was pretty convincing. He said he knew my family since he recruited my sister … and I saw it as a precaution just in case I didn't get any scholarship or financial help for college. I could do a lot with this money. My family has never been rich, and that money seems like so much…. I had my senior fees coming up and I can't expect my mother to pay for it. My senior trip, my prom. And it feels really good to have the money."

When I asked her if there were other reasons for her decision, she admitted that the events of 9/11 were also important.

> I have family in New York … and I sure didn't feel protected and I felt like [by joining the military] I would be protecting those I love. And even those I don't know, I felt like I could be doing something to help…. The women in my family have had to be protected by bad men [all their lives] and I don't want to have to do that. I could protect myself and not have to depend on others, but [my friends and family] can depend on me. Protecting people and standing up for what you believe in and being a good example of what the U.S. can be. I think our armed forces embody that.

The desire to feel protected and to equip oneself with important skills and financial resources to ensure independence fills Jasmín's narrative of education, her future, and her family indebtedness. They are also themes emerging from many Chicago Puerto Rican women's life histories and explain, in part, Aida's ambivalent embrace of military service for her daughters. For both Aida and Jasmín, the military's promise of money is clearly a means to ultimately achieve a better education and financial security….

Jasmín's decision, however, is also informed by her quest for autonomy and power that will allow the women in her family to depend on her rather than have to be "protected by bad men." Military service provides some possibilities not only for her, but specifically the women in her family whose options are often circumscribed because of limited education and low-wage employment.

Young women's participation in JROTC, therefore, is informed by social and economic need, although public officials are often quick to seize upon the deeply gendered concern with sexuality to promote the benefits of military service for young women. Public officials, for example, deploy the language of security and protection to advance the idea that military service protects "at risk" youth from danger and instills critical values allegedly absent in poor families, namely discipline, honor, and the value of hard work. Some JROTC officers and elected officials share parental concern with young women's sexuality and suggest that JROTC might be a new way to battle teen pregnancy. In Jackson, Mississippi, for example, state senator Robert Johnson recently applauded the increasing number of women entering into JROTC programs saying, "Jackson has more unwed mothers than just about any city of its size in the nation. We're talking about second- and third-generation single parents. The people criticizing JROTC are not the people living in these communities, because if they were, they would know that the people making the biggest difference, doing the most grass-roots work, are people with military backgrounds."

Regardless of whether the senator's statement is true, what is striking about his analysis is the explicit way in which he and others connect women's military service with sexual discipline, and how poor and working families also see these benefits of military service as well as the possibility it provides for sustaining precarious household economies and leading the way out of poverty. There are clearly a number of success stories of how this has happened, and these narratives are part of a larger American dream ideology that continues to animate impoverished Latina/o families.

CONCLUSION

With high poverty rates, low levels of educational attainment, and high drop-out rates among Latinas/os, military service becomes an appealing option for many Latinas/os whose life chances and economic options are increasingly circumscribed. These realities are not lost on military recruiters and the Department of Defense and the Pentagon, who have increased spending for JROTC programs in urban schools and target Latinos for military enlistment with the goal of boosting "the Latino numbers in the military." ... Like their male counterparts, young Latinas are particularly vulnerable to military appeals to their sense of family obligation as well as notions of pride, discipline, loyalty, honor, and citizenship as they consider military service as one of many economic strategies to make ends meet. This discussion, therefore, is an attempt to analyze Latinas/os' choice to participate in military programs within a broader cultural and political-economic context. What is needed, however, is more historically informed ethnographic research that engages with these complex issues of military service, poverty, race, gender, and notions of citizenship and patriotism in a way that honors the real lived experiences and struggles of our informants, but also challenges conventional notions of community based on militarized (and masculinist) notions of home and nation, to build, instead, communities of justice, equal opportunity, and solidarity.

NOTES

1. Latino Institute (1995).
2. John Betancur et al. (1993) point out that despite Cubans' and South Americans' economic success, their wages according to the 1980 census still approximated those of Puerto Ricans and Mexicans rather than that of whites. The 1990 census, however, paints a very different picture, emphasizing the growing gap between Puerto Ricans and other Latino groups in terms of average incomes, employment rates, and poverty levels.
3. See Chicago Urban League et al. (1994, 1995a, 1995b).
4. Chicago Public Schools, CPS at a Glance, February 2005. Available at www.cps .k.12.il.us/AtAGlance.html. The three full-time military academies, Chicago Military Academy, Carver Area High School, and Austin Community Academy, all opened since 1999 (www.cps.k12.il.us/AtAGlance.html).
5. Ana Beatriz Cholo, Military Marches into Middle Schools, *Chicago Tribune*, 26 July 2002; Chicago Public Schools, JROTC Program Book, nd; Education to Careers (ETC) FY 2003–2004, available at www.cps.k12.il.us/AboutCPS/Financial_ Information/FY2004_Final/CPS_Unit/Education/Education_to_Careers.pdf.
6. Elsewhere (Pérez 2002), I have documented how Latina/o youth (especially those in rapidly gentrifying neighborhoods) are implicated in the policing of urban space aimed at curbing, for example, gang activity. Although Chicago's anti-loitering ordinance was declared unconstitutional in 1999, some scholars and activists have highlighted the "ongoing attempts to legalize harassment and street sweeps of youth," particularly youth of color who are regarded as dangerous and who allegedly "need to be locked up or removed from public space" (Lipman 2003: 95).
7. All names are pseudonyms.
8. Lincoln's Challenge is a federally funded youth program for at-risk youth between the ages of sixteen and eighteen whose quasi-military training in "discipline, esprit-de-corps, leadership, and teamwork is producing rapid and effective change in individual behaviors and attitude." Lincoln's Challenge is the largest National Guard Youth Challenge program in the nation (www.lincolnschallengeacademy.org/ challenge/challenge.htm).
9. Racism and Conscription in the JROTC, *Peace Review*, September 2002; Class Warfare, *Time*, 4 March 2002, p. 50.

REFERENCES

American Friends Service Committee 1999. *Trading Books for Soldiers: The True Cost of JROTC*. Philadelphia, PA: AFSC.

Betancur, John J., Teresa Cordova, and Maria de los Angeles Torres 1993. Economic Restructuring and the Process of Incorporation of Latinos into the Chicago

Economy. In *Latinos in a Changing U.S. Economy*. Rebecca Morales and Frank Bonilla, eds. New York: Sage. Pp. 109–132.

Chicago Urban League, Latino Institute, and Northern Illinois University 1994. The Changing Economic Standing of Minorities and Women in the Chicago Metropolitan Area, 1970-1990. Chicago, IL: Chicago Urban League, Final Report.

Chicago Urban League, Latino Institute, and Northern Illinois University 1995a. When the Job Doesn't Pay: Contingent Workers in the Chicago Metropolitan Area. In *The Working Poor Project*. Chicago, IL: Chicago Urban League.

Chicago Urban League, Latino Institute, and Northern Illinois University 1995b. Jobs That Pay: Are Enough Jobs Available in Metropolitan Chicago? In *The Working Poor Project*. Chicago, IL: Chicago, Urban League.

Enloe, Cynthia 2000. *Maneuvers: The International Politics of Militarizing Women's Lives*. Berkeley: University of California Press.

Latino Institute 1994. *A Profile of Nine Latino Groups in Chicago*. Chicago IL: Latino Institute.

Latino Institute 1995. *Facts on Chicago's Puerto Rican Population*. Chicago IL: Latino Institute.

Lipman, Pauline 2003. Cracking Down: Chicago School Policy and the Regulation of Black and Latino Youth. In *Education as Enforcement: The Militarization and Corporatization of Schools*. Kenneth J. Saltman and David A. Gabbard, eds. New York and London: Routledge Falmer. Pp. 81–101.

Pérez, Gina 2002. The other "Real World": Gentrification and the social construction of place in Chicago. *Urban Anthropology* 3 (1): 37–67.

Souza, Caridad 2002. Sexual identities of young Puerto Rican mothers. *Dialogo* 6 Winter/Spring: 33–39.

Trujillo, Carla 1991. Chicana Lesbians: Fear and Loathing in the Chicano Community. In *Chicana Critical Issues*. Norma Alarcon, Rafaela Castro, Emma Perez, Beatriz Pesquera, Adaljiza Sosa Riddel, and Patricia Zavella, eds. Berkeley: Third World Woman Press. Pp. 117–126.

Zavella, Patricia 1997. "Playing with Fire": The Gendered Constructions of Chicana/Mexicana Sexuality. In *The Gender/Sexuality Reader: Culture, History, Political Economy*. Micaela di Leonardo and Roger Lancaster, eds. New York: Routledge. Pp. 392–408.

51

Policing the National Body

Sex, Race, and Criminalization

JAEL SILLIMAN

American politicians, eager to garner support from the large middle-class, promote "family values," endlessly debate issues of abortion, and outdo each other as champions of "working families" (read "middle-class" families)....

Essentially, what we have is an America deeply divided across class and race lines. This makes it possible for mainstream America—its politicians and media—to ignore or rarely address issues of poverty, criminalization, and race that are pressing for communities of color. Incarceration rates for people of color are disproportionately high and assaults and searches by police, the Immigration and Naturalization Service (INS), and border patrol forces are daily occurrences in communities of color. This aggressive law enforcement regime is increasingly accepted by the mainstream as the price to be paid for law and order. A lead article in the February 2001 edition of the *New York Times Magazine* marks this decisive shift in public attitude in New York City from a more libertarian, turbulent, and nonconformist city towards a greater acceptance of aggressive law enforcement.

... The new wisdom in New York City—one of the bastions of liberalism in the country—is "We no longer believe that to solve crime we have to deal with the root causes of poverty and racism; we now believe that we can reduce crime through good policing."[1]

Aggressive law enforcement policies and actions are devastating women of color and their communities. Though there is a strong and growing law enforcement accountability movement,[2] the women's movement in general has not seen state violence as a critical concern.[3] The mainstream reproductive rights movement, consumed with protecting the right to abortion, has failed to respond adequately to the policing, criminalization, and incarceration of large numbers of poor people and people of color. It has not sufficiently addressed cuts in welfare and immigrant services that have made one of the most fundamental reproductive rights—the right to have a child and to rear a family—most tenuous for a large number of people.

The mainstream reproductive rights movement, largely dominated by white women, is framed around choice: the choice to determine whether or not to

SOURCE: From Jael Silliman and Anannya Bhattacharjee, eds., *Policing the National Body: Race, Gender, and Criminalization* (Boston: South End Press, 2002), pp. x–xxvi. Reprinted by permission of the South End Press.

have children, the choice to terminate a pregnancy, and the ability to make informed choices about contraceptive and reproductive technologies. This conception of choice is rooted in the neoliberal tradition that locates individual rights at its core, and treats the individual's control over her body as central to liberty and freedom. This emphasis on individual choice, however, obscures the social context in which individuals make choices, and discounts the ways in which the state regulates populations, disciplines individual bodies, and exercises control over sexuality, gender, and reproduction.[4]

The state regulates and criminalizes reproduction for many poor women through mandatory or discriminatory promotion of long-acting contraceptives and sterilization, and by charging pregnant women on drugs with negligence or child abuse. An examination of the body politics, the state's power of "regulation, surveillance and control of bodies (individual and collective)," elucidates the scope and venues through which the state regulates its populations "in reproduction and sexuality, in work and in leisure, in sickness and other forms of deviance and human difference."[5]...

Women of color have independently articulated a broad reproductive rights agenda embedded in issues of equality and social justice, while keenly tuned to the state's role in the reproduction and regulation of women's bodies. In an effort to protect their reproductive rights, they have challenged coercive population policies, demanded access to safe and accessible birth control, and asserted their right to economic and political resources to maintain healthy children. Through these demands, they move from an emphasis on individual rights to rights that are at once politicized and collectivized. African-American women leaders articulate the barriers to exercising their reproductive rights:

> Hunger and homelessness. Inadequate housing and income to provide for themselves and their children. Family instability. Rape. Incest. Abuse. Too young, too old, too sick, too tired. Emotional, physical, mental, economic, social—the reasons for not carrying a pregnancy to term are endless and varied.[6]

These leaders remind us that a range of individual and social concerns must be engaged to realize reproductive rights for all women.... A key concern among women of color and poor communities today is: the difficulty of maintaining families and sustaining community in the face of increasing surveillance and criminalization.... Particular communities and women within them are conceived and reproduced as threats to the national body, imagined as white and middle-class....

Poor women and women of color are criminalized. Bhattacharjee elaborates upon the prison system, police, INS, and border patrol forces to illustrate how they routinely undermine and endanger women's caretaking, caregiving, and reproductive functions. She shows how systemic and frequent abuses of reproductive rights and threats to bodily integrity are often overlooked by narrow definitions of reproductive rights and single-issue movements.... Greater coalition-building efforts between the women's movement, the reproductive rights movement, the immigrant rights movement, the violence against women movement, and the enforcement accountability movement are needed to break down barriers and to ensure the safety and self-determination of women of color....

CRIME AND PUNISHMENT IN THE UNITED STATES

The criminal justice system has become a massive machine for arrest, detention, and incarceration. The events of September 11, 2001 have intensified this trend. Citing "national security" and the "war against terrorism," President Bush has furthered the power of the criminal justice system to arrest noncitizens and to circumvent the court system. Immigrants and communities of color will bear the brunt of the intensified assault on civil liberties. In 1998, on any given day, there were approximately six million people under some form of correctional authority. The number of people in American prisons is expected to surpass two million by late 2001. Federal Judge U. W. Clemon, after a visit to the Morgan County Jail in Alabama, wrote in a blistering ruling, "To say the Morgan County Jail is overcrowded is an understatement. The sardine-can appearance of its cell units more nearly resemble the holding units of slave ships during the Middle Passage of the eighteenth century than anything in the twenty-first century."[7] Imprisonment is the solution currently proffered for drug offenses in minority communities and for the other social problems spawned by poverty. As welfare and service programs are gutted, the only social service available to many of America's poor is jail![8]

Despite this surge in incarceration rates, it is widely accepted that prisons encourage recidivism, transform the occasional offender into a habitual delinquent,[9] fail to eliminate crime, and ignore the social problems that drive individuals to engage in illegal actions. French historian and social critic Michel Foucault explains the political rationale behind what the terms the "production of delinquency," and its usefulness for those in power:

> [T]he existence of a legal prohibition creates around it a field of illegal practices, which one manages to supervise, while extracting from it an illicit profit through elements, themselves illegal, but rendered manipulable by their organization in delinquency.... Delinquency represents a diversion of illegality for the illicit circuits of profit and power of the dominant class.[10]

The building and maintenance of policing and prison systems is politically expedient and highly profitable. Prisons boost local economies. Fremont County, Colorado, home to thirteen prisons, promotes itself as the Corrections Capital of the World. In *Going Up the River,* Joseph Hallinan explains how prisons have become public works projects that require a steady flow of inmates to sustain them.[11] Corporations engage in bidding wars to run prisons, and the federal government boasts about saving money by contracting out prison management to the private sector.

People of color are disproportionately represented in the prison industrial complex. Bureau of Justice statistics indicate that in 1999, 46 percent of prison inmates were Black and 18 percent were Hispanic. In "Killing the Black Community," Judith Scully argues that the war on drugs is used to justify and exercise control over the Black community. She contends that the United States government has historically maintained control over the Black community by selectively enforcing the law, arbitrarily defining criminal behavior and incarceration, and failing to punish white people engaged in lawless acts against the Black community, as exemplified in the infamous Rodney King and Amadou Diallo verdicts. The war on drugs demonstrates how government officials employ legal tools as well as racial rhetoric and criminal theory to criminalize and destroy Black communities. Scully explores how the creation of drug-related crimes such as the "crack baby" demonizes Black motherhood and undermines Black child-bearing. Her essay exposes the institutional links between Blackness, suspicion, and criminality.

As strongly as class and race biases determine the criminal justice system, so does gender bias. Since 1980, the number of women in state and federal correctional facilities has tripled.[12] Amnesty International figures indicate that the majority of the over 140,000 women in the American penitentiary system are Black, Latina, and poor women, incarcerated largely for petty crimes.[13] For the same offense, Black and Latina women are respectively eight and four times more likely to be incarcerated than white women. The United Nations special report on violence against women in U.S. state and federal prisons noted the trend in prison management that emphasizes punishment rather than rehabilitation and a widespread reduction in welfare and support services within the criminal justice system....

The criminal justice system works collaboratively with government, corporate, and professional institutions to perform and carry out disciplinary functions deemed necessary to uphold the system of injustice. The recent exposés on racial profiling, discriminatory sentencing, and the compliance of hospitals, medical professionals, and private citizens in administering drug tests or reporting substance abuse among pregnant women are a few examples of the ways in which a range of actors are drawn in (sometimes reluctantly) to the surveillance and disciplinary system. For example, feminist lawyer Lynn Paltrow, executive director of National Advocates for Pregnant Women, describes how the Medical University Hospital in Charleston instituted a policy of reporting and facilitating the arrest of pregnant patients, overwhelmingly African-Americans, who tested positive for cocaine. African-American women were dragged out of the hospital in chains and shackles where the medical staff worked in collaboration with the prosecutor and police to see if the threat of arrest would deter drug use among pregnant women.[14]

Such violations are not going unchallenged. In *Ferguson vs. City of Charleston* (March 2001), a lawsuit engineered by Lynn Paltrow, the US Supreme Court agreed that Americans have the right, when they seek medical help, to expect that their doctor will examine them to provide diagnosis and treatment, and not search them to facilitate their arrest.[15]...

It is essential that we do not separate the more blatant forms of policing—videocameras within prisons that track every move of a prisoner, racial profiling, drug tests disproportionately administered in poor communities, and the raids on illegal immigrants crossing over to the United States—from the disciplinary apparatus being deployed across society. The wide use of differential forms of control and discipline is apparent in the ways in which the public has acquiesced to the policing and surveillance increasingly employed in everyday lives.

As a society we are no longer outraged when we hear that public schools in poor neighborhoods are routinely policed and equipped with metal detectors, and that students are hauled away in handcuffs for petty misdemeanors. The bodies being patrolled, segregated by race, determine the form of discipline applied. Perhaps this differential treatment explains why, in response to the spate of shootings in public schools across the country, parents in affluent neighborhoods have rallied and called for greater policing to ensure their children's safety. Middle-class parents invite policing to "protect their children." This contrasts sharply with the policing imposed in schools in poor communities and communities of color that criminalizes rather than protects. Though surveillance and policing differ according to whether they are there to protect or to criminalize, both kinds of interventions further extend state control over individual and collective bodies. The overt and insidious intrusions consolidate power in state and corporate entities.

BIOLOGICAL CONTROL

Women of color have been the target of biological control ideologies since the founding of America. In her book *Killing the Black Body,* Dorothy Roberts traces the history of reproductive rights abuses perpetrated on the Black community from slavery to the present. Roberts shows how control over reproduction is systematically deployed as a form of racial oppression and argues that the denial of Black reproductive autonomy serves the interests of white supremacy.[16] Others have documented the history of sterilization abuse in the Latina and Native American communities, indicating that population control has a long history in the United States. A Native American reproductive activist reports that on the Pine Ridge reservation today, pregnant women with drinking problems are put in jail, as it is the only holding place for them.[17]...

The potential to extend biological control has expanded exponentially. Advances in data collection and storage through the computer, Internet, and genetic revolutions have made surveillance systems more efficient and invasive. The nineteenth-century "panoptical gaze" made it possible for a prisoner to be seen at all times, and through that process the prisoner internalized surveillance.[18] The new technologies radically expand the ability to collect, process, and encode large amounts of information on ever smaller surfaces. This makes it possible for the human body to be manipulated and controlled in radically new ways—from within the body itself. This intensification takes control from a mental to a physiological realm.

Like policing, corporate intrusion into the private sphere is increasingly naturalized. A great deal of data on individuals is bought, sold, and traded. In this instance, it is information on the rich and middle-class that is particularly coveted. The Internet revolution has made it possible for corporate and security interests to track every move made by an individual on the web to determine a person's consumer preferences, interests, and purchases, in addition to getting credit card information. Such intrusive data collecting techniques regarding an individual's tastes and preferences for corporate niche marketing is rarely framed as a surveillance problem. Critics have discussed it sometimes as a privacy concern, but by and large these incursions are accepted as a market-driven intrusion into our life.

Emerging reproductive technologies, such as cloning, have the potential to blur the distinctions between genetically distinct and genetically determined individuals. This raises ethical questions regarding who would count as fully human with the attendant civil rights and liberties. Valerie Hartouni points to how standards of humanity get "partialized" (making some less human than others) in this process.[19] She fears that, in the current social context, such technologies will be used to manage and contain diversity and the proliferation of difference.[20] Other critics are concerned with the eugenic possibilities of cloning and similar practices that work on humans from the inside out. The commodification of human life and the disruption that such technologies could have on kinship structures and human relations are a source of grave concern.

SURVEILLANCE AND NATIONAL (IN)SECURITY

Whereas the prison and policing systems are supposed to protect the nation from dangers within, the military, Border Patrols, and INS are ostensibly designed to protect the public from danger and threats that emanate from outside the national body. The discretionary authority and budgets of these institutions have expanded exponentially since September 11, 2001.... An influx of immigrants, the rhetoric of explosive population growth in the Third World, and angry young men inside and outside the United States are manufactured as a threat to national security.[21]

Immigrants in the United States are constructed as a source of danger. This threat has been used to justify the allocation of billions of dollars to the enforcement programs of the INS that patrol the nation. At present, the INS has more armed agents with arrest power than any other federal law enforcement agency.[22] Mandatory detention provisions have made immigrants the fastest growing incarcerated population in the United States. Stringent controls and border security forces are positioned at strategic places along the United States—Mexico border. Military-style tactics and equipment result in immigrants undertaking more dangerous, isolated routes to cross over where the risks of death, dehydration, and assault are exponentially higher.

Immigrant rights organizations and the press record detailed accounts of immigrants risking their lives to make their way through tortuous terrain to find work in the United States. The Mojave Desert in Southern California with its inhospitable terrain has become one such death trap for immigrants.[23] This dangerous crossing is a dramatic example of the extreme risks that immigrants take to escape INS agents and provide adequately for their families.

Vigilante groups and private citizens in Arizona have taken it upon themselves to "help" patrol ranchers "hunt" Mexican undocumented immigrants. Jose Palafox reports on the roundup of over 3,000 undocumented immigrants on Roger and Don Barnett's 22,000-acre property in Douglas, Arizona, near the United States-Mexico border. Ranchers circulated a leaflet asking for volunteers to help patrol their land while having "Fun in the Sun." Immigrants were considered "fair game."[24]

While border patrol forces and private citizens tighten their grip on illegal immigration and regulate the mobility of poor workers, social services are being slashed for legal immigrants. The latter are often portrayed as a drain on national resources and a direct threat to low-wage workers in the United States, and female immigrants of color are particularly targeted for their family settlement and community building roles....

Even though many studies have shown that immigrant labor contributes to the economy, there is a widely held perception that they represent a financial drain. Despite the proverbial anti-immigrant sentiment, the United States economy depends on cheap immigrant labor. Very often undocumented workers perform jobs or are made to engage in activities usually considered too low-paying or too risky for citizens and residents. It has been estimated that 45,000 to 50,000 women and children are trafficked annually into and across the United States for the sex industry, sweatshops, domestic labor, and agricultural work. The INS compares the trafficking in women and children with the drug and weapons smuggling industry....

BEYOND ANALYSIS: BUILDING A MOVEMENT

The ... Committee on Women, Population, and the Environment (CWPE) [is] a multiracial alliance of feminist activists, health practitioners, and scholars committed to promoting the social and economic empowerment of women in a context of global peace and justice. We work toward eliminating poverty, inequality, racism, and environmental degradation. A crucial feature of our work ... is our identification and cultivation of a political common ground, given our widely ranging ethnic and national identities....

CWPE asserts that the oppression of women, not their reproductive capacities, needs to be eliminated and calls for drug treatment for women with substance abuse problems, decent jobs, educational opportunities, and mental health and child-care services. It is the lack of these services that deny human dignity and exacerbate conditions of poverty, social status, and gender discrimination ...

This [essay] is directed at activists, students, policy-makers, and scholars who seek to understand the connections between the criminalization of people of color and the poor, and social and population control. We seek to [foster] a dialogue that builds collaborations between social movements to move beyond single-issue and identity-based politics towards an inclusive political agenda across progressive movements.

NOTES

1. Quoting Myron Magnet, editor of *City Journal.*

2. The law enforcement agencies referred to include local and state police agencies; prison systems at the local, state and federal levels; the United States Border Patrol and Interior Enforcement Operations of the Immigration and Naturalization Service (INS); and the rapidly expanding INS detention system.

3. This is not true in other parts of the world. For example, state violence has been a critical issue for the contemporary women's movement in India. Radha Kumar in *The History of Doing* writes: "The issue of rape has been one that most contemporary feminist movements internationally have focused on, firstly because sexual assault is one of the ugliest and most brutal expressions of masculine violence towards women, because rape and the historical discourse around it reveal a great deal about the social relations of reproduction, and thirdly because of what it shows about the way in which the woman's body is seen as representing the community. In India, it has been the latter reason which has been the most dominant in the taking up of campaigns against rape." (Delhi: Kali for Women, 1993), 128.

4. Foucault, in *The History of Sexuality*, refers to this form of state control as "biopower."

5. Nancy Scheper-Hughes and Margaret Lock, "The Mindful Body: A Prolegomenon to Future Work in Medical Anthropology," *Medical Anthropology Quarterly*, Vol. *1* (1987), 6–41.

6. "We Remember: African American Women are for Reproductive Freedom" (1998); statement signed and distributed by leaders in the African-American community including Byllye Avery, Reverend Willie Barro, Donna Brazil, Shirley Chisholm, and Dorothy Heights in support of keeping abortion safe and legal.

7. David Firestone, "Alabama's Packed Jails Draw Ire of Courts, Again," *New York Times*, May 1, 2001, A1.

8. Eve Goldberg and Linda Evans, "The Prison Industrial Complex and the Global Economy," *Political Environments* (Fall 1999/Winter 2000), 47.

9. Dr. James Gilligan, a Harvard psychotherapist who has worked on prisons and recidivism, argues that punishing violence with imprisonment does not stop violence because it continues to replicate the patriarchal code. See *Violence: Reflections on a National Epidemic* (New York: Vintage, 1997).

10. Michel Foucault, *Discipline and Punish* (New York: Vintage, 1995), 280.

11. Joseph Hallinan, *Going Up The River: Travels in a Prison Nation* (New York: Random House, 2001).

12. Jennifer Yanco, "Breaking the Silence: Women and the Criminal Justice System," *Political Environments*, No. 7 (Fall 1999/Winter 2000), 20.

13. This document, "The UN Special Report on Violence Against Women" by Special Rapporteur Radhika Coomaraswamy, is available from United Nations Publications at www.un.org.

14. Lynn Paltrow, "Pregnant Drug Users, Fetal Persons, and the Threat to *Roe v. Wade*," *Albany Law Review*, Vol. *62*, No. 3 (1999), 1024.

15. This decision affirms the Fourth Amendment to the US Constitution that protects every American, including those who are pregnant and those with substance abuse problems, from warrantless, unreasonable searches.

16. Dorothy Roberts, *Killing the Black Body* (New York: Vintage, 1998).

17. Native American Health Education Resource Center, SisterSong Native Women's Reproductive Health and Rights Roundtable Report (Lake Andes, SD: January 2001), 17.

18. Foucault, *Discipline and Punish*, 195–228.

19. For a rich discussion of this set of issues, see Valerie Hartouni, "Replicating the Singular Self," in *Cultural Conceptions: On Reproduction Technologies and the Remaking of Life* (Minneapolis, MN: University of Minnesota Press, 1997), 110–132.

20. Ibid., 119.

21. For more on the subject see Betsy Hartmann, "Population, Environment, and Security: A New Trinity" in Jael Silliman and Ynestra King, eds., *Dangerous Intersections* (Cambridge, MA: South End Press, 1999), 1–23.

22. Maria Jimenez, "Legalization Then and Now: An Eighty Year History," *Network News* (Oakland, CA: National Network for Immigrant and Refugee Rights, Summer 2000), 6.

23. Ginger Thompson, "The Desperate Risk of Death in a Desert," *New York Times*, October 31, 2000, A12.

24. Jose Palafox, "Welcome to America: Arizona Ranchers Hunt Mexicans," *Network News* (Summer 2000), 4.

52

The Color of Justice

MICHELLE ALEXANDER

Imagine you are Emma Faye Stewart, a thirty-year-old, single African American mother of two who was arrested as part of a drug sweep in Hearne, Texas. All but one of the people arrested were African American. You are innocent. After a week in jail, you have no one to care for your two small children and are eager to get home. Your court-appointed attorney urges you to plead guilty to a drug distribution charge, saying the prosecutor has offered probation. You refuse, steadfastly proclaiming your innocence. Finally, after almost a month in jail, you decide to plead guilty so you can return home to your children. Unwilling to risk a trial and years of imprisonment, you are sentenced to ten years probation and ordered to pay $1,000 in fines, as well as court and probation costs. You are also now branded a drug felon. You are no longer eligible for food stamps; you may be discriminated against in employment; you cannot vote for at least twelve years; and you are about to be evicted from public housing. Once homeless, your children will be taken from you and put in foster care.

A judge eventually dismisses all cases against the defendants who did not plead guilty. At trial, the judge finds that the entire sweep was based on the testimony of a single informant who lied to the prosecution. You, however, are still a drug felon, homeless, and desperate to regain custody of your children.

Now place yourself in the shoes of Clifford Runoalds, another African American victim of the Hearne drug bust. You returned home to Bryan, Texas, to attend the funeral of your eighteen-month-old daughter. Before the funeral services begin, the police show up and handcuff you. You beg the officers to let you take one last look at your daughter before she is buried. The police refuse. You are told by prosecutors that you are needed to testify against one of the defendants in a recent drug bust. You deny witnessing any drug transaction; you don't know what they are talking about. Because of your refusal to cooperate, you are indicted on felony charges. After a month of being held in jail, the charges against you are dropped. You are technically free, but as a result of your arrest and period of incarceration, you lose your job, your apartment, your furniture, and your car. Not to mention the chance to say good-bye to your baby girl.

This is the War on Drugs. The brutal stories described above are not isolated incidents, nor are the racial identities of Emma Faye Stewart and Clifford

SOURCE: Excerpt from *The New Jim Crow* - Copyright © 2010 by Michelle Alexander. Reprinted by permission of The New Press. www.thenewpress.com.

Runoalds random or accidental. In every state across our nation, African Americans—particularly in the poorest neighborhoods—are subjected to tactics and practices that would result in public outrage and scandal if committed in middle-class white neighborhoods. In the drug war, the enemy is racially defined. The law enforcement methods described [here] have been employed almost exclusively in poor communities of color, resulting in jaw-dropping numbers of African Americans and Latinos filling our nation's prisons and jails every year. We are told by drug warriors that the enemy in this war is a thing—drugs—not a group of people, but the facts prove otherwise.

Human Rights Watch reported in 2000 that, in seven states, African Americans constitute 80 to 90 percent of all drug offenders sent to prison. In at least fifteen states, blacks are admitted to prison on drug charges at a rate from twenty to fifty-seven times greater than that of white men. In fact, nationwide, the rate of incarceration for African American drug offenders dwarfs the rate of whites. When the War on Drugs gained full steam in the mid-1980s, prison admissions for African Americans skyrocketed, nearly quadrupling in three years, and then increasing steadily until it reached in 2000 a level *more than twenty-six times* the level in 1983. The number of 2000 drug admissions for Latinos was twenty-two times the number of 1983 admissions. Whites have been admitted to prison for drug offenses at increased rates as well—the number of whites admitted for drug offenses in 2000 was eight times the number admitted in 1983—but their relative numbers are small compared to blacks' and Latinos. Although the majority of illegal drug users and dealers nationwide are white, three-fourths of all people imprisoned for drug offenses have been black or Latino. In recent years, rates of black imprisonment for drug offenses have dipped somewhat—declining approximately 25 percent from their zenith in the mid-1990s—but it remains the case that African Americans are incarcerated at grossly disproportionate rates throughout the United States.

There is, of course, an official explanation for all of this: crime rates. This explanation has tremendous appeal—before you know the facts—for it is consistent with, and reinforces, dominant racial narratives about crime and criminality dating back to slavery. The truth, however, is that rates and patterns of drug crime do not explain the glaring racial disparities in our criminal justice system. People of all races use and sell illegal drugs at remarkably similar rates. If there are significant differences in the surveys to be found, they frequently suggest that whites, particularly white youth, are more likely to engage in illegal drug dealing than people of color. One study, for example, published in 2000 by the National Institute on Drug Abuse reported that white students use cocaine at seven times the rate of black students, use crack cocaine at eight times the rate of black students, and use heroin at seven times the rate of black students. That same survey revealed that nearly identical percentages of white and black high school seniors use marijuana. The National Household Survey on Drug Abuse reported in 2000 that white youth aged 12–17 are more than a third more likely to have sold illegal drugs than African American youth. Thus the very same year Human Rights Watch was reporting that African Americans were being arrested and imprisoned at unprecedented rates, government data revealed that blacks were no more likely to be guilty of drug crimes than whites and that white youth were actually the *most likely* of any racial or ethnic group to be guilty of illegal drug possession and sales. Any notion that drug use among blacks is more severe or dangerous is belied by the data; white youth have about three times the number of drug-related emergency room visits as their African American counterparts.

The notion that whites comprise the vast majority of drug users and dealers—and may well be more likely than other racial groups to commit drug crimes—may seem implausible to some, given the media imagery we are fed on a daily basis and the racial composition of our prisons and jails. Upon reflection, however, the prevalence of white drug crime—including drug dealing—should not be surprising. After all, where do whites get their illegal drugs? Do they all drive to the ghetto to purchase them from somebody standing on a street corner? No. Studies consistently indicate that drug markets, like American society generally, reflect our nation's racial and socioeconomic boundaries. Whites tend to sell to whites; blacks to blacks. University students tend to sell to each other. Rural whites, for their part, don't make a special trip to the 'hood to purchase marijuana. They buy it from somebody down the road. White high school students typically buy drugs from white classmates, friends, or older relatives. Even Barry McCaffrey, former director of the White House Office of National Drug Control Policy, once remarked, if your child bought drugs, "it was from a student of their own race generally." The notion that most illegal drug use and sales happens in the ghetto is pure fiction. Drug trafficking occurs there, but it occurs everywhere else in America as well. Nevertheless, black men have been admitted to state prison on drug charges at a rate that is more than thirteen times higher than white men. The racial bias inherent in the drug war is a major reason that 1 in every 14 black men was behind bars in 2006, compared with 1 in 106 white men. For young black men, the statistics are even worse. One in 9 black men between the ages of twenty and thirty-five was behind bars in 2006, and far more were under some form of penal control—such as probation or parole. These gross racial disparities simply cannot be explained by rates of illegal drug activity among African Americans.

What, then, does explain the extraordinary racial disparities in our criminal justice system? Old-fashioned racism seems out of the question. Politicians and law enforcement officials today rarely endorse racially biased practices, and most of them fiercely condemn racial discrimination of any kind. When accused of racial bias, police and prosecutors—like most Americans—express horror and outrage. Forms of race discrimination that were open and notorious for centuries were transformed in the 1960s and 1970s into something un-American—an affront to our newly conceived ethic of colorblindness. By the early 1980s, survey data indicated that 90 percent of whites thought black and white children should attend the same schools, 71 percent disagreed with the idea that whites have a right to keep blacks out of their neighborhoods, 80 percent indicated they would support a black candidate for president, and 66 percent opposed laws prohibiting intermarriage. Although far fewer supported specific policies designed to achieve racial equality or integration (such as busing), the mere fact that large majorities of whites were, by the early 1980s, supporting the antidiscrimination principle reflected a profound shift in racial attitudes. The margin of support for colorblind norms has only increased since then.

This dramatically changed racial climate has led defenders of mass incarceration to insist that our criminal justice system, whatever its past sins, is now largely fair and nondiscriminatory. They point to violent crime rates in the African American community as a justification for the staggering number of black men who find themselves behind bars. Black men, they say, have much higher rates of violent crime; that's why so many of them are locked in prisons.

Typically, this is where the discussion ends.

The problem with this abbreviated analysis is that violent crime is *not* responsible for the prison boom. As numerous researchers have shown, violent crime rates have fluctuated over the years and bear little relationship to incarceration rates—which have soared during the past three decades regardless of whether violent crime was going up or down. Today violent crime rates are at historically low levels, yet incarceration rates continue to climb.

Murder convictions tend to receive a tremendous amount of media attention, which feeds the public's sense that violent crime is rampant and forever on the rise. But like violent crime in general, the murder rate cannot explain the prison boom. Homicide convictions account for a tiny fraction of the growth in the prison population. In the federal system, for example, homicide offenders account for 0.4 percent of the past decade's growth in the federal prison population, while drug offenders account for nearly 61 percent of that expansion. In the state system, less than 3 percent of new court commitments to state prison typically involve people convicted of homicide. As much as a third of state prisoners are violent offenders, but that statistic can easily be misinterpreted. Violent offenders tend to get longer prison sentences than nonviolent offenders, and therefore comprise a much larger share of the prison population than they would if they had earlier release dates. The uncomfortable reality is that convictions for drug offenses—not violent crime—are the single most important cause of the prison boom in the United States, and people of color are convicted of drug offenses at rates out of all proportion to their drug crimes.

These facts may still leave some readers unsatisfied. The idea that the criminal justice system discriminates in such a terrific fashion when few people openly express or endorse racial discrimination may seem far-fetched, if not absurd. How could the War on Drugs operate in a discriminatory manner, on such a large scale, when hardly anyone advocates or engages in explicit race discrimination? ... Despite the colorblind rhetoric and fanfare of recent years, the design of the drug war effectively guarantees that those who are swept into the nation's new undercaste are largely black and brown.

This sort of claim invites skepticism. Nonracial explanations and excuses for the systematic mass incarceration of people of color are plentiful. It is the genius of the new system of control that it can always be defended on nonracial grounds, given the rarity of a noose or a racial slur in connection with any particular criminal case. Moreover, because blacks and whites are almost never similarly situated (given extreme racial segregation in housing and disparate life experiences), trying to "control for race" in an effort to evaluate whether the mass incarceration of people of color is really about race or something else—anything else—is difficult. But it is not impossible.

A bit of common sense is overdue in public discussions about racial bias in the criminal justice system. The great debate over whether black men have been targeted by the criminal justice system or unfairly treated in the War on Drugs often overlooks the obvious. What is painfully obvious when one steps back from individual cases and specific policies is that the system of mass incarceration operates with stunning efficiency to sweep people of color off the streets, lock them in cages, and then release them into an inferior second-class status. Nowhere is this more true than in the War on Drugs.

The central question, then, is *how* exactly does a formally colorblind criminal justice system achieve such racially discriminatory results? Rather easily, it turns out. The process occurs in two stages. The first step is to grant law enforcement officials extraordinary discretion regarding whom to stop, search, arrest, and charge for drug offenses, thus ensuring that conscious and unconscious racial beliefs and stereotypes will be given free reign. Unbridled discretion inevitably creates huge racial disparities. Then, the damning step: Close the courthouse doors to all claims by defendants and private litigants that the criminal justice system operates in racially discriminatory fashion. Demand that anyone who wants to challenge racial bias in the system offer, in advance, clear proof that the racial disparities are the product of intentional racial discrimination—i.e., the work of a bigot. This evidence will almost never be available in the era of colorblindness, because everyone knows—but does not say—that the enemy in the War on Drugs can be identified by race. This simple design has helped to produce one of the most extraordinary systems of racialized social control the world has ever seen.

53

Rape, Racism, and the Law

JENNIFER WRIGGINS

The history of rape in this country has focused on the rape of white women by Black men. From a feminist perspective, two of the most damaging consequences of this selective blindness are the denials that Black women are raped and that all women are subject to pervasive and harmful sexual coercion of all kinds....

THE NARROW FOCUS ON BLACK OFFENDER/
WHITE VICTIM RAPE

There are many different kinds of rape. Its victims are of all races, and its perpetrators are of all races. Yet the kind of rape that has been treated most seriously throughout this nation's history has been the illegal forcible rape of a white woman by a Black man. The selective acknowledgement of Black accused/white victim rape was especially pronounced during slavery and through the first half of the twentieth century. Today a powerful legacy remains that permeates thought about rape and race.

During the slavery period, statutes in many jurisdictions provided the death penalty or castration for rape when the convicted man was Black or mulatto and the victim white. These extremely harsh penalties were frequently imposed. In addition, mobs occasionally broke into jails and courtrooms and lynched slaves alleged to have raped white women, prefiguring Reconstruction mob behavior.

In contrast to the harsh penalties imposed on Black offenders, courts occasionally released a defendant accused of raping a white woman when the evidence was inconclusive as to whether he was Black or mulatto. The rape of Black women by white or Black men, on the other hand, was legal; indictments were sometimes dismissed for failing to allege that the victim was white. In those states where it was illegal for white men to rape white women, statutes provided less severe penalties for the convicted white rapist than for the convicted Black one. In addition, common-law rules both defined rape narrowly and made it a difficult crime to prove....

SOURCE: From Jennifer Wriggins, "Rape, Racism, and the Law" *Harvard Women's Law Journal* 6 (Spring 1983): 103–141. Reprinted by permission of *Harvard Journal of Law and Gender*.

After the Civil War, state legislatures made their rape statutes race-neutral, but the legal system treated rape in much the same way as it had before the war. Black women raped by white or Black men had no hope of recourse through the legal system. White women raped by white men faced traditional common-law barriers that protected most rapists from prosecution.

Allegations of rape involving Black offenders and white victims were treated with heightened virulence. This was manifested in two ways. The first response was lynching, which peaked near the end of the nineteenth century. The second, from the early twentieth century on, was the use of the legal system as a functional equivalent of lynching, as illustrated by mob coercion of judicial proceedings, special doctrinal rules, the language of opinions, and the markedly disparate numbers of executions for rape between white and Black defendants.

Between 1882 and 1946 at least 4,715 persons were lynched, about three-quarters of whom were Black. Although lynching tapered off after the early 1950s, occasional lynch-like killings persist to this day. The influence of lynching extended far beyond the numbers of Black people murdered because accounts of massive white crowds torturing, burning alive, and dismembering their victims created a widespread sense of terror in the Black community.

The most common justification for lynching was the claim that a Black man had raped a white woman. The thought of this particular crime aroused in many white people an extremely high level of mania and panic. One white woman, the wife of an ex-Congressman, stated in 1898, "If it needs lynching to protect woman's dearest possession from human beasts, then I say lynch a thousand times a week if necessary." The quote resonates with common stereotypes that Black male sexuality is wanton and bestial, and that Black men are wild, criminal rapists of white women.

Many whites accepted lynching as an appropriate punishment for a Black man accused of raping a white woman. The following argument made to the jury by defense counsel in a 1907 Louisiana case illustrates this acceptance:

> Gentlemen of the jury, this man, a nigger, is charged with breaking into the house of a white man in the nighttime and assaulting his wife, with the intent to rape her. Now, don't you know that, if this nigger had committed such a crime, he never would have been brought here and tried; that he would have been lynched, and if I were there I would help pull on the rope.[1]

It is doubtful whether the legal system better protected the rights of a Black man accused of raping a white woman than did the mob. Contemporary legal literature used the term "legal lynching" to describe the legal system's treatment of Black men. Well past the first third of the twentieth century, courts were often coerced by violent mobs, which threatened to execute the defendant themselves unless the court convicted him. Such mobs often did lynch the defendant if the judicial proceedings were not acceptable to them. A contemporary authority on lynching commented in 1934 that "the local sentiment which would make a lynching possible would insure a conviction in the courts." Even if the mob was not overtly pressuring for execution, a Black defendant accused of raping a white woman faced a hostile, racist legal system. State court submission to mob pressure is well illustrated by the most famous series of cases about interracial rape, the Scottsboro cases of the 1930s. Eight young Black men were convicted of what the Alabama Supreme Court called "a most foul and revolting crime," which was the rape of "two defenseless white girls." The defendants were summarily sentenced to death based on minimal and dubious evidence, having been denied effective assistance of counsel. The Alabama Supreme Court upheld the convictions in opinions demonstrating relentless determination to hold the defendants guilty regardless of strong evidence that mob pressure had influenced the verdicts and the weak evidence presented against the defendants. In one decision, that court affirmed the trial court's denial of a change of venue on the grounds that the mobs' threats of harm were not imminent enough although the National Guard had been called out to protect the defendants from mob executions. The U.S. Supreme Court later recognized that the proceedings had in fact taken place in an atmosphere of "tense, hostile, and excited public sentiment." After a lengthy appellate process, including three favorable Supreme Court rulings, all of the Scottsboro defendants were released, having spent a total of 104 years in prison.

In addition, courts applied special doctrinal rules to Black defendants accused of the rape or attempted rape of white women. One such rule allowed juries to consider the race of the defendant and victim in drawing factual conclusions as to the defendant's intent in attempted rape cases. If the accused was Black and the victim white, the jury was entitled to draw the inference, based on race alone, that he intended to rape her. One court wrote, "In determining the question of intention, the jury may consider social conditions and customs founded upon racial differences, such as that the prosecutrix was a white woman and defendant was a Negro man."[2] The "social conditions and customs founded upon racial differences" which the jury was to consider included the assumption that Black men always and only want to rape white women, and that a white woman would never consent to sex with a Black man.

The Georgia Supreme Court of 1899 was even more explicit about the significance of race in the context of attempted rape, and particularly about the motivations of Black men. It held that race may properly be considered "to rebut any presumption that might otherwise arise in favor of the accused that his intention was to obtain the consent of the female, upon failure of which he would abandon his purpose to have sexual intercourse with her."[3] Such a rebuttal denied to Black defendants procedural protection that was accorded white defendants....

The outcome of this disparate treatment of Black men by the legal system was often the same as lynching—death. Between 1930 and 1967, thirty-six percent of the Black men who were convicted of raping a white woman were executed. In stark contrast, only two percent of all defendants convicted of rape involving other racial combinations were executed. As a result of such disparate treatment, eighty-nine percent of the men executed for rape in this country were Black. While execution rates for all crimes were much higher for Black men than for white men, the differential was most dramatic when the crime was the rape of a white woman.

The patterns that began in slavery and continued long afterwards have left a powerful legacy that manifests itself today in several ways. Although the death penalty for rape has been declared unconstitutional, the severe statutory penalties for rape continue to be applied in a discriminatory manner. A recent study concluded that Black men convicted of raping white women receive more serious sanctions than all other sexual assault defendants. A recent attitudinal study found that white potential jurors treated Black and white defendants similarly when the victim was Black. However, Black defendants received more severe punishment than white defendants when the victim was white.

The rape of white women by Black men is also used to justify harsh rape penalties. One of the few law review articles written before 1970 that takes a firm position in favor of strong rape laws to secure convictions begins with a long quote from a newspaper article describing rapes by three Black men, who at 3 a.m. on Palm Sunday "broke into a West Philadelphia home occupied by an eighty-year-old widow, her forty-four-year-old daughter and fourteen-year-old granddaughter," brutally beat and raped the white women, and left the grandmother unconscious "lying in a pool of blood." This introduction presents rape as a crime committed by violent Black men against helpless white women. It is an image of a highly atypical rape—the defendants are Black and the victims white, the defendants and victims are strangers to each other, extreme violence is used, and it is a group rape. Contemporaneous statistical data on forcible rapes reported to the Philadelphia police department reveals that this rape case was virtually unique.[4] Use of this highly unrepresentative image of rape to justify strict rape laws is consistent with recent research showing that it is a prevalent, although false, belief about rape that the most common racial combination is Black offender and white victim.[5]

Charges of rapes committed by Black men against white women are still surrounded by sensationalism and public pressure for prosecution. Black men seem to face a special threat of being unjustly prosecuted or convicted. One example is Willie Sanders.[6] Sanders is a Black Boston man who was arrested and charged with the rapes of four young white women after a sensational media campaign and intense pressure on the police to apprehend the rapist. Although the rapes continued after Sanders was incarcerated, and the evidence against him was extremely weak, the state subjected him to a vigorous twenty-month prosecution. After a lengthy and expensive trial, and an active public defense, he was eventually acquitted. Although Sanders was clearly innocent, he could have been convicted; he and his family suffered incalculable damage despite his acquittal....

From slavery to the present day, the legal system has consistently treated the rape of white women by Black men with more harshness than any other kind of rape....

This selective focus is significant in several ways. First, since tolerance of coerced sex has been the rule rather than the exception, it is clear that the rape of white women by Black men has been treated seriously not because it is coerced sex and thus damaging to women, but because it is threatening to white men's power over both "their" women and Black men. Second, in treating Black offender/white victim illegal rape much more harshly than all coerced sex experienced by Black women and most coerced sex experienced by white women, the legal system has implicitly condoned the latter forms of rape. Third, this treatment has contributed to a paradigmatic but false concept of rape as being primarily a violent crime between strangers where the perpetrator is Black and the victim white. Finally, this pattern is perverse and discriminatory because rape is painful and degrading to both Black and white victims regardless of the attacker's race.

THE DENIAL OF THE RAPE OF BLACK WOMEN

The selective acknowledgement of the existence and seriousness of the rape of white women by Black men has been accompanied by a denial of the rape of Black women that began in slavery and continues today. Because of racism and sexism, very little has been written about this denial. Mainstream American history has ignored the role of Black people to a large extent; systematic research into Black history has been published only recently. The experiences of Black women have yet to be fully recognized in those histories, although this is beginning to change. Indeed, very little has been written about rape from the perspective of the victim, Black or white, until quite recently. Research about Black women rape victims encounters all these obstacles.

The rape of Black women by white men during slavery was commonplace and was used as a crucial weapon of white supremacy. White men had what one commentator called "institutionalized access" to Black women. The rape of Black women by white men cannot be attributed to unique Southern pathology, however, for numerous accounts exist of northern armies raping Black women while they were "liberating" the South.

The legal system rendered the rape of Black women by any man, white or Black, invisible. The rape of a Black woman was not a crime. In 1859 the Mississippi Supreme Court dismissed the indictment of a male slave for the rape of a female slave less than 10 years old, saying:

> [T]his indictment can not be sustained, either at common law or under our statutes. It charges no offense known to either system. [Slavery] was unknown to the common law ... and hence its provisions are inapplicable.... There is no act (of our legislature on this subject) which embraces either the attempted or actual commission of a rape by a slave on a female slave.... Masters and slaves can not be governed by the same system or laws; so different are their positions, rights and duties.[7]

This decision is illuminating in several respects. First, Black men are held to lesser standards of sexual restraint with Black women than are white men with white women. Second, white men are held to lesser standards of restraint with

Black women than are Black men with white women. Neither white nor Black men were expected to show sexual restraint with Black women.

After the Civil War, the widespread rape of Black women by white men persisted. Black women were vulnerable to rape in several ways that white women were not. First, the rape of Black women was used as a weapon of group terror by white mobs and by the Ku Klux Klan during Reconstruction. Second, because Black women worked outside the home, they were exposed to employers' sexual aggression as white women who worked inside the home were not.

The legal system's denial that Black women experienced sexual abuse by both white and Black men also persisted, although statutes had been made race-neutral. Even if a Black victim's case went to trial—in itself highly unlikely—procedural barriers and prejudice against Black women protected any man accused of rape or attempted rape. The racist rule which facilitated prosecutions of Black offender/white victim attempted rapes by allowing the jury to consider the defendant's race as evidence of his intent, for instance, was not applied where both persons were "of color and there was no evidence of their social standing."[8] That is, the fact that a defendant was Black was considered relevant only to prove intent to rape a white woman; it was not relevant to prove intent to rape a Black woman. By using disparate procedures, the court implicitly makes two assertions. First, Black men do not want to rape Black women with the same intensity or regularity that Black men want to rape white women. Second, Black women do not experience coerced sex in the sense that white women experience it.

These attitudes reflect a set of myths about Black women's supposed promiscuity which were used to excuse white men's sexual abuse of Black women. An example of early twentieth century assumptions about Black women's purported promiscuity was provided by the Florida Supreme Court in 1918. In discussing whether the prior chastity of the victim in a statutory rape case should be presumed subject to defendant's rebuttal or should be an element of the crime which the state must prove, the court explained that:

> What has been said by some of our courts about an unchaste female
> being a comparatively rare exception is no doubt true where the pop-
> ulation is composed largely of the Caucasian race, but we would blind
> ourselves to actual conditions if we adopted this rule where another race
> that is largely immoral constitutes an appreciable part of the population.[9]

Cloaking itself in the mantle of legal reasoning, the court states that most young white women are virgins, that most young Black women are not, and that unchaste women are immoral. The traditional law of statutory rape at issue in the above-quoted case provides that women who are not "chaste" cannot be raped. Because of the way the legal system considered chastity, the association of Black women with unchastity meant not only that Black women could not be victims of statutory rape, but also that they would not be recognized as victims of forcible rape.

The criminal justice system continues to take the rape of Black women less seriously than the rape of white women. Studies show that judges generally impose harsher sentences for rape when the victim is white than when the victim is Black. The behavior of white jurors shows a similar bias. A recent study found that sample white jurors imposed significantly lighter sentences on defendants whose victims were Black than on defendants whose victims were white. Black jurors exhibited no such bias.

Evidence concerning police behavior also documents the fact that the claims of Black rape victims are taken less seriously than those of whites. A ... study of Philadelphia police processing decisions concluded that the differential in police decisions to charge for rape "resulted primarily from a lack of confidence in the veracity of Black complainants and a belief in the myth of Black promiscuity."

The thorough denial of Black women's experiences of rape by the legal system is especially shocking in light of the fact that Black women are much more likely to be victims of rape than are white women.[10] Based on data from national surveys of rape victims, "the profile of the most frequent rape victim is a young woman, divorced or separated, Black and poverty stricken."...

CONCLUSION

The legal system's treatment of rape both has furthered racism and has denied the reality of women's sexual subordination. It has disproportionately targeted Black men for punishment and made Black women both particularly vulnerable and particularly without redress. It has denied the reality of women's sexual subordination by creating a social meaning of rape which implies that the only type of sexual abuse is illegal rape and the only form of illegal rape is Black offender/white victim. Because of the interconnectedness of rape and racism, successful work against rape and other sexual coercion must deal with racism. Struggles against rape must acknowledge the differences among women and the different ways that groups other than women are disempowered. In addition, work against rape must go beyond the focus on illegal rape to include all forms of coerced sex, in order to avoid the racist historical legacy surrounding rape and to combat effectively the subordination of women.

NOTES

1. *State v. Petit*, 119 La., 44 So. (1907).
2. *McQuirter v. State*, 36 Ala., 63 So. 2d (1953).
3. *Dorsey v. State*, 108 Ga., 34 S.E. (1899).
4. Out of 343 rapes reported to the Philadelphia police, 3.3% involved Black defendants accused of raping white women; 42% involved complaints of stranger rape; 20.5% involved brutal beatings; 43% involved group rapes.
5. In answer to the question, "Among which racial combination do most rapes occur?" 48% of respondents stated Black males and white females, 3% stated white males and Black females, 16% stated Black males and Black females, 33% stated white males

and white females. Recent victim survey data contradict this prevalent belief; more than four-fifths of illegal rapes reported to researchers were between members of the same race, and white/Black rapes roughly equaled Black/white rapes.

6. Suffolk Superior Court indictment (1980).

7. *George v. State*, 37 Miss. (1859).

8. *Washington v. State*, 38 Ga., 75 S.E. (1912).

9. *Dallas v. State*, 76 Fla., 79 So. (1918).

10. Recent data from random citizen interviews suggest that Black women are much more likely to be victims of illegal rape than are white women.

54

Interpreting and Experiencing Anti-Queer Violence

Race, Class, and Gender Differences among LGBT Hate Crime Victims

DOUG MEYER

Several studies of hate crime victims have documented the ways in which lesbian women and gay men determine that violence is based on their sexuality (Herek et al., 1997; Herek et al., 2002). These studies, however, make no reference to race, class, and gender. In this [reading], I employ an intersectionality framework to explore how lesbian, gay, bisexual, and transgender (LGBT) people determine that violence is based on their sexuality or gender identity. Employing an intersectionality framework reveals how LGBT people's violent experiences differ along the lines of race, class, and gender. Furthermore, an intersectionality approach facilitates our understanding of the ways in which LGBT people interpret and experience hate-motivated violence. To accomplish these research goals, I designed a qualitative research project in which I interviewed 44 people who experienced violence because they were perceived to be lesbian, gay, bisexual, or transgender.

... Studies of hate crime victims have revealed the degree to which hate motivated violence can have traumatic psychological effects. Unfortunately, the focus on the psychological effects of hate-motivated violence has left other research questions unexplored. In particular, studies of hate crime victims have overlooked important sociological research questions. We know little, for example, about the ways in which race, class, gender, and sexuality structure victims' experiences of hate-motivated violence. Intersectionality, a theoretical framework that has been highly influential in other bodies of literature, has remained absent from studies of hate crime victims.

Intersectionality theory denotes the ways in which institutional power structures such as race, class, gender, and sexuality simultaneously structure social relations. It conceptualizes these institutional power structures as distinct but mutually reinforcing systems of oppression. Conceptualizing systems of oppression as distinct prevents researchers from collapsing one system into another.

SOURCE: Meyer, Doug, "Interpreting and Experiencing Anti-Queer Violence: Race, Class, and Gender Differences among LGBT Hate crime victims" *Race, Gender, and Class* 15: 262–282. Copyright © 2008. Reprinted by permission.

Racism, for example, cannot be reduced to gender oppression and heterosexism cannot be understood as a product of patriarchy. Furthermore, because systems of oppression are interlocking, social scientists must account for multiple forms of social inequality to understand patterns of behavior. Thus, they cannot understand gendered patterns of activity without also understanding raced, classed, and sexualized ones....

BRINGING AN INTERSECTIONALITY APPROACH TO HATE CRIME RESEARCH

Although intersectionality has remained absent from studies of hate crime victims, it has been featured in other areas of the literature. Barbara Perry (2001), an eminent hate crime scholar, incorporates elements of intersectionality into her theoretical account of hate crime. She argues that traditional criminological theory has failed to account for hate-motivated violence. According to traditional criminological theory, hate crime occurs because relatively powerless individuals are unable to achieve society's goals through socially acceptable means.... Perry argues that this theoretical framework cannot fully account for hate crime because perpetrators are often relatively privileged members of society. Instead, she argues that a more satisfying explanation must incorporate power relations: hate crime should be understood as a social control mechanism rooted in institutional power structures. She conceptualizes hate crime as an outgrowth of systems of oppression; it is one of the ways in which perpetrators maintain social hierarchies. For instance, racially-motivated violence affirms White privilege, while anti-lesbian and anti-gay violence reinforces the subordination of women and the cultural devaluation of homosexuality (Perry, 2001).

Perry has advanced a persuasive sociological theory of hate crime. She reveals the cultural context in which hate-motivated violence can flourish and she documents the ways in which institutional power structures may lead to hate crime. Similarly, groundbreaking studies of organized hate groups and white supremacist discourse have contributed to intersectionality scholarship. They have revealed the factors leading to women's involvement in racist activism and they have demonstrated how white supremacist discourse reinforces patterns of social inequality. By doing so, they have revealed some of the dynamics that contribute to hate crime. While these studies have produced numerous contributions, they have focused on the causes and the perpetrators of hate-motivated violence more than the victims.

... While most studies of hate crime victims have examined the psychological effects of hate-motivated violence, some social scientists have begun to explore other research questions.... Rather than accepting abstract definitions of hate crime as given, this research approach allows victims to explain their understanding of hate-motivated violence. By doing so, it privileges the voice of victims and it constructs their understanding of hate crime as significant.

Herek and colleagues (Herek et al., 1997; Herek et al., 2002), focusing on anti–lesbian and anti–gay victimization, have revealed some of the ways in which hate crime victims interpret and understand their violent experiences. They have found that lesbian women and gay men often examine their perpetrators' statements to determine that violence is based on their sexuality. In these situations, lesbian women and gay men confront explicit, unambiguous homophobic remarks. In other, less common situations they identify incidents as anti–lesbian or anti–gay by relying on contextual cues. For instance, victims perceive violence as rooted in homophobia when it occurs near a gay-identified location or when it occurs after public displays of affection between same-sex couples.

In this [reading], I employ an intersectionality approach to expand upon studies that have examined how lesbian women and gay men determine that violence is based on their sexuality. As I argue throughout this [reading], systems of oppression affect how LGBT people make this determination. Thus, employing an intersectionality framework improves our understanding of hate crime by revealing some of the ways in which LGBT people's violent experiences differ along the lines of race, class, gender, and sexuality.

METHODS

To examine how victims experience hate-motivated violence, I designed a qualitative research project in which I interviewed 44 people who experienced violence because they were perceived to be lesbian, gay, bisexual, or transgender. … During the interview, I asked participants to describe their violent experiences and their response to the violence. To understand how they perceived the violence, I asked about their perpetrators' motivations ("As you look back on this incident, why do you think he/she/they used violence?"). I then asked detailed, follow-up questions about their understanding of why the violence had occurred.

Because I wanted to examine the intersection of multiple systems of oppression, it was important to have a diverse sample of LGBT people. Thus, I recruited participants from a wide range of advocacy and service organizations, many of which provide services for LGBT people of color. I interviewed 17 women, 17 men, and 10 transgender people. All of the men identified as gay and 13 of the women identified as lesbian; two women identified as heterosexual and two as bisexual. Eight of the transgender people identified as male-to-female (MTF), one as female-to-male (FTM), and one as intersexed. Participants ranged from 20 to 62 years old; the median age was 41. Twenty-one participants identified as Black, 13 as White, eight as Latino, and two as Asian. In terms of educational background, five participants had dropped out of high school, 14 had a high school diploma, six had taken some college, 16 had a college degree, and three had a postgraduate degree. The interviews lasted from approximately one to three hours….

RESULTS

Determining that Anti-Queer Violence is Rooted in Multiple Systems of Oppression

One of the ways in which queer people determined that violence was based on their sexuality was by examining what their perpetrators said about gender. When examining the intersection of gender and sexuality, queer people often determined that violence directed against their gender identity was rooted in homophobia. When doing so, they acknowledged societal processes that conflate gender nonconformity with homosexuality. Dorothy, a 49-year-old White lesbian woman, addressed this dynamic when arguing that violence directed against her gender identity was also rooted in homophobia. The violence occurred when she was attacked by three men on the street. One of the men punched her and stole her purse. The men called her a "bitch" and said she had "no business being on the street." Dorothy believed that her gender nonconformity marked her as visibly "out." As Dorothy explained, she was dressed "very aggressively—suit and tie." Although the perpetrators never mentioned homosexuality, she perceived the violence as homophobic: "I wasn't doing anything, but it was obvious that I was a lesbian. That's why they attacked me. They hated gay people."

Dorothy's experience was common among respondents. They were routinely victimized for violating gender norms. In some of these situations, perpetrators referred to the victim's gender nonconformity but not the victim's homosexuality; queer people frequently defined such violence as homophobic. When queer people made this determination, they perceived attempts to punish their gender performance as attempts to regulate their sexuality. For instance, Paul, a 57-year-old White gay man, perceived violence directed against his gender identity as an attempt to punish him for publicly identifying as gay. He was attacked by three strangers on the street. His perpetrators told him "you're not a woman" after pushing and hitting him. Paul thought that he was targeted because of the way he walks—"very feminine," as he described it. Moreover, he saw a relationship between how he performed gender and how his sexuality was perceived: "I think it happened because I'm gay. They didn't like that I'm feminine because it showed that I'm gay."

The examples above are a few among many. They illustrate that queer people sometimes perceived violence as related to their sexuality even when perpetrators did not explicitly address it. When queer people perceived gender-based forms of violence as homophobic, they acknowledged the cultural intersection of gender and sexuality. They recognized that in the United States one's gender display is often understood as indicative of one's sexuality—that is, conformist gender displays are associated with heterosexuality and nonconformist ones are associated with homosexuality. As a result, attempts to punish gender nonconformity could be perceived not only as attempts to enforce gender conformity but also as attempts to restrict homosexuality.

While some queer people highlighted the importance of gender and sexuality in structuring their experiences of violence, others argued that their violent experiences could not be reduced to these two aspects of their identity. These arguments were particularly common among queer people of color. Many queer people of color highlighted the role of racism, as well as homophobia and sexism, in structuring their violent experiences. Kevin, a 62-year-old Black gay man, maintained that his violent experiences could not be separated from his race: "I've experienced violence because I'm black *and* gay. When the police beat me up, they called me a fag … I would be surprised if they had done the same thing to a White gay guy, though." Here, Kevin argues that violence directed against his sexual identity was also rooted in racism. He highlights the significance of race in structuring forms of anti-queer violence and he suggests that if he were White and queer, he might not experience homophobic violence to the same degree or in the same way.

… Although queer people of color frequently determined that their violent experiences were at least partially rooted in homophobia, they usually thought that violence was based on more than their sexuality. When doing so, they advanced arguments that seemed to borrow from intersectionality. Page, a 45-year-old Latina woman, argued that her violent experiences could not be explained by only a few factors: "It's much more complicated politically than 'I'm a woman so this happened.' Things are just not necessarily about any one category, misogyny or homophobia or whatever." Similarly, Aisha, a 53-year-old Black lesbian woman, argued that her violent experiences could not be reduced to a few aspects of her identity: "I'm a Black lesbian woman who works in a job where mostly men work. Change any of those things and [the violence] would not have gone down in the same way." Highlighting how multiple systems of oppression structured her violent experiences, Aisha argued that homophobic violence can only be fully understood within the context of a racist, male-dominated, and capitalist society.

EXPRESSING UNCERTAINTY: DIFFERENCES
ALONG RACIAL LINES

Until this point, I have emphasized situations in which queer people determined that violence was at least partially rooted in homophobia. Many queer people, however, found it impossible to determine whether violence was based on their sexuality. They frequently responded to questions concerning why they thought the violence had occurred with phrases such as "I don't know" or "I'm not sure." In these situations, they expressed uncertainty as to whether violence was based on their sexuality.

Queer people most typically responded with a sense of uncertainty for two reasons: (1) the violence occurred in situations in which the perpetrator insulted many aspects of the victim's identity; or (2) the violence occurred in situations in which the perpetrator said very little about the victim's sexuality. These two situations are, in some sense, opposites. The latter occurred when perpetrators said very little; the former occurred when they said a lot. In both of these situations, however, victims struggled to make sense of their violent experiences.

Queer people of color were more likely than White gay men to express uncertainty as to the cause of their violent experiences. This difference reflects the reasons I have outlined above: queer people of color often faced situations in which many aspects of their identities were attacked and they frequently encountered situations in which their perpetrators did not mention homosexuality. For these reasons, queer people of color often found it more difficult than White gay men to determine whether violence was rooted in homophobia.

When queer people of color experienced violence in which their perpetrators did not mention homosexuality, it was usually intra-racial violence—that is, the victim and the perpetrator were the same race. In contrast, when queer people of color felt as if many aspects of their identity had been attacked, the violence was most typically interracial. Thus, patterns of activity reveal that these two situations were raced—they differed with regard to the racial make-up of the victim and the perpetrator.

Queer people of color had the most difficulty determining whether violence was based on their sexuality or gender identity when their perpetrators were White. In such situations, they often felt as if multiple aspects of their identity had been attacked. Dominique, a 23-year-old Black transgender woman, described this dynamic rather succinctly: "When I'm called a fag or a freak by a White person, I have a hard time telling if they hate me because I'm trans or because I'm Black." White perpetrators often mixed homophobic or transphobic insults with racist ones. This blurring of racist and homophobic insults made it difficult for queer people of color to determine whether violence was based on their sexuality. In these situations, they could not be certain that violence was rooted in homophobia because racism may have played an equal or even more significant role.

… While queer people of color sometimes found it difficult to determine whether interracial violence was based on their sexuality, White gay men almost always argued that their violent experiences were rooted in homophobia. Responses such as "it happened because of my sexuality" or "it happened because I'm gay" were common among White gay men. Even when perpetrators mentioned race, White gay men usually determined that interracial violence was based on their sexuality. For instance, Greg, a 43-year-old White gay man, believed that he was attacked by two Latino men because of his sexuality. His attackers called him a "fag" and told him to "take that White shit somewhere else." He explained his perpetrators' motivations in a rather matter-of-fact way: "Oh, I think it happened because I'm gay. What else could be the reason?" In stark contrast to the uncertainty expressed by some queer people of color, White gay men almost always determined that interracial violence was based on their sexuality.

… Queer people of color were more likely than White gay men to report violence in which their sexuality was not explicitly addressed. Some queer people of color focused on race when explaining why their perpetrators did not mention homosexuality. For instance, Cole, a 33-year-old Black gay man, explained his perpetrators' actions in racial terms:

INTERVIEWER: So, they didn't use any homophobic slurs?

COLE: Slurs? Well, you see, in the Black Community, it's a little different. They don't always say "faggot." They'll say "too sweet mother fucker" or they'll just call you a sissy … It's more about you being weak than being gay … For them to call me a "faggot" would have meant that homosexuality exists, so they'd rather just beat me and not say anything.

Here, the intersection of race, gender, and sexuality seems particularly stark. Cole's argument suggests that Black queer people frequently encounter violence in which their perpetrators focus on gender nonconformity rather than homosexuality. As a result, Black queer people may often confront violence in which their perpetrators do not explicitly address homosexuality.

DIFFERENCES AMONG LESBIAN WOMEN OF COLOR AND GAY MEN OF COLOR

While gay men of color sometimes experienced violence in which their perpetrators did not explicitly mention homosexuality, lesbian women of color encountered such situations even more frequently. Gay men of color most typically reported violence in which their perpetrators used homophobic insults such as "homo" or "faggot." In contrast, lesbian women of color reported more violent incidents in which their perpetrators did not use homophobic insults. In some of these situations, their perpetrators used misogynistic insults rather than homophobic ones. For instance, Leslie, a 50-year-old Black lesbian woman, was spat on and called a "bitch" by a man on the street. She thought that the violence might have been rooted in homophobia because it occurred when she was with her girlfriend. However, she found it difficult to determine whether the violence was based on her sexuality, since it also seemed to be rooted in sexism: "I don't know if it happened because I'm lesbian … It could have happened just because I'm a woman, but it seems like it happened because I'm gay, too. I don't know why they chose me and not [my girlfriend], though."

Lesbian women of color sometimes found it difficult to distinguish between misogynistic and homophobic forms of violence. Judy, a 43-year-old Latina lesbian woman, encountered discourse that was simultaneously misogynistic and homophobic when she was sexually assaulted. The sexual assault occurred at a party when she was 20. A man grabbed her breasts and tore open her t-shirt. When he could not forcibly remove her pants because Judy was holding onto her belt, he told her, "All I want to do is fuck you and I bet you'll come back straight." As he continually tried to remove her belt, he called her a "bitch" and a "whore." Before he could remove her belt, another woman entered the room where the sexual assault had occurred. The two women yelled and threw items at him. Shortly thereafter, he left the party.

Judy's experience illustrates the difficulty of unpacking misogynistic and homophobic forms of violence from one another. It's difficult to determine where the line for one begins and the other ends. Would Judy have been sexually assaulted had she been a heterosexual woman? Was her attacker trying to punish her for what he saw as a deviant sexuality? Did he actually believe that he could make her "come back straight," as he stated? Or was her homosexuality merely the most readily available discourse that he could draw upon to justify his own behavior while simultaneously shifting blame onto her? These questions, it seems to me, may be impossible to answer. It seems unlikely that even the attacker could fully explain all of his unconscious thoughts and feelings at the moment. While these questions may be unanswerable, my research suggests that lesbian women of color often ask themselves such questions as they struggle to make sense of their violent experiences. Indeed, Judy had asked herself many of these questions following the sexual assault. Examining the violence approximately 23 years later, she concluded: "I can't be sure if it occurred because of my sexuality or just because I'm a woman. Both probably played a role."

Lesbian women of color most frequently expressed uncertainty because their perpetrators did not use homophobic insults. Gay men of color, in contrast, more frequently encountered violence in which their perpetrators used many homophobic insults. This dynamic can be explained in part by patterns of victimization: heterosexual men perpetrated most of the anti-queer violence reported by respondents. Given this pattern of victimization, one would expect that male perpetrators would use homophobic insults more frequently against gay men than lesbian women. Using homophobic insults against gay men allows male perpetrators to distance themselves from homosexuality. It allows heterosexual men to construct themselves in opposition to the deviant men—the "fags" or "homos"—whom they attack. Thus, since heterosexual men appear to perpetrate most anti-queer violence, lesbian women of color might encounter fewer homophobic insults than gay men of color. In other words, the gender of the victim and the perpetrator affect the degree to which victims confront homophobic insults.

THE EFFECTS OF SOCIAL CLASS ON WHETHER QUEER PEOPLE EXPRESSED UNCERTAINTY

Social class also affected how queer people determined that violence was based on their sexuality. Middle- and upper-class queer people usually expressed more willingness than low-income queer people to examine whether their violent experiences were rooted in homophobia. Eva, a 46-year-old Black transgender woman, described herself as middle-class, but also said that she had very little money when she experienced transphobic violence several years prior to the interview. She described the effects of social class on her willingness to determine whether violence was based on her gender identity:

> "I didn't want to think if it was a hate crime. I didn't have heat. I didn't have heat! ... How was I supposed to sit around and spend time thinking about whether I had been bashed?"

Eva's experience suggests that working-class and low-income queer people may have more pressing concerns than determining whether violence is rooted in bias. She indicates that poverty hinders the willingness of queer people to determine whether violence is based on their sexuality or gender identity. Indeed, many queer people who were living in poverty began to wonder over the course of the interview whether more of their violent experiences were rooted in homophobia than they had previously thought. Nevada, a 36-year-old White person who identified as intersexed and lived in a homeless shelter at the time of the interview, conveyed this feeling: "I had never thought about all of this as related to my sexuality. Maybe it was now that I think about it." Nevada's response was common among low-income queer people. They often began the interview by describing a violent incident that they thought was rooted in homophobia or transphobia. As the interview progressed, they frequently described more violent incidents. When describing these incidents, they sometimes said that they had not thought about whether their violent experiences were based on their sexuality or gender identity prior to the interview. Conversely, middle- and upper-class queer people seemed to have considered prior to the interview whether their experiences were rooted in homophobia. They frequently responded with phrases such as "I've thought about that before" or "I've thought about this a lot" when explaining whether they thought their violent experiences were based on their sexuality or gender identity. Thus, social class affects the degree to which queer people are willing to determine whether violence is based on their sexuality or gender identity.

DISCUSSION

Previous studies that have documented the ways in which lesbian women and gay men determine that violence is based on their sexuality have made no reference to race, class, and gender (Herek et al., 1997; Herek et al., 2002). As I have argued throughout this [reading], these systems of oppression affect how queer people determine that violence is based on their sexuality. Previous studies, then, have overlooked some of the ways in which lesbian women and gay men make this determination....

Race ... structured how queer people determined that violence was rooted in homophobia. White gay men generally expressed certainty as to the cause of their violent experiences—that is, they usually believed that violence had occurred because of homophobia. Conversely, queer people of color sometimes found it difficult to determine whether violence was based on their sexuality.

They often felt as if multiple aspects of their identity had been attacked and they frequently encountered violence in which their perpetrators said very little about homosexuality. For these reasons, queer people of color were more likely than White gay men to express uncertainty as to the cause of their violent experiences. Thus, the degree to which queer people are willing to determine that violence is based on their sexuality differs along racial lines.

If queer people of color find it more difficult than White gay men to determine whether violence is based on their sexuality, then hate crime statutes may primarily serve to protect the interests of White gay men. Hate crime statutes, which increase criminal sanctions against hate crime perpetrators, benefit victims who are willing to define violence as bias-motivated. Victims who cannot classify violence as bias-motivated will be less likely to report it as a hate crime and, consequently, less likely to have it prosecuted as one. As a result, victims who find it easiest to determine that violence is rooted in bias will benefit disproportionately from hate crime statutes....

... Considering the experiences of lesbian women of color suggests that they may find it particularly difficult to pursue hate crime legislation. As my results suggest, lesbian women of color often confront violence in which their perpetrators do not use homophobic insults. Because perpetrators' hate speech is often used to prosecute hate-motivated violence, hate crime statutes may rarely serve the interests of lesbian women of color.

... Social class further complicates the ways in which victims pursue hate crime statutes. The experiences of working-class and low-income queer people suggest that they may not pursue hate crime statutes because of the financial demands of their lives. Financial anxieties, in other words, make it more difficult for victims to pursue hate crime statutes. As a result, middle- and upper-class victims may benefit disproportionately from hate crime legislation.

Considering the experiences of queer people of color reveals that racism makes possible certain forms of homophobic violence and homophobia makes possible some forms of racist violence. As a result, queer people of color face situations that neither heterosexual people of color nor White LGBT people must confront. Of course, queer people of color are not a monolithic group.... While some queer people of color argued that both racism and homophobia were implicated in anti-LGBT violence, others expressed uncertainty as to the cause of their violent experiences. Although I have tried not to ignore these differences, my primary focus has been to examine the obstacles confronted by queer people of color and to explore how their experiences may differ from the experiences of White gay men....

Because of the exploratory nature of this research project, I have highlighted victims' voices as much as possible.... Hate crime researchers should ... continue to explore how victims' experiences differ along lines of race, class, gender, and sexuality. Indeed, as I have shown throughout this [reading], an intersectionality approach can provide a better understanding of the ways in which hate crime statutes concern the lives of all LGBT people.

REFERENCES

Herek, G.M., Cogan, J.C., & Gillis, J.R. (2002). Victim experiences in hate crimes based on sexual orientation. *Journal of Social Issues*, 58(2): 319–339.

Herek, G.M., Gillis, J.R., & Cogan, J.C. (1999). Psychological sequelae of hate crime victimization among lesbian, gay, and bisexual adults. *Journal of Consulting and Clinical Psychology*, 67(6): 945–951.

Herek, G.M., Gillis, J.R., Cogan, J.C., & Glunt, E.K. (1997). Hate crime victimization among lesbian, gay, and bisexual adults: Prevalence, psychological correlates, and methodological issues. *Journal of Interpersonal Violence*, 12(2): 195–215.

Perry, B. (2001). *In the name of hate: Understanding hate crimes*. New York: Routledge.

PART IV

✳

Pulling It All Together

MARGARET L. ANDERSEN AND
PATRICIA HILL COLLINS

Upon reaching the end of this book, students often want to know "What can I do?" As the editors of this volume who compile and revise it every couple of years, we know there is not a simple answer to this question. Once people know about the social injustices brought about by race, class, and gender, they may feel overwhelmed by the possibility of changing society. Some may still think that social inequities happen to other people, not to them, and they tune out. We understand that developing and then acting on an intersectional analysis of race, class, and gender is a lot to expect from our readers, most of whom are undergraduate students. For some, reading this book will be the first time they have even thought about such things. Others will have experienced some of what is written about here, but perhaps they will not before have thought beyond the particulars of their own lives. We know that developing an inclusive perspective and then deciding what to do about it is a complex process—one without simple solutions or ways of thinking. Thus, we have developed this last section of the book to examine what it means to pull together a new way of thinking—and acting—that recognizes and engages the complexity of race, class, gender, and the various other social factors that together make up this complex system we have called a *matrix of domination*.

In Part IV we ask "What would things look like were you to take an intersectional view of society?" This includes having the vision to think comprehensively and from multiple points of view as well as analyzing the various contexts where people negotiate their way through the structures of society. Vision is important to imagining social justice, and here we include articles that

467

show how people articulate a vision that is informed by race, class, and gender. Our goal is not to provide a roadmap about how to bring about change. Instead, we want to show you some possibilities.

One of the best places to find people with vision is among people who are young. High school and college students are often the ones who imagine new possibilities and who follow through with social action. But even here, race, class, and gender relations encourage us to see some youth as having more important things to say than others. We open this part with an article from Jessica Taft's *Rebel Girls: Youth Activism and Social Change across the Americas*. Taft's "rebel girls" fight back against pressures to silence them. They see the version of individual empowerment offered to girls as limited and instead claim the social change goals of activism. These two forms of action—empowerment and activism—are often seen as being the same. But they are not and we include this article because it helps us broaden our vision about what is possible.

Is there something in your life that you care about so much that it would spur you to work for social justice? Is it your family? your children? your faith? your neighborhood? a concern for a social issue? or an ethical framework that requires not just talk but action? Most people think that people who work for social justice must be somehow extraordinary like Martin Luther King Jr. or other heroic figures. But most people who engage in social activism are ordinary, everyday people who decide to take action about something that touches their lives. Like Taft's rebel girls, their quest for individual empowerment may push them toward broader forms of activism.

For Eisa Nefertari Ulen ("Tapping Our Strength") religious identity as a U.S. Muslim is the foundation both for her individual empowerment as well as for her activism. Ulen urges us to build cross-cultural, multi-ethnic connections among people. In a time period when simplistic and stereotypical views of people defined as "other" pervade the dominant culture, Ulen expresses the need to embrace the diverse experiences and perspectives of people, while simultaneously using that diversity to build a more inclusive and just society.

The effort to generate a more just society develops in many social contexts—schools, homes, communities, churches, and other locales. In "'Whosoever' Is Welcome Here: An Interview with Reverend Edwin C. Sanders II," Gary David Comstock describes how African American minister Sanders challenged the exclusionary practices of Southern African American churches by building a new church that included everyone. In this organization a small group of people came together in Nashville, Tennessee, in search of a community of worship that did not exclude anyone and was welcoming of everyone. From its small beginnings, the Metropolitan Interdenominational Church grew and is welcoming of people of different races, classes, genders, sexualities, and religious traditions.

Possessing a vision of equality and an ethical framework that helps people see how and where diversity can be achieved, but also illustrates how this process can be filled with contradictions, Janani Balasubramaniam is a staunch advocate for sustainability, but questions the race and class politics of this social movement. She reminds us that people can share a vision yet must also struggle with the contradictions of race, class, and gender in order to bring their vision about. Whether social actors are located inside the institutions they wish to change or whether they stand outside its boundaries, the strategies they select reflect the opportunities and constraints of each specific site. Working from within organizations can mean trying to change the institutional policies and practices that either overtly discriminate based on race, class, and gender or that have unjust outcomes. Other people and groups work outside formal social institutions to effect social change. The familiar boycotts, picketing, public demonstrations, leafleting, and other direct-action strategies long associated with social movements of all types typically constitute actions taken outside an institution, and designed to pressure it to change. Although activities such as these can be trivialized in the media, it is important to remember that direct action from outsider locations represents one important way to work for social justice.

Robin D. G. Kelley's article "How the New Working Class Can Transform Urban America" shows people engaging in just such action. Kelley details some of the social actions being taken by coalitions of Latino, Black, Asian America, and some White workers. In an economy where jobs and employment have moved to the front of the line as important social issues, protecting the economic status of working people has become a hot topic. Many people think that the union movement is on the decline, mainly because they see unions as primarily composed of White, male industrial workers. Kelley describes the organizing practices of the national Justice for Janitors unionization movement and the Bus Riders Union in Los Angeles, two organizations that have continued to grow. Kelley's article points to the multiple strategies used by these grassroots movements that draw upon traditional organizing principles but that also respond to the new challenges of economic change. Kelley shows how applying a race, class, and gender framework to all efforts for social change is important for grassroots movements and organizations that are marginalized by dominant institutions.

Throughout this book we have examined how entrenched power relations of race, class, and gender affect us all in the United States. Dominant social institutions use ideologies to obscure the individual and collective political actions of everyday individuals who oppose social injustices. Latinos, African Americans,

Native Americans, new immigrant groups, women, gays and lesbians, and members of the poor and working class have long challenged the institutional structures that constrain them. Thus, social institutions are both sites that reproduce social inequalities of race, class, gender, ethnicity, and sexuality and sites where people have challenged these same injustices. Workplaces, families, schools, the media, the military, and the criminal justice system—some of the social institutions we examined in Part III—are all potential sites of change.

Yet working for a more just society requires looking beyond both the borders of the United States as well as beyond what already exists. Charlotte Bunch's article discusses how the existing politics of race, class, and gender leave us with a restricted view of women's rights. Bunch encourages us to look beyond U.S. borders to envision how human rights encompass women's rights. Such a shift in our thinking moves us in the direction of inclusive thinking that is essential to race, class, and gender studies. In her work, we see ways to think relationally and to see new possibilities that move us in the direction of a broader, social justice vision.

Whether working for individual empowerment or social activism, we must learn to see beyond what is in order to imagine what is possible. For example, what type of environmental policies would result if White working-class women and women of color were central in the environmental movement's decision-making processes? If all religious institutions welcomed all people, how might families and communities be changed? How might immigrant communities be changed if all service workers were paid an adequate wage? If working-class people's need for affordable, reliable transportation were central to public policy, would more money be devoted to mass transit and less to highway repair? If men and women truly learned to work together, how might economic security be better provided for all? Thinking inclusively about race, class, and gender stimulates this type of vision. All sites of change contain emancipatory possibilities, if only we can learn to imagine them.

55

We Are Not Ophelia

Empowerment and Activist Identities

JESSICA K. TAFT

In January 2004, in a speech to the International Women's Health Coalition, the then secretary-general of the United Nations Kofi Annan told the assembled group "when it comes to solving many of the problems of this world, I believe in girl power."[1] Girls, according to Annan, need to be educated to take up the mantle of civic responsibility and humanitarian leadership. Meanwhile, the Nike Foundation, a charity wing of the giant athletic gear corporation, focuses all of its funding on adolescent girls in the "developing world" in order to "empower impoverished girls by expanding their opportunities, capabilities, and choices.[2] Countless organizations, books, Web sites, and after-school programs around the world state that their mission is to "empower girls." Furthermore, an additional panoply of programs and initiatives focus on enhancing the civic engagement of all youth, both girls and boys. Taken together, these programs and statements suggest the high levels of transnational corporate, non-governmental, and philanthropic interest in girls' empowerment and girls' civic identities. Whether it be expressed in concerns about boosting girls' self-esteem or the potential for reducing poverty through girls' education, the figure of the empowered girl and the notion of girls' empowerment are powerful and pervasive features in our current discourses of girlhood.

But what, exactly, does it mean to empower girls? What kind of power do empowered girls possess and demonstrate in these imaginings? And what are the implications of this version of empowerment for teenage girls' critical political participation? In this [paper], I briefly deconstruct and analyze the current institutionalized models of girls' empowerment and civic engagement, contrasting them with the activist practices, strategies, and social change visions of the girls who are featured in this book. I'll argue that empowerment, as it is currently articulated, is quite distinct from activism. Girls' empowerment is all too often focused on incorporating girls into the social order as it stands, rather than empowering them to make any meaningful changes to it.... Girls in this study, on the other hand, are interested in substantial social change, they are "empowered," but they are critical of the narrow versions of empowerment usually offered to them. They are not the kind of empowered girls usually celebrated

SOURCE: Taft, Jessica, "We Are Not Ophelia" from *Rebel Girls: Youth Activism and Social Change Across the Americas*. Copyright © 2011 New York University Press. Reprinted with permission.

by the media. They are not likely to turn up on Oprah, nor to be featured in popular teen magazines for their "positive social contributions." Their empowerment is too rebellious, too critical, too political. Girl activists embody an alternative kind of empowered girl citizen and a different understanding of empowerment, both of which will be made visible throughout this [paper].

This [paper] also explores how girl activists construct and define their identities as *activists,* as opposed to empowered individuals.... I ask what is activism and who is an activist? Definitions of activism can be drawn from people's explicit commentary on the subject and extrapolated from their political practices. Therefore, I pair girls' statements about the meaning of activism with concise descriptions of the activism I encountered in my research in order to offer an overall picture of these young women's social movement activity....

ACTIVISM AND EMPOWERMENT

My account of girls' activist identities is, of course, shaped by my own research practices and criteria for inclusion in this project. Consequently, it is also worth addressing some of those choices here. I conducted preliminary research before my arrival in each metropolitan area, gathering information about local social movement organizations and youth activist groups. I contacted as many of these adult and youth groups as possible before and during my time in each city, asking for their assistance in locating teenage girl activists. Then, I followed the suggestions and leads of these local movement participants, seeking out girl activists by visiting the recommended locations, organizations, and events. When choosing which leads to follow and which groups or organizations to contact, I focused only on those that could be considered part of left and/or progressive social movements, broadly conceived. I did not include organizations and events that primarily emphasized individual growth and personal development, government-centered political participation, or community service. Instead, following agreed-upon sociological definitions of social movements, I looked to spaces and groups engaged in collective, non-governmental, and change-oriented political activities. Once I had made contact with some girl activists in these locations, I used a snowball sampling approach, asking the girls themselves for their help finding other activist teens. This means that for each city there are a few tendencies and groupings of girls, each branching out from an original set of contacts.

Unlike much social movement research that narrows its focus by selecting a single movement for analysis, my research is not movement-specific.... I sought out girl activists across progressive movements and, in doing so, found that a majority of the girls that I met in one movement context were also often involved in one or more other political activities and organizations.

When I encountered teenage girls in these social movement spaces, I would ask them if they were interested in participating in the study and if they would self-identify as activists. Only those girls who acknowledged and claimed an activist identity or who said that they were somewhere on the route to becoming activists were then interviewed. Girls who replied that they were not really active, were just stopping by this single event to see a friend, or who were not regular or frequent participants in any kind of collective political project were not interviewed. Thus, both ongoing involvement in social movement activities and/or organizations and an activist self-identification (including as someone "becoming an activist") were necessary conditions in my own determination of who was or was not a girl activist.

Rather than simply relying on my own criteria to create a definition of girl activist identity, I wanted to know how the girls themselves understood and constructed activist identities both through their narratives and their political practices. Central to their understandings is the claim that activism, first and foremost, is about a desire to create change, to make the world into a different and better place. According to one interviewee from Mexico, Sicaru, activism is "the interest in trying to change things, not just staying like this." Ixtab, another Mexican teen, stated that activism is "doing something to change the situation." Ella, from the San Francisco Bay Area, described an activist as "someone who spends time trying to make the world a better place." Of course, people have conflicting ideas about what would make the world a better place. Changes are not neutral, and they are assessed based on particular values and beliefs about what makes a good community or society. Girl activists have a variety of visions and hopes for the world.

... Activists concern themselves not only with their own well-being but also with that of others and of communities. Lucia, from Mexico City, argued that "an activist is someone who ... struggles for people—someone who is involved and who wants things to get better, not just for themselves, but for people who they know and people who they don't know." In this view, activists are people who act for the good of communities, not just the good of individuals. Activists view entrenched social structures and systemic problems as presenting barriers to community well-being, a view which then necessitates collective action. Ana, from Buenos Aires, argued, therefore, that an activist is "someone who sees what is really happening and who has a critical vision about this.... An activist is someone who can do many things ... and who doesn't just have the capacity (because everyone has the capacity), but who also has the desire, who wants to change what they see." It is worth noting here that this public-spirited commitment is something that girls think anyone can develop. By emphasizing community well-being and a critical understanding of the social, political, and economic problems facing communities, these girls draw links between activism and a particular way of seeing and understanding the world—a sociological, rather than individualized, understanding. The changes they imagine are about creating a world that is better for many people, not just improving their own abilities to deal with and overcome the problems they see in the world.

Girl activists, located within different communities, political contexts, and struggles, address a breathtaking range of social problems and contemporary political issues in their activism. Because schools and education are particularly powerful forces in girls' lives, many of their organizations focus on improving these institutions. In all five locations I encountered girls organizing and agitating for more decision-making authority in their schools and more influence over educational policy and curriculum. They also work toward racial equity in education, for safer schools, for low-cost student transportation, and against the privatization of public education. In addition to student rights and student issues, girl activists engage in projects focused on the well-being and rights of children and youth. Several of the young activists were involved in campaigns that address teenagers' reproductive rights and access to quality health care and health information. Several others were active in labor organizing and worker's rights campaigns, including those oriented toward the needs and rights of child and youth workers.

Girl activists participate in numerous local campaigns around issues of community development, including struggles against gentrification, unemployment and hunger, challenges to destructive building proposals (shopping malls and highways that would disrupt important ecosystems and community spaces), demands for corporate accountability and solutions to ongoing corporate environmental health and safety violations, and projects that aim to build alternative, community-controlled social service institutions. These local campaigns, particularly in Latin America, are complemented by activism aimed toward the global institutions that structure local problems. This includes mobilization against free trade agreements and the institutions of neoliberal globalization including the World Trade Organization, the International Monetary Fund and the World Bank, anti-corporate campaigns, resistance to privatization, to structural adjustment and other neoliberal economic policies, struggles for immigrant and refugee rights and against displacement, and other related goals that could be identified as part of a global justice or alter-globalization agenda.

Girl activists are also specifically concerned with the problem of violence, both state-based and individual. Many of the girls I interviewed are active in a loosely structured anti-war movement and speak out about U.S. imperialism, the war in Iraq, land mines, and/or militarized violence. A few are also involved in trying to address the Israeli occupation of Palestine. Other girls engage in activism around prison issues and the rights of imprisoned people, making improvements to the California juvenile justice system, against police brutality and racial profiling, and against torture. Still others are part of campaigns that address political repression and violence against activists or that mobilize for the freedom of political prisoners. Finally, some of the young women organize against hate crimes, violence against GLBTQ (gay, lesbian, bisexual, transgendered, queer) individuals, and gender-based violence.

Alongside this almost dizzying array of specific concerns, most girl activists expressed their desires for widespread and systemic social change. Many of them have big dreams and expansive hopes for an entirely different world. This was the case for Celia, a seventeen-year-old who was so interested in the political struggles happening in Latin America that she decided to leave her native Italy to spend a year as an exchange student in Venezuela. An activist in both countries, she sees herself as part of "a movement of youth who imagine something better for the world. Or, of youth who haven't lost the hope of changing the future of the world, who want to make a world with equality, with justice." Girls in all five locations spoke out and acted against inequality and oppression in its multiple forms: racism, sexism, homophobia, ableism, ageism, economic inequality, and international or global inequalities. They want "a world where all of us are equal, where there is no discrimination" (Victoria, Caracas). In addition to the ideals of justice and equality, young women spoke about working to create a world based on the values of ecological sustainability, solidarity, community, democracy, love, liberty, human rights, peace, and indigenous people's rights to land and sovereignty.

... A defining feature of activism is its emphasis on social change. In striking contrast, programs for girls' empowerment tend to focus primarily on personal change. Girls' empowerment is frequently presented by globalized media institutions and a variety of girls' organizations as a process by which girls learn to develop their own personal power, in particular, the power to make choices and construct their own individual identities.

... This emphasis on changing the self can be seen in some of the mission statements of key organizations for girls around the world. These organizations regularly identify a variety of social problems that produce barriers to the happiness and success of girls (unequal education, sexualized media cultures, inadequate access to contraception, etc.), but their solutions are primarily oriented toward improving girls' individual ability to cope with these problems, rather than removing or changing the problems themselves....

Girls' empowerment as a discourse of self-change and individual growth is not, however, identical for all girls. Rather, it is a discourse that takes on a variety of racialized and located forms. Programs for more privileged, middle-class North American girls tend to focus on helping these girls to navigate the treacherous waters of falling self-esteem and aggressive peers in order to become empowered. Low-income or "at risk" girls in North America are instead told that they must overcome the supposed dangers of their upbringing and make "healthy choices." Meanwhile, according to a variety of the development organizations and policy makers, girls in the Global South (including but not limited to Latin American girls) need to be empowered in order to free themselves from the constraints of their patriarchal national cultures and excessive machismo. This version of the empowerment discourse not only replicates heavily criticized ideas about third-world women as victims of "local" patriarchies whose potential contributions are stifled by "traditional" ideas, but also provides discursive support for the idea that Latin America's young women experience substantial benefits from the "opportunity" to work....

"Empowering girls" has very positive connotations, invoking ideas of gender equity, community improvement, and a brighter future. But, when examined more closely, it can also be simply another way to present meritocratic notions of personal growth and individual opportunity. As an entirely individualized project of self-creation and transformation, this version of girls' empowerment weaves together the language of liberal feminism and gender equity, colonialist images of third-world women who need to be saved, and neoliberal ideologies of individual responsibilities and self-production. Individual empowerment makes no references to social and political rights, to economic justice, to equality, or to changing the overall contexts and conditions of girls' lives, but only discusses girls' individual strength and resilience. Any girl then who does not "succeed" is just not empowered enough.

This does not mean that the idea of individualized empowerment for girls should be dismissed completely. The organizations in both North America and Latin America that are concerned with the well-being of girls are, of course, engaged in valuable work. Acknowledging the importance of girls and expressing an interest in empowering them is certainly much better than simply ignoring girls and the very real challenges that they face in their lives. By focusing on psychology, self-reliance, healthy choices, and individual achievements, however, this approach to girls' empowerment encourages girls to think of their lives in these terms, often at the expense of a more sociological or political analysis. As girls learn to assess their lives through the language of self-esteem, healthy decision making, and individual opportunities, they are more likely to see their problems as personal troubles, rather than as issues of public concern. If their problems are not seen as publicly relevant, they are also much less likely to engage in social action to remedy them. By only teaching skills for facing barriers as self-made individuals, not removing them, this model of empowerment implies that society, the public, and the community are unchanging arenas. The empowered girl, in this model, thus has the power to remake herself, to define and reconstruct her own individual identity, but not to make social change. The young women who are a part of this study, on the other hand, do not want to merely overcome the problems they see around them, but to change the conditions of their lives and the lives of those around them. These young women are motivated not to simply remake themselves as "successful individuals," but to remake the world as a more just and equitable place. Girls' activism and girls' empowerment have very different goals....

NOTES

1. Kofi Annan, "No Development Tool More Effective Than Education of Girls." United Nations Information Service, 2004, http://www.unis.unvienna.org/unis/pressrels/2004/sgsm9118.html

2. Nike Foundation, "Who We Are." http://www.nike.com/nikebiz/nikbuzfoundation/who.jhtml, accessed Mar. 20, 2007.

56

Tapping Our Strength

EISA NEFERTARI ULEN

I walk with women draped in full-length fabric. We swirl through the delicate smell of incense and oil filling air around the mosque. Even as the mad Manhattan streets overflow with noise, we sisters rustle past the crisp ease of brothers in pressed cotton tunics and loose-fitting pants, past tables of overgarments and woven caps, Arabic books, and Islamic tapes. Our scarves flap and wave in bright color or sober earth tones above an Upper East Side sidewalk that is transformed, every Friday, into a bazaar—the sandy souk reborn on asphalt.

Worshippers walk through a stone gate, along a path, and into a room unadorned yet filled with spiritual energy. Women and men lean to remove their shoes where rows of slippers, sandals, heels, and sneakers line the entrance hall. White walls bounce light onto the high ceiling. Men sit in rows along the carpeted floors, shifting in silence as we file past them, up the stairs and to a loft where other women sit and wait. "As salaam u alaikum." "Wa alaikum as salaam." "Kaifa halak—how are you, sister?" "Al humdilillah—praise be to God—I am well."

Soon the imam's voice begins to resonate in the hushed rooms. Contemplative quiet focuses communal piety. The cleric speaks in Arabic, then in English, building ideas about a complete way of life. About an hour later, when he concludes, a chanting song, lyrical poetry, calls the Muslims to prayer. We women stand tall, shoulder to shoulder, forming ranks, facing Mecca, kneeling down and then forward in complete submission to Allah, our faces just tapping the prayer rug.

And our ranks are growing. Islam is the second most popular religion in the world, with over one billion Muslims forming a global Umma (community) that represents about 23 percent of the world's population. In virtually every country of Western Europe, Islam is now the second religion after Christianity. There are approximately six million Muslims in the U.S. today, and about 60 percent or so are immigrants. About 30 percent of the remaining American-Muslim population are African-American, with U.S.-born Latinos and Asian-Americans making up the 10 percent difference. There are now more Muslims in the U.S. than Jews, and the numbers of new shahadas and Muslim immigrants continue to rise. Islam is even changing the way the United States sounds, as the azan converges with church bells, calling Muslims to prayer five times a day.

SOURCE: *Shattering the Stereotypes: Muslim Women Speak Out*, edited by Fawzia Afzal-Khan, pp. 42–50. Copyright 2005 by Eisa Nefertari Ulen. Reprinted with the permission of Olive Branch Books, an imprint of Interling Publishing Group, Inc.

Words of worship are filling the air with Arabic all across America. This country will increasingly need to explore gender, generation, politics, and plurality from an Islamic perspective. The veiled lives of American-Muslim women, so often garbled into still passivity, pulsate with social ramifications.

So how will pluralistic America shift and groan under the weight of this new diversity? What happens when uber-girrrl in spiked heels and spiked hair turns the corner on her urban street and peers into the wide eyes of a woman whose face is covered with cloth? What happens when uber-girrrl's daughter brings home a friend who has two mommies—and one daddy? Under what terms do we launch that dialogue of encounter? Are women who insist on wearing hijab unselfconsciously oppressed, or—particularly in the land that gave birth to wet T-shirt contests—are they performing daily acts of resistance by covering their hair? In the West, where long blond tresses signify a certain power through sexuality and set the standard for beauty, are veiled women the most daring revolutionaries? In workplaces, where anything less than full assimilation is dismissed, are women who quietly refuse to uncover actually storming the gates for our own liberation? Is liberation possible within the veil? American feminists would do well to engage these and other questions, and then again to engage what may seem the easy answers.

I am peering outward, clustered with sisters in scarves. And I am also aligned with women sporting spikes. I am a Muslim woman. I am also a womanist, a feminist rooted in the traditions of Sistah Alice Walker. I run those mad Manhattan streets, contribute my own voice to the cacophony; I also sit in focused silence, shifting space to embrace the presence of my many-hued sisters. I celebrate the sanctity of varicolored flesh, of difference, that is celebrated in Islam. And I am also an ardent advocate of Black Empowerment, of Uplift, of Pride—I am a Race Woman.

When non-Muslims ask how a progressive womanist sistah like me could convert to Islam, I tell them Muslim women inherited property, participated in public life, divorced their husbands, worked and controlled the money they earned, even fought on the battlefield—1,400 years ago. When Muslims ask how a woman who submits to Allah like me could still be feminist, I tell them the same thing and add that modern realities too often fall short of the Islamic ideal. The Qur'an was revealed because Arabs were burying their newborn daughters in the sand, because Indians were burning their wives for dowry, because Europeans were keeping closet mistresses in economic servitude, because Africans were mutilating female genitals, and because the Chinese word for woman is slave. Obviously, these forms of violence against women continue, very often at the hands of Muslim women and men. The presence of patriarchal, sanctioned assaults against women and girls anywhere in the global Ummah horrifies me, particularly because I recognize these atrocities as anti-Islamic.

Contrary to popular opinion among non-Muslims, the Qur'an rejects the sexist propaganda that Eve is the first sinner who tempted Adam and led him to perdition. Both are held responsible for their exploits in the garden. Muslims believe Islam is the perpetuation—the refinement—of monotheistic religion and admonishes the persecution of people on the basis of gender, race, and class.

Non-Muslims often confuse sexist individuals or groups with an entire religious system—and a cross-racial, multicultural swath of the world's population gets entangled in the inevitable stereotypes.

This political irresponsibility is dangerous especially because Islam is the sustenance more and more women feel they need. Indeed, while American women were publicly calling for more foreplay a few decades ago, Islam sanctioned equal pleasure in spouses' physical relationship—again, 1,400 years ago. Muslim men actually receive Allah's blessing when they bring their wives to orgasm.

Knowing many Muslim women are cutting their daughters, I celebrate Islam and teach to raise awareness about the culturally manifested, pre-Islamic practice of female genital mutilation, which is falsely considered an Islamic practice worldwide. I feel the same passionate need for widespread truth and empowerment when I read about women across this country knifing their breasts and hips and faces in the name of Western-inspired beautification. Fellow feminists too often allow difference to impede a coalition based on these virtual duplications, on this cross-cultural torture. My Muslim sisters too often think feminism is a secular evil that would destroy the very foundations of our faith. This is all a waste. While we allow difference to divide us, women everywhere are steadily slicing into their own flesh—and into the flesh of emergent women around them.

Any concerted efforts to link and liberate on the part of American feminists must proceed from factual knowledge of the veiled "other." Immigrant Muslim women must begin to align themselves with non-Muslim-American women, even as they maintain their *deen* (religion) and their home culture. Increasing conversion in this country demands U.S. citizens interested in women's freedom begin to understand why women like myself have chosen Islam.

I became Muslim because the Qur'an made sense, because my mind and spirit connected. Islam is a thinking chick's religion. Education is more than just a privilege in Islam; it is a demand. Qur'anic exhortations to reflect and understand highlight each Muslim's duty to increase in knowledge, a key component of this *deen,* this religion, where science especially supports a better understanding of spirit. Islam makes no distinction between women and men and access to knowledge, though some men would deny women's Islamic right to education, just as men in America have historically denied women the opportunity to learn.

The more I think about feminism and Islam, the more compatibility emerges from the dust of difference, and the more potential I see. I want to reconcile the great gulf that all my sisters—American non-Muslims who can't get past the *hijab* and American Muslims who can't get past secularism—see when they peer (usually past) "the other." I understand my Muslim sisters' trepidation, because first wave feminism's relationship with African-Americans lacked a cohesion born of acute commitment and fell victim to white supremacist techniques. Likewise, second wave feminism fell victim to the science of divide and conquer. Daring individuals pushed past division, though. They leaped over the great gulfs system controllers contrived to separate abolition and suffrage, Black Power and Women's Lib, and embraced a decidedly universal freedom. I am still slightly shocked when Muslims and non-Muslims claim I cannot be both Muslim and feminist. I am leaping the great abyss dividing submission (to Allah) and resistance (to patriarchy) in this increasingly complex place called America. We must build bridges.

We must build cross-cultural, multi-ethnic bridges. Now, especially now, I ask my African-American sistahs to remember our legacy of domestic terrorism, of white sheets streaking by on horseback, of strange fruit, of Black men burned alive, of four little girls. Now, more than ever, we must remember the centuries of domestic terrorism in this country, but we must also remember this: that countless white women were our sisters in fighting the horror and pain their fathers and brothers wrought. We must remember this, too: We must remember the white men appalled by the terrorism of white supremacists, the white men who battled their own souls. We must allow these memories to help form a link connecting non-Muslims with Muslims in the country today. I am asking women to remember today. I am asking you.

I am ready to do this important new work. Islam fuels my momentum. I empower myself when I wash and wrap for prayer. I transform out of a space belonging to big city chaos and into a space conjuring inner peace. I renew. With the ritual Salaat, I generate serenity. I can create and channel strong energy as I pray.

Although I only cover for prayer, I deeply admire women who choose to outwardly manifest their connection to the Divine within. I want more non-Muslims to understand veiled Muslim women and respect them for celebrating Islamic creed, for resisting overwhelming economic forces in this country, for not succumbing to the images captured in high fashion gloss. By living in constant alignment with faith, they challenge the misogynist systems that compel too many Western women and girls to binge, purge, and starve themselves. For these pious sisters, plain cloth is the most meaningful accessory they could ever wear. To me, American-Muslim women who choose to cover undeniably act out real life resistance to the hyper-sexualization of girls and women in the West. In the context of consumerist America, women who cover express power of intellect over silhouette, of mind over matters of the flesh.

Because I move in non-Muslim circles, I hear too many of my fellow feminists focus on *hijab,* urging complete unveiling as the key to unleashing an authentic liberation. For them, scarves strangle any movement toward Muslim women's emancipation. I ask them to just imagine 1,400 years—generations—of women moving without bustles, hoops, garters, bustiers, corsets, zippers, pantyhose, buckles, belts, pins, and supertight micro-minis. The way I see it, Western men wear comfortable shoes and slacks while women are pinned, underwired, heeled, and buttoned to psychological death. We American women still strangle ourselves every day we get up and get dressed for work.

Ah, you say—but even in their loose garb, Muslim women are still, so, so … passive. But Muslim women are not silent, not sitting still. We do not require American pity. We take the very best America has to offer. We are moving our bodies. As American women wow the world with unleashed athletic excellence in the WNBA, women's soccer, bobsledding, and pole vaulting, Muslim-American women are running and kicking along that mainstream—in full *hijab* or not. For Muslim immigrants who hail from nations that denied women access to physical movement, this country has freed them to pilot their own bodies. Many American Muslims are destroying the cultural forces that chained them while remaining true to the essence of Islam.

I remember hearing Sister Ama Shabazz, a bi-racial Muslim educator and lecturer (her mother is Japanese-American and her father is African-American) urging a large group of Muslim women to take swimming classes and learn CPR to satisfy the Shariah (Islamic law) not to defy it. (Anecdotes can be so helpful sometimes.) A friend of mine, whose mother immigrated to this country from Colombia, wears long loose clothing to the gym three days a week, then washes to cover and go home. An African-American girlfriend of mine roller-blades through her Bedford Stuyvescent[*sic*] neighborhood, full *hijab* blowing through her own body's wind.

These women fiercely assert Islam even though they have felt American hands tug at their clothing, especially since 9/11. They are obviously Muslim even though non-Muslims hurl offensive epithets or gestures at them. They do not cower. "It takes a warrior to be a Muslim woman," says another friend, a New Yorker born and bred Dominican. I agree with her.

Yet there is so much promise in the future: I know two Muslim high school students of mixed Iraqi/Indian heritage who play tennis in traditional whites and have earned black belts in karate. One even coaches the boys' basketball team. Interestingly, their immigrant mother could not wear blue jeans because her father forbade it. Certainly some men are still using women to assert a political agenda via Islam. I recoil when I see young Muslim girls in full *hijab* while their brothers skip beside them in shorts and t-shirts. I think about the women I know who cover themselves and their daughters for the wrong reason, and then I remember I know some women who wear push-up bras for the same wrong reason: to please men.

We must recognize that the similarities in our oppression as women far out-weigh the differences in the ways that oppression manifests. And to do that we must fuse our stories. Like African-American women who have fought to wear locks and braids and naturals on the job—or have fought to use relaxers without the ultra-righteous disparaging them as loser sell-outs—Muslim women have had to fight to wear *hijab* here. For everything from job security to an American passport photo, Muslim women have been asked to uncover. At root, we are all denied our right to represent an authentic self by these predominant cultural and social forces.

The last place we women of all faiths need to suffer the indignity of judgment based solely on outward appearance is in the company of other women. Right now, half of American non-Muslim women encourage other women to be free by being naked, and the other half desperately tries to get women and girls to cover up. Meanwhile, the men simply get dressed in the morning. Likewise, Muslim women who wear *hijab* are automatically considered unhappy, while men who wear turbans and long loose clothing are just considered Muslim.

Non-Muslim women need to stop telling Muslim women their traditional Islamic garb symbolizes oppression. Muslim women need to open themselves to coalitions with women in mini-skirts. Only then will we work successfully toward a world where all women can truly wear what they feel. Ultimately, societies grant men much more freedom in clothing. Perhaps this is the point from which our discussion should launch.

We must begin to think more critically, and honestly, about media representations of all women. While Muslim immigrants need to reconsider the East's portrayal of American women as loose and wild, feminists need to check their sweeping generalizations about the seemingly inherent violence and suppression the media projects as Islam.

Since the 1970s America has slowly shifted evil empire status from the former Soviet Union to the site of underground power, where black, slick, liquid energy fuels America's Middle East policy. Americans have been taught to fear the Arab world so that America can easily justify killing Arabs. Images of Middle Eastern men in the state of *jihad* demonized the people of an entire region. But the only legitimate Islamic war is a war waged in self-defense—the other guy must be the clear aggressor. And the direct translation of *jihad* is struggle, while the primary focus of that struggle is within. We must remember this as we watch our evening news, as we watch bombs fall from U.S. planes. We must remember this as we vote.

American feminists should not join the Pentagon and media in denigrating veiled women and our faith as archaic, out-of-touch, regressive. This is part of the propaganda of fear America needs to perpetuate in order to maintain world dominance. This country takes the very universal problem of sexism and often presents it as if it were exclusively a Muslim issue, as when non-Muslims degrade Islam for allowing men to marry up to four wives, even as American men practice their own kind of polygamy—via mistresses, madams, and baby-mammas. Who are we to judge? After all, while the United States has never had a woman president, Pakistan, Turkey, and Bangladesh—all Muslim countries—have had female heads of state.

Of course, as Jane Smith of the Hartford Seminary says, "I think you'd have to be blind not to see things going on in the Islamic world—and in the name of Islam—that are not Islamic." Muslim women would do well to remember that the Hadith (sayings of Prophet Muhammad) have been interpreted to give men powerful social advantages over women, and that there are men, and women complicit in their own oppression, who use Islam as justification for misogyny.

Certainly the 9/11 attacks were not Islamic. Islam means peace. We greet each other with peace. Islam is no more violent than Christianity. Yet there have been American Christian networks formed to throw bombs—at abortion clinics. When have you ever heard the term Christian terrorist? We Americans do not profile white men with crew cuts, but a white man in a crew cut bombed the federal office building in Oklahoma City. Certainly we should not denigrate Christianity—and Christians—because of the few who would use their faith as a justification for violence. Why has it been easy for white Americans to turn on their own darker brothers?

Maybe we simply need to understand each other. Certainly America needs to begin the work necessary to understand Islam. El Hajj Malik El Shabazz said in his letter from Mecca, "America needs to understand Islam, because it is the one religion that erases from its society the race problem." Back at the Islamic Cultural Center in Manhattan, when the congregational Jummah prayer concludes, chatter fills the once hushed room as women and men prepare to leave. They step off the carpet, slip on their shoes, and the women readjust their *hijabs*. Vari-colored Indian and Pakistani sisters toss beautifully brilliant cloth sari-style. Olive-skinned Arab sisters check the pin securing the cotton scarves underneath their chins. Deep brown African sisters toss oversized lightweight cloth in their handbags, revealing their artfully wrapped gelees. And some women—of all colors and nationalities—take their *hijab* off completely, now that prayer has ended. This dynamic diversity might just be what the next wave of American feminism needs.

Muslim women and men are active forces in many different struggles, just as Western feminists struggle against misogynist forces. As a Muslim brother of mine once reminded me, race is just a smokescreen. Gender is just a smoke-screen. Religion is just a smokescreen. These are tools of the oppressor used to separate and slay as he takes.

We have so much work to do. I chose Islam for the wonder of the word, because I believe in the five pillars of the faith, because I love Allah and justice. I have been blessed to bear witness to women's realities in what people think of as two different worlds, and I have seen that those realities are essentially the same.

I bear witness to the woman beaten by her lover in the street outside my Brooklyn apartment—and to the woman tied to a Nigerian whipping post. I bear witness to the woman forced to strip to survive in Atlanta—and the woman forced to cover to survive in Afghanistan. I bear witness to the ever-increasing legions of women caught in this country's prison industrial complex, often because of their associations with husbands and male lovers—and I bear witness to the women struggling against inequity in interpretations of the Shariah in Islamic courtrooms. How do we measure a veiled woman's pain? Does it weigh more or less than the trauma in an American woman's eyes? Should we compare and contrast the horror, brutality, and hard smack against a woman's cheek? I simply ask that we warrior women, Muslim and non-Muslim, stand shoulder to shoulder, forming ranks, bending forward to carry all our sisters, Muslim and non-Muslim, tapping our collective strength.

57

"Whosoever" Is Welcome Here

An Interview with Reverend Edwin C. Sanders II

GARY DAVID COMSTOCK

Reverend Edwin C. Sanders II is the founding pastor of Metropolitan Inter-denominational Church, which is located in a working class neighborhood of small houses in Nashville, Tennessee. Reverend Sanders is African American and the congregation is predominantly African American. Lesbian/bisexual/gay/transgendered people are welcome and encouraged to participate in the life of the church. I attended a packed Sunday morning service in June 1998 and interviewed Rev. Sanders in the afternoon.

GARY COMSTOCK: How did the church get started?

REVEREND SANDERS: I had been the Dean of the Chapel at Fisk University. I left in 1980 in a moment of controversy. A new president had come, and we weren't able to mesh. I left and had no-where to go. I didn't have a plan. And instantly there were folks who were part of the chapel experience there at Fisk who wanted to organize a new church. I felt no spiritual interest whatsoever in organizing a new church, but about seven months later I felt like I clearly heard the voice say-ing, "This is something to do." I'm glad it worked out that way, because I think if I had done it directly after Fisk it would have been born out of a reaction and we probably wouldn't have developed the kind of identity, sense of mission, and direction the way we did. There were twelve people who came together and said they wanted to do this.

One of my good friends, Bill Turner, is a sociologist, and we were at Fisk together. Bill has a theory he advanced that institutions—and he built his theory around black institutions—cannot break out of the mold from which they were born. There was something about the way an institution is framed in its beginning, and no matter what you do you don't escape it. I thought that was absurd, but in time I came to think that he had something.

SOURCE: From Eric Brandt, ed., *Dangerous Liaisons: Blacks, Gays, and the Struggle for Equality* (New York: The New Press, 1999), pp. 142–157. Reprinted by permission of the author.

The congregation was a mix of white and black men and women from all kinds of denominational backgrounds, and that mix turned out to be significant because from the beginning people identified us as being inclusive at least across racial terms and definitely inclusive and equal in gender terms....

In that original group we had also a young man—one of my very dear friends—who was gay. Don was living a bisexual lifestyle at that point, but mainly to keep up appearances for professional purposes. Another one of my dearest friends went through a major mental breakdown at the time we were starting the church. I felt it very important not to abandon him and to include him. So we had a guy going through major psychological problems, somebody who was gay, and we had the racial mix.

We did not have the class mix at the beginning. Most of the folks were associated in some way with the academic community. Nothing like we have now. Today we have an unbelievable mix of people. I mean there are people who are doctors, lawyers, dentists and business people, and we also have a lot of folks who are right off the street, blue collar workers, in treatment for drugs and alcohol, going through a lot of transition. And that mix has evolved. Although we said in the beginning that's what we wanted to be, in actuality the current mix goes beyond that of the original twelve members.

The presence of Don, the one gay black male in the original congregation, had a lot to do with our current mix of people, and it had a lot to do with how we got so involved with HIV/AIDS ministry. The church began in '81 and he died in '84. It's almost hilarious when I think about it because I'm so involved in HIV/AIDS work now, but when he died I remember he was real sick, we didn't know what was going on, and they told us he died of toxicosis. I remember saying what in the world is that? I researched it and found out that it had something to do with cat and bird droppings, but Don didn't have cats. It was AIDS. You'd hear people talking about this strange disease because then it was 1984, but what that meant for us was that we got involved before it became a publicly recognized and discussed issue. Don's presence and death immediately pushed us in ministering to and being responsive to folks with AIDS and in dealing with the issue of homosexuality.

The other thing I was going to tell you which is kind of funny is the name. I will never forget when I told one of

my friends we were going to name the church Metropolitan Interdenominational. He said are you sure you want to
do that. I said what do you mean, you know, I was pretty
naive. He said all the gay churches across the country are
called Metropolitan churches. I said that's where I feel the
Lord leads me, I feel Metropolitan. In my mind what that
meant was that Nashville happens to have been the first
metropolitan government in the United States. It's the first
place where the county and city combined, so everything
in and around Nashville is referred to as metropolitan
government. But my friend said to me that everybody is
going to instantly say you're the gay church. I said so be
it and we went on with it. Like I said, there was a hand
bigger than mine at work....

COMSTOCK: What was the turnaround point for the class mix? How and
when did it happen?

REV. SANDERS: It was real clear. Early on what happened was we got
involved in prison ministry—actually directed a ministry
called the Southern Prison Ministry for a while. Going in
and out of the prisons we started to develop relationships
that translated to folks coming out of prison and getting
involved in the church. After I had done the prison
ministry for a while it became crystal clear to me that
80 percent of the people I was dealing with in prison were
there for alcohol and drug related issues. I started reassessing
this whole issue of drugs and alcohol and decided that's an
area we had to begin to focus our ministries. So, I got
involved, did the training, and became certified as a
counselor. I would venture to say that 25 percent of the
people in this church are folks that I first met in treatment.
Thirty to 35 plus percent of this congregation are folks
who are in treatment from alcohol and drug use. This is a
place where a lot of folks know they can come. We've got
all these names, you know, the drug church, the AIDS
church, we get those labels. But it's all right because that's
what we do. We hit those themes a lot.

I've learned some things over the years that I pretty
much hold to and this is one of them ... We don't say we
are a black church, we don't say that we are a gay church
or a straight church, we don't say that we are anything
other than a church that celebrates our oneness in Christ.
I'm convinced that has turned out to be the real key to
being able to hold this diverse group of folks together.
I must admit I'm a person that has a negative thing about
the word diversities. I don't use it much. We don't

celebrate our differences, we celebrate our oneness. That ends up being an avenue for a lot of folks being attracted and feeling comfortable here. New folks say I got here and I just felt like no one was looking at me strange, no one was treating me different, I was just able to be here....

COMSTOCK: I was impressed by the informality of the service today. People seem to relax and fit in. The choir is not so focused on performance that they're not interacting with the congregation. And your own manner is informal. It's a style that lets people in. You aren't just saying all people are welcomed here, you actually do something that lets them feel at home. I was also struck by the openness of the windows—the plants inside and the view into the park....

REV. SANDERS: ... We have the sense of informality. People are arriving from the time we welcome the guests at the beginning until just after the sermon when there's maximum presence in the audience. People come in slow, and we give folks an opportunity to leave early. We even say it in the bulletin, we just say if you need to leave just leave quietly. And we know folks do. There's a lot of folks who want to hear the choir or they want to hear the sermon, but they don't want all the rest of it. So we do it that way. It works out. One of the real hooks for just about everyone at Metropolitan is the fellowship circle at the end. We actually have a few Jews and a couple of Muslims who worship here regularly. The Muslim family does a very interesting thing which helped me to appreciate the significance of the fellowship circle at the end. We offer communion—the Eucharist, the Lord's supper—every Sunday, and they stay until we get to that part of the service. Then they go outside or into the vestibule, but they come back in. I remember when Omar first started attending, I said to him it's interesting to me that you don't leave at that point. He said, "No, no, no, I wouldn't miss the fellowship circle." And it made me realize that the communion of the fellowship circle is more important than the bread and the wine. Probably the real communion is what we do when we stand there at the end and sing "We've Come Too Far to Turn Back Now." We do that every week. It's our theme. That's a very significant moment. I've heard people tell me when they have to be late that they rush to get here just to catch the end of it. One woman said to me it was enough for her if she just got here for the fellowship circle.

A lot of the informality is very intentional. For instance, the only thing at Metropolitan that's elevated is the altar. Nothing else is above ground level. None of the seating is differentiated. In most churches the ministers have different, higher, bigger chairs than everyone else. We don't do that. We sit in the same seats the other folks do. My choir director is always telling me we'd get better sound if we could elevate the back rows. I say no, everybody's got to be on the same level and got to sit in the same seats. We do a lot of symbolic things like that.

COMSTOCK: Do other clergy give you much flack for working with needle exchange, welcoming gay folks, and working on issues that they may see as too progressive?

REV. SANDERS: They do. But let me tell you something. I have been able to have a level of involvement with ministers in this community that probably has brought credibility to what we do. I would like to think that we have maintained our sense of integrity especially as it relates to our consistency in ministry. Folks tend to respect that, so even when they disagree, they also look more seriously and harder at what it is they're questioning....

COMSTOCK: "Whosoever" is from John 3:16–17: "For God so loved the world that he gave his only Son, that whosoever believes in him should not perish but have eternal life." Has it been an expressed theme from the beginning?

REV. SANDERS: It's been our theme for the last seven or eight years....
I think we're growing in this inclusivity all the time. The language issue was big for us. We try to use inclusive language. If somebody else comes into the church that is not into that, it sticks out like a sore thumb. Does our inclusivity also mean that there is a real tolerance for folks who are perhaps not where we think folks should be in terms of issues like inclusive language, issues that relate to sexuality, gay and lesbian issues? I think the answer to that question is yes, you have to make room for those folks too. That's a real struggle. Another one of our little clichés, and we don't have a lot of them, is we say we try to be inclusive of all and alienating to none. Not being alienating is a real trick. It's amazing how easy you can alienate folks without realizing it, in ways that you're just not aware of....

COMSTOCK: What would you do if some new people had trouble with cross dressing, transgendered, lesbian or gay people? How do you get them to stay and deal with it rather than leave?

REV. SANDERS: We tracked that issue a couple of times. Let me tell you what's happened. Most folk are here for a while before some of it settles in, before they start to notice everything.

It's amazing to me how people get caught up in Metropolitan. They'll join and get involved and then they'll go through membership class, that's usually when it starts to hit them, that they say to themselves, "Oh oh, I'm seeing some stuff that I'm not sure about here." But what we've discovered is that what seems to help us more than anything else is that folks end up remembering what their initial experience had been when they first came here. We even have one couple, this is my favorite story to tell, who has talked about coming here on one of those Sundays when we were hitting the theme of inclusive hard. And they left here and said what kind of church is that? But then a couple of weeks later they said let's go back over there again. They came back and then they didn't come again for about two or three months. They were visiting other churches. They said when they got down to thinking about all the churches, they had felt warmth and connection here. They said you know that church really did kind of work; and they ended up coming back. When folks are here there's a warmth they feel, there is a connection they feel, there's a comfort zone in which they'll eventually deal with the issues that might be their point of difference. We've seen folks move in their thinking and there's some folks that have not been able to do it. I've got one young man—I really do think he just loves being here—who says, "I just can't fathom this gay thing. Why do you insist on it." I said, "You know, you're the one who's lifting this up."

As I said, we try to make sure that when any issue is brought up, it's in the course of things. It's not like we stop and have gay liberation day, just like we don't celebrate Black History month. We don't focus on or celebrate these identities or difference, but yet if you're around here you can't help but pick up on the church's support for these issues and differences. Today, for instance, we sang "Lift [Every] Voice and Sing," which is known as the black national anthem. But we call it "Our Song of Liberation," and I've tried to help people to understand that....

I have a lot of divinity students who serve as my pastoral assistants, and most of them are women. This place has become a real refuge for black women in the ministry. There are not a lot of clergy opportunities for them, so I've tried to figure out how to incorporate them into the life and ministry of the church. They actually run a lot of our ministries. And most of them are extremely well prepared academically. They're more prepared than I am to do the work and they do it well. When they started coming we suddenly became a magnet, a place where there was this rush of folks out of divinity school to come and be a part of ministry here. At one level I probably seem like I'm extremely freewheeling and loose, but I'm probably a lot more intentional about how things are going than folks realize, especially as they relate to the focus of ministry here at the church. One of my real concerns was not having the time to orient the young ministers, to bring them into a full awareness of what makes this place click. What I have discovered is that if I get the inclusive piece established in the beginning with them, I don't have to worry about the rest of it as much. If they buy into understanding that inclusivity is at the heart of this church, I don't have to worry about keeping my eye on them all the time....

COMSTOCK: You know that the example here of accepting and welcoming lesbians and gay men is an anomaly in the Church. You provide inspiration and hope, an example of something positive that is actually happening. Most gay people feel alienated from their churches, but I've found that African-American more often than white gay people emphasize the importance of religion in their family, community, and history and say that most of the pain and sadness in their lives centers around the Church. They claim that being rejected by the black church is especially devastating because this institution has been, and continues to be, the only place where they can take real refuge from the racism they experience in the society. Clearly that's not happening here at Metropolitan.

REV. SANDERS: We realized when we started doing our HIV/AIDS ministry that there are organizations that were established by gay white men who had done a good job of developing services, but we kept seeing there was something that was not happening for gay and lesbian African Americans. And we realized it was community. No matter how much they tried to get that in other communities, there is the whole thing of cultural comfort. They were looking for the context where there were people who looked like them, who were extensions of their family supporting them. It became clear to me that what we needed to do was to establish a place where people could literally come and where people had a sense of community. Consequently, what we call the Wellness Center here is more than anything else just a place where you can come and feel at home. It has sofas and tables and chairs. Folks sit around.

It's a place where there is a community that is an insulating, supporting place to come to. And although we provide some direct services for those folks, more than anything else I think the greatest service we provide is a safe place, a comfort zone, a community, a way to be connected to community. So I know real clearly what you're talking about, and I've heard it spoken too. It's one of the real issues for African Americans who are gay and lesbian. So our simple response has been to try to create a space where folks will have a certain level of comfort.

The problem I often run into, which speaks to what you were asking about, is that I think the greatest trouble we have sometimes is folks in the gay and lesbian community want to celebrate their sense of life and lifestyle more than Metropolitan lends itself to. In the same way, I have folks who don't understand why we don't do more things that are clearly more defined as being Afrocentric. There's a strong Afrocentric movement now within religious circles, and I end up dealing with them the same way I do with the folks who want to have ways in which they celebrate being gay and lesbian. My greatest struggle ends up being with folks who want to lift that up more, and I say the only thing we lift up is the basis of our oneness. I think the Church has been a place where folks have not been able to find community, when they have been rejected. The African American church is a pretty conservative entity, always has been. It's probably even better at holding up the conventions of American Christianity than other institutions. Consequently folks can draw some pretty hard lines, and that's what we've been dealing with at Metropolitan. But the other side of the Black church is that it is known as being an institution of compassion. So at the same time you have folks giving voice to some conservative ideas and practices, you can also appeal to a tradition and practice of compassion. That's why in our HIV/AIDS ministry we've been able to bring other churches into the loop. We're trying to get thirty churches to be involved in doing education and intervention on HIV/AIDS. We've been able to engage them on the level of compassion. Once we get them involved at that level then, we see folks open their eyes to other issues. African American churches have been very effective in compassion ministry for years around issues of sickness, death and dying. What we're trying to do is get folks to put as much emphasis on living and not just being there to minister to sickness and dying.

Another thing is the contradictions in the Church. The unspoken message that says it's all right for you to be here, just don't say anything, just play your little role. You can be in the choir, you can sit on the piano bench, but don't say you're gay. We had an experience here in Nashville which I'm sure could be evidenced in other places in the world. You know how a few years ago in the community of male figure skaters there was a series of deaths related to HIV/AIDS. The same thing happened a few years ago with musicians in the black churches. At one point here in Nashville there were six musicians who died of AIDS. In every instance it was treated with a hush. Nobody wanted to deal with the fact that all of these men were gay black men, and yet they'd been there leading the music for them. It's that contradiction where folks say yeah you're here but don't say anything about who you really are, don't be honest and open about yourself. I believe that the way in which you get the Church to respond is to continue to force the issue in terms of the teachings of Christ, to be forthright in seeing how the issue is understood in relationship to Jesus Christ. One thing I've learned about dealing with inner city African Americans is that you have to bring it home for them in a way that has some biblical basis. I'm always challenged by this, but I'm always challenging them to find a place where Jesus ever rejected anyone. I don't think anyone can find it. I don't think there's anybody that Jesus did not embrace.

58

Sustainable Food and Privilege

Why is Green Always White (and Male and Upper Class)

JANANI BALASUBRAMANIAN

When asked to name the heroes of food reform and sustainable agriculture, who comes to mind? Michael Pollan, Joel Salatin, Eric Schlosser, Peter Singer, Alice Waters maybe? Notice any patterns? The food reform movement is predicated on rather shaky foundations with regard to how it deals with race and other issues of identity, with its focus on a largely white and privileged American dream.

Still, what could be better than a return to family farms and home-cooking, which many of these gurus champion? The images are powerfully nostalgic and idyllic: cows grazing on sweet alfalfa, kids' mouths stained red with fresh heirloom tomato juice, and mom in the kitchen rolling out dough for homegrown-apple pie. But this is not an equal-access trip down memory lane. While we would like to think the American dream of social communion around food is a universal one, this assumption glosses over the very real differentials in gender, class, race, ethnicity, and nationality that were enabled and exacerbated by specific communities (white plantation owners, for example) through the use of food.

This is not to say that activists in the sustainable food movement are unconcerned with issues of identity, but that their rhetoric tends to disallow discussions on race, history, and food in a number of ways. First, Pollan and others situate the current state of American consumption in a patriarchal paradigm. These writers speak about a disappearance of food culture that for the most part accompanies male privilege. For example, Pollan, in an article for *The New York Times* on cooking and entertainment aptly titled "Out of the Kitchens, onto the Couch," explores the relationship between second-wave feminism and the gender politics of cooking. He argues that Betty Friedan's *The Feminine Mystique* convinced women to regard their housework, specifically cooking, as drudgery. Friedan did not, in fact, construct this sentiment herself; she merely observed the existent trends in white women's attitudes about food and housewifery. Pollan goes on to describe how Julia Child inspired his mother and other women like her, empowering them to channel their creativity into the kitchen. This is apt praise for the lively and engaging cook, but can Pollan not drive home the point that Americans need to cook more often without guilting American feminists?

SOURCE: Janani Balasubramanian, *Sustainable Food and Privilege: Why Green is Always White (and Male Upper-Class)*. Copyright © Janani Balasubramanian. Reprinted with permission.

Second, the emphasis on the local food economy, though admirable, has certain anti-global and overly nationalist undertones. Let us take the example of Joel Salatin, owner of Polyface Farms, featured in many of Pollan's books, as well as the movies *Food Inc.* and *Fresh!* Salatin is an ex-lawyer, of considerable means, who moves to the countryside, establishes a dynamic, organic, solar-powered farm, and sells top-quality animal products at top-quality dollar. If the nation is truly to scale up sustainable foods, we cannot fixate on the early image of the American farmer as white, male, and conservative. Instead, we must acknowledge (as USDA statistics tell us) that the face of farming is changing, and women and people of color will continue to grow in number as stewards of sustainable agriculture. Furthermore, we need to consider the real impact of foods we purchase, rather than mindlessly buying produce labeled "local" and "organic." The United States supports a lot of global agriculture through its food purchases, and this is a relationship we should not break off entirely. True, we can do more to support efficient, environmentally friendly purchasing, but we should also not be too hasty to reject globalization.

Finally, the major voices in food are not talking about race and class as often as they should. Food justice is fundamentally a race and class issue. Schlosser's *Fast Food Nation* elucidates labor practices that disproportionately affect people of color, but does not engage the issue of race specifically. Partly, this stagnancy is a matter of perception: after all, activists of color like Bryant Terry and Winona La Duke do brilliant work in their communities with regards to food justice. For some reason, however, their work goes largely underappreciated.

All social movements need a variety of voices, but I argue that food reform requires this diversity even more urgently because it is so universal in its reach. And if we can reach all those voices, then think of all the activists we will have as allies—feminists, anti-racists, interfaith leaders, and so on—interested and involved because food justice speaks to the needs of their communities and their call for action (activists: this is on you too—get on board!). As consumers of this kind of liberal rhetoric, we need to demand that the powers and big hitters in the food world diversify their representations. The food movement can only grow more powerful for it.

59

How the New Working Class Can Transform Urban America

ROBIN D. G. KELLEY

Corporate downsizing, deindustrialization, racist and sexist social policy, and ... the erosion of the welfare state do not respect the boundaries between work and community, the household and public space. The battle for livable wages and fulfilling jobs is inseparable from the fight for decent housing and safe neighborhoods; the struggle to defuse cultural stereotypes of inner city residents cannot be easily removed from the intense fights for environmental justice. Moreover, the struggle to remake culture itself, to develop new ideas, new relationships, and new values that place mutuality over materialism and collective responsibility over "personal responsibility," and place greater emphasis on ending all forms of oppression rather than striving to become an oppressor, cannot be limited to either home or work.

Standing in the eye of the storm are the new multiracial, urban working classes. It is they, not the Democratic Party, not a bunch of smart policy analysts, not corporate benevolence, who hold the key to transforming the city and the nation.... But my point is very, very simple.

MAKING OF THE NEW URBAN WORKING CLASS

...The pervasive imagery of the "underclass" makes the very idea of a contemporary urban working class seem obsolete. Instead of hardworking urban residents, many of whom are Latino, Asian-Pacific Islanders, West Indian immigrants, and U.S.-born African Americans, the dominant image of the "ghetto" is of idle black men drinking forty-ounce bottles of malt liquor and young black women "with distended bellies, their youthful faces belying the fact that they are often close to delivering their second or third child." White leftists nostalgic for the days before identity politics allegedly undermined the "class struggle" often have trouble seeing beyond the ruddy-faced hard hats or European immigrant factory workers who populate social history's images of the American working class. The ghetto is the last place to find American workers.

SOURCE: "How the New Working Class Can Transform Urban America" by Robin Kelley from *Yo' Mama's Disfunktional: Fighting the Culture Wars in Urban America* by Robin D. G. Kelley, pp. 125–153. Copyright © 1997 Beacon Press.

If you are looking for the American working class today, however, you will do just as well to look in hospitals and universities as in the sooty industrial suburbs and smokestack districts of days past. In Bethlehem, Pennsylvania, for example, once a stronghold of the steel industry, nursing homes have become the fastest growing source of employment, and the unions that set out to organize these workers have outgrown the steelworkers' union by leaps and bounds. The new working class is also concentrated in food processing, food services, and various retail establishments. In the world of manufacturing, sweatshops are making a huge comeback, particularly in the garment industry and electronics assembling plants, and homework (telephone sales, for example) is growing. These workers are more likely to be brown and female than the old blue-collar white boys we are so accustomed to seeing in popular culture....

Organizing the new immigrant labor force is perhaps the fundamental challenge facing the labor movement. For one, a substantial proportion of immigrant workers is employed by small ethnic firms with little tolerance for labor unions. Besides obvious language and cultural barriers, union leaders are trying to tackle the herculean task of organizing thousands of tiny, independent, sometimes transient firms. Immigrants are also less represented in public sector jobs, which tend to have a much higher percentage of unionized employees. (Indeed, the heavy concentration of native-born black people in public sector jobs partly explains why African Americans have such a high unionization rate.) The most obvious barrier to organizing immigrant workers, however, has been discriminatory immigration policy.... The 1986 Immigration Reform and Control Act imposed legal sanctions against employers of "aliens" without proper documents. Thus, even when unions were willing to organize undocumented workers, fear of deportation kept many workers from joining the labor movement.

Unions are also partly to blame for the state of labor organizing among immigrant workers. Until recently, union leaders too often assumed that Latino and Asian workers were unorganizable or difficult to organize— arguments that have been made about women and African American workers in the past.... Even when union organizers were willing to approach undocumented workers, they often operated on the assumption that immigrants were easily manipulated by employers, willing to undercut prevailing wages, or were "target workers" whose goal was to make enough money to return to their place of origin....

Several decades of grassroots, rank-and-file efforts on the part of workers of color [have tried to] re-orient unions toward issues of social justice, racism, sexism, and cultural difference within their ranks. Indeed, one of the most significant labor-based social justice movements emerged out of the Service Employees International Union (SEIU).... Launched in 1985, Justice for Janitors sought to build a mass movement to win union recognition and to address the needs of a workforce made up primarily of people of color, mainly immigrants. As Sweeney explained, "The strategy of Justice for Janitors was to build a mass movement, with workers making clear that they wanted union representation and winning 'voluntary recognition' from employers. The campaigns addressed the special needs of an immigrant workforce, largely from Latin America. In many cities, the janitors' cause became a civil rights movement—and a cultural crusade." Throughout the mid- to late 1980s, Justice for Janitors waged several successful "crusades" in Pittsburgh, Denver, San Diego, Los Angeles, and Washington, D.C. With support from Latino leaders and local church officials, for example, their Denver campaign yielded a sudden growth in unionization among janitors and wage increases of about ninety cents an hour.

From its inception, Justice for Janitors has been deeply committed to anti-racism and mass mobilization through community-based organizing and civil disobedience. Heirs of the sit-down strikers of the 1930s and the Civil Rights movement of the 1950s and 1960s, they have waged militant, highly visible campaigns in major cities throughout the country. In Los Angeles, for example, Justice for Janitors is largely responsible for the dramatic increase in unionized custodial employees, particularly among workers contracted out by big firms to clean high-rise buildings. The percentage of janitors belonging to unions rose from 10 percent of the workforce in 1987 to 90 percent in 1995. Their success certainly did not come easily. Indeed, the turning point in their campaign began around 1990, when two hundred janitors struck International Service Systems (ISS), a Danish-owned company. A mass march in Century City, California, in support of the strikers generated enormous publicity after police viciously attacked the demonstrators. Overall, more than sixty people were hospitalized, including a pregnant woman who was beaten so severely she miscarried. An outpouring of sympathy for the strikers turned the tables, enabling them to win a contract with ISS covering some 2,500 janitors in Southern California....

In Washington, D.C., Justice for Janitors led the struggle of Local 82 of the SEIU in its fight against U.S. Service Industries (USSI)—a private janitorial company that used nonunion labor to clean downtown office buildings. But because they conceived of themselves as a social movement, they did not stop with the protection of union jobs. In March 1995, Justice for Janitors organized several demonstrations in the district that led to over 200 arrests. Blocking traffic and engaging in other forms of civil disobedience, the protesters demanded an end to tax breaks to real estate developers as well as cutbacks in social programs for the poor. As union spokesperson Manny Pastreich put it, "This isn't just about 5,000 janitors; it's about issues that concern all D.C. residents—what's happening to their schools, their streets, their neighborhoods." Many people who participated in the March demonstrations came from all over the country, and not all were janitors. Greg Ceci, a longshoreman from Baltimore, saw the struggle as a general revitalization of the labor movement and a recognition, finally, that the unions need to lead a larger fight for social justice. "We need to reach out to the workers who have been ignored by mainstream unions. We need to fight back, and I want to be a part of it."

Indeed, D.C.'s Justice for Janitors is made up precisely of workers who have been ignored—poor women of color. Black and Latino women make up the majority of its membership, hold key leadership positions, and have put their bodies on the line for the SEIU as well as the larger cause of social justice. Twenty-four-year-old Dania Herring is an example of Justice for Janitors' new leadership cadre. A mother of four and resident of one of the poorest neighborhoods in the southeast section of the District, whose husband (at the time of the demonstrations) was an unemployed bricklayer, Herring had quit her job to become a full-time organizer for Local 82 and Justice for Janitors. Yet, these women were so militant that members of Washington's District Council dismissed them as "hooligans." "Essentially, they use anarchy as a means of organizing workers," stated African American councilman Harold Brazil. "And they do that under the mantle of justice—for janitors—or whoever else they want to organize."

In the end, their challenge to the city and to USSI paid off. In December 1995, the National Labor Relations Board (NLRB) concluded that USSI had "a history of pervasive illegal conduct" by threatening, interrogating, and firing employees they deemed unacceptable, especially those committed to union organizing. African American workers, in particular, had suffered most from the wave of firings, in part because USSI, like other employers, believed immigrant workers were more malleable and less committed to unionization because of their tenuous status as residents. The pattern was clear to Amy Parker, a janitor who had worked under USSI for three years. "Before USSI got the contract to clean my building, there were 18 African-Americans. Now, I am the only one." After three steady years cleaning the same building, she earned only $5.50 an hour. The NLRB's decision against USSI, therefore, was a substantial victory for Local 82 of the SEIU. It meant that USSI could no longer discriminate against the union and it generated at least a modicum of recognition for Justice for Janitors.

Such stories have been duplicated across the country. Justice for Janitors, in short, is more than a union. It is a dynamic social movement made up primarily of women and people of color, and it has the potential to redirect the entire labor movement....

LABOR/COMMUNITY STRATEGIES AGAINST CLASS-BASED RACISM

... Urban working people spend much if not most of their lives in their neighborhoods, in their homes, in transit, in the public spaces of the city, in houses of worship, in bars, clubs, barber-shops, hair and nail salons, in various retail outlets, in medical clinics, welfare offices, courtrooms, even jail cells. They create and maintain families, build communities, engage in local politics, and construct a sense of fellowship that is sometimes life sustaining. These community ties are crucial to the success of any labor movement....

Working people live in communities that are as embattled as the workplace itself. Black and Latino workers, for example, must contend with issues of police brutality and a racist criminal justice system, housing discrimination, lack of city services, toxic waste, inadequate health care facilities, sexual assault and domestic violence, and crime and neighborhood safety. And at the forefront of these community-based movements have been women, usually struggling mothers of all ages dedicated to making life better for themselves and their children— mothers who, as we have seen, have become the scapegoats for virtually every- thing wrong with the "inner city." In cities across the United States, working- class black and Latina women built and sustained community organizations that registered voters, patrolled the streets, challenged neighborhood drug dealers, defended the rights of prisoners, and fought vigorously for improvements in housing, city services, health care, and public assistance....

Whereas most community- and labor-based organizations limit their focus to an issue or set of issues, even when they are able to see the bigger picture, once in a while there are movements that attempt to fight on all fronts. Such organi- zations, where they do exist, are often products of the best elements of Third World, feminist, and Black Liberation movements. Rather than see race, gender, and sexuality as "problems," they are, instead, pushing working-class politics in new directions....

One of the most visible and successful examples of such a broad-based radical movement is the Labor/Community Strategy Center based in Los Angeles. The leaders of the Strategy Center have deep roots in social move- ments that go back to Black Liberation and student activism of the 1960s, urban antipoverty programs, farmworkers' movements, organized labor, and popular left movements in El Salvador and Mexico. They have been at the forefront of the struggle for clean air in the Harbor area of Los Angeles—a region with a high concentration of poor communities of color. They have worked closely with Justice for Janitors, providing crucial support to Local 660 of the SEIU....

The Strategy Center's most important campaign during the last few years has been the Bus Riders Union (BRU), a multiracial organization of transit- dependent working people who have declared war on race and class inequality in public transportation.... Public transportation is one of the few issues that touches the lives of many urban working people across race, ethnic, and gender lines. And ultimately, equal access to affordable transportation is tied to equal access to employment opportunities, especially now that manufacturing and other medium- to high-wage jobs have migrated to suburban rings or industrial parks. The combination of rising fares and limits on services has sharply curtailed the ability of workers to get to work. The cost of transportation for many families is a major part of their monthly budget. The average working- class commuter in Los Angeles rides the bus between sixty and eighty times a month, including for work, family visits, medical care, and shopping. For transit-dependent wage earners, bus fare could potentially comprise up to one-fourth or more of their total income! For many poor people even a small increase makes riding prohibitive. They might visit family less often, skip grocery shopping, or even miss work if they do not have bus fare. Forcing low-income riders off the buses with higher fares moves the Los Angeles

County Metropolitan Transit Authority (MTA) out of the business of transporting the urban poor.

After three years of conducting research and riding the buses to recruit members, in 1994 the Strategy Center and the Bus Riders Union, along with the Southern Christian Leadership Conference and the Korean Immigrant Workers' Advocates as coplaintiffs, filed a class action suit against the MTA on behalf of 350,000 bus riders, the vast majority of whom are poor people of color. Represented by the National Association for the Advancement of Colored People's Legal Defense and Educational Fund, the plaintiffs challenged proposals to raise bus fares from $1.10 to $1.35; eliminate monthly bus passes, which cost $42 for unlimited use; and introduce a zone system on the blue line commuter rail that would increase fares by more than 100 percent for half of the passengers.... The plaintiffs argued that the MTA's policies violate Title VI of the Civil Rights Act of 1964, which provides that no person "shall on the ground of race, color, or national origin, be excluded from participation in, be denied the benefits of, or be subjected to discrimination under any program or activity receiving Federal financial assistance."

The evidence that the MTA *has* created a separate and unequal transit system is overwhelming. On one side is the underfunded, overcrowded bus system which carries 94 percent of the MTA's passengers (80 percent of whom are African American, Latino, Asian-Pacific Islander, and Native American) yet receives less than one-third of the MTA's total expenditures. As a result, riders of color experience greater peak-hour overcrowding and tend to wait at neighborhood bus stops that are benchless and unsheltered. On the other side is the lavish commuter rail system connecting (or planning to connect) the predominantly white suburbs to the L.A. downtown business district. Although commuter line passengers make up 6 percent of the ridership, the system receives 70.9 percent of the MTA's resources. To make matters worse, the fare hike and proposed elimination of the monthly bus passes were to be accompanied by massive cutbacks in bus service.

... The proposed rail system has already created a two-tiered riding public— luxury trains for the white middle class, buses for the colored poor. Furthermore, the MTA paid for its Metrolink (train) lines with revenues earned from poor and working-class bus riders. Monies that should have been earmarked to improve bus service instead went to a partially finished, impractical rail system that will cost approximately $183 billion and only serve 11 percent of the population. It doesn't take a high IQ to realize that poor black and brown riders who have to crowd onto South Central, Inglewood, and East Los Angeles buses have been subsidizing the commuter rail. The average commuter rail rider is a white male professional with a household income of $64,450, who receives a $58 transportation subsidy from his employer and owns a car. By contrast, a typical bus rider is a person of color with no car and whose household income falls below $15,000. Put another way, the MTA's subsidy for bus #204, which runs through predominantly black and Latino neighborhoods along Vermont Avenue, will remain 34 cents per passenger, while the projected subsidy for the proposed Azusa and Torrance rail lines will be $55.64 and $52.87, respectively. Of course, the MTA will use revenues from sales tax and fare-box receipts to pay for this, but it is also dependent on federal funding: hence the violation of Title VI of the Civil Rights Act.

Like the Student Non-Violent Coordinating Committee in the 1960s, Strategy Center and BRU organizers take litigation *and* mass organizing seriously, but they are very clear which component drives the process. They have held legal workshops for members, conducted hours and hours of background research, prepared the initial strategy document on which the case is based, and carefully examined and re-examined all legal documentation pertaining to the suit. At the same time, they have worked hard to ensure that "the class" has a voice in the class action suit. In addition to collecting detailed declarations from riders, their main core of organizers are also plaintiffs—that is, bus riders....

The Bus Riders Union has been incredibly successful. Besides building a truly multicultural, multiracial mass movement, the BRU ... wrested a settlement from the MTA that imposed a two-year freeze on bus fares, a substantial reduction in the price of monthly and biweekly bus passes, a 5-percent increase in the number of buses in the MTA fleet in order to reduce overcrowding, the creation of new bus lines with the intention of eradicating "transportation segregation," and the creation of a joint working group in which the BRU would play a major role in drafting a five-year plan to improve bus service for all. Hence a solidly antiracist movement resulted in a victory for all riders, including transit-dependent white workers, the disabled, the elderly, and students.

While the settlement marks a critical landmark in the history of civil rights struggle, Strategy Center organizers understand that the specific legal challenge is less important than the social movement that grows out of it.... Through their own struggles, in the courts and city council chambers, on the streets and buses of Los Angeles, the Bus Riders Union has advanced a radical vision of social justice.... Lacking such a vision, many of us have not been able to see why the struggle over public transportation is so important for working people, or how antiracism benefits the entire working class.... The bus campaigns powerfully demonstrate that the issues facing the vast majority of people of color, for the most part, are working-class issues. Thus, unless we understand the significance of the "billions for buses" campaign or the historic class action suit against the MTA, we will never understand the deeper connections between welfare, workfare, warfare, and bus fare ... or bus riders and unions, for that matter.

In this tragic era of pessimism and defeat, these grassroots, working-class radical movements go forward as if they might win.... The women and men who have built and sustained these organizations do not have the luxury not to fight back. In most cases, they are battling over issues basic to their own survival—decent wages, healthy environment, essential public services. At the heart of these movements are folks like my mother and the many other working-class women of color ... who have borne the brunt of the material and ideological war against the poor. They understand, better than anyone, the necessity of fighting back. And they also understand that change does not come on its own. As Dr. Martin Luther King, Jr., so eloquently explained:

> A solution of the present crisis will not take place unless men and women work for it. Human progress is neither automatic nor inevitable. Even a superficial look at history reveals that no social advance rolls in on the wheels of inevitability. Every step toward the goal of justice requires sacrifice, suffering, and struggle; the tireless exertions and passionate concern of dedicated individuals. Without persistent effort, time itself becomes an ally of the insurgent and primitive forces of irrational emotionalism and social destruction. This is no time for apathy or complacency. This is the time for vigorous and positive action.

> The time is now.

60

Women's Rights as Human Rights

Toward a Re-Vision of Human Rights

CHARLOTTE BUNCH

The concept of human rights is one of the few moral visions ascribed to internationally. Although its scope is not universally agreed upon, it strikes deep chords of response among many. Promotion of human rights is a widely accepted goal and thus provides a useful framework for seeking redress of gender abuse. Further it is one of the few concepts that speaks to the need for transnational activism and concern about the lives of people globally. The Universal Declaration of Human Rights,[1] adopted in 1948, symbolizes this world vision and defines human rights broadly. While not much is said about women, Article 2 entitles all to "the rights and freedoms set forth in this Declaration, without distinction of any kind, such as race, colour, sex, language, religion, political or other opinion, national or social origin, property, birth or other status."...

Since 1948 the world community has continuously debated varying interpretations of human rights in response to global developments. Little of this discussion, however, has addressed questions of gender, and only recently have significant challenges been made to a vision of human rights which excludes much of women's experiences. The concept of human rights, like all vibrant visions, is not static or the property of any one group; rather, its meaning expands as people reconceive of their needs and hopes in relation to it. In this spirit, feminists redefine human rights abuses to include the degradation and violation of women. The specific experiences of women must be added to traditional approaches to human rights in order to make women more visible and to transform the concept and practice of human rights in our culture so that it takes better account of women's lives....

I. BEYOND RHETORIC: POLITICAL IMPLICATIONS

Few governments exhibit more than token commitment to women's equality as a basic human right in domestic or foreign policy. No government determines its policies toward other countries on the basis of their treatment of women, even when some aid and trade decisions are said to be based on a country's human

SOURCE: Bunch, Charlotte. 1990. "Women's Rights as Human Rights: Toward a Re-Vision of Human Rights." *Human Rights Quarterly* 12: 486–498. Copyright © 1990 John Hopkins University Press. Reprinted with permission.

rights record. Among nongovernmental organizations, women are rarely a priority, and Human Rights Day programs on 10 December seldom include discussion of issues like violence against women or reproductive rights. When it is suggested that governments and human rights organizations should respond to women's rights as concerns that deserve such attention, a number of excuses are offered for why this cannot be done. The responses tend to follow one or more of these lines: (1) sex discrimination is too trivial, or not as important, or will come after larger issues of survival that require more serious attention; (2) abuse of women, while regrettable, is a cultural, private, or individual issue and not a political matter requiring state action; (3) while appropriate for other action, women's rights are not human rights per se; or (4) when the abuse of women is recognized, it is considered inevitable or so pervasive that any consideration of it is futile or will overwhelm other human rights questions. It is important to challenge these responses....

The most insidious myth about women's rights is that they are trivial or secondary to the concerns of life and death. Nothing could be farther from the truth: sexism kills. There is increasing documentation of the many ways in which being female is life-threatening. The following are a few examples:

—Before birth: Amniocentesis is used for sex selection leading to the abortion of more female fetuses at rates as high as 99 percent in Bombay, India; in China and India, the two most populous nations, more males than females are born even though natural birth ratios would produce more females.

—During childhood: The World Health Organization reports that in many countries, girls are fed less, breast fed for shorter periods of time, taken to doctors less frequently, and die or are physically and mentally maimed by malnutrition at higher rates than boys.

—In adulthood: The denial of women's rights to control their bodies in reproduction threatens women's lives, especially where this is combined with poverty and poor health services. In Latin America, complications from illegal abortions are the leading cause of death for women between the ages of fifteen and thirty-nine.

Sex discrimination kills women daily. When combined with race, class, and other forms of oppression, it constitutes a deadly denial of women's right to life and liberty on a large scale throughout the world. The most pervasive violation of females is violence against women in all its manifestations, from wife battery, incest, and rape, to dowry deaths, genital mutilation, and female sexual slavery. These abuses occur in every country and are found in the home and in the workplace, on streets, on campuses, and in prisons and refugee camps. They cross class, race, age, and national lines; and at the same time, the forms this violence takes often reinforce other oppressions such as racism, "able-bodyism," and imperialism. Case in point: in order to feed their families, poor women in brothels around US military bases in places like the Philippines bear the burden of sexual, racial, and national imperialism in repeated and often brutal violation of their bodies.

... Violence against women is a touchstone that illustrates the limited concept of human rights and highlights the political nature of the abuse of women.

… Victims are chosen because of their gender. The message is domination: stay in your place or be afraid. Contrary to the argument that such violence is only personal or cultural, it is profoundly political. It results from the structural relationships of power, domination, and privilege between men and women in society. Violence against women is central to maintaining those political relations at home, at work, and in all public spheres.

Failure to see the oppression of women as political also results in the exclusion of sex discrimination and violence against women from the human rights agenda. Female subordination runs so deep that it is still viewed as inevitable or natural, rather than seen as a politically constructed reality maintained by patriarchal interests, ideology, and institutions. But I do not believe that male violation of women is inevitable or natural. Such a belief requires a narrow and pessimistic view of men. If violence and domination are understood as a politically constructed reality, it is possible to imagine deconstructing that system and building more just interactions between the sexes.

The physical territory of this political struggle over what constitutes women's human rights is women's bodies. The importance of control over women can be seen in the intensity of resistance to laws and social changes that put control of women's bodies in women's hands: reproductive rights, freedom of sexuality whether heterosexual or lesbian, laws that criminalize rape in marriage, etc. Denial of reproductive rights and homophobia are also political means of maintaining control over women and perpetuating sex roles and thus have human rights implications. The physical abuse of women is a reminder of this territorial domination and is sometimes accompanied by other forms of human rights abuse such as slavery (forced prostitution), sexual terrorism (rape), imprisonment (confinement to the home), and torture (systematic battery). Some cases are extreme, such as the women in Thailand who died in a brothel fire because they were chained to their beds. Most situations are more ordinary like denying women decent education or jobs which leaves them prey to abusive marriages, exploitative work, and prostitution.

This raises once again the question of the state's responsibility for protecting women's human rights. Feminists have shown how the distinction between private and public abuse is a dichotomy often used to justify female subordination in the home. Governments regulate many matters in the family and individual spheres. For example, human rights activists pressure states to prevent slavery or racial discrimination and segregation even when these are conducted by nongovernmental forces in private or proclaimed as cultural traditions as they have been in both the southern United States and in South Africa. The real questions are: (1) who decides what are legitimate human rights; and (2) when should the state become involved and for what purposes.…

The human rights community must move beyond its male defined norms in order to respond to the brutal and systematic violation of women globally. This does not mean that every human rights group must alter the focus of its work. However it does require examining patriarchal biases and acknowledging the rights of women as human rights. Governments must seek to end the politically and culturally constructed war on women rather than continue to perpetuate it.

Every state has the responsibility to intervene in the abuse of women's rights within its borders and to end its collusion with the forces that perpetrate such violations in other countries.

II. TOWARD ACTION: PRACTICAL APPROACHES

The classification of human rights is more than just a semantics problem because it has practical policy consequences. Human rights are still considered to be more important than women's rights. The distinction perpetuates the idea that the rights of women are of a lesser order than the "rights of man," and, as Eisler describes it, "serves to justify practices that do not accord women full and equal status."[2] In the United Nations, the Human Rights Commission has more power to hear and investigate cases than the Commission on the Status of Women, more staff and budget, and better mechanisms for implementing its findings. Thus it makes a difference in what can be done if a case is deemed a violation of women's rights and not of human rights.

… I have observed four basic approaches to linking women's rights to human rights. These approaches are presented separately here in order to identify each more clearly. In practice, these approaches often overlap, and while each raises questions about the others, I see them as complementary.…

1. Women's Rights as Political and Civil Rights. Taking women's specific needs into consideration as part of the already recognized "first generation" political and civil liberties is the first approach. This involves both raising the visibility of women who suffer general human rights violations as well as calling attention to particular abuses women encounter because they are female. Thus, issues of violence against women are raised when they connect to other forms of violation such as the sexual torture of women political prisoners in South America. Groups like the Women's Task Force of Amnesty International have taken this approach in pushing for Amnesty to launch a campaign on behalf of women political prisoners which would address the sexual abuse and rape of women in custody, their lack of maternal care in detention, and the resulting human rights abuse of their children.…

The political and civil rights approach is a useful starting point for many human rights groups; by considering women's experiences, these groups can expand their efforts in areas where they are already working. This approach also raises contradictions that reveal the limits of a narrow civil liberties view. One contradiction is to define rape as a human rights abuse only when it occurs in state custody but not on the streets or in the home. Another is to say that a violation of the right to free speech occurs when someone is jailed for defending gay rights, but not when someone is jailed or even tortured and killed for homosexuality. Thus while this approach of adding women and stirring them into existing first generation human rights categories is useful, it is not enough by itself.

2. Women's Rights as Socioeconomic Rights. The second approach includes the particular plight of women with regard to "second generation" human rights such as the rights to food, shelter, health care, and employment. This is an approach favored by those who see the dominant Western human rights tradition and international law as too individualistic and identify women's oppression as primarily economic.

… It focuses on the primacy of the need to end women's economic subordination as the key to other issues including women's vulnerability to violence. This particular focus has led to work on issues like women's right to organize as workers and opposition to violence in the workplace, especially in situations like the free trade zones which have targeted women as cheap, nonorganized labor. Another focus of this approach has been highlighting the feminization of poverty or what might better be called the increasing impoverishment of females. Poverty has not become strictly female, but females now comprise a higher percentage of the poor.…

3. *Women's Rights and the Law.* The creation of new legal mechanisms to counter sex discrimination characterizes the third approach to women's rights as human rights. These efforts seek to make existing legal and political institutions work for women and to expand the state's responsibility for the violation of women's human rights. National and local laws which address sex discrimination and violence against women are examples of this approach. These measures allow women to fight for their rights within the legal system. The primary international illustration is the Convention on the Elimination of All Forms of Discrimination Against Women.[3]

The Convention has been described as "essentially an international bill of rights for women and a framework for women's participation in the development process … [which] spells out internationally accepted principles and standards for achieving equality between women and men."[4] Adopted by the UN General Assembly in 1979, the Convention has been ratified or acceded to by 104 countries as of January 1990. In theory these countries are obligated to pursue policies in accordance with it and to report on their compliance to the Committee on the Elimination of Discrimination Against Women (CEDAW).…

The Convention outlines a clear human rights agenda for women which, if accepted by governments, would mark an enormous step forward. It also carries the limitations of all such international documents in that there is little power to demand its implementation. Within the United Nations, it is not generally regarded as a convention with teeth, as illustrated by the difficulty that CEDAW has had in getting countries to report on compliance with its provisions. Further, it is still treated by governments and most non-governmental organizations as a document dealing with women's (read "secondary") rights, not human rights. Nevertheless, it is a useful statement of principles endorsed by the United Nations around which women can organize to achieve legal and political change in their regions.

4. *Feminist Transformation of Human Rights.* Transforming the human rights concept from a feminist perspective, so that it will take greater account of women's lives, is the fourth approach. This approach relates women's rights and human rights, looking first at the violations of women's lives and then asking how the human rights concept can change to be more responsive to women.…

This transformative approach can be taken toward any issue, but those working from this approach have tended to focus most on abuses that arise specifically out of gender, such as reproductive rights, female sexual slavery, violence against women, and "family crimes" like forced marriage, compulsory heterosexuality, and female mutilation. These are also the issues most often dismissed as not really human rights questions. This is therefore the most hotly contested area and requires that barriers be broken down between public and private, state and nongovernmental responsibilities....

Yet most women experience abuse on the grounds of sex, race, class, nation, age, sexual preference, and politics as interrelated, and little benefit comes from separating them as competing claims. The human rights community need not abandon other issues but should incorporate gender perspectives into them and see how these expand the terms of their work. By recognizing issues like violence against women as human rights concerns, human rights scholars and activists do not have to take these up as their primary tasks. However, they do have to stop gate-keeping and guarding their prerogative to determine what is considered a "legitimate" human rights issue.

As mentioned before, these four approaches are overlapping and many strategies for change involve elements of more than one. All of these approaches contain aspects of what is necessary to achieve women's rights. At a time when dualist ways of thinking and views of competing economic systems are in question, the creative task is to look for ways to connect these approaches and to see how we can go beyond exclusive views of what people need in their lives. In the words of an early feminist group, we need bread and roses, too. Women want food and liberty and the possibility of living lives of dignity free from domination and violence. In this struggle, the recognition of women's rights as human rights can play an important role.

NOTES

1. Universal Declaration of Human Rights, *adopted* 10 December 1948, G.A. Res. 217A(III), U.N. Doc. A/810 (1948).

2. Riane Eisler, "Human Rights: Toward an Integrated Theory for Action," *Human Rights Quarterly* 9 (1987): 297. See also Alida Brill, *Nobody's Business: The Paradoxes of Privacy* (New York: Addison-Wesley, 1990), 291.

3. Convention on the Elimination of All Forms of Discrimination Against Women, G.A. Res. 34/180, U.N. Doc. A/Res/34/180 (1980).

4. International Women's Rights Action Watch, "The Convention on the Elimination of All Forms of Discrimination Against Women" (Minneapolis: Humphrey Institute of Public Affairs, 1988), 1.

Index